To: Geoff

wishing you the very best

Leo Beranek

April 2, 2004

Concert Halls and Opera Houses sets forth the harvest of six decades of intensive study of acoustics for music performance. It is a comprehensive (and indispensable) aid to architects, musicians and design teams who tackle the incredibly daunting task of creating new performance spaces.

— RUSSELL JOHNSON
Acoustics and Theater Consultant, ARTEC Consultants, Inc.,
New York

This book assembles architectural and acoustical data on 100 spaces for music and rank-orders over two-thirds according to their acoustical quality as judged by musicians and music critics. It gives comprehensive knowledge of room acoustics and offers a basic foundation for acoustical research long into the future.

— HIDEKI TACHIBANA
Professor of Acoustics, University of Tokyo, Japan

Beranek has created a new Rosetta Stone for the languages of music, acoustics, and architecture. Lovers of music everywhere will welcome this extraordinary work for its scope, depth, and ease of reading, and for heightening our understanding and enjoyment of the musical experiences that so enrich our lives.

— R. LAWRENCE KIRKEGAARD
Acoustical Consultant, Chicago, Illinois

Beranek's latest reference work is an essential volume in every auditorium designer's library. It will also bring information and pleasure to all with an interest in music, acoustics and architecture. In our offices, a common response to a question in an acoustical design session is "Let's check in Beranek."

— ROB HARRIS
Director, Arup Acoustics, Winchester, Hampshire, England

Concert Halls and Opera Houses is the definitive work on the architectural acoustic design of classical music spaces. With presentation of 100 halls, it illustrates various levels of acoustical quality. Written for the lay reader it deserves to be in every school of music, architecture and science and with every musician and music lover.

— CHRISTOPHER JAFFE
Founding Principal, Jaffe Holden Acoustics, Norwalk, CT

Concert Halls and Opera Houses

Springer
New York
Berlin
Heidelberg
Hong Kong
London
Milan
Paris
Tokyo

Concert Halls and Opera Houses

Music, Acoustics, and Architecture

Second Edition

LEO BERANEK

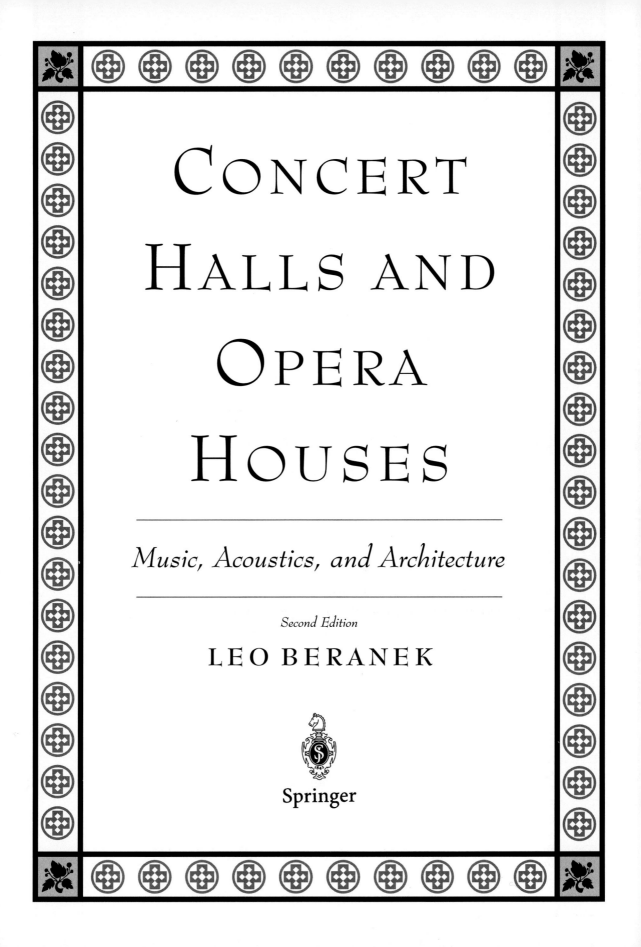

Springer

LEO BERANEK
975 Memorial Drive
Suite 804
Cambridge, MA 02138-5755
USA
beranekleo@ieee.org

LIBRARY OF CONGRESS CATALOGING-IN-PUBLICATION DATA
Beranek, Leo Leroy, 1914–
Concert halls and opera houses : music, acoustics, and architecture / Leo Beranek.—2nd ed.
        p.   cm.
Rev. ed. of: Concert and opera halls, 1996.
Includes bibliographical references and index.
ISBN 0-387-95524-0 (alk. paper)
1. Architectural acoustics.   2. Music halls.   3. Theaters.   I. Beranek, Leo Leroy, 1914–
NA2800.B39   2002
725'.81—dc21   2002070734

ISBN 0-387-95524-0   Printed on acid-free paper.

An earlier edition of this book was published by the
Acoustical Society of America in 1996.

Text design by Steven Pisano.

Printed in the United States of America.

9 8 7 6 5 4 3 2 1   SPIN 10882373

www.springer-ny.com

Springer-Verlag   New York Berlin Heidelberg
A member of BertelsmannSpringer Science + Business Media GmbH

Preface

The first question any lover of classical music usually asks an acoustician is, "Which are the best halls in the world?" The response—the three halls rated highest by world-praised conductors and music critics of the largest newspapers were built in 1870, 1888, and 1900—always prompts the next query: "Why are those so good while many halls built after 1950 seem to be mediocre or failures?" You will find answers to these questions in this book, the result of a half-century's research into the very complex field of acoustics of halls for music.

The dialog re-enacted above bears a close resemblance to another illustration that typically troubles music lovers. They frequently ask, "Why is a Stradivarius violin so good and so many built since then not in the same league?" In this case, we know that Antonio Stradivari, working at the turn of the eighteenth century, employed the utmost skill, a good ear, and perhaps a little luck to capture the dozens of details that make up a great violin. Subsequent violinmakers have learned that only by producing close copies of his masterpieces can they expect their instruments to be highly acceptable.

Indisputably, the acoustics of halls for music are more diverse than those of violins. As this book will explain in depth, halls for music encompass a broader range of different types with very different acoustics, and one should always remember that composers often wrote music for a particular concert hall or opera house. Consequently, a given composition usually sounds best when performed in its intended acoustics. For instance, Gregorian chants were written for performance in large churches with high reverberance; a small, quiet church never comes close to doing it justice. As Chapter 1 discusses, compositions of different musical periods—Baroque, Classical, or Romantic, for example—sound best in halls whose reverberation times vary from medium low to relatively high. Can one hall serve all purposes? Halls with variable acoustics are among those treated here.

Since we can, today, identify the acoustical characteristics of the finest halls in existence, we could create an unerring duplicate of any one of the several best and thus reproduce its acoustics exactly. Why not do so? Because building committees generally select architects not to make exact copies of a great hall but to do something original and visually inspiring, *with the hope that the halls will have excellent sound.* Most architects will not argue with that approach. Who would be awarded an architectural prize for the construction of an exact copy? Consequently, the acoustical consultant is faced with a dilemma. To have the best acoustics, the hall should be close in design to one of the great halls—and should yield similar electro-acoustic data when measured. So the consultant usually follows a subtle path, pushing for as many similarities as possible and making recommendations, where differences occur, of features—often novel—that *may* salvage the new design.

For every new hall, with its untried acoustics, opening night may become a trial by fire. Of course, the local orchestra and conductor may do all in their power to adapt their playing style to the new acoustics, as the history of the Philadelphia Academy of Music in Chapter 1 illustrates. But well-traveled music critics, often in attendance only this once, may judge the acoustics of the new venue against those of the four or five top-ranked old halls, and opening night reviews may set the reputation of the hall, negatively, for years to come. On occasion, these assessments turn out to be unjust, failing to account for how a hall's acoustics may be adjusted over time or the possible misuse of the hall that first night. Such bad fortune befell one important hall that was designed for a standard-sized orchestra playing the kinds of compositions that make up the bulk of the repertoire of today's symphonic concerts. For the opening night, however, the conductor chose a new composition, with a double-sized orchestra and a chorus of several hundred. The stage had to be extended to over twice its normal size, and the choristers in the back row stood on bleachers so high that their heads threatened to touch the stage ceiling, thus amplifying their voices unevenly. In some parts of the composition, the musicians created unusually loud sound effects, in one case by hitting a suspended three-meter section of railroad track with a sledge hammer. Nearly everybody in the audience went home with a headache. The music critic's response? The hall was at fault.

Following the first chapters, which establish a base for understanding the effects of acoustics on composers, performers, and listeners, and guiding the reader to a common vocabulary, the bulk of this book, Chapter 3, contains the write-ups, photographs, drawings, and architectural details on 100 existing halls in 31 countries. Thirty of the halls are completely new. Although the remainder appeared in earlier books by the author, the materials have been updated wherever necessary. The later chapters present the relation of a hall's acoustics to its age, shape, type of seats, and the materials used for the walls and ceiling. The sequence of events that led to Boston Symphony Hall's excellent acoustics, which opened in 1900, is covered in detail—although it went through a troubled first few years because the leading local music critic considered the predecessor hall as better. Detailed discussions also appear for balcony, box, stage, and pit designs. All the known electro-acoustical measurements on 100 existing halls are examined and compared with the rank orders of 58 concert halls and 21 opera houses that were obtained from interviews and questionnaires. Finally, the optimal electro-acoustical results are presented for concert halls and opera houses used for today's repertoires.

Three appendices supplement the chapters: the first gives definitions of all of the major acoustical and architectural terms and symbols used in the book; the second provides the electro-acoustical data available on the 100 halls; and the third presents in tabular form much of the dimensional and electro-acoustical data for the 100 halls.

CREDITS AND ACKNOWLEDGMENTS

Hundreds of persons are responsible for the material presented in this book: conductors, music critics, composers, musicians, orchestra and opera directors, hall managers, architects, acousticians, and musical friends.

The largest contributors of new information were Takayuki Hidaka, Noriko Nishihara, and Toshiyuki Okano, devoted members of the acoustical department at the Takenaka Research and Development Institute in Chiba, Japan. They are responsible for the electro-acoustical data in this book on twenty-two concert halls and seven opera houses in nine nations of Europe, the Americas, and Japan. Other major contributions came from experts at Mueller-BBM of Munich/Planegg, Germany; the National Research Council of Canada; the Technical University of Denmark; ARTEC Consultants of New York; Kirkegaard Associates of Chicago; Jaffee Holden Acoustics of Norwalk, Connecticut; Michael Barron of Bath University; Jordan Acoustics of Denmark; Arup Acoustics of Winchester, Hampshire, U.K.; Sandy Brown Associates of London; Nagata Acoustics of Tokyo; InterConsult Group of Trondheim, Norway; Garcia-BBM of Valencia; ACENTECH (successor to Bolt Beranek and Newman) of Cambridge, Massachusetts; Cyril Harris of New York; and Albert Yaying Xu of Paris. Others too numerous to name here also provided invaluable information for this volume.

For the "biographies" of the 100 halls, the architectural drawings for 64 were produced by Richard Shnider, 30 by the late Wilfred Malmlund, and 6 by Daniel Chadwick. Important editorial assistance on the first two chapters was rendered by Ondine E. Le Blanc.

To all of the above, I owe my deepest thanks.

LEO BERANEK
October 2003

Contents

CHAPTER TWO

THE LANGUAGE OF MUSICAL ACOUSTICS 19

Concert Halls and Opera Houses

JAPAN

MALAYSIA

MEXICO

NETHERLANDS

NEW ZEALAND

NORWAY

Concert Halls and Opera Houses

ONE

MUSIC

AND

ACOUSTICS

WHAT ARE GOOD ACOUSTICS?

Every concert hall and opera house has its own distinct acoustics. Any music lover, of course, feels the effect of a hall's acoustical design, often without realizing its importance—unless he or she has made a practice of listening to music in many different venues. Consequently, an attempt to determine which acoustical qualities concertgoers prefer usually elicits recollections about the particular concerts that gave a listener the deepest pleasure. For that individual, a number of elements come together to create that pleasure—the composition, the conductor, the orchestra, and the hall must in combination be excellent to produce a memorable listening experience. For the music professional, however, whether a conductor, a performer, or an acoustical engineer, it is vital to distinguish among these ingredients and to understand what each contributes to the totality.

Although that task requires a precise language and a correspondingly concrete understanding of acoustics, most musicians and music lovers can agree in general on what makes a "good" concert hall. It must be so quiet that the very soft (*pp*) passages are clearly audible. It must have a reverberation time long enough to carry the crescendos to dramatic very loud (*ff*) climaxes. The music must be sufficiently clear that rapidly moving violin passages do not melt into a general "glob." The hall should have a spacious sound, making the music full and rich and apparently much "larger" than the instrument from which it emanates. It must endow the music with a pleasant "texture," an indescribable, but hearable, quantity that can be demonstrated by electronic measurements. The bass sounds of the orchestra must have "power" to provide a solid foundation to the music. Finally, there should be no echoes or "source shift"; that is to say, all or part of the orchestra should not seem to originate at some side or ceiling surface.

In an effort to pin down, in physical terms, the qualities that make up this optimum listening experience, I have interviewed well over a hundred conductors and music critics about their acoustical preferences. Those interviews have resulted in the rank-ordering of 58 concert halls according to their acoustical quality for today's symphonic repertoire (see Chapter 4). Physical measurements of the acoustics were made in these halls using the latest in electronic equipment. An analysis of the subjective responses made it apparent that of all acoustical elements, *reverberation* time comes to a conductor's mind first, because it directly affects his ability to achieve the clarity, fullness of tone, and phrasing consistent with his interpretation of the composer's intention.

Nearly all of my interviewees preferred to perform or listen to music of the classical and romantic periods in rectangular halls with reverberation times that fell into a specific window: between 1.7 and 2.1 secs, measured at mid-tones with the hall fully occupied. A shorter reverberation time, however, can provide a different benefit: halls with reverberation times at 1.5 sec or below seem to hone the overall performance quality of their home orchestras. At the Philadelphia Academy of Music (1.2 secs), the Cleveland Severance Hall (1.5 sec before recent revisions; now 1.6 secs), and Chicago Orchestra Hall (1.2 sec before recent revisions; now 1.75 secs), for example, an orchestra must play precisely because no lingering reverberation smoothes over slight imperfections. Many think that the resident orchestras in those three halls, at least in the years before the recent renovations, ranked highest in performance in the United States. Let us now review the question of good acoustics as viewed by performers, composers, and well-traveled listeners.

ACOUSTICS AND THE PERFORMERS

Conductors

Herbert von Karajan, music director of the Berlin Philharmonic Orchestra, the Vienna Philharmonic, the Vienna Staatsoper, and the Salzburg Festival for various overlapping time periods from 1955 until his death in 1989, was keenly aware of the acoustics of different music halls, as any well-traveled conductor must be. He also did not retreat from sharing his opinions. In 1943, for example, he wrote a damning criticism of the new Berlin Staatsoper: "I was obliged [after four concerts] to conclude that it is so constituted that a realization of the kind of performance I am expected to produce is not possible" [Karajan letter to H. Tietjen, 12 May 1943, Osborne, Random House, *Karajan,* 1998, p. 164]. In 1956, he wrote to the competition judges of the prospective Philharmonie Hall in Berlin with his support for a particularly innovative design. "Of all the designs submitted," he said, "one [by Architect Hans Scharoun] seems to stand out above the others; which is founded on the principle that the performers should be in the middle. . . . It seems to me . . . that this arrangement with the orchestra centrally placed will be better suited than any known hall to the musical style of the Berlin Philharmonic" [Osborne, *Karajan,* p. 476]. With these well-placed words, Karajan helped break the tradition of the European rectangular hall as the only accepted performing space for symphonic music.

In this situation, the judges most certainly valued Karajan's opinion as a performer, but other factors also unquestionably influenced them—factors that are evident in the completed hall. The space is visually breathtaking, at least on a first visit, particularly if one enters at one of the upper "vineyard" levels (See Hall No. 57, Chap. 3). The city also, undoubtedly, sought something different, something to attract attention. The choice did result in certain trade-offs, however: many of the seats in a surround hall are acoustically inferior to those in the great traditional shoebox halls. Of course, the loss is balanced by a gain: listeners behind the stage trade acoustical quality for the opportunity to see the conductor face-on and to feel closer to the performers.

The Berlin experiment, and other innovations in performance space around the world, demonstrates nothing if not that different halls provide different acoustics almost always with a degree of give and take, of value lost and found. Karajan's preferences notwithstanding, conductors typically learn to play in and value the acoustics of their home space, thus adapting themselves and their orchestras to one hall's "strengths" and "weaknesses." At Philadelphia's Academy of Music, both Leopold Stokowski (conductor from 1912 to 1936) and his successor, Eugene Ormandy (from 1936 to 1980), taught the players of the Philadelphia Orchestra to stretch out the endings of notes so as to simulate the effects of hall reverberation; the violinists even had to learn to bow out of unison. These techniques would seem to defy logic, but in fact both conductors knew the same thing: the Academy had "dry" acoustics, which robbed music of the usual fullness it would derive from a hall's reverberation.*

In the 1930s, Stokowski in Philadelphia and Serge Koussevitzky of the Boston Symphony Orchestra became objects of study for some students of acoustics at Harvard University. Isolating the different conducting techniques that each man used, the students deduced that each had adopted his approach to achieve his individual musical style in the distinct acoustics of his home hall.

Stokowski was known for his emphasis on orchestral color—bar to bar and phrase to phrase—as well as on a long, rich, flowing melodic line. The dry, clear, warm acoustics of the Academy, however, lend high definition and rather low fullness of tone to compositions performed in it. The Harvard study suggested the bridge between his style and this space, surmising that Stokowski had developed an orchestral technique that rounded and prolonged the attack and release of each tone, tended to blend successive notes, and gave the performance a flowing silky tone. He required his violists to practice free bowing to assure smooth orchestral texture. The violas, cellos, and bases were coached to produce a smooth, rich foundation to the full ensemble. Stokowski's individual style became especially apparent when he conducted in other halls, albeit perhaps not with the same effect as in the Academy of Music.

Koussevitzky also emphasized orchestral color, but the lively acoustics of Boston's Symphony Hall favored that goal. The Harvard acoustics group observed that Koussevitzky made his attacks and releases more abrupt than Stokowski's, and he depended on the hall to enhance the fullness of tone. The bowing techniques of the cellos and basses were not so critical, since the reverberation of the hall elongates

---

* It was said that Karajan emulated Stokowski's bowing techniques. Walter Legge, who was the classical-recording manager of the British gramophone company EMI, put it this way in 1953, "[Karajan and I] worked together for years on the theory that no entrance must start without the string vibrating and the bow already moving, and when you get a moving bow touching an already vibrating string, you get a beautiful entry. But if either of those bodies is not alive and already moving, you get a click, and Karajan had already calculated all that" (Osborne, *Karajan,* p. 307).]

and rounds the tones. Koussevitzky's violin tone—ringing, brilliant, and loud—came easily in the acoustics of Symphony Hall, so much so that even slight imprecisions could go unnoticed—something that Koussevitzky would not have tolerated had they been perceptible. He loved to build up a dramatic conclusion to an allegro finale, an effect particularly suited to the acoustics of Symphony Hall.

Stokowski and Koussevitzky were themselves well aware of the different acoustics of their two halls. Each of them developed his technique to achieve his greatest perfection in his own hall and then strongly preferred that hall to all others. Koussevitzky is known to have said, "the Academy of Music is good, but not nearly as good as Symphony Hall." And Stokowski said, "Symphony Hall has good but not outstanding sound. The Academy of Music is the best concert hall in America. It has natural clear sound."

Audiences also sensed the differences in the two men's techniques. An annual visit to Boston by the Philadelphia Orchestra prompted comments about the "oversmooth, too-silky tone." In Philadelphia, the tone of the reciprocating Boston Symphony Orchestra was described as "crisp, too clearly molded, and sometimes slightly imprecise." Both reactions are eminently understandable. The result was that many years ago the two orchestras ceased visiting each other's hall—the reason given then was that the halls were unable to sell enough seats in either city. And today? The Boston Symphony Orchestra does not perform in Philadelphia's Academy of Music and the Philadelphia Orchestra only comes to Boston's Symphony Hall about once in five years as part of a local bank's "Celebrity Series."

Maestro James De Preist, Music Director of the Oregon Symphony, has had extensive experience in three of America's important East Coast halls, including the Academy of Music. Asked for his thoughts on the different reverberation times in these performing spaces, he wrote:

> A native Philadelphian, I grew up immersed in the splendid, luxurious velvet of the Philadelphia Orchestra. Hearing concerts in the highest perches of the Academy of Music remains one of the most vivid memories of my teen years. Not until years later, when I made my conducting debut with the Orchestra, did I realize how hard the musicians had to work to produce the "Philadelphia Sound."
>
> Every concert hall has its individual acoustics, which demand different adjustments and accommodations of the orchestra that plays there. The conductor in particular develops the performance of a composition in accordance with those acoustics—adjusting tempo, dynamics, structural spacing—to fashion an interpretation. The Academy of Music, with its dry acoustics [short reverberation time] that absorbs and even chokes sound, impedes leisurely tempi and long soaring lines. In response, the orchestra's conductors have required endless bow arms of the players and directed them to attack and

sustain notes in order to make the music "sing." That effort to compensate for the dryness of the hall, of course, had produced the orchestra's trademark opulence—its rich blending sonorities.

I discovered the challenges of the Academy when I began working with the Philadelphia Orchestra. It was in 1972 and I had chosen Schoenberg's ultra-romantic *Pelleas and Melisande* because I knew it would suit the rich, expansive sonority of the orchestra. Only when I stepped on stage for the first rehearsals did I realize that the dry acoustics of the hall undermined my plans for dynamics and tempi. However, the players' experience and intimate knowledge of the hall allowed us to realize the printed dynamics of the composition. More than a decade later, albeit much more experienced with my orchestra and my hall, I encountered similar difficulties during rehearsals of Mahler's *Fifth Symphony:* diminuendi evaporated too abruptly and the exquisite *pp* and *ppp* string playing, which sounded fine to me on the podium, was but a hollow vapor in the hall.

Because the Philadelphia Orchestra repeated this same program at Carnegie Hall, we had an opportunity to compare the effects of the two vastly different halls on the same orchestration—and to find out how the Philadelphians could exploit their distinctive sound in a more helpful acoustic. Anthony Gigliotti, the orchestra's longtime principal clarinet, felt that over time the Academy's acoustic became drier and drier, and by comparison the sound of the orchestra bloomed during its visits to Carnegie. The Mahler performance bore out his appraisal—in Carnegie Hall I could encourage the softest pianissimi and allow the hall to play a role in sustaining the notes. Although the adjustments were quite subtle—really a matter of recalibrating dynamics—Academy subscribers who attended the Carnegie Hall program claimed that they had never before heard the full splendor of The Philadelphia Orchestra.

A few years ago I had the chance to compare the effects of two very different acoustics in two halls on the playing techniques of their home orchestras—Boston Symphony Hall and Philadelphia Academy of Music. The Boston Symphony Orchestra, playing in its legendary Hall, offered a Brahms 2nd *Symphony* that was lean, angular and sharply etched with splendid clarity: a crisp and highly articulated Brahms, which was ideal for the Hall's long reverberation time. A great hall tends to be permissive and inviting in the process of music making. A few months later I conducted the Philadelphia Orchestra in the Academy of Music. As with Boston, I invited the orchestra to repose into the Brahms of their tradition—a more rounded and expansive performance. The results were shaped as much by the acoustics of the hall as by any other factor. Both of these magnificent orchestras are products of their environments as their respective sonic profiles witness.

As with every orchestra, the Philadelphia developed its sound in response to the acoustics of its home, and although some might view the blanketing effect of the Academy as a drawback, it has in fact demanded of its musicians, lushness, a distinctive expansiveness, that would never have been realized in a more forgiving hall.

Performers

Conductors experience acoustics in relation to the orchestra as a whole; performers, and especially soloists, need to think about the sound of their particular instrument in a space. Depending on the instrument he or she plays, a performer may react differently to an acoustical environment. As noted before, violins can suffer from a space that is too reverberant or too dry, and thus prefer a fine balance. Isaac Stern has explained this in detail:

> Reverberation is of great help to a violinist. As he goes from one note to another the previous note perseveres and he has the feeling that each note is surrounded by strength. When this happens, the violinist does not feel that his playing is bare or "naked"—there is a friendly aura surrounding each note. You want to hear clearly in a hall, but there should also be this desirable blending of the sound. If each successive note blends into the previous sound, it gives the violinist sound to work with. The resulting effect is very flattering. It is like walking with jet-assisted takeoff.

Although both are keyboards, organs—especially pipe organs—and pianos illustrate just how divergent the needs of different instruments can be. Pianists appear to be satisfied with spaces less reverberant, "dryer," on the whole, than those preferred by other instrumentalists. One rarely hears a pianist complain of a short reverberation time with the same dissatisfaction as a violinist, probably because, when chordal change is slow, the pianist has a sustaining pedal that can prolong notes. Because there is a technique for merging a tone with its successor, and because the piano is itself loud and reverberant, the pianist depends more on his performance than on the hall to create the desired effects. Music for the pipe organ, on the other hand, needs special consideration. Since the organ has no sustaining pedal, the tone stops very soon after a key is released. The performer can, with considerable effort, achieve some fullness of tone, but technique alone can never fully substitute for reverberation.

As a result, much music for the organ has developed such that it depends on reverberation. E. Power Biggs, a prominent twentieth-century organist, wrote:

> An organist will take all the reverberation time he is given, and then ask for a bit more, for ample reverberation is part of organ music itself. Many of Bach's organ works are

designed actually to exploit reverberation. Consider the pause that follows the ornamented proclamation that opens the famous *Toccata in D minor*. Obviously, this is for the enjoyment of the notes as they remain suspended in air. In harmonic structure, Mendelssohn's organ music is tailored to ample acoustics, for the composer played frequently in the great spaces of St. Paul's Cathedral in London. Franck's organ music, like that of Bach, frequently contains alternation of sound and silence, and depends for its effect on a continuing trajectory of tone. In general, a reverberation period of at least two seconds, and preferably more, is best for organ and organ music.

The performance spaces that dominated Europe, and North America, since the Renaissance have shaped, or were sometimes shaped by, the styles of music that prevailed through different periods in modern history. Today, this acoustical give-and-take continues to influence efforts to match compositions with appropriate halls. So although the pipe organ itself cannot fake reverberation, the great composer of organ music thought in terms of highly reverberant spaces and thus created compositions that still depend upon the acoustic of architecture at a particular moment in history.

ACOUSTICS AND MUSICAL PERIODS

The conductor and performer, presented with a composition and a space in which to play it, meet the challenge of making that piece shine in that environment, no matter the disadvantages of the match. The composer, on the other hand, has often enjoyed the advantage of creating a piece of music with a particular space, or kind of space, already in mind. Obviously, the reverberation a composer imagined while working depended on the architecture that dominated musical performance in his surroundings. Consequently, we can look back upon a history of musical periods with particular and sometimes very divergent acoustical demands.

Baroque Period

In spite of differences in the music written between 1600 and approximately 1750 by European composers scattered, the term *Baroque* gives us a convenient designation for the contrapuntal style from that period, a style best exemplified by Bach and Handel in the north of Europe and Corelli and Vivaldi in Italy. This century and a half witnessed the evolution of music from an unaccompanied choral song to a more highly rhythmic, harmonic-thematic balance in which voice and instrument frequently combined and the parts were not all of equal melodic interest.

The spacing of instrumental colors and the emphasis on contrast also typified Baroque music; with each movement confined to a fixed palette, the variety occurred only from one movement to another.

Acoustically, the Baroque period developed in two very divergent performance spaces, one highly reverberant and the other with high definition and low fullness of tone. Wholly familiar with these acoustical environments, the Baroque composer wrote music to suit them. Secular music, or Baroque orchestral music, generally found its audience in the dry acoustic—in the rectangular ballroom of a palace, for instance, which had considerable intimacy, or in small theaters that replicated these private spaces. In either case, these were relatively small rooms with hard, reflecting walls and, when occupied, a reverberation time longer than that of a conventional living room, yet low—less than 1.5 sec. In the small theaters, such as Munich's Altesresidenz Theater (opened in 1753), the music sounded intimate because of the many nearby sound-reflecting surfaces; when full, it had an especially short reverberation time. Even today, we prefer to listen to this highly articulated music in a small space with fairly low reverberation time.

The spaces available for Baroque sacred music covered an unusually diverse range, presenting composers with certain challenges and possibilities. Bach may be the composer who best exploited that range. Because most of the important churches of the eighteenth century were very large and highly reverberant, listeners continued to prefer these for the musical forms of earlier times, such as the plain-chant. On the other hand, much of the sacred music of this period was written for performance in private royal or ducal chapels with low reverberation times; to these rooms we owe the brisk tempos of Bach's early fugues. Furthermore, during this same period, Europe saw the spread of converted and newly built Lutheran churches, in which the congregation occupied galleries as well as the main floor; the acoustics thus created were quite moderate—considerably less reverberant than that of the medieval cathedral. Hope Bagenal, the leading architectural acoustician of England in the mid-twentieth century, wrote,

> The reducing of reverberation in Lutheran churches by the inserted galleries, thus enabling string parts to be heard and distinguished and allowing a brisk tempo was the most important single fact in the history of music because it lead directly to the *St. Matthew Passion* and the *B-Minor Mass.*

Bach composed many choral works specifically for this unique environment and created many of his large works, including the *B-Minor Mass* and the *St. Mathews Passion,* during his tenure as cantor of the Thomaskirche in Leipzig. Drawing on extant representations of the original Thomaskirche—lithographs and descriptions of the tapestries, altars, and other art works—we can estimate its acoustical qualities with a great deal of accuracy. The original building probably had a reverberation time of about 1.6 sec at mid-frequencies with a full congregation and of a little over 2 sec when partly full—a dry environment, as we would call it today, for ecclesiastical organ and choral music. Bach was the supreme master of counterpoint, the art of combining different melodic lines in a musical composition, and this environment made hearing the harmonic relationship between the lines ideal.*

Classical Period

From 1750 until roughly 1820, European audiences enjoyed music written in the "Classical" style. During this relatively short span of time, a wider secular appeal gave a new impetus to the composer. Although the church and the court still commissioned works throughout most of the eighteenth century, the growing interest of music publishers, entrepreneurs, and purveyors of public entertainment increased the composer's influence and imposed changing demands on him. Haydn, Mozart, and Beethoven created their great symphonies in this period.

From an acoustical point of view, the Classical symphony and sonata together constituted the most important development of the period. These forms typically synthesized a number of independent musical ideas—some related, some contrasting—into a single unit. The way in which the ideas were put together—the structure of the music—sometimes became even more important than the musical material itself. Characteristic of the Classical period is a diminished emphasis on the contrapuntal style, the new direction following the operatic idea of accompanied melody rather than the interweaving of equal parts that characterizes a Brandenburg Concerto. As Leonard Bernstein explains the Classical style in his *The Joy of Music* [Simon and Schuster, New York, pp. 232–233 (1959)]:

> Counterpoint is melody, only accompanied by one or more additional melodies, running along at the same time. . . . [T]his music is difficult for us to listen to. . . . Today, we are used to hearing melody on top with chords supporting it underneath like pillars—melody and harmony, a tune and its accompaniment.

Overall, the changes from Baroque to Classical resulted in a bigger sound, in which fullness and depth gained ground over the clarity and briskness that Bach had exemplified. In the Classical symphony, a wave of strings carried the main part of

---

*After it acquired a higher ceiling during a nineteenth-century renovation, the reverberation time at mid-frequencies with full audience increased to about 1.9 sec [L. Keibs and W. Kuhl, "Zur Akustik der Thomaskirche in Leipzig," *Acustica,* Vol. 9, pp. 365–370 (1959)].

the melodic material, augmented by woodwind passages as these became more prominent. Bringing in the full orchestra emphasized larger movements.

These new sounds developed concomitantly with the expanding audience. Growing in number throughout the eighteenth century, public concerts were performed in London and Paris; by the end of the eighteenth century, public concerts became highly popular, owing mainly to historical and sociological developments but perhaps also to the new musical style. Significantly, Haydn composed twelve of his symphonies between 1791 and 1795 especially for Salomon's series of concerts at the Hanover Square Room in London. At the turn of the century, concert music as public entertainment spread across the continent, appearing in Leipzig, Berlin, Vienna, Stockholm, and elsewhere.

While building in this period did reflect the growing popular appeal of orchestral performance, the architecture lagged behind the new potential of the music. The first real concert halls, built in the last half of the eighteenth century, still showed the influence of the court halls: they were almost all rectangular, seated audiences of 400 or fewer, and had relatively short reverberation times. The Holywell Music Room in Oxford, England, completed in 1748 and recently restored, seats about 300 and has a reverberation time of about 1.5 sec at mid-frequencies. Vienna's Redoutensaal, which stood in Beethoven's time, seated an audience of 400 people and had a reverberation time, with full audience, of about 1.4 sec at mid-frequencies. The Altes Gewandhaus, which stood in Leipzig from 1780 to 1894, also seated 400 and had a reverberation time, when fully occupied, of not more than 1.3 sec at mid-frequencies.

Toward the middle of the nineteenth century, the popular appeal of orchestral concerts became manifest in the construction of the first large halls specifically designed for concerts. These halls also had much longer reverberation times. For example, the old Boston Music Hall, which opened in 1863, retained the rectangular shape of the earlier halls described above, but it seated 2,400 persons and had a reverberation time of over 1.8 sec, with full audience. The "Neues" Gewandhaus in Leipzig, completed in 1886 (and destroyed in World War II) embodied the same change on a slightly smaller scale: also rectangular, it held an audience of 1,560 and had a reverberation time of 1.6 sec. The longer reverberation times of the best of these large concert halls added to their fullness of tone and, hence, to the dramatic value of the music, while at the same time their narrow rectangular shapes, which provided early reflections from the side walls, preserved the clarity necessary for Classical music.

Today the preferred reverberation times for music of the Classical period appear to be in the range of 1.6 to 1.8 sec, which is reasonably consistent with the acoustics of the Leipzig, Oxford, and Vienna halls of that time. Beethoven's symphonies, particularly his later ones, showed the immense scope of his imagination—he wrote almost as though he anticipated the large reverberant halls that would be built in the next 150 years.

Romantic Period

For the next hundred years a succession of composers—from Schubert and Mendelssohn, through Brahms, Wagner, Tchaikovsky, Richard Strauss, Ravel, and Debussy—created a body of music that, together with the Classical symphonies, make up the preponderant part of today's orchestral repertoire. From Haydn onward, each generation of composers increased the size and tone color of the orchestra and experimented with the expressive possibilities of controlled definition. The music no longer required listeners to separate out each sound they heard to the same extent that Baroque and Classical music had. In some Romantic compositions, a single melody might be supported by complex orchestral harmonies; in others, a number of melodies interweave, their details only partly discernible in the general impression of the sound; and in some musical passages no melody seems to emerge, only an outpouring of sound, perhaps rhythmic or dramatic, often expressive or emotional.

The music of this period, as normally performed, thrives in an acoustical environment that provides high fullness of tone and low definition. Conductors and musicians confirm the experience of recording engineers that these qualities are achieved with a relatively long reverberation time, about 1.9 to 2.1 sec, and a small ratio of sound that arrives directly from the performing group or from nearby side walls to the reverberant sound energy that follows. That finding reflects the preferences of performers and concertgoers today, and not surprisingly it matches the choices made in concert hall construction in the last half of the nineteenth century.

Composers of this period sometimes wrote with a specific concert hall in mind. Wagner composed *Parsifal* expressly for his Festspielhaus in Bayreuth, Germany, for example, and Berlioz composed his *Requiem* for Les Invalides in Paris. In the last half of the nineteenth century, halls built specifically for the performance of concert music reflected the composers' desire for acoustics with high fullness of tone. The Grosser Musikvereinssaal in Vienna, for example, completed in 1870, has a reverberation time at mid-frequencies of about 2 sec when the hall is fully occupied. The hall is small enough for *ff* orchestral effects to sound very loud, and its narrowness, which emphasizes the early sound reflections from sidewalls, lends both significant definition and spaciousness to the music. The Concertgebouw of Amsterdam, which was completed in 1887, also has a reverberation time of 2.0 sec at mid-frequencies, but as it is wider it emphasizes the early sound reflections less; therefore, the music played in it emerges with less clarity and more fullness of tone. The Concertgebouw has excellent acoustics for music of the late Romantic period.

Twentieth Century

Since the 1880s, concert going has grown into a well-established cultural activity in Europe and the Americas. More recently, an impressive surge of interest in concert-hall music has swept Japan, and Tokyo might soon become the world capital of concert-hall music. Japan boasts a large number of symphony orchestras and music conservatories, as well as widespread public patronage; over eighty concert halls have gone up across the country just since World War II. Tokyo alone has more than ten symphonic orchestras, of which two, perhaps, are world-class; descriptions of nine Tokyo halls appear in Chapter 3 of this book.

The building as well as renovation of performance spaces at this time brings with it the challenge of accommodating earlier styles and meeting the developments of more recent compositions. The concert music of the twentieth century presents audiences with a great deal of variety, some of it reflecting earlier styles and some of it largely unprecedented. Many contemporary composers have written for large orchestras along lines similar to those developed in the second half of the nineteenth century, though with the addition of new harmonies, new instruments, and new effects. Some of these works evince a return to the clearer sound of pre-Classical periods, which requires smaller instrumental combinations and sometimes calls for instruments directly patterned on those of that earlier epoch. At the same time, new musical innovations emerge every year, including experiments with sound from sources other than conventional instruments—the electronics laboratory, the tape machine, and the computer all furnish the components of a new music or at least novel sounds.

Daniel Pinkham, an active composer in New England, illustrates the dilemma faced by a composer today working primarily in the "Baroque" environment of King's Chapel in Boston (1749). He writes,

> Music that I have composed for King's Chapel is in a style which might sound muddy when performed in a reverberant concert hall, but which sounds at its best in that rather dry environment, which transmits the details of each line with crystalline clarity while still providing a useful blend for the various lines. When I was preparing my *Easter Cantata for Chorus, Brass and Percussion*, the rehearsals were held in Jordan Hall, Boston, which is fairly live. For the actual performance that followed in a TV studio, I found that the only way to cope with the dead acoustics was to permit the percussion instruments to ring as long as they would, and this gave the *whole sound* the impression of

adequate reverberation. As a result of this experience, I have written my *Concertante No. 3 for Organ and Percussion Orchestra* so that the after-ring of the percussion following each phrase is deliberately carried over into the beginning of the next phrase; in a dead hall this will compensate for the lack of reverberation, while in a live hall it may either enhance the reverberant sound of the room or the percussion ring may be curtailed at the will of the performers to minimize confusion. On the other hand, I have found great difficulty, even with highly experienced musicians in performing in a live hall some music that had originally been written for the dead acoustics of the TV studio.

To meet the needs of modern concert music, a contemporary hall would best accommodate a variety of styles. Music of the transparent, "intellectual" type wants a hall with relatively high definition, of the kind required for Bach. New halls now exist that fulfill this requirement; often referred to as "hi-fi" halls, they combine high definition with reverberation times at or near 1.4 sec at mid-frequencies when occupied. Modern music of a more passionate or sentimental quality sounds best in a hall with high fullness of tone and low definition, the same profile typical of many late-nineteenth-century spaces.

A real need persists, however, for halls that can straddle a range of musical styles—halls with variable acoustics. Some attempts at such designs appear in Chapter 3 of this book, including halls that make use of devices that reduce reverberation times, such as retractable curtains, and devices that increase reverberation times, such as operable doors that permit the addition of large reverberant chambers. A few major halls are experimenting or have experimented with electronic augmentation—an art and craft that is slowly gaining in acceptance—but this effort still applies mostly to halls that need to correct for basic acoustical deficiencies.

European Opera

Of all the types of halls for music ever designed, the European opera house has proved the most consistent over time. From at least 1700 on, the horseshoe-shaped theater has been built with rings of boxes one atop the other and a crowning gallery of low-priced seats. The form reached its apogee in the Teatro alla Scala, the well-loved La Scala of Milan, completed in 1778. Its horseshoe design has cropped up in nearly every important city in Europe and on other continents as well. This ubiquitous opera house, with its relatively short reverberation time (1.2 sec), has allowed composers a unique creative privilege: to write opera with one kind of acoustics in mind.

The acoustical requirements of opera differ from those for orchestral concerts. Because the vocal part functions, like speech, as a form of communication, listeners should be able to hear the words clearly. To preserve a libretto's intelligibility, especially at the tongue-twisting musical speeds of Mozart and Rossini, the performance space must provide a relatively short reverberation time, so that the reverberation from one sound or chain of sounds will not mask successive syllables.

Many opera houses satisfy these requirements, especially in Europe, where audiences usually hear an opera in their own language. The opera houses of Naples San Carlo, Paris Garnier, London Royal Opera, Vienna Staatsoper, Munich Staatsoper, and Academy of Music in Philadelphia exemplify this type of acoustic. In these houses, the singers' voices reach the audience with clarity and sufficient loudness, and the orchestra sounds clean and undistorted. This balance between orchestra and vocalists is assisted by the acoustical design as well as by the conductor's control of the orchestra. In the Americas and in Japan, where audiences typically listen to an opera in the language of its original composition, the need for precision becomes less important. Consequently, in these houses, the reverberation can be longer—more attuned to the music than to following the libretto. One hears this acoustic at, for example, Teatro Colón in Buenos Aires (1.6 sec), Metropolitan Opera in New York (1.5 sec), and New National Theater in Tokyo (1.5 sec).

The works of Richard Wagner constitute the one exception to the European standard: Wagner broke with the tradition of Baroque-like opera and evolved a style closely related to the traditions of the Romantic period, albeit still wholly distinctive. From his pen flowed some of the most unusual and stirring of operas, "musical dramas" as he called them. Wagner's rich Romantic music is best supported by high fullness of tone and relatively low definition. His orchestral passages, with their relatively low speed, sound best in a hall with a long reverberation time, approximately 2 sec at mid-frequencies; however, in order for the libretto to remain intelligible the reverberation time needs to be somewhat shorter, in the vicinity of 1.6 sec.

In an effort to achieve the perfect acoustical environment for his musical style, Wagner designed his own opera house—the Festspielhaus at Bayreuth, Germany—a house that combines a relatively long reverberation, 1.6 sec at mid-frequencies, fully occupied, with a thoroughly blended orchestral tone. Although performances here use a very large orchestra (100 to 130 pieces), proper balance is maintained between singers and orchestra by means of a sunken and covered pit, which, through a slot in the cover, also imparts a mysterious quality to the music.

Acoustics and Listeners

In the end, any study of performance hall acoustics has one goal: to perfect the audience's enjoyment. All the constituent elements—the architecture, the composition, the conductor, the performers, everything—come together in the listener's experience of live music. And listeners take in the hall as a whole, rather than as a diagrammatic breakdown of reflective and absorbing surfaces. For example, as mentioned regarding the Berlin Philharmonie above, a layout that impedes the acoustics of a hall may still please the audience, since it can add to the visual experience.

In an effort to make sense of what listeners prefer in a concert hall, one naturally gravitates toward those institutions with outstanding reputations on the assumption that audiences intuitively recognize superior acoustics. Although the ingredients that create a hall's reputation may seem to have little to do with acoustics—positive reviews, high-profile guest composers, etc.—many are nonetheless traceable, at some level, to the physical dynamic of the space.

Music critics are, in a sense, audience members with exceptional experience and a heightened awareness for the details of a performance. They can be of vital importance when it comes to gauging a hall's reputation for its acoustical quality. Every critic knows that the acoustics of a hall form the conduit between the performing body and the listener and, hence, shape what the latter hears. Every critic also knows that a musical composition was probably composed for performance in a particular acoustical environment. This knowledge, then, contributes to the critic's assessment of a performance.

Critics have a significant influence on a hall's reputation when the institution holds its gala opening or inaugural performance after major renovations. At that opening concert, the critic probably asks him or herself an array of questions all of which reveal something about the new building's acoustics. Are the tone qualities of and the balance among the different sections of the orchestra to my liking? Is the sound as good or better than that in the hall(s) that I regularly attend? Is the hall overly bright or reverberant or too clinical or dry? Did the principal piece sound as good as the best performance of that composition that I have ever heard? If not, was the orchestra at fault or the acoustics? Were there any echoes or disturbing sound reflections? Was there any acoustic distortion?

Of course, a careful critic realizes that it may be too early for pronouncements: it takes months for an orchestra to adjust to the acoustics of a new hall and later "tuning" of the hall may improve its sound (several famous halls received low critical acclaim based on their first concerts, but rose to their present high estimation over the years). Nonetheless, the critic must analyze the music heard that night to produce the review, the wording of which will affect, possibly substantially, the hall's reputation. Taken together, the judgments of a number of writers average out, with the most weight going to the most prestigious publication with the widest circulation.

Undeniably, a hall also wins large, regular, and satisfied audiences when it can feature visiting musical dignitaries season after season, as well as a strong home orchestra. Carnegie Hall is famous not only because its acoustics are favorable but also because the great orchestras and performers of the world appear there. In this, the relationship between good acoustics and good seasons is symbiotic, as demonstrated above: the best performers will gravitate to the best halls because that acoustic enhances their sound. The combination produces the excellent performance that leaves listeners profoundly happy.

The conductors and soloists—the most high-profile performers—make judgments based, in part, on factors that the audience is not aware of, such as the response to their ears of the stage enclosure, the ability of the musicians to hear each other, and the early reflections from the auditorium that reach their ears. If every conductor says that Hall "X" is one of the great halls of the world, their combined opinion will outweigh any results of questionnaires that might be addressed to that hall's concertgoers.

From these dense and unscientific phenomena—reputation, listening pleasure, etc.—the acoustical engineer must find a way to tease out acoustical quality and to relate that to how listeners respond. Unfortunately, we do not have recent studies to go on: no comprehensive, modern, laboratory-controlled tests of listeners' preferences for a concert hall's acoustical characteristics. Nonetheless, many halls are undergoing renovations, with the principal object of improving their acoustics, typically by increasing their reverberation times. Orchestra Hall in Chicago, for example, now has a reverberation time of 1.75 sec, up from 1.2 sec in 1994. For halls that generally feature standard orchestral repertoires, those changes indicate that the mid-frequency reverberation times should be in excess of 1.6 sec, but optimally between 1.8 and 2.1 sec with the hall fully occupied.

A recent in-depth study of 24 opera houses has attempted to determine which acoustical characteristics most affect conductors' judgments of acoustical quality in those venues (Hidaka and Beranek, 2000). The survey obtained responses from 21 well-known opera conductors and six or more opinions on 21 houses. The questionnaires contained a rating scale for each house, with the options "Poor" and "One of the Best" at the two extremes and "Passable," "Good," and "Very Good" in between. The average ratings of the houses by the conductors extended from halfway between "Passable" and "Good" to half-way between "Very Good" and "One of the Best," with the Buenos Aires Opera Colón at the top and the very large Tokyo NHK Hall at the bottom.

The reverberation times of the six best opera houses ranged from 1.24 sec (Milan, La Scala) to 1.6 sec (Dresden, Semperoper) with the mean value for all of

1.4 sec. The best halls seated less than 2,500 persons, and thus were not too wide, creating a sense of "intimacy," which is provided only if there are sound-reflecting surfaces near the proscenium. The loudness of the singers' voices in the audience area depended on the type of scenery on stage and whether the singers stood near the apron of the stage or farther back. In the best houses, the singers' voices were judged as "clearer" than the sound of the orchestra, because singers are more exposed. The conductors were also asked to give separate grades for the acoustics as heard in the audience and as heard in the pit. In only three houses (Munich, Prague, and New York) was the sound in the pit (the conductor's position) judged significantly better than the sound in the audience. Finally, the "texture" of the sound in the best houses rated above that in the least good houses (see Chapter 2 for a definition of texture and the usual method of measurement).

With this general introduction, we shall move in the next chapter to the terminology used in musical acoustics and to an understanding of the acoustical factors that are believed, at this stage of acoustical knowledge, to determine the quality of a concert hall or opera house.

TWO

The Language

of

Musical

Acoustics

Definitions and Explanations of Selected Terms

In recent years a common language of acoustics has developed out of the dialogue between musicians and acousticians. Not every musician is familiar with all the terms, nor do all acousticians agree on the definitions applied in this text. Nevertheless, the list compiled here results from a studied compromise, compiled from interviews with musicians and concert aficionados and from literature about concert hall and opera house acoustics.

A third group, music critics, provide a useful foil. Since they need to convey their reactions to a lay public in vivid terms, they use a subjective language that is rarely amenable to precise definition. They may describe the acoustics of a new concert hall as "overbearing," "ravishing," "shimmering," or "shattering." While such words cannot serve as guides to successful acoustical design, they do have a purpose. Because the critic is a perceptive, experienced listener, both the musician and the acoustician must pay attention to this language, gleaning from it information that facilitates their own professional goals.

The 25 terms defined here cover all the important aspects of music performed in the acoustics of a closed space (a chart at the end of this chapter provides an overview). Accompanying each definition is an explanation of the physical characteristic that affects the defined acoustical parameter. Please keep in mind that these definitions and the discussions in this book concern concert halls and opera houses seating more than 700 persons, with occasional references to churches and halls for chamber music. The acoustics of small rooms and broadcast studios, which typically have problems at the low frequencies, are not treated here.

Reverberation and Fullness of Tone

Reverberation, to the acoustician, is the continuation of a musical sound in a hall after the instrument that produced it ceases to sound. This "after-ring" is particular to music played indoors; outside, a note has no reverberation—sounds simply become weaker as they travel outward. Imagine that a violinist on stage in a concert hall plays a single note. The acoustical wave that radiates outward from the violin in all directions encounters the surrounding surfaces—walls, balcony fronts, ceiling, and the audience area. A listener, as shown in Fig. 2.1, hears first the direct sound wave, followed, after a brief interval, by a succession of "first" reflections (designated in Fig. 2.1 as R_1, R_2, R_3, etc.). Those first reflections will go on to encounter more surfaces, including the audience "surface," and more waves of reflections ensue. In one second, in fact, because sound travels so rapidly, the single note from that violin can impinge on and be reflected from a room's surfaces some 20 times. Of course, the sound loses some energy at each encounter—particularly at the audience area—so that what the listener hears gradually dies down to inaudibility.

In the work that acousticians do in concert halls, "reverberation time" (RT) holds a key position. In general terms, the reverberation time of an enclosed space refers to the amount of time it takes for a loud sound in that space to become completely inaudible after it is stopped. In more technical terms, reverberation time is the number of seconds it takes for a loud tone to decay 60 decibels after being

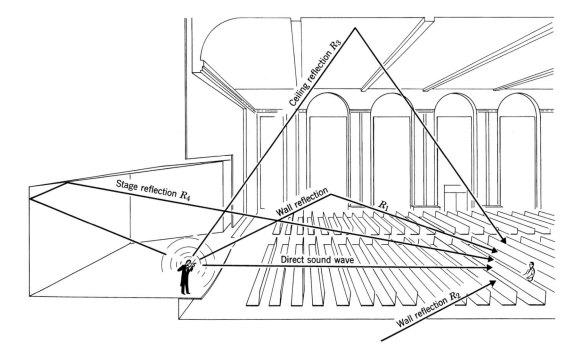

Ceiling reflection R_3

Stage reflection R_4

Wall reflection

R_1

Direct sound wave

Wall reflection R_2

ℱIGURE 2.1. An illustration of the paths that sound travels from player to listener, including the direct sound and four early reflected sound waves out of many that may occur in the early time period of 80 msec. These additional reflections may occur from balcony faces, rear wall, niches, and any other surfaces, including the audience "surface."

stopped, also known as the "sound decay process" (Fig. 2.2). A sound created in an unoccupied concert hall, with hard interior surfaces and without sound-absorbing materials (carpets, for example), will continue to reverberate for three or more seconds after cessation; that is to say, the room has a reverberation time of three or so seconds, unoccupied. The addition of an audience, which reflects but also absorbs sound, will cause the reverberation to die out sooner. The most famous halls have reverberation times, fully occupied, at middle frequencies (the frequencies between 350 and 1400 Hz; see definition below), in the range of 1.8 to 2.0 sec.

Reverberation is in itself neither desirable nor undesirable; it is one of the components available to the composer (and the performer) for producing a musical effect. Because the reverberant sound fills the spaces between new notes as they are played, it provides the "fullness of tone" that musicians may employ or restrain, as

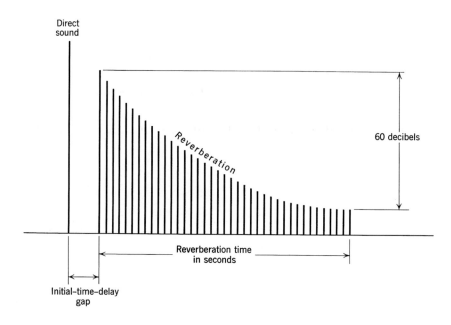

𝒯IGURE 2.2. A diagram of sound decay at a listener's ears. Here the direct sound and its strength appears as the vertical line at the left of the diagram, followed by the initial-time-delay gap (ITDG). The ITDG is followed by a succession of sound reflections that decrease in amplitude as they suffer loss at each reflection from surfaces in the room. The reverberation time of the room is defined as the length of time required for the reverberant sound to decay 60 decibels. This time is usually measured from −5 dB to −35 dB and multiplied by a factor of 2 to give the equivalent of a 60 dB decay. Starting the measurement after the sound has dropped 5 decibels eliminates the ITDG from the result. In this drawing, the length of the ITDG is exaggerated, shown as about 200 msec, nearly 10 times the length found in the best concert halls, which have reverberation times of about 2.0 secs, fully occupied.

their needs demand. Beginning in the sixteenth century, composers wrote choral and liturgical music for performance in both large highly reverberant cathedrals and smaller, less reverberant churches or chapels. In every period, composers have written chamber music for small groups of instruments played in small rooms, where the reverberation times are short. Between 1700 and 1850 much orchestral music was written for groups that performed in larger but reasonably narrow public halls and in ballrooms in the palaces of Europe. The reverberation times in these spaces reached a maximum of 1.5 sec at mid-frequencies. Similarly, opera was usually performed in relatively small, and narrow, horseshoe-shaped opera houses, such as the Teatro di San Carlo in Naples, Italy, which has a reverberation time of only about 1.2 sec.

This array of venues, with their distinct acoustical personalities, has allowed composers to conceive their musical works with particular ranges of reverberation times in mind. Ideally, every performance of a work finds an environment suited to its composition—or an orchestra skilled at negotiating discrepancies between composition and hall. When a work is performed in a hall with an unsuitable reverberation time, a sophisticated listener will quickly realize the mismatch. As the organist E. Power Biggs once said, "The listener immediately senses something wrong when he hears one of the organ works composed for performance in a cathedral played in a small college auditorium."

Direct Sound, Early Sound, Reverberant Sound

Direct sound is the first sound a listener hears coming from an instrument on the stage—that is, the sound that travels directly from instrument to listener. In Figs. 2.2 and 2.4, it appears as the vertical line at the left of the graph. The term *early sound* encompasses the direct sound plus all the reflections that reach a listener's ears in the first 80 milliseconds (msec) after arrival of the direct sound (a millisecond is equivalent to one-thousandth of a second; 100 msec is a tenth of a second). The reverberant sound includes all the reflections that arrive after 80 msec. In the paragraphs that follow, frequent references appear to the ratio of the energy in the early sound to that in the reverberant sound.

Early Decay Time (EDT) (Also Early Reverberation Time)

When a musician or ensemble plays rapidly, only the early part of the sound decay process remains audible between successive notes (see "speed" below). "Early decay time" (EDT) designates that initial phase of sound decay. More specifically, it is the exact amount of time it takes for a sound from a musical note to decay 10 decibels after it is cut off, multiplied by a factor of 6. Why the factor of 6? Ten decibels of decay occurs in a time period roughly equivalent to one-sixth of the time required for 60 decibels of decay, defined above as the reverberation time. The factor 6 allows a direct comparison between EDT and RT (both in seconds). The descriptions of the concert halls in Chapter 4 include a correlation between each hall's subjective rank-ordering, gathered from interviews and questionnaires, and its EDT and RT measurements. The results make it apparent that EDT indicates acoustical quality better than RT does, because notes played by violinists in symphonic compositions usually follow each other very rapidly.

Speed of Successive Tones

When a musician performs, the speed at which he or she plays has a vital relationship with the acoustics of the hall. In particular, the speed at which successive tones follow one another interacts with reverberation time and thus shapes what the audience hears. The two graphs in Fig. 2.3 demonstrate this process: each graph shows two successive tones played on a musical instrument in different acoustical environments. If the reverberation time is long, as in Graph A (3 secs), the second tone falls beneath the reverberant sound and becomes inaudible. In other words, music played at this speed in a hall with high reverberation time will have little clarity (see "Definition," below). Graph B shows the same two notes in relation to a reverberation time of 1 sec. In this case, the second tone comes through clearly, as does its ensuing reverberation.

The faster the musicians play, the more quickly notes pile up under even short reverberation times. Were the speed of successive notes as represented in Graph B to increase, the second tone would move to the left because the time between the two notes becomes shorter. If the speed is great enough the second tone will fall below the reverberation line for the first tone even for a 1-sec reverberation time and will become inaudible. If there is no reverberation (RT = 0), the notes will stand out clearly whether played fast or slow, just as with music outdoors.

We must emphasize, however, that for any given tempo, the number of notes squeezed into a musical measure may be large, particularly for stringed instruments, thus the separation between notes is a fraction of a second, and the strengths of the successive notes are not greatly different. Thus the presentation in Fig. 2.3 of two successive notes 15 decibels apart in strength, and played very slowly, demonstrates principle only, not actuality.

Definition (or Clarity)

The terms "definition" and "clarity" are synonyms for the same musical quality. They name the degree to which a listener can distinguish sounds in a musical performance. Definition is discernable in two forms: *horizontal,* related to tones played in succession, and *vertical,* related to tones played simultaneously. In either case, definition results from a complex of factors, both musical and acoustical—a certain piece of music, played in a certain way, in a certain acoustical environment.

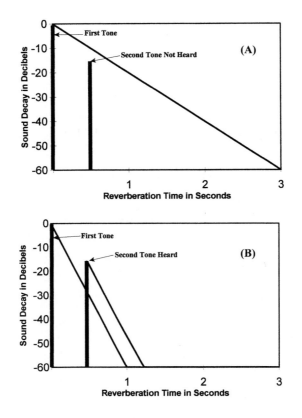

𝓕IGURE 2.3. A demonstration of the way a long reverberation time "masks" (hides or drowns out) a musical note. Graph A shows the first note starting at the left side of the graph and dying out (−60 dB) 3 secs later. A second note (very weak with a strength that is 15 dB less than the first note) is sounded 0.5 sec later. It is masked by the falling reverberation of the first note. Graph B shows that if the reverberation time is shorter (1 sec), the second tone and its falling reverberation will remain audible. The decaying of the first note "bridges" the gap between the two notes, which adds "fullness of tone" to music.

"Horizontal definition" refers to the degree to which sounds that follow one another stand apart. The composer can specify certain musical factors that determine the horizontal definition, such as tempo, repetition and number of tones in a phrase, and the relative loudness of successive tones. The performer can vary the horizontal definition with phrasing and tempo. Both, however, have to work within the acoustical qualities of the hall. Horizontal definition increases both as the length of the reverberation time decreases and as the ratio of the loudness of the early sound to that of the reverberant sound increases.

Horizontal definition is usually defined by acousticians as the ratio *expressed in decibels* of the strength of the early sound to that of the reverberant sound. Thus, if the definition, in decibels, is a positive quantity, the early sound dominates. If negative, the strength of the reverberant sound dominates. If zero, they are alike.

"Vertical definition" refers to the degree to which notes that sound simultaneously are heard separately. Like horizontal definition, vertical definition depends on the score, the performers, and the auditory acuteness of the listener, as well as on the acoustics of the room. The composer specifies the vertical definition by choosing which tones sound at once, the instrument(s) for each tone, and the relation among the tones; the latter can vary, for instance, depending on whether the composition is hymnlike, chordal, contrapuntal, or simply an accompanied melody. Performers can alter the vertical definition by varying the dynamics of their simultaneous sounds and by the precision of their ensemble. The performance space shapes vertical definition vis-à-vis acoustical factors such as balance among the sounds of the various instruments as they reach the audience; the degree to which the tones from the different instruments in the stage enclosure blend together; the relative response of the hall at low, middle, and high frequencies; and the ratio of the energy in the early sound to that in the reverberant sound.

To communicate a piece of music faithfully to an audience, the musicians and conductor need to discern and follow the degree of definition that the composer intended; success in that endeavor requires a thoughtful choice of hall. Gregorian chant—with its slow melodic lines that build and recede gradually—is best performed with little horizontal definition, preferably in a cathedral-like room with a very long reverberation time and lots of reverberant energy compared to the early sound energy. Bach's *Toccata in D Minor* for organ needs a reverberation time of at least 3 sec in order to realize its full sonorities. At the other end of the spectrum, a piano concerto by Mozart—with its rapid solo passages and the delicate interplay of piano and different orchestral voices—needs considerable horizontal and vertical definition. It should be performed in a room that has a relatively short reverberation time and that allows for a larger amount of energy in the early sound relative to that in the reverberant sound. Mozart, after listening to a performance of *Die Zauberflöte* from various locations in an unoccupied opera house, wrote in October 1791: "By the way, you have no idea how charming the music sounds when you hear it from a box close to the orchestra—it sounds much better than from the gallery." Mozart learned what later generations have confirmed, that even in a hall with a relatively short reverberation time, his style of music sounds best at a location where the amount of early energy exceeds that of reverberant energy.

Resonance

In speaking of concert hall acoustics, a musician may use the terms "reverberance" and "resonance" interchangeably, but an acoustician employs each to signify very specific—and distinct—phenomena. The time-worn musical example of resonance is the story of the soprano who hits a certain high note and shatters a crystal stem-glass. The explanation? The frequency (tone) of her loud high note coincided with the frequency of a "natural mode of vibration" of the glass, thereby inducing a vibration vigorous enough to break the glass.

Every body or object, even those that appear to be wholly inanimate, is constantly in motion on a molecular level. By the same token, every object vibrates, quite naturally, at certain frequencies. Resonance refers to the specific coincidence of two phenomena: the glass resonates (1) because an external sound, vibration, or other force—the soprano's high note—matches (2) one of the glass's natural modes of vibration and causes it to vibrate vigorously at exactly that frequency. When you strike the head of a tympani with a mallet the tones you hear come from one or several of its natural modes of vibration: the air column inside the tube beneath the bar of a xylophone will resonate when the frequency of the bar matches the air column's natural mode of vibration.

For technical discussions, it is also necessary to distinguish resonance from "augmentation of loudness," although musicians may conflate the two. For example, when the string on a violin is set in motion by the bow, the body of the violin will amplify the string's tone: the violin's large surface area radiates sound many times better than does the small area of the string. The amplification does *not* occur because the string's frequency matched the frequency of a natural mode of vibration of the body. An even louder tone is generated, however, when a natural mode of vibration of the violin body coincides with the frequency of the tone generated by the vibration of the string.

Intimacy or Presence and Initial-Time-Delay Gap

A small room has visual intimacy—people in the room see the walls and other objects relatively nearby. By the same token, a hall can have "acoustical intimacy" if sounds seem to originate from nearby surfaces. (Professionals in the recording and broadcast industries would say that such a hall has "presence.")

This phenomenon matters to the acoustical engineer because it arises from the specific physics of a space. The degree of musical intimacy in a space corresponds to how soon after the direct sound the first reflection reaches the listener's ears. As described above, any listener in a performance hall hears first the direct sound, which travels in a straight line from an instrument to the person's ears in just a few hundredths of a second. The reflected sound waves follow just behind. If the time difference between the direct sound and first reflection, also known as the *initial-time-delay gap* (ITDG) (Figs. 2.2 and 2.4), is short, the hall sounds intimate. Acoustical measurements show that in the center of the main floor of the best-liked halls the ITDG remains at or below 25 msec. In a lower-grade hall, the ITDG exceeds 35 msec; in a poor hall, it reaches 60 msec or greater.

Usually the first reflection arrives from a sidewall or a balcony front. Thus, for a low ITDG, a hall should be narrow and have near-parallel sidewalls. In a fan-shaped hall, the early reflections are reflected into the back corners, so that the first reflection a listener in the center of the main floor hears will probably come from a high ceiling, which means that the ITDG may greatly exceed 30 msec. In that case, hanging reflecting panels or "saw-toothed" panels along the sidewalls can be used to guide early sound to the listener and to reduce the ITDG to the 20-msec region.

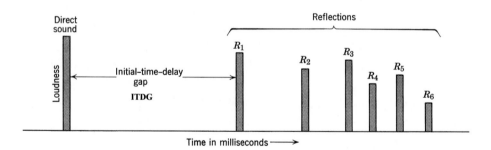

FIGURE 2.4. A "reflection pattern" diagram, useful for measuring the initial-time-delay gap (ITDG). It illustrates the sequence in which sounds travel to a listener's ears and the relative loudness of each sound. The vertical line at the left shows the sound that travels directly from the performer to the listener, the first sound to arrive. After the ITDG, reflections from the walls, ceiling, stage enclosure (see Fig. 2.1), and other reflecting surfaces arrive in rapid succession. The height of the direct sound plus the height of the series of reflections suggests the loudness of the sound.

Liveness and Mid-Frequencies

A "live" hall has a long reverberation time; a hall with a short reverberation time is called "dead" or "dry." "Liveness," a largely subjective term, corresponds generally to the reverberation times at the frequencies between 350 and 1400 Hz (for orientation, the note A above middle C on the piano keyboard has a frequency of 440 Hz) because a person's hearing is more sensitive in that region. The average of the reverberation times in this frequency range, called the *reverberation time at mid-frequencies,* constitutes a part of the description for each hall in the next chapter. In a concert hall, when a tone is sounded and then suddenly turned off, the reverberation time at mid-frequencies remains audible for about 1.5 to 2.2 sec. In an opera house the range is generally 1.2 to 1.6 sec.

Spaciousness

When a cello, for example, plays in a relatively narrow, shoebox-shaped concert hall, the sound appears to the listener to emanate from a space substantially wider than the actual width of the cello. This phenomenon, known as spaciousness, comes primarily from lateral reflections in the hall, such as reflections from sidewalls and side-balcony fronts. Properly measured, the degree of spaciousness ranks as one of the primary reasons why one concert hall or opera house sounds better than another.

A listener takes sound into the brain binaurally—that is, through the two ears that are separated from each other by the head; this configuration and the way the brain interprets the resulting transmission of information from the two ears is responsible for the perception of spaciousness. When a listener faces a source of music outdoors, the sound that reaches him or her will be identical at both ears and thus bereft of spaciousness; however, when a sound comes from one side, one ear hears it slightly sooner than does the other. Furthermore, the sound that arrives at the second ear will be of lower strength because the intervening head shadows it slightly. In a hall, where many reflections converge at various angles on the audience in the first 80 msec, a substantial portion of those heard will be lateral (Fig. 2.1). If the lateral reflections combine at each listener's two ears with the right degrees of difference, the spaciousness that results will cause the hall to be judged as having superior acoustics. Lesser spaciousness than this optimum amount results in reduced acoustical quality. Acoustical engineers measure spaciousness with instrumentation that yields a quantity called Binaural Quality Index (BQI), detailed in Chapter 4. Incidentally, the presence of strong bass sounds will also add somewhat to a hall's spaciousness, but bass strength is treated separately below.

Warmth

In general, "warmth" of music in a concert hall is directly related to whether the bass sounds are clearly *audible* when the full orchestra is playing. In technical terms, "warmth" in music is determined by the strength of the bass tones, simply measured by a sound level meter at various audience seats in a hall during which time a standardized loudspeaker radiates a 125-Hz tone from the stage. Problems with warmth can arise at either extreme. Musicians sometimes describe as "dark" a hall that has too strong a bass, which only occurs because the high frequencies become too attenuated by draperies, carpets, or other sound-absorbing materials in the hall. Conversely, the music will lack warmth if the walls or the ceiling surfaces—or both—are constructed of thin wood paneling, which soaks up low-frequency sounds. The pipes and swell boxes of a pipe organ or overly thick upholstery on the seats can also absorb low frequencies in significant amounts.

Listener Envelopment

"Listener envelopment" refers to the degree to which the reverberant sound seems to surround the listener—to come from *all* directions rather than from limited directions. In Boston's Symphony Hall, where the acoustics are judged excellent, the reverberation appears to originate in the entire upper-hall space and arrives at the audience from above, ahead, and behind. By contrast, any person seated on the main floor of a hall with a steep audience balcony at the back will perceive the reverberation as coming primarily from the front of the hall. Although the electronic measures for listener envelopment are too involved to discuss here, one can estimate it with a visual inspection of a hall. Just observe whether the sound waves have the freedom to travel around the overhead surfaces at the front, sides, and rear of the hall (imagine the freedom that a billiard ball has to bounce around the four sides of a pool table); note the presence of any significant irregularities and ornamentation on the sidewalls, ceiling, and balcony fronts, which will help spread the sound reflections to all surfaces.

Strength of Sound and Loudness

The word *loudness* as employed by acousticians bears a strong resemblance to the lay person's use of the term—the subjective perception of the volume or force of a sound at one's ears. But acousticians use "strength of sound," measured in decibels, to formulate loudness with much greater precision. Measuring a sound's strength with a microphone and a calibrated sound-level meter, engineers look at loudness in two stages of its arrival at a listener's ears: early and reverberant. Loudness of the early sound comprises the sound that comes directly from the source and the energy received from the early reflections in the first 80 msec. Loudness of the reverberant sound is defined as the total sound energy that reaches a listener in the period following the first 80 msec. Parenthically, if the strength of a sound increases or decreases by about 10 decibels, its loudness as one hears it is doubled or halved, respectively.

From these definitions, one can estimate different levels of loudness from hall to hall, depending on the physical design of each. A sound emitted uniformly in a concert hall seating 1,000 listeners would be louder than that in a hall seating 3,000 to 5,000 if both halls have the same reverberation time. (Note that in the usual concert hall at mid-frequencies the audience area actually absorbs up to 80 percent of the total sound radiated by the orchestra.) Thus, the total sound power that the orchestra emits is divided almost equally among the number of listeners, and each of 1,000 listeners will receive three times the amount of sound energy (5 decibels more) as each of 3,000 listeners. Music also sounds somewhat louder in a highly reverberant hall than in a dead hall, even though both may seat the same number of listeners.

Timbre and Tone Color

"Timbre" is the quality of sound that distinguishes one instrument from another or one voice from another. "Tone color" describes balance between the strengths of low, middle, and high frequencies, and balance between sections of the orchestra. The acoustical environment in which the music is produced affects tone color. If the hall either amplifies or absorbs the treble sound, brittleness or a muffled quality may mar the music. If the stage enclosure or the main ceiling directs certain sounds only toward some parts of the hall and not toward others, the tone color will be affected differentially.

Acoustical Glare

If the sidewalls of a hall or the surfaces of hanging panels are flat and smooth, and are positioned to produce early sound reflections, the sound from them may take on a brittle, hard, or harsh quality, analogous to optical glare. This "acoustical glare" can be prevented either by adding fine-scale irregularities to those surfaces or by curving them. In the eighteenth and nineteenth centuries, baroque carvings or plaster ornamentation provided fine-scale irregularities on sound-reflecting surfaces which reduced glare very effectively.

Brilliance

A bright, clear, ringing sound, rich in harmonics, is described as "brilliant." A brilliant sound has prominent treble frequencies that decay slowly. In a concert hall, the high frequencies, to some extent, will diminish naturally because they are absorbed in the air itself through which the waves travel. A serious lack of brilliance, however, arises primarily from the presence of carpets, draperies, or any significant amount of sound-absorbing materials. By contrast, sound may become overly brilliant with improper additions of electronic amplification.

Balance

Good "balance" entails both the balance between sections of the orchestra and balance between the orchestra and a vocal or instrumental soloist. Some of the ingredients that combine to create good balance are acoustical and others are musical. A performance can lose balance if the stage enclosure or some other surface near the players overemphasizes certain sections of the orchestra or if it fails to support the soloists adequately. Beyond that, balance is in the hands of the musicians, their seating, and the conductor's control of the players. In an opera house, balance between singers and orchestra is achieved by the stage design, the early reflective surfaces provided near the stage to assist the singers' voices, the pit design, and, again, the conductor's control of the orchestra.

Blend

"Blend" describes a mixing of the sounds from the various instruments of the orchestra such that the listener finds them harmonious. Blend depends partly on the placement of the orchestra, which should be spread neither too wide nor too deep. Blend also depends heavily on the design of the sound-reflecting surfaces close to the stage, such as those of a stage enclosure.

Ensemble

"Ensemble" refers to the ability of the performers to play in unison—to initiate and release their notes simultaneously so that the many voices sound as one. Orchestral ensemble depends on the ability of the musicians to hear (and perhaps to see) their fellow performers. The sound-reflecting surfaces near and above the performers should carry the sound from the players on one part of the stage to those on other parts. Risers on the stage are often used to enable the musicians to see each other better (they can also allow the audience on the main floor to see the players at the rear of the orchestra).

Immediacy of Response (Attack)

From the musician's standpoint, a hall should give the performers the feeling that it responds immediately to a note. This can depend largely on the hall's "immediacy of response": the manner in which the first reflections from surfaces in the hall arrive back at the musician's ears. If the reflections occur too long after the note is sounded, the players will hear them as an echo. Conversely, if the musicians hear reflections only from the nearby surrounding stage walls, they will have no sense of the hall's acoustics.

Texture

"Texture" refers to the listeners' subjective impression of the music based on the patterns in which the sequence of early sound reflections arrives at their ears. In an excellent hall those reflections that arrive soon after the direct sound follow in a more-or-less uniform sequence. In other halls there may be a considerable interval between the first and the following reflections. High-quality texture requires a large number of early reflections, uniformly but not precisely spaced apart, allowing no one to dominate.

Echoes

"Echo" describes a delayed reflection sufficiently loud to annoy the musicians on stage or the listeners in the hall. Ceiling surfaces that are very high or that focus sound into one part of the hall may create echoes. Echoes may also result from a long, high, curved rear wall whose focal point is near the front of the audience or on the stage. Echoes are more obtrusive in halls with short reverberation times.

Dynamic Range and Background Noise Level

"Dynamic range" is the spread of sound levels over which music can be heard in a hall. This range extends from the low level of background noise produced by the audience and the air-handling system up to the loudest levels produced in the performance. All extraneous sources of noise—including traffic and aircraft noise—must be avoided in order to obtain a wide dynamic range.

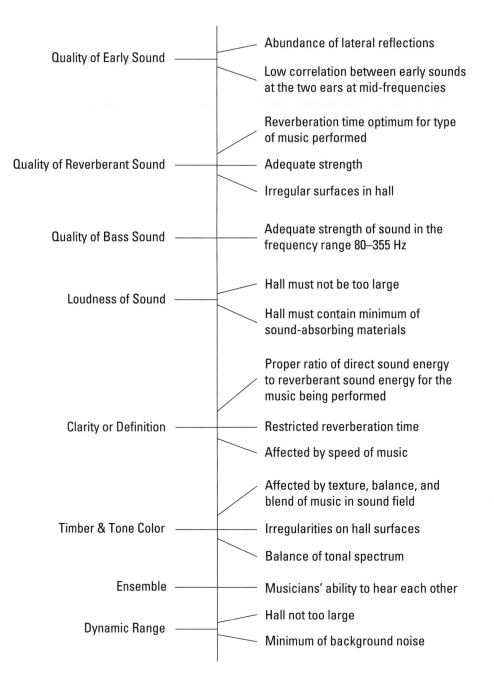

Quality of Early Sound ——— Abundance of lateral reflections

Low correlation between early sounds
at the two ears at mid-frequencies

Reverberation time optimum for type
of music performed

Quality of Reverberant Sound ——— Adequate strength

Irregular surfaces in hall

Quality of Bass Sound ——— Adequate strength of sound in the
frequency range 80–355 Hz

Loudness of Sound ——— Hall must not be too large

Hall must contain minimum of
sound-absorbing materials

Proper ratio of direct sound energy
to reverberant sound energy for the
music being performed

Clarity or Definition ——— Restricted reverberation time

Affected by speed of music

Affected by texture, balance, and
blend of music in sound field

Timber & Tone Color ——— Irregularities on hall surfaces

Balance of tonal spectrum

Ensemble ——— Musicians' ability to hear each other

Hall not too large

Dynamic Range ——— Minimum of background noise

𝓕IGURE 2.5. A chart showing the interrelations between the audible factors
of music and the acoustical factors of the halls in which the music is performed.

Detriments to Tonal Quality

Like a fine instrument, a concert hall can have fine tonal quality—the faithful transmission of the sounds from the instruments without any added (or subtracted) sounds, distortion, or shift of source. Tone quality can suffer any number of injuries, such as a rattle in a metallic surface or metal bars in the front of an organ that resonate in unison with certain musical notes. Sometimes a special kind of distortion will intrude—some architectural or decorative aspect of the hall may add a rasping sound to the orchestral music. Fine-scale irregularities on reflecting plane surfaces alleviate this type of distortion. A "shift of source," heard at certain seats in even some of the best halls, can also distort the tonal quality. If a particular sound-reflecting surface focuses a large amount of sound toward one part of the audience, the listeners there will hear the sound as emanating from that surface rather than as coming directly from the orchestra.

Uniformity of Sound in Audience Areas

The same music, tonal quality, etc., should reach every listener. The quality of a listener's experience will suffer if he or she is seated under a deep overhanging balcony or at the sides of the front rows in the hall. In certain locations, reflections may overemphasize one section of the orchestra or produce echoes, muddiness, or lack of clarity. Musicians sometimes speak of "dead spots" where the music is not as clear or as live as it is in other arts of the hall. Acousticians reserve that term only for locations where the music is especially weak.

Summary of the Musical Qualities Affected by Acoustics

Figure 2.5 summarizes the interrelations between the musical qualities heard in a concert hall and the acoustical factors that affect those qualities. This chart, together with the definitions of this chapter, covers the known interrelations among the qualities of music performed in a concert hall. Much of the information also applies to opera houses, but there are sufficient differences to require a separate discussion (see Chapter 5).

THREE

One Hundred Concert Halls and Opera Houses

These 100 Halls

In the world of the performing arts are some thousands of rooms used for musical performances. Because no two of these spaces are exactly alike, each gives the music played in it a different sound. While doing the research for this book, I have enjoyed the opportunity to visit many halls all over the world, sampling the enormous range of acoustical environments they provide. As much as I appreciate the uniqueness of each space, I cannot agree with those who would call acoustics a mere matter of taste, denying the existence of "good" and "bad" sound. Were that the case, acoustics would stand alone as the one thing in the world not possessed of different degrees of quality.

Even music lovers who frequent live performances can be unaware of the factors that shape the sound they enjoy. Most listeners seldom hear music in halls other than the one or two they regularly attend, and those visits engage their senses on many levels at once—visual and tactile as well as auditory. Similarly, the professionals responsible for creating concert halls and opera houses typically prioritize concerns other than the acoustics. Architects design for clients, and either may have specific goals in mind. The owner presents the architect with a list of desired or requisite features—such as seating capacity, size of the stage, lobby areas, ventilation, and lighting—that can be tallied up once the structure is completed. The architect may wish to build a monument that the public will travel far to see and that will win international awards. Either through lack of knowledge or interest, architects and owners may fail to build for, arguably, the most important feature of a hall for musical performance: how the acoustics of such a creation will or should sound.

The one hundred halls described on the following pages provide a tour of the world, as it were, via its acoustical achievements. Musicians and music aficionados may gain from this book an understanding of why some halls have earned better reputations as venues for music than others. By defining and measuring how the physical elements of a room shape its acoustics, I also have endeavored to make sound quality a concrete goal for the individuals who are planning or renovating performance spaces.

$\mathscr{P}$OCHEÉ CODE FOR ARCHITECTURAL DRAWINGS

Stage floor area

Remote wall surfaces in longitudinal section

Special sound–absorbing material

Regions outside auditorium boundary

1

Aspen, Colorado

Benedict Music Tent

*L*ocated in the Rocky Mountains, Aspen, Colorado, is a quaint Victorian town, albeit with a cornucopia of cultural offerings, including music, dance, theater, painting, and films. It boasts many shops and galleries as well as excellent living and dining facilities. The Aspen Music Festival and School were established in 1949. It is a foremost training ground for pre-professional musicians and offers studies of classical music for nine weeks each summer. There are five orchestras staffed by the students and their instructors, four of which perform under guest conductors and one that trains student conductors.

The Music Festival comprises more than 200 events annually, including orchestral concerts, chamber music, opera, contemporary music, and lectures. The principal festival performances occupy three venues, the Opera House, Concert Hall, and Music Tent. Aspen differs from its sister institution, the Tanglewood Music Center in Western Massachusetts, mainly in that the premier orchestra is composed of the instructors and the most accomplished students, while at Tanglewood the main performing entity is the Boston Symphony Orchestra. Tanglewood has a premier student orchestra and its own concert hall.

The Benedict Music Tent was dedicated in 2000. With 2,050 seats, it has the outdoor appearance of a circus tent, but closer inspection reveals a hurricane- and snow-resistant structure augmented to give it an acoustical ambience that rivals that of excellent concert halls. The main supporting structure is a circular grid, 100 ft (30.5 m) in diameter, formed from intersecting, elliptical-shaped steel trusses that, in turn, rest on four large, telescoping columns, 45 feet (13.7 m) in height, two of which stand at the front corners of the stage, while the other two are in the middle part of the audience. This structural "disk" supports an upper "roof" of sound-reflecting wood and glass panels. They are protected from the elements by the tent material itself, which is Teflon-coated glass-fiber fabric.

The "flying" sound-reflecting surfaces immediately above the stage are made from 3 × 3-in. (7.6-cm squares) welded wire mesh secured to curving steel trusses

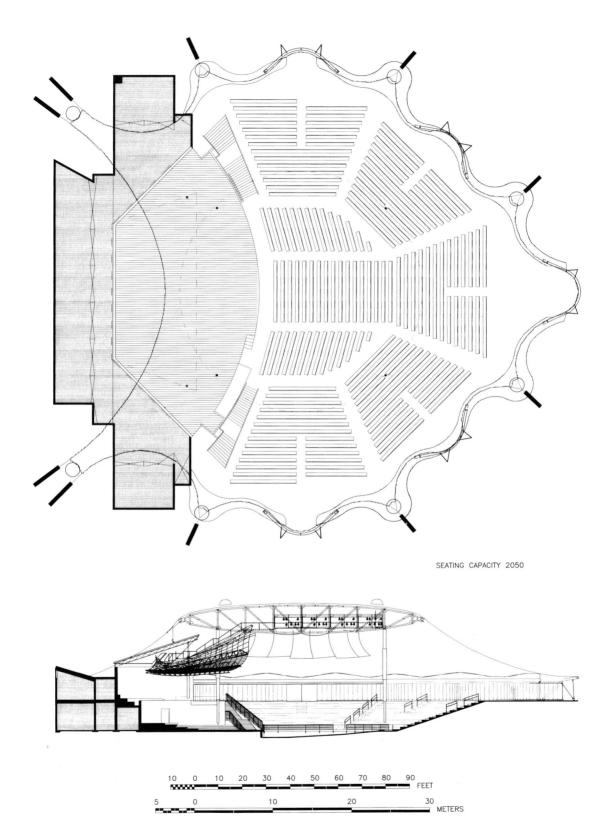

SEATING CAPACITY 2050

10   0   10   20   30   40   50   60   70   80   90
                                                    FEET

5   0   10   20   30
                                                    METERS

that are attached above to the structural disk. The wire mesh panels are covered with No. 10 duct canvas, which has been sprayed with two coats of polyurethane finish. Finally, the periphery of the tent has operable wooden louvers that are used to exclude wind, rain and sun, or opened to permit the audience and orchestra to view the surrounding mountains.

Conductors and audiences have spoken highly of the acoustics, whether the performance involves orchestra and chorus, chamber group, or soloists.

ARCHITECTURAL AND STRUCTURAL DETAILS

Uses: Summer symphony and chamber orchestra, soloists, lectures. **Ceiling:** Center section, partly glazing, partly tongue-and-groove decking, all covered by the outer tent which is Teflon-coated glass-fiber fabric. **Canopy above stage:** Panels composed of No. 10 duct canvas with two coats of spray-applied polyurethane finish stretched over 3 × 3-in. (7.6-cm) welded wire mesh, all secured to curving steel trusses that are anchored to the hard shell above. **Low side and rear walls:** Operable steel-framed vertical louvers, of which lower 66% is 0.75 in. (1.9 cm) plywood, all wrapped in dyed acrylic fabric. **Stage enclosure:** Veneered plywood on two layers of 0.75-in. substrate affixed to 12-in. (30 cm)-thick grout-filled CMV block. **Audience floors:** Bare concrete. **Stage floor:** Tongue-and-groove wooden flooring, over two layers of 0.75-in. plywood, mounted on rafters spaced 16 in. (40.6 cm) apart, over airspace. **Stage height:** 3 ft (0.9 m) above floor at first row. **Carpet:** None. **Seats:** Molded plywood benches with cast aluminum supports and closed-cell foam cushions wrapped in dyed acrylic fabric.

ARCHITECT: Henry Teague Architects. ACOUSTICAL CONSULTANT: Kirkegaard Associates. PHOTOGRAPHS: Tim Hursley, Little Rock.

TECHNICAL DETAILS

V = 700,000 ft³ (19,830 m³)	S_a = 10,373 ft² (964 m²)	S_A = 12,880 ft² (1,197 m²)
S_o = 5,240 ft² (487 m²)	S_T = 14,816 ft² (1,377m²)	N = 2050
H = 44 ft (13.4 m)	W = 136 ft (41.5 m)	L = 98 ft (30 m)
D = 97 ft (29.5 m)	V/S_T = 47.2 ft (14.4 m)	V/S_A = 54.3 ft (16.6 m)
V/N = 341.5 ft³ (9.67 m³)	S_A/N = 6.28 ft² (0.58 m²)	H/W = 0.32
L/W = 0.72		

Note: $S_T = S_A$ + 180 m²; (see definition of S_T in Appendix 1).

NOTE: The terminology is explained in Appendix 1.

2

Baltimore

Joseph Meyerhoff Symphony Hall

The Joseph Meyerhoff Symphony Hall was built only for concert use. It is a beautiful building, with a unique shape, panorama of glass and brick, and grand staircases. It was opened in 1982 to generally favorable reviews by music critics. Home of the Baltimore Symphony Orchestra, it is one of America's important concert halls. There are no flat walls, no ninety-degree angles; instead at each level there are curved randomly shaped "bumps" that reflect and diffuse the sound field. The ceiling is covered with 52 convex disks that emulate the coffered ceilings of classical halls. The seating capacity is 2,467 and the reverberation time, at mid-frequencies, fully occupied, is 2.0 sec, which is optimum for today's symphonic repertoire.

From the beginning, the orchestra was not happy with the sound on stage. The players cited excessive loudness, poor ensemble conditions, and some early reflections strong enough to be distracting both to the conductor and the musicians. In 1990, a large array of QRD sound diffusers was installed around the periphery of the stage, with early reports that "they have had the positive effect of clearing up the sound on stage." However, after several seasons' use the QRD's were removed because of the general feeling that the diffused sound made it more difficult for the orchestra to play in good ensemble.

In 2001 a major renovation of the performing area was completed. The original stage ceiling was altered and an array of reflecting panels was hung below from rows of battens. The average size of each reflector is 5.75 ft (1.75 ms) square; each is adjustable in angle, both fore–aft and sideways; and the rows can individually be changed in height, as can the individual panels. Risers have been added to the stage, both to improve players' ability to see each other and better to project the sound to the audience. Some changes have been made in the sidewalls at the main floor level to remove harshness of tone, and absorbing material has been added in a few places on the rear wall to reduce faint echoes.

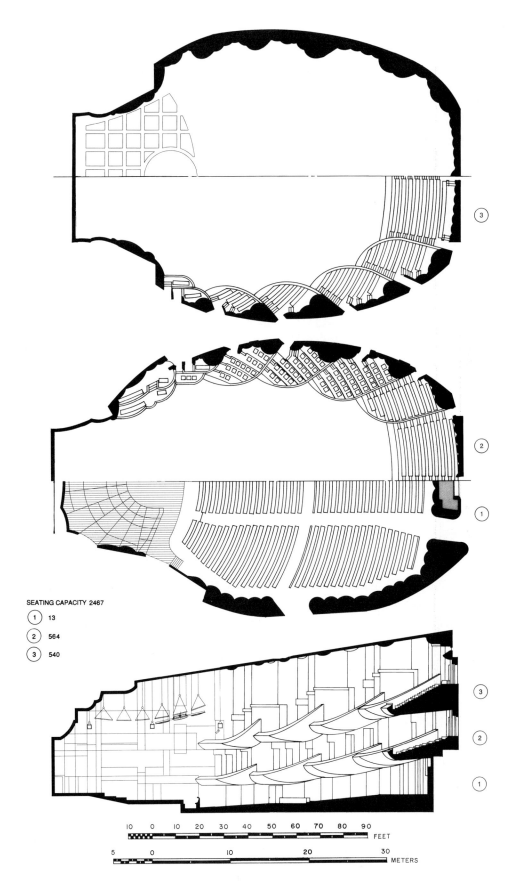

SEATING CAPACITY 2467

(1) 13
(2) 564
(3) 540

10 0 10 20 30 40 50 60 70 80 90
FEET

5 0 10 20 30
METERS

Acoustically, the sound is a little louder than that in Boston Symphony Hall, and the bass is a little stronger, but the high level of the balconies at the rear of the hall and the resulting need to slope the ceiling downward prevent the around-the-room reverberation that is a trademark of the Boston hall.

ARCHITECTURAL AND STRUCTURAL DETAILS

Uses: Concerts, recitals, and conferences. **Ceiling:** Plaster over concrete planks. **Walls:** Plaster over concrete block; plaster on balcony faces. **Floors:** Cast-in-place concrete slab. **Carpets:** On aisles of main floor and balconies—0.33-in. (0.83 cm)-thick, directly affixed to concrete. **Stage sidewalls:** Masonry construction. **Stage canopy:** Each reflector is spherically curved with a 22 ft (6.7 m) radius. **Stage floor:** Wood, tongue-and-groove over planks. **Stage height:** 39 in. (1 meter) height above main floor at first row of seats. **Seating:** Backrest is 0.38-in. (1 cm)-thick molded plywood. Top of seat bottom and front of backrest are upholstered, porous fabric over open-cell foam. Armrests are wooden.

ORIGINAL ARCHITECT: Pietro Belluschi, Inc. ASSOCIATE ORIGINAL ARCHITECT: Jung/Brannen Associates, Inc. RENOVATION ARCHITECT: RTKL Associates, Dallas. ORIGINAL ACOUSTICAL CONSULTANT: Bolt Beranek and Newman, Inc. RENOVATION ACOUSTICAL CONSULTANT: Kirkegaard Associates. PHOTOGRAPHS: Upper photograph, courtesy Meyerhoff Symphony Hall; lower photograph, Richard S. Mandelkorn, courtesy of Jung/Brannen.

TECHNICAL DETAILS

$V = 760{,}000$ ft³ (21,530 m³)	$S_a = 12{,}870$ ft² (1,196 m²)	$S_A = 16{,}000$ ft² (1,487 m²)
$S_o = 2{,}467$ ft² (229 m²)	$S_T = 17{,}940$ ft² (1,667 m²)	$N = 2{,}467$
$H = 59$ ft (18 m)	$W = 96$ ft (29.3 m)	$L = 116$ ft (35.4 m)
$D = 123$ ft (37.5 m)	$V/S_T = 42.4$ ft (12.9m)	$V/S_A = 47.5$ ft (14.5 m)
$V/N = 308$ ft³ (8.73 m³)	$S_A/N = 6.49$ ft² (0.60 m²)	$S_a/N = 5.22$ ft² (0.48 m²)
$H/W = 0.61$	$L/W = 1.21$	ITDG $= 22$ msec

Note: $S_T = S_A + 1{,}940$ ft² (180 m²); see Appendix 1 for definition of S_T.

NOTE: The terminology is explained in Appendix 1.

3

Boston

Symphony Hall

*S*ymphony Hall, built in 1900, is rectangular in shape with a high, horizontal coffered ceiling and two wraparound balconies. On entering the hall, one encounters two strong architectural features: the stage with its back wall devoted to a row of gilded organ pipes, and the upper walls of the hall with their niches, in front of which stand replicas of Greek and Roman statues. The combination of shades of gray and cream paint, gilded proscenium frame and balcony fronts, red-plush balcony rails, black leather seats, and red carpets would place this hall architecturally in the middle of the nineteenth century, although it was built fifty years later. In some ways it resembles the Vienna Grosser Musikvereinssaal; nevertheless, it is different, primarily because it seats 2,625 people compared with 1,680 for the Vereinssaal. During May, June, and December each year, tables are installed on the main floor for "pops" concerts and the capacity is reduced to 2,369.

The sound in Symphony Hall is clear, live, warm, brilliant, and loud, without being overly loud. The hall responds immediately to an orchestra's efforts. The orchestral tone is balanced, and the ensemble is excellent.

Conductors von Karajan, Bernstein, De Preist, Leinsdorf, and many others have agreed it is the "most noble of American concert halls." Encomiums such as "one of the world's greatest halls," "when this hall is fully occupied the sound is just right—divine," "an excellent hall, there is none better," are expressed by players and leaders.

Seven crucial design features were responsible for the immediate success of this hall. The shoebox shape, which was modeled after the (old) Leipzig, Gewandhaus; the proper reverberation time, determined by Sabine's formula, which set the ceiling height; the preservation of bass both by avoiding large areas of wood, which followed the decision to make the building fireproof, and by choice of seats with a minimum of upholstering; limiting its width and length to insure intimacy; providing sound diffusion by niches, statues, and ceiling coffers; and, finally, by a stage

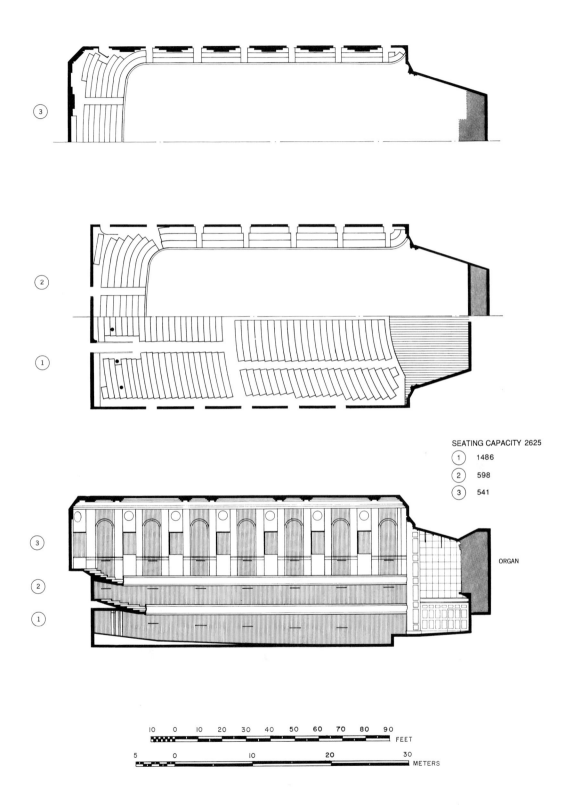

SEATING CAPACITY 2625

① 1486

② 598

③ 541

ORGAN

10 0 10 20 30 40 50 60 70 80 90
FEET

5   0   10   20   30
METERS

house that blends the orchestral sound beautifully and projects the music uniformly throughout the audience.

Whoever has the good fortune to travel to many halls, finds Boston Symphony Hall possessing features that hardly exist elsewhere. Most compelling is the high upper space that allows the sound to travel around the room so that it is heard as coming from the sides, front, and back. Balconies, in other halls, that rise to the ceiling at the rear are the chief barriers to such sound.

ARCHITECTURAL AND STRUCTURAL DETAILS

Uses: Orchestra and soloists. **Ceiling:** 0.75-in. (1.9 cm) plaster on metal screen. **Walls:** 30% plaster on metal lath, 50% on masonry backing, and 20% of 0.5–1-in. (1.25–2.5 cm)-thick wood, including the stage walls; balcony fronts, open-pattern cast iron. **Floors:** Base floor is flat concrete with parquet wood affixed; in winter concert season, sloping floor of 0.75-in. boards on 4 × 4-in. (10 × 10-cm) framing members—the airspace beneath varies from zero at the front to 5 ft (1.52 m) at the rear; balcony floors, wood supported above concrete. **Carpets:** Thin, on main aisles. **Stage enclosure:** Wood paneling about 0.5 in. thick, but from the stage floor up to a height of about 14 ft (4.3 m), is about 1 in. thick. **Stage floor:** 1.5-in. wooden planks over a large airspace with 0.75-in. flooring on top. **Stage height:** 54 in. (1.37 m). **Seating:** The front and rear of the backrests and the top of the seat bottoms are leather over hair; the underseats and the arms are of solid wood.

ARCHITECT: McKim, Mead, and White. ACOUSTICAL CONSULTANT: Wallace C. Sabine. REFERENCES: Richard P. Stebbins, *The Making of Symphony Hall,* Boston Symphony Orchestra, Inc. (2000).

TECHNICAL DETAILS

Concerts

V = 662,000 ft³ (8,750 m³)	S_a = 11,360 ft² (1,056 m²)	S_A = 14,750 ft² (1,370 m²)
S_o = 1,635 ft² (152 m²)	S_T = 16,385 ft² (1,523 m²)	N = 2,625
H = 61 ft (18.6 m)	W = 75 ft (22.9 m)	L = 128 ft (39 m)
D = 133 ft (40.5 m)	V/S_T = 40.4 ft (12.3 m)	V/S_A = 44.9 ft (13.7 m)
V/N = 252 ft³ (7.14 m³)	S_A/N = 5.62 ft² (0.52 m²)	S_a/N = 4.33 ft² (0.40 m²)
H/W = 0.81	L/W = 1.71	ITDG = 15 msec

NOTE: The terminology is explained in Appendix 1.

4

Buffalo

Kleinhans Music Hall

he Kleinhans Music Hall is a large venue for today's symphony-loving
public, seating 2,839, but its well-proportioned lines and primavera wood
interior render an immediate feeling of intimacy, warmth, and comfort. Buffalo's
audiences are proud of the hall and most find no fault with its acoustical properties.
The seats are luxuriously upholstered and widely spaced. The balcony is enormous,
yet it leaves one with the feeling of an intimate space. Controversy has surrounded
its acoustics largely because its parabolic shape brings the music directly to the
listener, emphasizing the early sound, at the expense of the reverberant sound and
giving the hall the reputation of being too dry. The music is loud because of this
early-sound emphasis and the bass is strong. The reverberation time at mid-
frequencies, fully occupied, is estimated at 1.5 sec, below optimum, but above the
RT in Philadelphia's Academy of Music, where the famed Philadelphia sound originated. Originally, the ventilation noise was excessive; that has been eliminated.

I have on several occasions attended concerts in the Kleinhans Music Hall.
The brilliance of the string tone is excellent, particularly on the main floor, and the
sound is warm with rich full bass. In certain parts of the main floor, the sound takes
on a more reverberant character. I have classified this as "stage liveness" rather than
"hall liveness." Isaac Stern praised the hall, and he was usually critical of a dry
sound. He said that on stage the sound was quite good, and he felt a sense of
immediacy and support. Izler Solomon said, "I conducted there for many years.
The acoustics are good, perhaps not quite as good on the stage as on the main floor
of the auditorium."

The principal difference between this hall and the shoeboxes of Boston, Vienna, and Amsterdam, besides the reverberation time, which for those halls is about
2.0 sec, is that there is no upper space for the reverberation to form and to reach
the listener from all directions.

In defense of its designers, it was one of the first halls built between 1900 and
1950 and there was only one good model in the United States to follow, Boston.

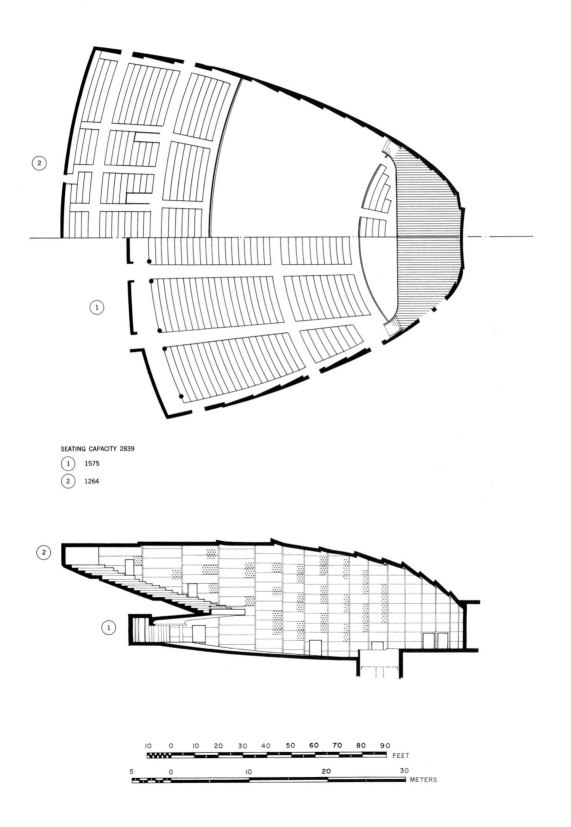

SEATING CAPACITY 2839

1 1575
2 1264

10 0 10 20 30 40 50 60 70 80 90
FEET

5 0 10 20 30
METERS

The architect wanted to deviate from this nineteenth-century shape. Not enough was known acoustically to give the architect the guidance about the need for more early sound. A serious problem was the lack of knowledge about the large absorbing power of an audience, which caused the lower than expected reverberation time. Buffalo is fortunate that the hall has come out so well.

ARCHITECTURAL AND STRUCTURAL DETAILS

Uses: Orchestra, soloists, glee club, and lectures. **Ceiling:** 0.75-in. (2 cm) painted plaster. **Sidewalls:** 0.75-in. plaster on metal lath on which is pasted a linen cloth over which are cemented 0.06-in. (0.16 cm) flexible wooden sheets. **Rear walls:** 0.75-in. plaster on metal lath on which is cemented a heavy woven monk's cloth. **Floors:** Concrete. **Carpet:** The main floor seating area and aisles are fully carpeted. **Stage enclosure:** The orchestra enclosure is permanent and is made of 0.75-in. (2 cm) plywood irregularly supported on 2 × 2-in. (5 × 5-cm) furring strips held, in turn, by a hollow-tile structure. Lighting coves overhead open to a high attic and permit air to filter out. **Stage floor:** 1.1-in. (2.8 cm) wooden planks on 0.75-in. subfloor over a large airspace. **Stage height:** 42 in. (107 cm). **Added absorbing material:** On each wall are located ten vertical strips of thin, perforated asbestos, backed in places by sound-absorbing material. **Seating:** Front of the backrest and the top of the seat are upholstered; seat-bottoms and armrests are solid.

ARCHITECT: Eliel Saarinen and F. J. and W. A. Kidd. ACOUSTICAL CONSULTANTS: C. C. Potwin and J. P. Maxfield. PHOTOGRAPHS: Courtesy of Kleinhans Music Hall Management, Inc. REFERENCES: J. P. Maxfield and C. C. Potwin, "A modern concept of acoustical design," pp. 48–55; and "Planning functionally for good acoustics," pp. 390–395, *J. Acous. Soc. Am.* **11**, (1940).

TECHNICAL DETAILS

V = 644,000 ft³ (18,240 m³)	S_a = 17,000 ft² (1,580 m²)	S_A = 21,000 ft² (1,951 m²)
S_o = 2,200 ft² (205 m²)	S_T = 22,940 ft² (2,131 m²)	N = 2,839
H = 44 ft (13.4 m)	W = 129 ft (39.3 m)	L = 123 ft (37.5 m)
D = 144 ft (43.9 m)	V/S_T = 28.1 ft (8.56 m)	V/S_A = 30.7 ft (9.35 m)
V/N = 227 ft³ (6.24 m³)	S_A/N = 7.4 ft² (0.69 m²)	S_a/N = 6.00 ft² (0.556 m²)
H/W = 0.34	L/W = 0.95	ITDG = 32 msec
Note: $S_T = S_A$ + 1,940 ft² (180 m²); see Appendix 1 for definition of S_T.		

NOTE: The terminology is explained in Appendix 1.

5

Chicago

Orchestra Hall in Symphony Center

This oddly shaped hall, which reopened 4 October 1997, has undergone extensive renovations designed to improve the acoustics and to adapt the stage shape to present-day orchestra seating. Viewed from the stage, the visual appearance of the audience areas has changed little from the 1904 design by Chicago's music giant, Theodore Thomas, who conducted the Chicago Symphony Orchestra from 1891 to 1905. But viewed from the audience, the revision is breathtaking. Three meters above and around the narrowed and deepened stage is a "U-shaped" terraced seating area, three rows deep. Resting on the rear wall of this seating area, 9 ms above the stage, is a hemispherical, sound-transparent dome beneath which hangs an unusually attractive glass-and-steel acoustical canopy. The strikingly beautiful interior overall attests to the faultless task of the designers: all silver and cream, new but not garish, shining yet balanced.

Blair Kamin, the *Chicago Tribune*'s architecture critic, wrote (3 October 1997):

> . . . the debut of the renovated and expanded Orchestra Hall merits a full-throated "bravo" for pumping new lifeblood into both the arts and the heart of Chicago . . . One of the arts that it invigorates is architecture . . . it reinterprets the 93-year-old original in an adventurous conversation between past and present . . . the concert hall itself has been made larger while losing none of its former visual intimacy. The most prominent addition to the hall, a steel-and-glass acoustical canopy . . . now floats elegantly above the stage, lacy as a doily . . . on the whole, this is a sparkling transformation.

Negative criticism of the 1904 Orchestra Hall began the day it opened. Although the orchestral music was clear and well balanced, the sound was dry, meaning the reverberation time was too short. Listeners traveling to the oft-praised halls in Boston, Vienna, and Amsterdam were dismayed by the lack of support for the orchestral tone; missing was a "singing tone" that made those halls famous. Only by an increase in reverberation time, accompanied by an increased number of strong, early sound reflections, could the hall be made competitive.

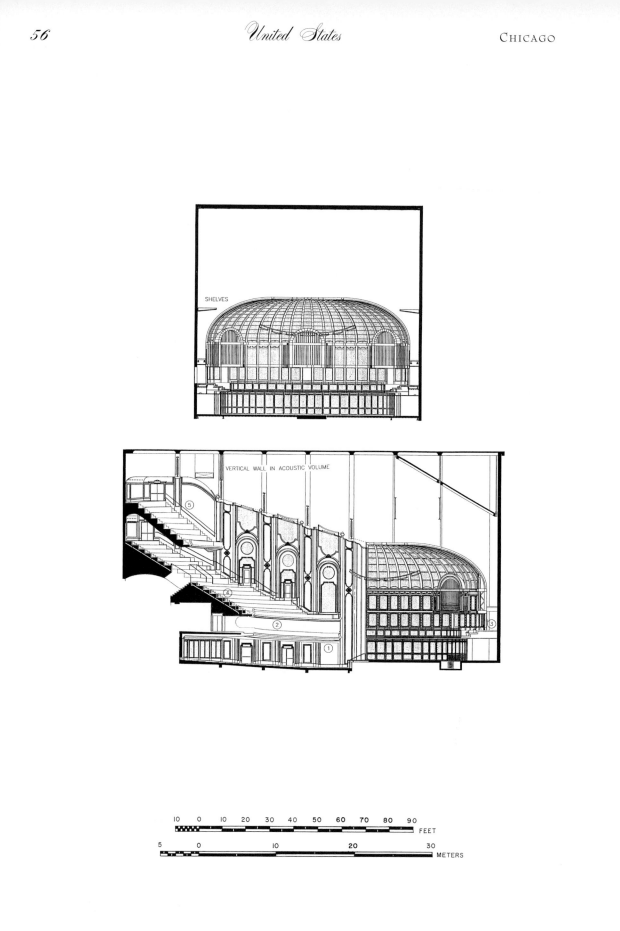

SHELVES

VERTICAL WALL IN ACOUSTIC VOLUME

| 10 | 0 | 10 | 20 | 30 | 40 | 50 | 60 | 70 | 80 | 90 |
FEET

| 5 | 0 | | 10 | | 20 | | 30 |
METERS

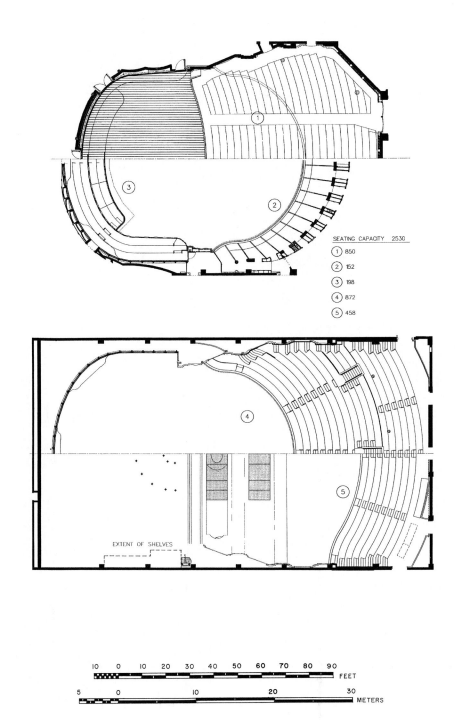

SEATING CAPACITY 2530

1) 850

2) 152

3) 198

4) 872

5) 458

EXTENT OF SHELVES

10 0 10 20 30 40 50 60 70 80 90
                                              FEET

5   0   10   20   30
                                              METERS

In the 1997 renovation the (acoustical) ceiling was raised 15.8 ft (4.6 m). It is flat and thick, reaching a height that is 37 ft (11.3 m) above the over-stage hemisphere (see the drawing). The visual ceiling is now more open so that it freely communicates with the space above. The new cubic volume is 50 percent greater than the previous value of 18,000 m³ making the reverberation time, fully occupied and at the middle range of frequencies, about 1.6 sec, compared to the previous 1.2 sec. There are other changes acoustically, the most important being the suspended glass-and-steel canopy which enables the players to hear themselves. Another is the new seats on which the upholstering is half as thick, which means that they absorb less bass sound. Suspended panels above the hemispherical orchestra enclosure reflect sound to the various areas of the audience and back to the players.

Very importantly, the front part of the hall is now narrower by about 7.6 m, which creates side-reflecting surfaces for sending highly desirable, early, lateral sound to listeners on the main floor. The width of the stage is reduced by about 3.6 m. Orchestra Hall can again boast of a pipe organ. The hall was lengthened, both to make room for the three-row terrace seating and to increase the stage depth by about 3 m. The hall now seats 2,530 persons.

The general evaluation by Chicago listeners and most music critics is largely favorable—the sound has been improved significantly. James R. Oestreich, *The New York Times,* in The Living Arts Section, October 6, 1997, wrote:

> The results, to judge on Saturday from a choice seat in the lower balcony during the long concert . . . are remarkable. The strings, for a change, could hold their own against those mighty Chicago winds. The low strings, especially, had a new warmth and solidity, and the whole bass and baritonal range of the orchestra provided a firmer basis and a mellower foil for the sound above it, which has always been brash and brilliant.

A music critic of the *Chicago Tribune* in 2002 has denigrated the new acoustics, "[The changes] have robbed the upper strings of shimmer and warmth even as they added depth and spaciousness to the low frequency sound, flattering the cellos and basses." The acoustical consultant responded, ". . . we are very pleased with much of what was accomplished in the renovation, but, along with others, are not yet completely satisfied with the results . . . further refinements that will achieve everyone's goals can be accomplished within reasonable means."

At two center seats in the front balcony and two on the main floor, where I sat during two concerts, the sound was very good. Some listeners report that the sound on the main floor under the boxes/balcony overhang is less good. Observe that the sound will never exactly duplicate that of Boston's and Vienna's famous halls, because the steep balconies in Orchestra Hall prevent the sound from reverberating in spaces behind the listeners.

ARCHITECTURAL AND STRUCTURAL DETAILS

Uses: Orchestra, chamber music, soloists, chorus, organ, lectures. **Upper ceiling:** Precast concrete with sound-diffusing shaping, 6–14 in. (15–36 cm). **Visual ceiling:** Plaster ribs occupy 5%; perforated metal between ribs 42% open; above the stage the perforated metal is 95% open. **Stage canopy:** 0.5-in. (1.25-cm) laminated glass panels of double-convex shape hung from a steel structure, 70% solid. **Walls above visual ceiling:** Sides, plastered brick, 12-in. (30-cm); Rear stage, 16 in. (41 cm). **Walls below visual ceiling:** Plaster applied to brick, 12–20 in. (30–50 cm). Majority of rear walls and box rear walls are QRD diffusers behind perforated metal with limited areas of 4-in. (10-cm) plaster. Balcony side walls are plaster and perforated metal. Rear of first balcony and rear sides and underside of upper balcony are plaster 2–4 in. (5–10 cm). **Stage walls, all behind perforated metal:** Center of choral seating is diffusive 4-in. (10-cm) plaster. Side walls of choral seating are concrete block 12 in. (30 cm), bowed in plan. Rear walls of stage are a combination of wood paneling and perforated metal over plaster. **Floors** (audience floors everywhere): Painted concrete with thin carpet in aisles; choral seating area, tongue-and-groove wood; stage floor, tongue-and-groove maple, 1 in. (2.5 cm), over two layers plywood, 0.75 in. and 1.13 in. (1.5–2.9 cm)—all over a 1-in. (2.5-cm) airspace formed by sleepers and neoprene pads. **Orchestra risers:** Wood, which telescope and retreat onto castered carriages for removal from stage.

ARCHITECT: 1904, D. H. Burnham; 1997, Skidmore, Owings & Merrill, LLP. ACOUSTICAL CONSULTANT: Kirkegaard & Associates. PHOTOGRAPHS: Jon Miller, Hedrich-Blessing. REFERENCE: Kevin Hand and Celeste Bernard, *Chicago Tribune,* Sunday, September 28, 1997, Arts and Entertainment Section.

TECHNICAL DETAILS

V = 953,000 ft³ (27,000 m³)	S_a = 11,406 ft² (1,060 m²)	S_A = 12,471 ft² (1,159 m²)
S_o = 2,884 ft² (268 m²)	S_T = 14,408 ft² (1,339 m²)	N = 2,530
H = 89 ft (27.1 m)	W = 72 ft (21.9 m)	L = 80 ft (24.4 m)
D = 105 ft (32 m)	V/S_T = 66.1 ft (20.2 m)	V/S_A = 76.4 ft (23.2 m)
V/N = 377 ft³ (10.7 m³)	S_A/N = 4.93 ft² (0.46 m²)	S_a/N = 4.50 ft² (0.420 m²)
H/W = 1.24	L/W = 1.11	ITDG = 36 msec

Note: S_T = S_A + 1,940 ft² (180 m²); see Appendix 1 for definition of S_T.

NOTE: The terminology is explained in Appendix 1.

6

Cleveland

Severance Hall

*S*everance Hall is a handsome classical building, located in the University Circle area of Cleveland. Completed in 1931, the hall was designed during the period when some acousticians adhered to the philosophy that the acoustics of the audience area should be "dead" and that the stage enclosure alone should be live. In 1958, as part of a program of rehabilitation, most of the draperies and carpets in the hall were removed, and a stage enclosure of contemporary style, with relatively low ceiling, was installed. With fewer sound-absorbing materials present, the reverberation time increased noticeably. Certainly, the hall, with only 2,101 seats, was favorable to the production of good music—having assisted two famous conductors, George Szell and Christoph von Dohnanyi, in bringing the Cleveland group world acclaim as the equal to today's best symphonic orchestras.

In 1999, a second general rehabilitation took place. Lobbies and musicians' facilities were enlarged, exhibition space and dining facilities indoors and outdoors were added, and underground parking was provided. The audience space was not changed; except new audience seats were installed. The older stage enclosure was scrapped and replaced by one that not only blends with the 1931 ornamental Art Deco style, but also has brought about a significant improvement in the acoustics. In addition, the original pipe organ, long buried behind the orchestra shell, was restored. There was always considerable unused space to the sides and above the 1958 stage enclosure and that now is employed to house the refurbished organ and to add cubic volume above and to the sides of the semi-transparent stage enclosure. The reverberation time, with full occupancy, at mid-frequencies has been increased throughout the hall from 1.5 to 1.6 sec and the modified on-stage acoustics apparently pleases the players. Visually, the new stage is stunning. Behind the upper sidewalls are concealed doors that can be closed or opened by degrees to adjust the acoustics to fit the preferences of the musicians and conductors.

One of the most satisfying concerts that I have attended in recent years has convinced me that the stage modifications were successful acoustically. Allan Kozinn

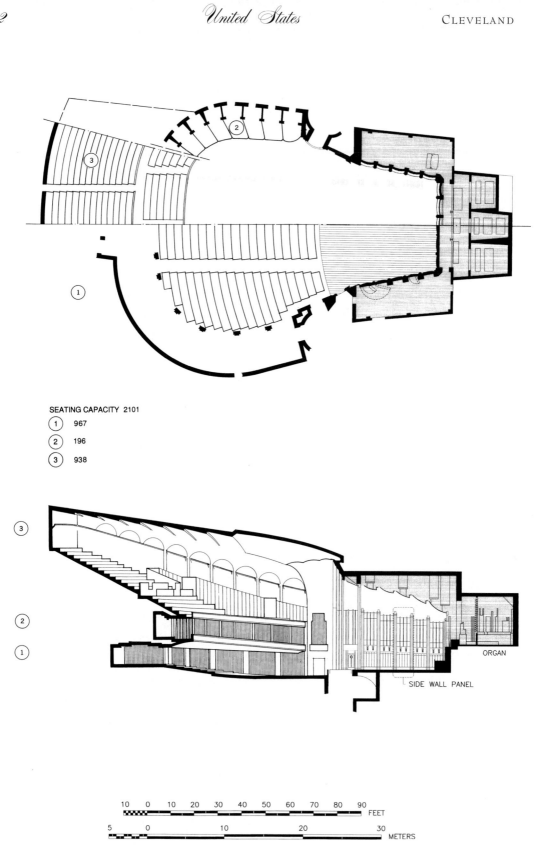

SEATING CAPACITY 2101

1 967

2 196

3 938

ORGAN

SIDE WALL PANEL

10 0 10 20 30 40 50 60 70 80 90
FEET

5 0 10 20 30
METERS

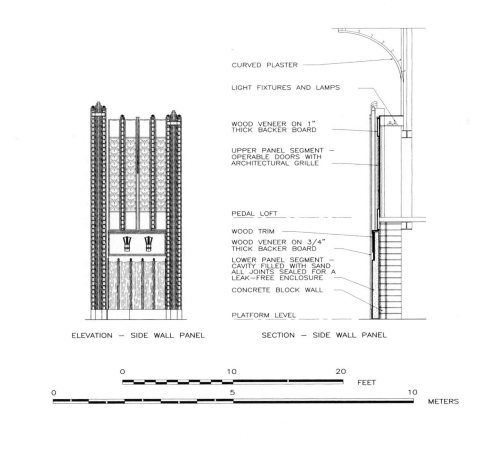

CURVED PLASTER

LIGHT FIXTURES AND LAMPS

WOOD VENEER ON 1"
THICK BACKER BOARD

UPPER PANEL SEGMENT —
OPERABLE DOORS WITH
ARCHITECTURAL GRILLE

PEDAL LOFT

WOOD TRIM

WOOD VENEER ON 3/4"
THICK BACKER BOARD

LOWER PANEL SEGMENT —
CAVITY FILLED WITH SAND
ALL JOINTS SEALED FOR A
LEAK—FREE ENCLOSURE

CONCRETE BLOCK WALL

PLATFORM LEVEL

ELEVATION — SIDE WALL PANEL   SECTION — SIDE WALL PANEL

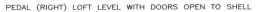

0   10   20
                              FEET
0   5   10
                              METERS

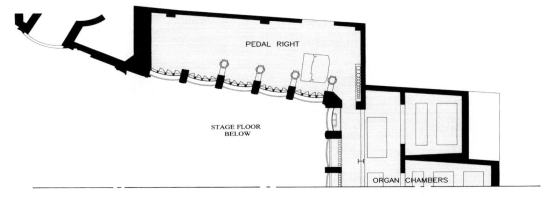

PEDAL RIGHT

STAGE FLOOR
BELOW

ORGAN CHAMBERS

PEDAL (RIGHT) LOFT LEVEL WITH DOORS OPEN TO SHELL

0   10   20   30   40   50
                                        FEET
0   5   10   20
                                        METERS

in *The New York Times,* January 10, 2000, said, "Less than 20 minutes into the program it was evident that there was no cause for worry."

$\mathscr{A}$RCHITECTURAL AND STRUCTURAL DETAILS

Uses: Symphonic music, chamber music, and recitals. **Ceiling:** Plaster on wire lath. **Stage ceiling:** Five irregular sections, separated by air spaces. **Walls:** Plaster, except for doors and space dividers. **Stage walls:** Each side has five bays besides the original bay with an entry door; from the bottom of each bay upward, 14 in. (36 cm) solid, then solid, sand-filled section followed by lighting bays; next is a 14-in. hard section, above which are acoustically transparent screens, backed by operable doors; then another 14-in. solid section above which is a curved plaster section; the rear wall is the same except for organ pipes. **Floors:** Concrete covered with vinyl tile. **Carpet:** None. **Stage floor:** Two layers wood, over sleepers on joists over airspace. **Stage height:** 27 in. (69 cm) above front of main floor. **Seats:** Upholstered on seat backs and seat bottoms, no perforations beneath seat bottom.

ARCHITECTS: original, Walter and Weeks; renovation, 1999, David Schwartz. ACOUSTICAL CONSULTANT: original, Dayton C. Miller; 1999, Jaffe Holden Acoustics, Inc. ORGAN CONSULTANT, 1999, Schoenstein & Company. PHOTOGRAPHS: Steve Hall of Hendrich Blessing.

TECHNICAL DETAILS

$V = 575{,}000$ ft³ (16,290 m³)	$S_a = 10{,}000$ ft² (930 m²)	$S_A = 13{,}000$ ft² (1,210 m²)
$S_o = 2{,}313$ ft² (215 m²)	$S_T = 14{,}960$ ft² (1,390 m²)	$N = 2{,}101$
$H = 55$ ft (16.8 m)	$W = 90$ ft (27.4 m)	$L = 108$ ft (32.9 m)
$D = 135$ ft (41.2 m)	$V/S_T = 38.4$ ft (11.72 m)	$V/S_A = 44.2$ ft (13.5 m)
$V/N = 273.7$ ft³ (7.75 m³)	$S_A/N = 6.19$ ft² (0.58 m²)	$H/W = 0.61$
$L/W = 1.2$		

Note: $S_T = S_A + 180$ m²; see definition of S_T in Appendix 1.

NOTE: The terminology is explained in Appendix 1.

7

Costa Mesa, California

Orange County Performing Arts Center, Segerstrom Hall

Opened in 1986, Segerstrom Hall, seating 2,903 for full orchestra, is a multi-use auditorium. The acoustical design called for early, laterally reflected sound to all seating areas. These reflections are produced by walls created at the edges of four seating trays; by large tilted panels that float within the volume near the front of the hall; and by tilted rear walls. Some of the tilted panels are pleated (so-called quadratic residue diffusers, QRD, discussed in Chapter 4), designed to spread a sequence of early sound reflections to each listener. They also decrease the intensity of overtones relative to fundamental tones, which gives the music a warm, mellow characteristic, and eliminates acoustic "glare" common to reflections from large smooth surfaces. All seats share the same reverberant sound field which is determined by the outer walls of the auditorium, including the volume behind the floating panels. The reverberation time can be shortened for rehearsals or speech events by curtains that are drawn out in varying lengths from suspended housings distributed throughout the ceiling area and from behind the large panel reflectors. At the stage end, the proscenium opening is high, chosen to make the performing orchestra be within the room, while its width is dictated by the owner's request for a stage area of 2,400 ft² (223 m²) for normal symphony concerts, increasing to 3,060 ft² (284 m²) for choral works or decreasing to 500 ft² (46 m²) for soloists. The orchestra enclosure is only 15 ft (4.6 m) deep, so that about half the players are in front of the curtain line. To provide them with cross-stage communication, three removable reflecting panels are located overhead.

The measured acoustical data show that the hall has met the design objectives of achieving sufficient early laterally reflected sound energy, an optimum early decay time (2.2 seconds, unoccupied), and excellent sound levels over the audience area. Segerstrom Hall differs from the classical halls in that the ratio of early sound energy compared to later reverberant energy is higher and the later reverberation time is shorter—1.6 sec. In the upper two trays, the later reverberant sound is not heard as surrounding the listener, as in a conventional rectangular hall, but rather is heard as a broadening of the source, a satisfactory result.

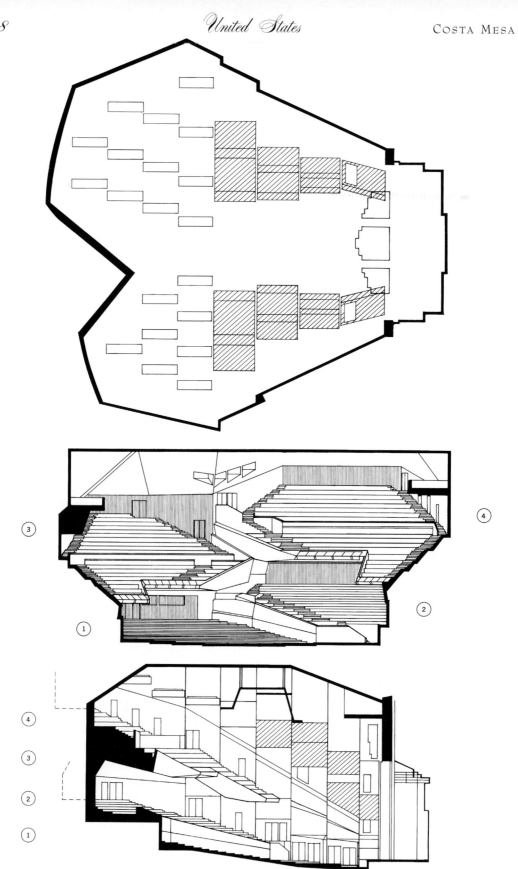

CENTERLINE SECTION

SEATING CAPACITY 2903 to 2994

1   1145

2   679

3   478

4   601

FEET   10   0   10   20   30   40   50   60   70   80   90

METERS   5   0   10   20   30

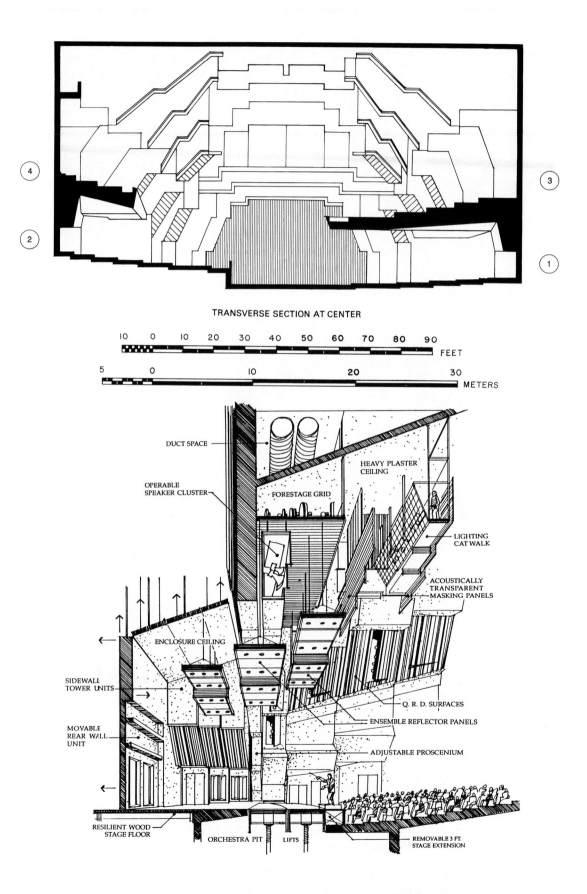

TRANSVERSE SECTION AT CENTER

DUCT SPACE

HEAVY PLASTER
CEILING

OPERABLE
SPEAKER CLUSTER

FORESTAGE GRID

LIGHTING
CAT WALK

ACOUSTICALLY
TRANSPARENT
MASKING PANELS

ENCLOSURE CEILING

SIDEWALL
TOWER UNITS

Q. R. D. SURFACES

ENSEMBLE REFLECTOR PANELS

MOVABLE
REAR WALL
UNIT

ADJUSTABLE PROSCENIUM

RESILIENT WOOD
STAGE FLOOR

ORCHESTRA PIT   LIFTS

REMOVABLE 3 FT.
STAGE EXTENSION

Architectural and Structural Details

Uses: About 25% each for classical music, musicals, and dance; remainder divided among opera, pops, and rentals. **Ceiling:** 2-in. (5-cm) high-density plaster; under-balcony soffits 1-in. (2.5-cm) plaster. **Walls:** 2-in. high-density plaster; balcony faces and surfaces between trays, same. **Sound diffusers:** Located in back corners of each seating area, chevron in shape, made of 0.75-in. (1.9-cm) wood; each about 70 ft² (6.5 m²) in area; depth 1–8 in. (20 cm); and open at perimeter. **Quadratic residue diffusers:** Twelve in number. Constructed of 0.5-in. (1.25-cm) plywood, totaling 1,550 ft² (144 m²); design frequency near 500 Hz based on prime number 7. *Floors:* Dense concrete. Vinyl tile cemented to concrete in seating areas. **Variable elements:** Proscenium opening variable between 68 and 52 ft horizontally and 42 to 32 ft vertically; variable part, 1-in. plywood. **Carpets:** 0.33-in. directly affixed to concrete, totaling about 3,630 ft² (340 m²). **Stage enclosure** (see sketch): Each side formed by three tall rolling towers; back wall section and the three ceiling sections are flown by rigging, all composed of 0.75-in. plywood laminated to 0.5-in. plywood; for large choral works, a 12-ft (3.7-m) insert increases the performing area; between lower and upper portions of the sidewalls, an intermediate 10-ft (3 m) length, 20 degrees from vertical, is built in. These sloped surfaces have a single-period QRD attached to them; chevron diffusers are attached to the lower vertical surfaces on the tower portions and are arrayed across the entire back wall of the enclosure; three large reflecting panels above the players, each about 50-ft (15-m) wide and 9-ft (2.74-m) deep at heights (from down-stage to upstage) of 33 ft, 30 ft, and 28 ft (10 m, 9.25 m, and 8.5 m) above stage level. **Stage floor:** Extends into auditorium on two pit lifts plus a 3-ft. extension; fixed floor is composed of 0.75-in. particle board on 1 × 4-in. (10-cm) tongue-and-groove subfloor, supported on 2 × 3-in. strips, 10 in. (25 cm) on centers—strips are separated from the flat concrete surface below by 0.325-in. (1-cm) neoprene pads. **Stage height:** 40 in. (1.02 m) above floor level at first row of seats. **Seating:** The backrest is 0.325-in. molded plywood. The front of the backrest and the seat top are upholstered. The seat bottom is unperforated. Seat arms are wooden.

Architect: Charles Lawrence. Associate Architects: Caudill-Rowlett-Scott and The Blurock Partnership. Acoustical Consultants: Joint Venture: Paoletti/Lewitz Associates, Jerald R. Hyde and Marshall/Day Associates. Photo-graphs: Courtesy of J. R. Hyde. Stage Sketch: John von Szeliski. References: D. Paoletti and J. R. Hyde, "An acoustical preview of OCPAC," p. 23, Sound & Video Contractor July 15, 1985; J. R. Hyde and J. von Szeliski, "Acoustics and Theater Design: Exploring New Design Requirements for Large Multi-Purpose Theaters," 12th International Congress on Acoustics, Vancouver, Canada (August 1986); A. H. Marshall and J. R. Hyde, "Some Practical Considerations in the Use of Quadratic Residue Diffusing Surfaces," Tenth International Congress on Acoustics, Sydney, Australia (1980); J. R. Hyde, "Segerstrom Hall in Orange County—Design, measurements and results after a year of operation," *Proceedings of Institute of Acoustics,* **10**, 155 (1988); J. R. Hyde, "Sound strength in concert halls: The role of the early sound field on objective and subjective measures," J. Acoust. Soc. Amer. **103**, 2748 (A), (1998).

TECHNICAL DETAILS

V = 981,800 ft³ (27,800 m³)	S_a = 16,190 ft² (1,504 m²)	S_A = 18,750 ft² (1,742 m²)
S_o = 2,400 ft² (223 m²)	S_T = 20,690 ft² (1,922 m²)	N = 2,903
H = 80 ft (24.4 m)	W = 136 ft (41.5 m)	L = 119 ft (36.2 m)
D = 144 ft (44 m)	V/S_T = 47.4 ft (14.5 m)	V/S_A = 60.6 ft (18.5 m)
V/N = 338 ft³ (9.58 m³)	S_A/N = 6.46 ft² (0.6 m²)	S_a/N = 6.46 ft² (0.600 m²)
H/W = 0.59	L/W = 0.88	ITDG = 31 msec
Note: S_T = S_A + 1,940 ft² (180 m²); see Appendix 1 for definition of S_T.		

NOTE: The terminology is explained in Appendix 1.

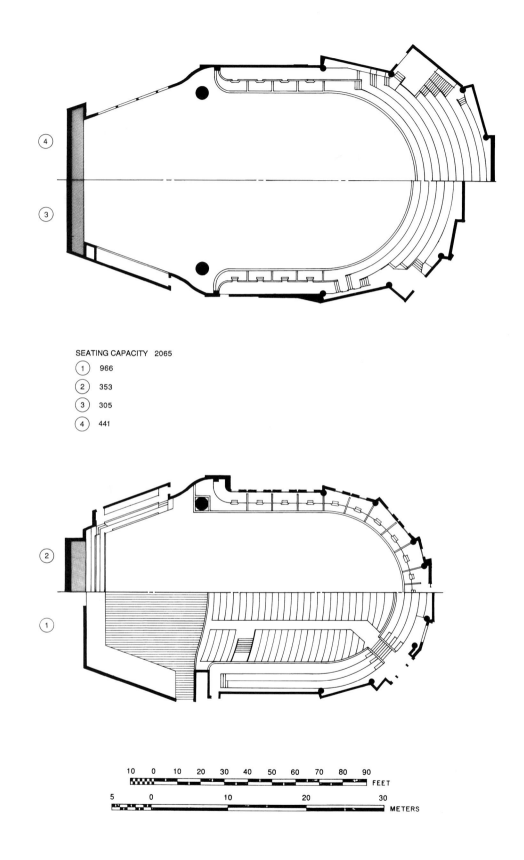

SEATING CAPACITY 2065

- (1) 966
- (2) 353
- (3) 305
- (4) 441

shall and J. R. Hyde, "Some Practical Considerations in the Use of Quadratic Residue Diffusing Surfaces," Tenth International Congress on Acoustics, Sydney, Australia (1980); J. R. Hyde, "Segerstrom Hall in Orange County—Design, measurements and results after a year of operation," *Proceedings of Institute of Acoustics,* **10**, 155 (1988); J. R. Hyde, "Sound strength in concert halls: The role of the early sound field on objective and subjective measures," J. Acoust. Soc. Amer. **103**, 2748 (A), (1998).

TECHNICAL DETAILS

V = 981,800 ft³ (27,800 m³)	S_a = 16,190 ft² (1,504 m²)	S_A = 18,750 ft² (1,742 m²)
S_o = 2,400 ft² (223 m²)	S_T = 20,690 ft² (1,922 m²)	N = 2,903
H = 80 ft (24.4 m)	W = 136 ft (41.5 m)	L = 119 ft (36.2 m)
D = 144 ft (44 m)	V/S_T = 47.4 ft (14.5 m)	V/S_A = 60.6 ft (18.5 m)
V/N = 338 ft³ (9.58 m³)	S_A/N = 6.46 ft² (0.6 m²)	S_a/N = 6.46 ft² (0.600 m²)
H/W = 0.59	L/W = 0.88	ITDG = 31 msec

Note: $S_T = S_A$ + 1,940 ft² (180 m²); see Appendix 1 for definition of S_T.

NOTE: The terminology is explained in Appendix 1.

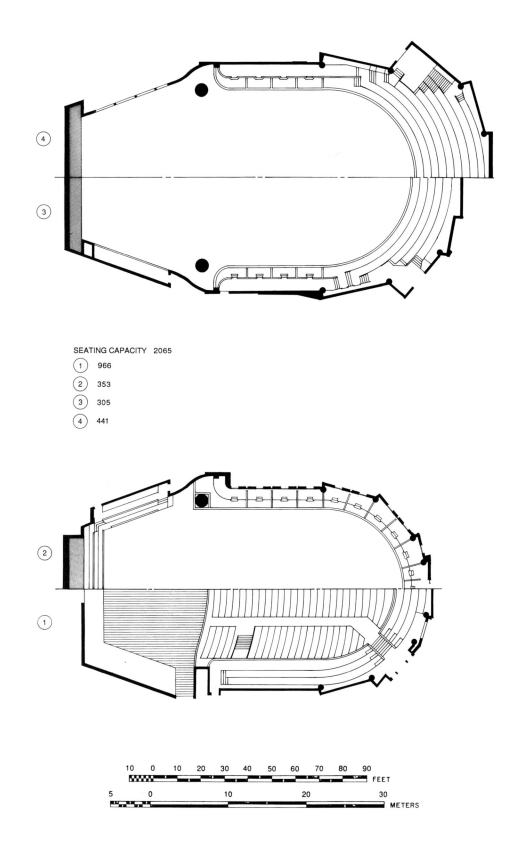

SEATING CAPACITY 2065

1 966

2 353

3 305

4 441

8

Dallas

Eugene McDermott Concert Hall in Morton H. Meyerson Symphony Center

The McDermott Hall in the Meyerson Symphony Center is a classical rectangular hall in the front two-thirds and a classical opera house in the remainder. Entrance is via circular, beige, Italian marble stairs that lead one into an asymmetrical lobby that surrounds the hall like an eighteenth-century glass ballgown. The concert hall, seating 2,065, is in direct contrast—an intimate, warm room in soft red and dark wood tones trimmed with bronze. On the occasion of its opening in 1989, a *New York Times* critic wrote, "One of the handsomest new rooms in which to hear music anywhere."

Acoustically, the unusual feature of the Meyerson is the 254,000 ft^3 (7,200 m^3) of partially coupled reverberation space wrapped around the perimeter of the hall above the highest audience level and concealed by an open-weave cloth. The chamber is opened and closed by seventy-four, 4-in. (10-cm)-thick, hinged, concrete doors, motor-operated by remote control. With them, the length of the reverberation time can be varied, and the four-part canopy can be raised to give the sound energy greater access to the chamber for large-scale works, or lowered for more intimate chamber and recital music.

During my two-concert, five-seat visit to the hall the canopy was raised high enough that its surface was parallel to the vision of listeners in the highest rows. It was obvious that the reverberation chamber doors, at least some of them, were open. The fidelity of tone, orchestral balance, and intimacy were excellent nearly everywhere. The spaciousness and the fullness of the sound was best on the main floor and in the seats in the curved rear one-third. On the sides the orchestra was less well balanced, one side or the other dominating over the cello in Shostakovich's *Concerto No. 1*. The effect of the chamber was hard to evaluate. Its presence was not obvious in all the seats, except after sudden-stop chords, when the sound seemed to reverberate for about 3.0 sec. In the running music, the early reverberation time sounded like 2.0 sec. Critical reviews have been favorable. The hall, being basically rectangular, combines a favorable design with an interesting addition of a lengthened

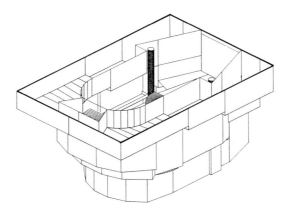

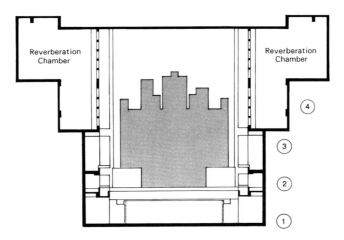

Reverberation Chamber

Reverberation Chamber

④

③

②

①

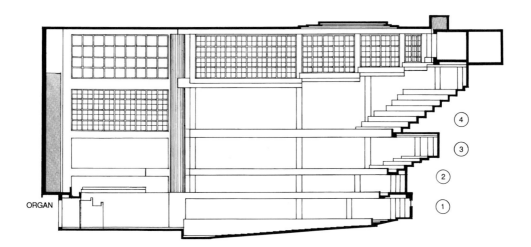

ORGAN

④

③

②

①

reverberation time. Since this lengthened RT is audible only with very slow tempos or after stop chords, the question is whether the extra construction cost is merited.

ARCHITECTURAL AND STRUCTURAL DETAILS

Uses: Symphonic and chamber orchestras, recitals, and soloists. **Ceiling:** 5.5-in. (14-cm) concrete with plaster skim coat. **Overstage canopy:** 6-in. (15-cm) laminated wood, 4,000 ft² (372 m²), over orchestra and front of audience; normal height 36–50 ft (11–15 m). **Walls:** Thin wood veneer on 0.5-in. (1.5-cm) particle board bonded by adhesive and plaster to 10-in. (25-cm) masonry, some areas of limestone; center canopy has 0.06-in. (0.16-cm) layer of felt to suppress acoustic glare. **Balcony fronts:** 2.0-in. (5-cm) plaster, on upper two fronts, added thin wood veneer. **Variable absorption:** Motor-operated curtains in storage pockets for covering 6,190 ft² (575 m²) of walls; in each pocket, one set of curtains is single-layer thin fabric, the other is multilayer, tightly woven, heavy velour. **Floor:** Terrazzo or painted concrete; no carpet. **Stage floor:** 0.5-in. (1.5-cm) tongue-and-groove wood on wood boards on joists. Floor beneath cellos and double basses has 2 in. (5 cm) of wood over 3 ft (0.9 m) of airspace. **Stage walls:** Lower 14 ft (4.3 m) is 2-in. (5 cm) wood doors that are selectively opened to expose a 12-ft (3.7-m)-deep airspace behind instruments. **Reverberant chamber:** 254,000 ft³ (7,200 m³), containing 4,844 ft² (450 m²) of variable sound absorption. **Stage height:** 42 in. (1.38 m). **Seating:** Molded plywood back and unperforated plywood bottom; upholstered seatback and seat with porous fabric over polyvinyl foam cushion; armrests, wood.

ARCHITECT: Pei Cobb Freed & Partners. ACOUSTICAL CONSULTANT: ARTEC Consultants, Inc. THEATER CONSULTANT: ARTEC Consultants, Inc. PHOTOGRAPHS: Courtesy of Pei Cobb Freed & Partners.

TECHNICAL DETAILS

V = 844,000 ft³ (23,900 m³) closed	S_a = 10,550 ft² (980 m²)	S_A = 12,500 ft² (1,161 m²)
S_o = 2,691 ft² (250 m²)	S_T = 14,440 ft² (1,341 m²)	N = 2,065
H = 86 ft (26.2 m)	W = 84 ft (25.6 m)	L = 101 ft (30.8 m)
D = 133 ft (40.5 m)	V/S_T = 58.4 ft (17.8 m)	V/S_A = 67.5 ft (20.6 m)
V/N = 409 ft³ (11.6 m³)	S_A/N = 6.05 ft² (0.56 m²)	S_a/N = 5.11 ft² (0.474 m²)
H/W = 1.02	L/W = 1.2	ITDG = 21 msec
Note: $S_T = S_A$ + 1,940 ft² (180 m²); see Appendix 1 for definition of S_T.		

NOTE: The terminology is explained in Appendix 1.

9

Denver

Boettcher Concert Hall

*D*enver is known as the gateway to the Rocky Mountains. At an altitude of 1 mile (1.6 km) it boasts 300 days of sunshine a year. In 1978 the opening of the Boettcher Concert Hall signaled a move toward the population's greater appreciation of the performing arts. The Colorado Symphony Orchestra performs from September through May and attracts leading soloists.

Boettcher, seating 2,750, is circular in design; with the orchestra near the center, no listener is more than 100 feet from the stage. Actually, only 311 seats are directly behind the orchestra, although a substantial number are to the sides. Modeled after the Philharmonie in Berlin, the Boettcher's seating is terraced—composed of trays—so that part of the sound from the orchestra is reflected from the tray-sides laterally to the listeners, which creates a desirable acoustic effect.

Most obvious to a visitor are the hall's 106 large, translucent disks hovering above the stage and parts of the audience, which guide the sound of the performers to all sections of the hall. The opening night music critics were enthusiastic about the acoustics—Schonberg of *The New York Times* wrote, "In mid-range and top frequencies the sound was beautiful—warm and intimate, well detailed, with plenty of color and presence . . . sauvity and smoothness of response. [Except for a few spots] the bass can be heard well enough. . . . listening to music there is a comfortable, enjoyable experience." Martin Bernheimer of the *Los Angeles Times* offered, "Pianissimos shimmer at Boettcher, the string tone is really lustrous, and the brass cut the air with clarion purity. . . . The acoustics are fine." At the hall's tenth anniversary in 1988, the *Denver Post* interviewed the hall's musical users and concluded, "Boettcher Hall is earning good grades for acoustics." These listeners sat facing the front of the orchestra.

On the other hand, those seated behind an orchestra that is situated near the center of a hall hear a poor rendition of a piano or a singer's voice, because the high tones are radiated forward, while only the lower ones project rearward. The orchestra sounds are also unbalanced, because the French horns face backward and

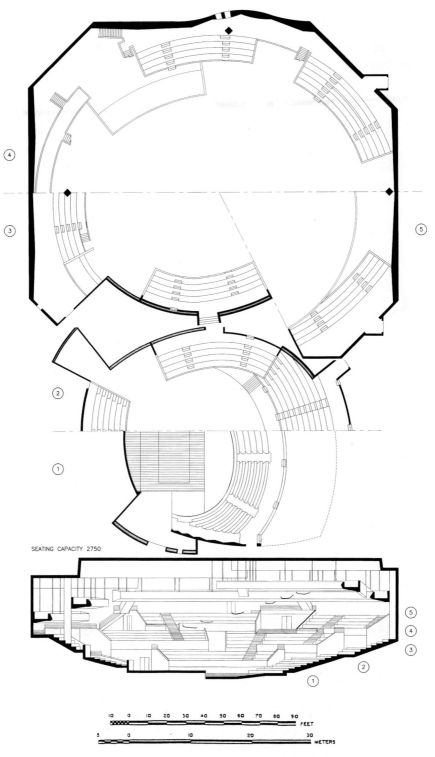

SEATING CAPACITY 2750

Modified Scale

the other brass instruments face forward. Placing a singer stage-right on risers behind the orchestra string section would help the voice problem. With that orientation, 90 percent of the audience would be forward of the singers, and those behind would be near enough to hear a reasonably well-balanced vocal sound. Further, years ago, the Baldwin Piano Co. developed a reversed piano lid that would resolve any directionality problems for that instrument. To many listeners, a distinct advantage to those rear seats is the experience of facing the conductor and being "part of the orchestra."

ARCHITECTURAL AND STRUCTURAL DETAILS

Uses: Primarily symphonic concerts, but otherwise general purpose. **Ceiling and side walls:** Three layers of gypsum board. **Stage floor:** Tongue-and-groove wood above wooden underlayer on sleepers with a moat below. **Seats:** Molded wood with upholstering on top of seat bottom 2 in. (5 cm) thick and that on front of seat back 1 in. (2.5 cm) thick.

ARCHITECT: Hardy Holzman Pfeiffer Associates. ACOUSTICAL CONSULTANT: Jaffe Holden Acoustics. REFERENCES: J. S. Bradley, Gary Madaras, and Chris Jaffe, "Acoustical characteristics of a 360° surround hall," *J. Acoust. Soc. Am.* **101**, 3135(A) (1997).

TECHNICAL DETAILS

$V = 1{,}321{,}760$ ft³ (37,444 m³)	$S_a = 18{,}290$ ft² (1,700 m²)	$S_A = 23{,}564$ ft² (2,190 m²)
$S_o = 2{,}561$ ft² (238 m²)	$S_T = 25{,}504$ ft² (2,370 m²)	$N = 2{,}750$
$H = 40$ ft (12.2 m)	$W = 174$ ft (53 m)	$L = 190$ ft (59 m)
$D = 100$ ft (30.5 m)*	$V/S_T = 51.8$ ft (15.8 m)	$V/S_A = 56.1$ ft (17.1 m)
$V/N = 481$ ft³ (13.6 m³)	$S_A/N = 8.57$ ft² (0.796 m²)	$S_a/N = 6.65$ ft² (0.62 m²)
$H/W = 0.23$	$L/W = 1.07$	ITDG $= 17$ msec

Note: $S_T = S_A + 1{,}940$ ft² (180 m²); see Appendix 1 for definition of S_T.

* With the stage extended, $D = 75$ ft (23 m).

NOTE: The terminology is explained in Appendix 1.

10

Fort Worth, Texas

Bass Performance Hall

The Bass Performance Hall, situated in the heart of downtown Fort Worth, is a neo-classic building with an outdoor motif of two trumpet-heralding angels. The hall, which seats 2,072 listeners, gives the feeling of the Teatro Colón in Buenos Aires, one of the world's best-loved opera/concert halls, while American musicians liken it to Carnegie Hall.

Bass Performance Hall in Fort Worth is a combination concert hall and opera house, also suitable for ballet, Broadway theater, and popular music. The principal problem with multi-purpose halls is the difficulty of adjusting the reverberation time to suit the different performances. In the Bass Hall two means are employed to overcome this problem. First, the "Concert Hall Shaper" shown in the drawings was employed to increase the cubic volume of the stage end of the hall without exposing the space to the sound absorption of scenery stored in the upper part of the stage house. Second, lower reverberation times for theater, voice, and contemporary music can be attained by deploying sound-absorbing curtains that can cover 6,600 ft² (613 m²) of wall space and by storing the "Shaper" against the upper rear wall, thus exposing the absorption of the normal house scenery in the stage house. The reverberation times at mid-frequencies, fully occupied, are 1.9 sec, optimum for symphonic concerts, and 1.6 sec, excellent for opera and contemporary music. Intimacy is preserved by the relatively small distances between the balcony fronts in the front of the hall.

Music critics have praised its acoustics from the time of its opening in early 1998, and the New York Philharmonic Orchestra, performing there in 1999 under the direction of Kurt Mazur, sent word via their executive director, Deborah Borda, that, "Its acoustics as well as the aesthetics make it truly an international gem." Kurt Mazur added, "A beautiful hall with perfect acoustics."

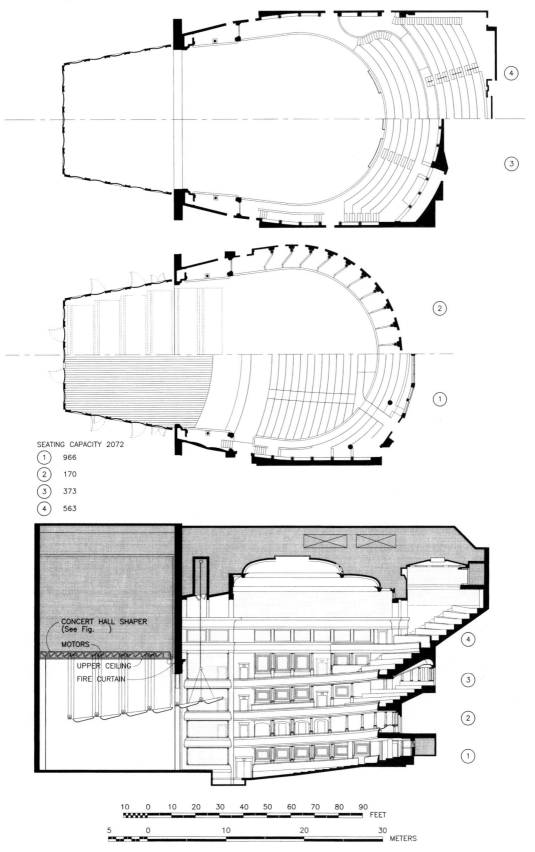

SEATING CAPACITY 2072
1 966
2 170
3 373
4 563

CONCERT HALL SHAPER
(See Fig.   )
MOTORS
UPPER CEILING
FIRE CURTAIN

10 0 10 20 30 40 50 60 70 80 90
FEET
5 0 10 20 30
METERS

ARCHITECTURAL AND STRUCTURAL DETAILS

Uses: Multi-use hall used by the Fort Worth Symphony, Opera, Ballet, and general performing arts functions. **Ceiling:** Dome area is 1-in. (2.5-cm) plaster and remaining area is 2–3 layers of 0.62-in. (1.58-cm) gypsum board. **Side walls:** 90% drywall laminated to masonry; 10%, plaster on masonry; balcony fronts are plaster. **Floors:** Concrete. **Stage house:** Hall shaper ceiling is series of reflectors made of 0.5-in. masonite and perforated brass covers on top; outer stage house walls are painted grout-filled masonry; upper shaper ceiling is 0.5-in. (1.25-cm) plywood; concert towers surrounding the players are plywood veneer over brass. **Carpet:** Thin, on aisles. **Height of the stage above the audience floor level:** 54 in. (1.37 m). **Seats:** Front of seat back and top of seat bottom upholstered; back of seat back wood; bottom of seat plastic. Armrests not upholstered.

ARCHITECT: David M. Schwarz Architects. ACOUSTICAL CONSULTANT: Jaffe Holden Acoustics. PHOTOGRAPHS: Len Allington.

TECHNICAL DETAILS

Concerts

V = 964,000 ft³ (27,300 m³)	S_a = 10,180 ft² (946 m²)	S_A = 13,150 ft² (1,222 m²)
S_o = 3,270 ft² (304 m²)	S_T = 15,085 ft² (1,402 m²)	N = 2,072
H = 82 ft (25 m)	W = 71 ft (21.6 m)	L = 94 ft (28.6 m)
D = 131 ft (40 m)	V/S_T = 63.9 ft (19.5 m)	V/S_A = 73.3 ft (22.3 m)
V/N = 465 ft³ (13.18 m³)	S_A/N = 6.35 ft² (0.59 m²)	S_a/N = 4.91 ft² (0.456 m²)
H/W = 1.156	L/W = 1.324	ITDG = 25 msec

Note: $S_T = S_A$ + 1,940 ft² (180 m²); see Appendix 1 for definition of S_T.

Opera

V = 652,000 ft³ (18,470 m³)	S_a = 9,500 ft² (883 m²)	S_A = 12,370 ft² (1,150 m²)
S_{pit} = 796 ft² (74 m²)	S_p = 2,320 ft² (215.6 m²)	S_T = 15,486 ft² (1,440 m²)
N = 1,960	V/S_T = 42.1 ft (12.83 m)	V/S_A = 52.7 ft (16.06 m)
V/N = 333 ft³ (9.42 m³)	S_A/N = 6.31 ft² (0.587 m²)	S_a/N = 4.85 ft² (0.45 m²)

NOTE: The terminology is explained in Appendix 1.

11

Lenox, Massachusetts

Seiji Ozawa Hall

The Seiji Ozawa hall opened July 7, 1994, with four conductors, two symphony orchestras, and a three-hour program that ended with the audience singing Randall Thompson's *Alleluia*. Modeled after Vienna's Grosser Musikvereinssaal, Ozawa Hall is a somewhat smaller "shoebox" seating 1,180 vs. 1,680. The hall is beautiful inside, with stucco walls painted a warm, off-white. The two tiers of balconies are faced with railings and gridded fronts in teak wood. The loge boxes, with the railings, become the most important architectural element when the hall is filled with people. Even the backrests of the seats pick up this gridded pattern. The ceiling is made of pre-cast coffers and the side walls are irregular to provide good sound diffusion.

The musicians sit on risers modeled so that they can see each other and be seen by the audience. High up, there are clerestory windows. Unusual, are high narrow windows above and behind the stage, northerly directed, that give one partial views of the sky changing from sunset to twilight to night.

The hall had to meet one other requirement, an audience seated outside on the lawn. Located in a natural shallow bowl, the rear wall can be completely opened, bringing the concert to outdoor audiences as large as 2,000, with sound augmented by an excellent sound system. (See lower photograph.)

The very thick concrete walls, doors, windows, and ceiling preserve the bass like no other hall, giving the music a rich sound and emphasizing the tones of the lower strings. The reverberation time, fully occupied with rear wall open, is 1.7 sec at mid-frequencies. At the lower frequencies, the reverberation is 2.2 sec and the bass ratio equals 1.32, which accounts for the warmth and richness of the bass sound. Because the hall is narrow and with plenty of surfaces for reflecting sound laterally to the audience, the subjective acoustical parameter called "spaciousness" also exceeds that in any larger hall, including that in its Viennese model. It contrasts with the nearby 5,000-seat Koussevitzky Music Shed, with greater loudness and a feeling of listening in a small group. Edward Rothstein, music critic of the *New*

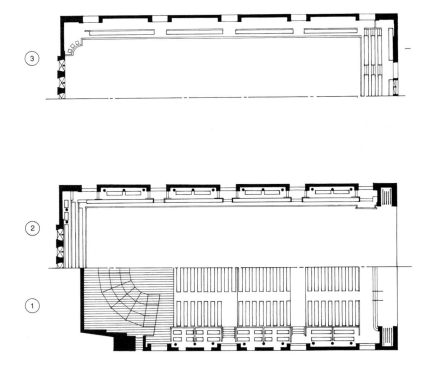

SEATING CAPACITY   1180

①   692

②   268

③   220

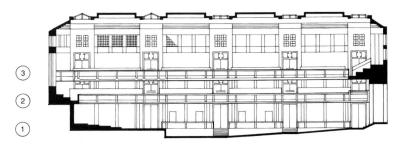

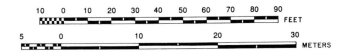

York Times, said, "[It] is precisely what a concert hall should be: a resonant, warm space that comes to life with sound."

ARCHITECTURAL AND STRUCTURAL DETAILS

Uses: Rehearsals, performances, and recording. **Ceiling:** 35 pre-cast concrete panels weighing 27,000 lbs (12,200 kg) each. **Main sidewalls:** 12-in. (30-cm) grout-filled concrete blocks with stucco covering. **Balcony fronts:** Lower part irregular to provide early lateral reflections; upper part open gridded wood. **Stage ceiling:** None; overhung slightly by balconies. **Stage sidewalls:** Removable wooden panels on stage right, and canted concrete block walls, behind wood trimmed skrim fabric panels on stage left. **Stage floor:** Tongue-and-groove maple wood strips glued to canvas and floated on bed of plastic and 1.12-in. (2.8-cm) plywood to approximate the looseness of classic, aged, tongue-and-groove stage floors and to prevent off-season buckling. **Stage height:** 42 in. (106 cm). **Audience floor:** Wooden tongue-and-groove boards affixed to 3.25-in. (8.26-cm) tongue-and-groove timber planking. **Carpet:** None. **Sound absorption:** Six large panels, 6 in. (15 cm) thick can be lowered by steps from the ceiling over the stage and absorbing curtains along the upper sidewalls can be dropped to reduce the empty reverberation time (with closed rear wall) down by 0.4 to 0.6 sec to the desired recording reverberation time. **Seating:** Seat back, open slatted; seat pans covered with thin cushions; structure, backrest, and armrests, solid wood.

ARCHITECT: William Rawn Associates, Architects, Inc. ACOUSTICAL CONSULTANT: Kirkegaard & Associates. PHOTOGRAPHS: Steve Rosenthal.

TECHNICAL DETAILS

V = 410,000 ft³ (11,610 m³)	S_a = 5,340 ft² (496 m²)	S_A = 7,955 ft² (739 m²)
S_o = 2,175 ft² (202 m²)	S_T = 9,895 ft² (919 m²)	N = 1,180
H = 49 ft (14.9 m)	W = 68 ft (20.7 m)	L = 94 ft (28.6 m)
D = 94 ft (28.6 m)	V/S_T = 41.4 ft (12.63 m)	V/S_A = 51.5 ft (15.7 m)
V/N = 347 ft³ (9.83 m³)	S_A/N = 6.74 ft² (0.63 m²)	S_a/N = 4.52 ft² (0.42 m²)
H/W = 0.72	L/W = 1.38	ITDG = 23 msec

Note: $S_T = S_A + 1,940$ ft² (180 m²); see Appendix 1 for definition of S_T.

NOTE: The terminology is explained in Appendix 1.

12

Lenox, Massachusetts

Tanglewood, Serge Koussevitzky Music Shed

anglewood, where the Boston Symphony Orchestra's Koussevitzky Music Shed is located, is an incredibly beautiful estate in the Berkshire Hills of Massachusetts. The Music Shed at Tanglewood boasts a unique position among concert halls. It is the only place that houses a very large audience, 5,121 listeners, under acoustical conditions that rival the best in America. And an additional 10,000 people seated on the lawns outside can enjoy the music that issues from the partly open sides of the shed and a superior sound amplification system. Even though the sides of the shed are open to a height of about 15 ft (4.6 m), the reverberation behaves like that in a regular auditorium.

The shed began its life using an inadequate stage enclosure that was moved indoors from a tent in 1938. The combination of a high reverberation time, and no surfaces for projecting the sound from the stage, led to a "muddy" acoustical environment. The sound lacked definition and the balance between early and late sound energy was poor. In 1954, the trustees engaged Bolt Beranek and Newman to undertake a study to improve the acoustical quality. Working began with the architect in 1958; a new orchestra enclosure, acoustic canopy, and rear-wall sound diffusing surface were designed, and the whole was dedicated in 1959.

The 50-percent-open low ceiling, comprising 26 non-planar triangular panels, varying in width from 7 to 26 ft (2.1 to 7.9 m), reflects about half of the early sound energy down onto the audience, arriving shortly after the direct sound, thus giving music the quality heard in classical rectangular halls. The upper volume of the hall receives sufficient energy to maintain an optimum ratio of early sound energy to later reverberant sound energy at the listeners' ears. On stage, the enclosure/canopy also contributes to excellent sectional balance and ease of ensemble playing.

Isaac Stern said after the 1959 season, "The new orchestra enclosure in the Tanglewood Music Shed is one of the most fantastically successful efforts to create brilliant, ringing sound with wonderful definition, despite the enormous size of this

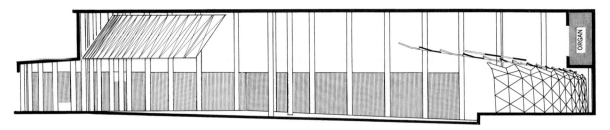

SEATING CAPACITY 5121

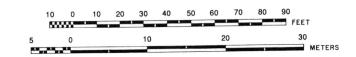

PANEL 11

ORGAN

6.250

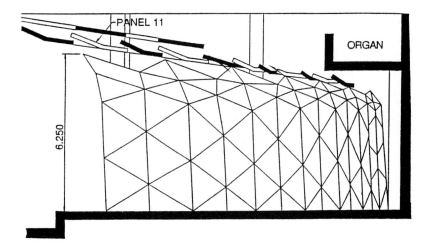

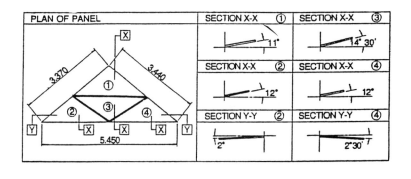

DETAIL OF PANEL 11

0   5 METERS

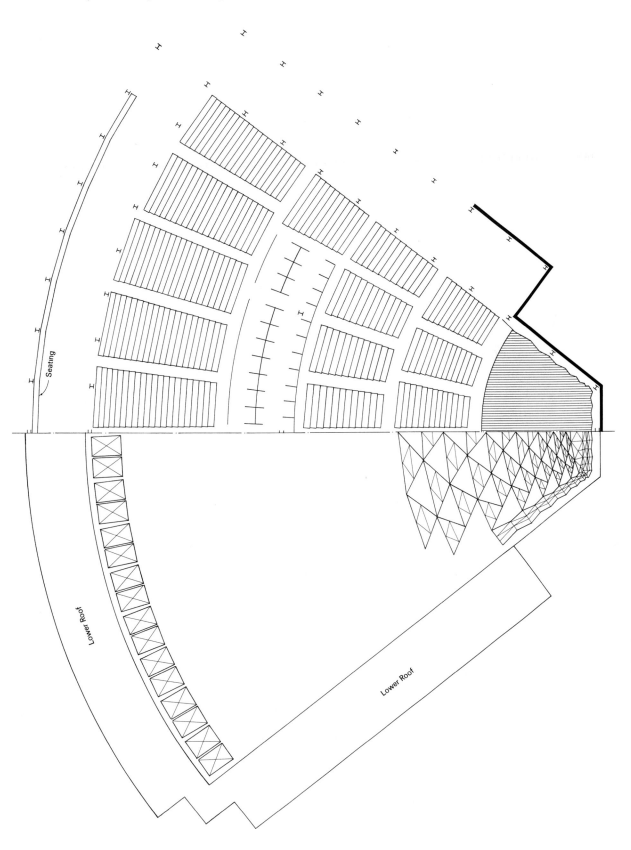

Seating

Lower Roof

Lower Roof

hall. It is particularly successful in providing an equal sound value wherever one sits. On stage there is a wonderfully live quality, and yet complete clarity for balancing with a large orchestra."

At the end of the summer season of 1959, Charles Munch, then the Music Director of the Boston Symphony Orchestra, wrote, "The new canopy has solved all the old problems of disproportion among the various elements of the orchestra. The greatest benefit has come to the strings and especially the violins which now can be heard in the shed with as much brilliance and clarity as in the best concert halls."

Pierre Monteux said in 1959, "What has been done is absolutely marvelous. Last year I could not hear the violins. This year the sound is marvelous."

Owning a summer home nearby, Eugene Ormandy, then Music Director of the Philadelphia Orchestra said (1962), "This year and also a year ago I sat in boxes at the center of the Shed and was amazed at the fine quality of the sound. The acoustical enclosure is wonderful. The brass is somewhat predominant in the large orchestra, though the conductor should be able to control it."

The acoustics of the completed hall corroborated our predictions. The Shed's canopy and enclosure were the first application of the results of a six-year program of studies. The low ceiling over the stage and front part of the audience (almost 50 percent open) produces the necessary short-initial-time-delay reflections in the hall, yet leaves the upper volume of the hall available for reverberation. In addition, the canopy contributes to excellent sectional balance for a large orchestra and improved clarity inside the shed.

Measurements of reflected sound from this type of panel array, in the laboratory, are in agreement with mathematical theory, which shows that lateral diffraction from the edges of the panels maintain an uniform energy flow even down to very low frequencies. This energy flow seems mostly to come from directly ahead because measurements in the *unoccupied* Shed reveal low values for the Binaural Quality Index BQI. But this apparent deficiency is more than made up for by the strength of the bass. The bass ratio, BR, is 1.45, far greater than Boston's 1.03.

It has been suggested that the optimum reverberation time, 1.9 sec at mid-frequencies, fully occupied, and the BR so enhance the effectiveness of this type of canopy that strong lateral reflections are not necessary to give a feeling of spaciousness. No confirming experiments have been performed to show that the favorable values of these two acoustical attributes are necessary to make this type of canopy work.

ARCHITECTURAL AND STRUCTURAL DETAILS

Uses: Symphony orchestra, chamber orchestra, and choral music. **Ceiling:** 2-in. (5-cm) wooden planks. **Side and rear walls:** Closed part of walls is 0.75-in. (1.9-cm) painted fiberboard; large area open to outdoors during concert season. **Floors:** Packed, treated dirt. **Stage enclosure:** 0.4–0.6-in. (1–1.5-cm)-plywood with modulations in shape and randomly and heavily braced. **Canopy:** Suspended by steel cables from the roof, consisting of a series of non-planar triangular plywood panels 0.4–0.6 in. thick, heavily framed, connected tip-to-tip (see the drawings). **Stage floor:** 1.25-in. (3.2-cm) wood over large airspace. **Stage height:** 33 in. (84 cm). **Sound-absorbing materials:** None. **Seating:** All wooden with metal arms, no cushions.

ORIGINAL ARCHITECT: Joseph Franz, engineer (1938), from preliminary design of Eliel Saarinen. In 1937, Saarinen consented to the use of his original drawings with such changes as were necessitated by the financial resources of the Berkshire Festival Corporation. RENOVATION ARCHITECT: Eero Saarinen and Associates (1959). ACOUSTICAL CONSULTANT: Bolt Beranek and Newman, now Acentech. PHOTOGRAPHS: Courtesy of Boston Symphony Orchestra. REFERENCE: F. R. Johnson, L. L. Beranek, R. B. Newman, R. H. Bolt, and D. L. Klepper, "Orchestra enclosure and canopy for the Tanglewood Music Shed," *Jour. Acoust. Soc. of Am.* **33**, 475–481 (1961).

TECHNICAL DETAILS

V = 1,500,000 ft³ (42,490 m³)	S_a = 24,000 ft² (2,230 m²)	S_A = 30,800 ft² (2,861 m²)
S_o = 2,200 ft² (204 m²)	S_T = 32,740 ft² (3,041 m²)	N = 5,121
H = 44 ft (13.4 m)	W = 200 ft (61 m)	L = 167 ft (50.9 m)
D = 163 ft (49.7 m)	V/S_T = 45.8 ft (13.97 m)	V/S_A = 48.7 ft (14.8 m)
V/N = 293 ft³ (8.29 m³)	S_A/N = 6.01 ft² (0.56 m²)	S_a/N = 4.69 ft² (0.435 m²)
H/W = 0.22	L/W = 0.835	ITDG = 19 msec

Note: $S_T = S_A + 1{,}940$ ft² (180 m²); see Appendix 1 for definition of S_T.

NOTE: The terminology is explained in Appendix 1.

13

Minneapolis

Minnesota Orchestra Association
Orchestra Hall

*O*rchestra Hall opened in October 1974. Designed primarily for The Minnesota Symphony Orchestra, which turned 100 in 2003, seats 2,450. The concert hall, one of two components in the complex, is made of concrete, brick, and oak. In the tradition of Vienna's Grosser Musikvereinssaal and Boston's Symphony Hall, it is basically rectangular, but with three wraparound balconies and a much less exposed wall above the top balcony. Missing also are the gilded or white statues of its famous predecessors.

Orchestra Hall must be seen to be believed. As one approaches, the exposed pipes come into view, blue on the exterior and yellow, green, and blue in the lobby. Inside the concert hall, a newcomer is struck by the playful cube-like shapes that cover the entire ceiling and the rear wall of the stage. In direct contrast is the militaristic formality of the balcony fronts and the stage side walls. These intriguing rectangular boxes are made of plaster and are randomly oriented to create a diffuse sound field—one in which the reverberant energy is uniform throughout the hall, giving the music a pleasant singing tone.

The balconies seat only 938 people, with movable cushioned chairs arranged in boxes along the sides and theater-style seats in the rear. Their shallowness assures the ticket buyer that no under-balcony dead spots exist. Great efforts were made to eliminate outside noises. The hall is separated by a 1-in. (2.5-cm) airspace from the structure which surrounds it on three sides. Adjacent to the entry doors on the three sides is a "ring corridor," with carpeted walls and floors designed to combat the noise of air-handling machinery, transformers, and office equipment. This second part of the building also contains rehearsal space, artistic and administrative offices, and the lobby.

Orchestra Hall's usable stage area is 2,174 ft^2 (202 m^2). The hall is sufficiently narrow that no central aisle is needed. A state-of-the-art sound system renders it usable for pops concerts, popular artists, seminars, and school functions. During the month of June, for pops, the seating configuration on the main floor

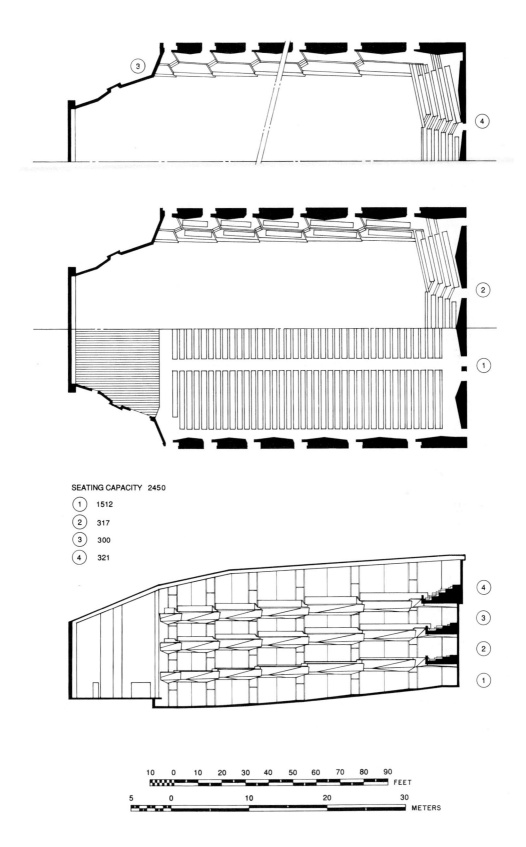

SEATING CAPACITY 2450

(1) 1512
(2) 317
(3) 300
(4) 321

is converted into terraced levels with tables and chairs, reducing the hall's capacity to 1,838.

ARCHITECTURAL AND STRUCTURAL DETAILS

Uses: Mostly orchestra, soloists, and popular artists. **Ceiling:** Plaster on wire lath. **Walls:** Side walls are hardwood on concrete block. Balcony wall is plaster on wire lath. **Floors:** The base floor is flat concrete with parquet wood affixed; during the winter concert season, a sloping floor is installed (see the drawing). **Carpets:** On main aisles downstairs, with no underpad. **Stage enclosure:** Most of the enclosure, including the ceiling, is thick wood. **Stage floor:** 1.5-in. (3.8-cm) wooden planks over a large airspace with 3/4-in. (1.9-cm) flooring on top. **Stage height:** 46 in. (117 cm) above floor level at first row of seats. **Seating:** The rear of the backrest is 0.5-in. (1.25-cm) molded plywood. The front of the backrest and the top of the seat bottom are upholstered over a polyurethane cushion. The underseat is unperforated. The arms are wooden.

ARCHITECT: Hammel Green and Abrahamson. ASSOCIATE ARCHITECT FOR DESIGN: Hardy Holzman Pfeiffer Associates. ACOUSTICAL CONSULTANT: Cyril M. Harris. PHOTOGRAPHS: Tom W. McElin and Banbury Studios.

TECHNICAL DETAILS

$V = 670,000$ ft³ (18,975 m³)	$S_a = 13,627$ ft² (1,266 m²)	$S_A = 16,942$ ft² (1,574 m²)
$S_o = 2,185$ ft² (203 m²)	$S_T = 18,882$ ft² (1,754 m²)	$N = 2,450$
$H = 54$ ft (16.5 m)	$W = 94$ ft (28.6 m)	$L = 125$ ft (38.1 m)
$D = 134$ ft (40.8 m)	$V/S_T = 35.48$ ft (10.82 m)	$V/S_A = 39.5$ ft (12 m)
$V/N = 273$ ft³ (7.74 m³)	$S_A/N = 6.91$ ft² (0.642 m²)	$S_a/N = 5.56$ ft² (0.516 m²)
$H/W = 0.57$	$L/W = 1.33$	ITDG $= 33$ ms

Note: $S_T = S_A + 1,940$ ft² (180 m²); see Appendix 1 for definition of S_T.

NOTE: The terminology is explained in Appendix 1.

Avery Fisher Hall

*A*very Fisher Hall is the successor to Philharmonic Hall, although the exterior and the lobbies to the building have remained unchanged. The original hall was opened in 1962 and suffered such criticism that a series of renovations were performed over the next several years. Then, through the generosity of Avery Fisher, a complete reconstruction was made, including relocating the balconies, lowering the ceiling somewhat, adding irregularities to all surfaces, installing a new stage enclosure, changing the contour of the main floor (beneath the seats), and selecting new seats. Also, the decor is completely different. The new hall is a classic rectangular shoebox, like Boston Symphony Hall, but its drawings more resemble those of Salt Lake and Minneapolis.

The reconstructed hall opened in 1976. It seats 2,742 persons. On entering the hall, three elements command one's attention. First, the sound reflectors on the walls and ceiling of the stage enclosure. Then, the side balconies, resembling boxes, each with a gold-leaf-covered, cylindrical front. And, third, the ceiling with "waves" to facilitate sound diffusion. The change in acoustics was welcome, and music critics, immediately after the reopening, placed Avery Fisher Hall on an approximate par with Carnegie Hall.

The stage is large for a regular concert hall, 2,230 ft² (207 m²) in area, 81,000 ft³ (2,295 m³) in volume and 41 ft (12.5 m) in depth. It was augmented in 1992 by the addition of two tiers of rounded sound diffusers on the side walls (see the photograph) and two diffusing elements in the ceiling. The published purpose was to reflect sound back to the musicians so that they could hear themselves and their colleagues better.

The reverberation time (occupied) is shorter than that in Boston Symphony Hall (1.75 compared to 1.9 sec), and the ratio of the low-frequency reverberation times to those at middle frequencies is appreciably less in Avery Fisher (0.93) than in Boston (1.03). There are still complaints about bass weakness, as is borne out by the acoustical data just quoted, and about the lack of intimacy in the hall.

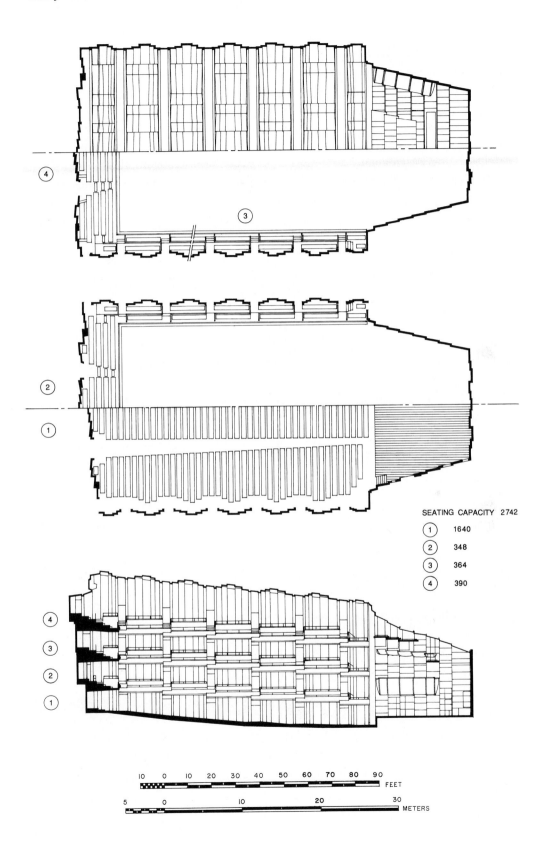

SEATING CAPACITY 2742

(1)	1640
(2)	348
(3)	364
(4)	390

10 0 10 20 30 40 50 60 70 80 90
FEET

5 0 10 20 30
METERS

Recently, Lincoln Center has been discussing the possibility of razing the hall's interior and replacing it with one having fewer seats.

ARCHITECTURAL AND STRUCTURAL DETAILS

Uses: Concerts, primarily. **Ceiling:** 1.5-in. (3.8-cm) plaster on metal lath. **Main and balcony side walls:** 0.75-in. (1.9-cm) painted plywood panel over minimum 1-in. (2.5-cm) compressed fiberglass (varies with angle of wall) screwed to grounds attached to 6-in. (15 cm) concrete block. **Balcony fronts:** Gold-leaf surface on 1-in. plaster. **Stage ceiling:** 0.75-in. (1.9-cm) natural-faced wooden plywood over which was poured 4-in. (10.2 cm) layered plaster ceiling. **Stage sidewalls:** 0.75-in. (2-cm) natural wooden plywood over 0.75-in. compressed fiberglass screwed to 10-in. (25.6-cm) concrete blocks. **Stage floor:** 0.6-in. (1.5-cm) tongue-and-groove oak wood strips over 0.75-in. (2-cm) plywood on 2 × 4-in. (5 × 10-cm) sleepers over original 3-in. (7.5-cm) wooden planks. **Stage height:** 42 in. (106 cm). **Audience floor:** To give the floor a new slope a new concrete floor was poured; 1.62-in. (4.1-cm) sleepers were laid on the floor; a 0.5-in. [1.25-cm plywood subfloor was attached and the finish flooring is 0.6-in. (1.5-cm)] oak wood boards. **Carpet:** Thin, on aisles. **Added sound absorption:** None. **Seating:** Seat backs, molded plywood; tops of seats and fronts of the backrests are upholstered with vinyl cushion, porous cloth covering; underseat, unperforated metal; armrests, wooden.

ARCHITECTS: For building exterior (1962), Harrison and Abramovitz; for concert hall (1976), Philip Johnson/John Burgee Architects; stage modifications (1992), John Burgee Architects. ACOUSTICAL CONSULTANTS: For the hall (1976), Cyril M. Harris; for the stage modification (1992), ARTEC Consultants, Inc. PHOTOGRAPHS: Sandor Acs and Norman McGrath.

TECHNICAL DETAILS

V = 720,300 ft³ (20,400 m³)	S_a = 12,798 ft² (1,189 m²)	S_A = 15,930 ft² (1,480 m²)
S_o = 2,180 ft² (203 m²)	S_T = 17,870 ft² (1,660 m²)	N = 2,742
H = 55 ft (16.8 m)	W = 85 ft (25.9 m)	L = 126 ft (38.4 m)
D = 135 ft (41.2 m)	V/S_T = 40.3 ft (12.29 m)	V/S_A = 42.2 ft (13.8 m)
V/N = 262 ft³ (7.44 m³)	S_A/N = 5.81 ft² (0.54 m²)	S_a/N = 4.67 ft² (0.434 m²)
H/W = 0.65	L/W = 1.5	ITDG = 30 msec

Note: $S_T = S_A$ + 1,940 ft² (180 m²); see Appendix 1 for definition of S_T.

NOTE: The terminology is explained in Appendix 1.

15

New York

Carnegie Hall

*I*saac Stern and Board Chairman James D. Wolfensohn are pictured in a 1991 Carnegie brochure embracing each other outside Carnegie Hall on December 15, 1986, beneath the caption: "Renovation and Re-Opening; The renovation . . . restored the auditorium to its 19th-century splendor." Immediately thereafter controversy erupted over the radically altered acoustics of this famous hall and persists to this day.

The lower photograph here shows the 1960 appearance of the stage proscenium. Note that a curtain hung over the upper half of the proscenium opening, even when the lower curtains were retracted. My description of the stage architecture, presented in *Music, Acoustics, and Architecture* (Wiley, New York, 1962) reads—

> **Stage enclosure:** Walls of the stage are plaster on wire lath; just prior to 1960, thin nylon draperies 30 ft [9 m] high were hung on either side [end] of the stage; in front of these draperies there were three folding screens, each with two panels. Each panel is made of 1/4-inch plywood on 9-ft-high by 4-ft-wide [2.74 m × 1.22 m] frames of 2 × 2 in. [5 cm] wood. Above the stage there are two frames, each about 6 ft [1.83 m] wide and 40 ft [12.2 m] long, sloped forward about 20°; painted canvas is stretched over the frames [these create an acoustic ceiling for the orchestra]. **Stage floor:** wood on *sleepers over concrete* [emphasis added].

Direct quotations from interviews with ten leading conductors of that day spoke of Carnegie with statements like: "In Carnegie Hall the music is clear as a bell. It has excellent high-frequency brilliance, but the middle- and low-frequency brilliance is not the best. There is a lack of body in the sound." "Carnegie Hall is only 'good' to 'fair.' From the podium there is a slightly damped feeling—especially on climaxes." "It is better than Philadelphia [Academy of Music], but worse than Boston [Symphony Hall]."

Irving Kolodin, revered music critic for the *Saturday Review,* said to me in December 1960, "I have a fondness for this hall. It is not a liability to any performance. Carnegie Hall lets the orchestra be itself. It doesn't impose its personality

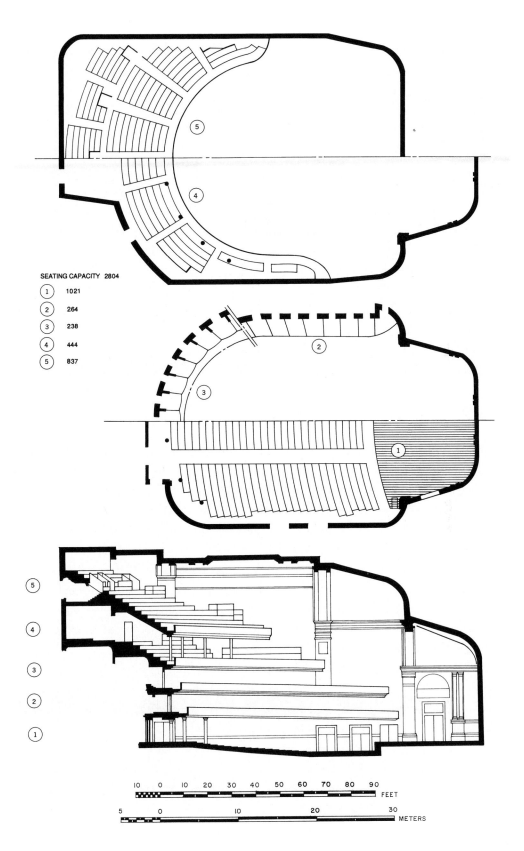

SEATING CAPACITY 2804

1 1021
2 264
3 238
4 444
5 837

10 0 10 20 30 40 50 60 70 80 90 FEET

5 0 10 20 30 METERS

on it. Whether from Europe or Philadelphia, an orchestra always sounds with its own style." We must remember that Kolodin occupied the same seat, one of the best, for this period.

Prior to September 8, 1966, new seats were installed, a new main floor was installed, the stage was extended 6 ft [1.83 m], the stage floor was renewed, and the center aisle was eliminated. Harold Schonberg wrote in the *New York Times,* November 10, 1966, "Ever since the new seats with their wooden backs were installed. . . . The auditorium is less mellow than it used to be. What we now are hearing is not necessarily a bad sound, but it certainly is a different sound." In *The New York Times* of May 3, 1982, Donald Henahan attempts to analyze the origin of Carnegie's "satisfying acoustics," while admitting, "its seating capacity of 2,800 is 69 greater than that of everybody's favorite American concert space, Symphony Hall in Boston . . ." Without mentioning the great changes to the stage in the late 1950s and again in 1966, he attributes the satisfactory acoustics to the architect William B. Tuthill.

Carnegie officials now have a very complete report on the acoustics of the hall, written by Tuthill in 1928, a year before his death, and published privately by his son in 1946, a copy of which came to the Carnegie archives in 1985 with an inscription to Mr. Stern. The booklet is entitled, *"Practical Acoustics, A study of the Diagrammatic Preparation of a Hall of Audience."* Tuthill is hardly modest; he starts by writing, "Practical acoustics. . . . becomes a demonstrably easy theme and task for a designer who by nature is scientifically endowed, is generally cultured and has been trained in all phases that may relate to the subject." Tuthill delineates some of his basic principles, as follows: (1) A hall should preferably not be longer than about 100 ft [30.5 m] [all seats in Carnegie are within this distance from the front of the stage, except those in the top balcony]. (2) The earliest reflections must be heard within about a ninth of a second after the direct sound [this requirement is met by all first reflections except those from the rear wall below the first balcony]. (3) If the arrivals of the earliest reflections are delayed longer than one-ninth of a second, there will be an echo, and (4) there should be no wood of any kind, except in the planking of the floors, to preserve reverberation in the room.

How successful was Tuthill's Carnegie Hall? Major Henry Higginson, then owner of the Boston Symphony Orchestra and chairman of the building committee for Boston Symphony Hall which opened in 1900, stated that he and the committee disliked Carnegie Hall's acoustics and looked to European halls as models. More conclusive has been the general response to the 1986 renovation that "restored the auditorium to its 19th-century splendor." Sad to say, there was a significant change in the orchestral sound. The brass and percussion strode forth in full volume. The famous mellow tone ceased to exist. The reflected sound took on a "hard" texture. Echoes from the rear wall of the main floor and from the rear stage wall troubled the audience at the front of that floor. On stage, the musicians were disturbed by focused ceiling echoes and by the same echoes as the audience. Those time-delayed reflections across-stage exceeded the direct sound heard from one section to another, and made ensemble playing difficult. Something had to be done.

In 1989, novel and effective renovations were instituted—with the understanding that the 19th-century visual appearance was not to be changed. The shape of the stage ceiling was slightly altered and a thin layer of sound-absorbing material was applied to it. In the three square panels on the back wall of the stage, felt was attached in thicknesses varying from 0.25 to 0.50 in. (see upper photograph). In the small rectangle and the semicircular decoration above each side-stage door, more felt was applied. Finally, to eliminate the high-frequency components in the reflections from the rear wall of the main floor, scientifically designed acoustic panels, about 28 ft (8.53 m) long (in three sections), were installed just below the soffit of the balcony above. They act to spread the high- and middle-frequency sound energy sideways, which improves spaciousness on the main floor, eliminates the echo from the lower rear wall, and masks the echoes from the box levels.

However, some grave negatives still exist. The worst acoustics I have ever heard in any concert hall exist today in the front rows of the top balcony. Without the pre-1986 curtain, the sounds from the back of the stage rise to the curved section of the ceiling between the proscenium and the flat upper ceiling and focus on these seats. At a 2001 concert of the Chicago Symphony with Yo Yo Ma as soloist, I could hardly hear the cello and the violins because of the overwhelming strength of the sound from the back of the stage—horns, percussion, brass, and some woodwinds. Even in the front half of the main floor, there is amplification of the sound from the rear of the orchestra owing to the shape of the stage ceiling.

There have been general complaints about a lack of bass since the 1986 renovations even on the main floor. This was caused by the removal of the hanging curtain, which permits the entire middle and rear sections of the orchestra to be amplified by the stage ceiling—making the basses and cellos sound less prominent. This deficiency led to the discovery of a solid layer of concrete beneath the stage floor. The immediate conclusion by music critics was that the concrete must cause a reduction in the strength of the basses and cellos that had not existed before the stage floor was rebuilt in 1986. Alan Kozinn in the September 14, 1995, issue of *The New York Times* wrote, "How the concrete got there—and who was responsible—remains a mystery. The architects and builders have no record of it." But it is clear from my 1962 description above that *the concrete was there in 1962 in exactly the form that is shown in the sketch* with the Kozinn article. The concrete was removed in 1995. At the first concert that followed, many believed that the bass strength was restored. Indeed, there may have been a difference, but that would also have resulted if the concrete had been removed in 1962 while the curtains were there. In various articles in *The New York Times* in the 1995–96 music season, the theme enunciated by Kozinn reoccurs, "There are still ways in which the pre-1986 sound was preferable. The old stage curtain, though less attractive, absorbed high frequencies and kept strings and winds from sounding strident, as they sometimes still do." Unless Carnegie is drastically rebuilt, e.g., eliminating focusing surfaces like that in the ceiling just outside the proscenium, the simplest answer is to restore the proscenium curtain and, possibly, the hanging panels above the orchestra.

ARCHITECTURAL AND STRUCTURAL DETAILS

Uses: Mostly orchestra, soloists, and chorus. **Ceiling:** 1-in. (2.5-cm) plaster on metal screen, 5% open for ventilation. **Walls:** 0.75-in. (1.9-cm) plaster on metal screen with small airspace between the plaster and the solid backing; beneath the first tier of boxes, the walls are plaster on solid backing; balcony fronts are plaster. **Floors:** Wood flooring on sleepers over concrete. **Carpets:** In the aisles on main floor and upper two tiers (3 and 4); carpet in the boxes. **Stage enclosure:** Walls of the stage are mostly plaster on masonry, but a few areas are plaster on metal lath; stage ceiling is vaulted, plaster resiliently suspended free of wall surfaces (other details in the text). **Stage floor:** Prior to 1995, the construction was wood on sleepers over concrete. In mid-1995, the concrete was removed, opening a large airspace underneath. **Stage height:** 48 in. (122 cm) above floor level at first row of seats. **Seating:** Irwin PAC chair with high-impedance mohair upholstery, perforated metal pans.

ARCHITECT: William B. Tuthill; for the 1986 and 1989 renovations, James Polshek. ACOUSTICAL CONSULTANTS: For the 1986 renovation, Abraham Melzer; for 1989 renovation, Kirkegaard & Associates. PHOTOGRAPHS: Lower photograph courtesy of Carnegie Hall in 1962; upper photograph, Michelle V. Agins *The New York Times* (1991).

TECHNICAL DETAILS

$V = 857,000$ ft³ (24,270 m³)	$S_a = 12,300$ ft² (1,145 m²)	$S_A = 17,220$ ft² (1,600 m²)
$S_o = 2,440$ ft² (227 m²)	$S_T = 19,160$ ft² (1,780 m²)	$N = 2,804$
$H = 78$ ft (23.8 m)	$W = 85$ ft (25.9 m)	$L = 108$ ft (32.9 m)
$D = 147$ ft (44.8 m)	$V/S_T = 44.7$ ft (13.63 m)	$V/S_A = 59.8$ ft (15.2 m)
$V/N = 306$ ft³ (8.65 m³)	$S_A/N = 5.71$ ft² (0.53 m²)	$S_a/N = 4.39$ ft² (0.408 m²)
$H/W = 0.92$	$L/W = 1.27$	ITDG $= 23$ msec
Note: $S_T = S_A + 1,940$ ft² (180 m²); see Appendix 1 for definition of S_T.		

NOTE: The terminology is explained in Appendix 1.

16

New York

Metropolitan Opera House

*I*n September 1966, the new Metropolitan Opera House opened with all the pomp and circumstance that a devoted New York opera public could muster. The renowned Metropolitan Opera Association regularly features on its roster some of the world's finest singers. Its pit orchestra is among the best in this country. Its tradition has been carefully nurtured by a proud New York society and by the nationwide audience devoted to its Saturday afternoon broadcasts.

Its 3,816 seats make the Met one of the largest houses in the world used exclusively for opera. Its shape is not like Milan's horseshoe La Scala, traditionally used as a model for most opera houses, though there are resemblances. The seating is on five tiers, plus a balcony, and the fronts of the tiers are flattened to make the viewing distance from the stage as short as possible. Its volume of 873,000 ft³ (24,724 m³) is more than double that of La Scala, making it necessary for singers to have strong voices in order to project. Its reverberation time, fully occupied, at mid-frequencies, is about 1.6 sec, as high as that in any other opera house, which aids in augmenting the loudness of the singer's voices.

The Met's stage is probably the most highly mechanized in the world. Scenes can float up from below on seven hydraulic lifts which occupy a 60 × 60-ft (18 × 18-m) area. Scenes can glide in from two side stages on large motorized wagons and from back-stage, which also boasts a 57-ft (17.4-m) diameter turntable. Scenes and people can pop up through traps in the lifts. The front-stage curtains and all hanging-scenery battens are motorized. The extensive controls for these mechanisms and the lighting are truly awe-inspiring.

The designing acousticians wisely tailored the box fronts and the tall flat areas to either side of the proscenium to create the necessary early lateral sound reflections to give the voices breadth and intimacy. The acoustics benefit from the large apron on the stage, which reflects the voices of the singers into the upper four tiers and the balcony. The sound is distributed reasonably equally to all seats throughout the house except those in the upper tiers nearest to the proscenium.

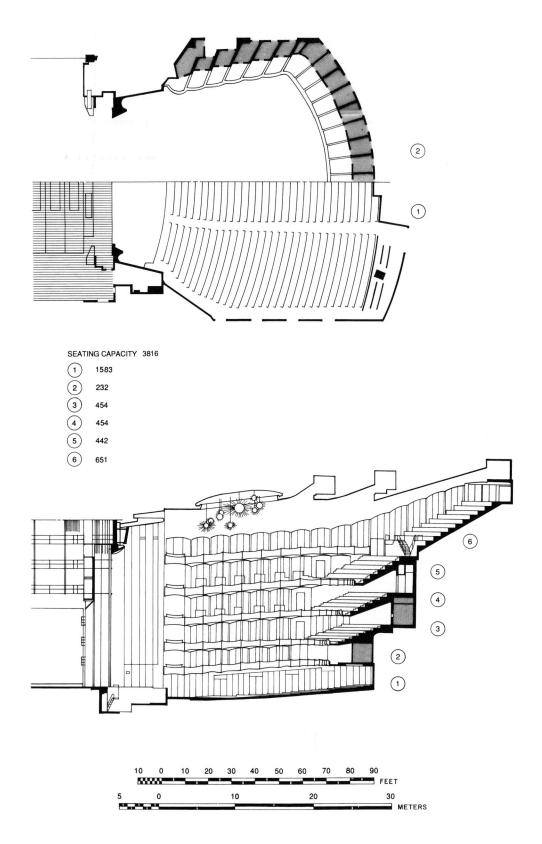

SEATING CAPACITY 3816

(1)	1583
(2)	232
(3)	454
(4)	454
(5)	442
(6)	651

10 0 10 20 30 40 50 60 70 80 90
FEET

5 0 10 20 30
METERS

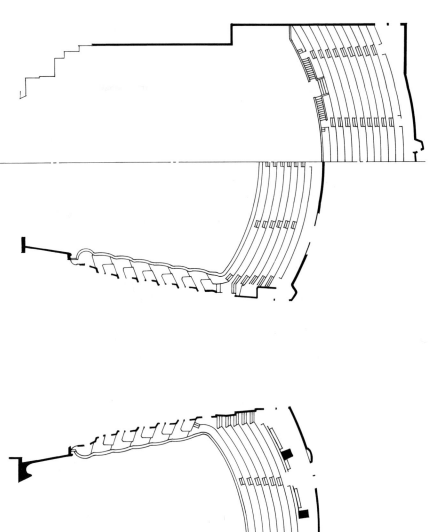

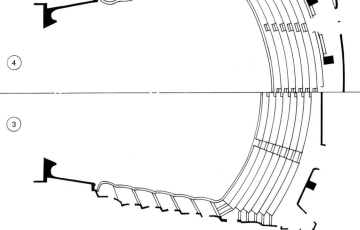

⌒ ARCHITECTURAL AND STRUCTURAL DETAILS

Uses: Opera. **Ceiling:** Plaster, covered with 4,000 rolls of 23 carat gold leaf. **Walls:** Plaster, some areas covered with 27,000 ft² (2,500 m²) of burgundy velour, some painted and some with wood paneling. **Floors:** Solid. **Carpets:** 18,000 ft² (1,670 m²). **Pit:** Floor is made of 2.5-in. (6.4-cm) wooden planks; the pit walls are of 0.75 in. (1.9 cm) wood. **Stage height:** 48 in. (122 cm) above floor level at first row of seats. **Seating:** Fully upholstered, except underseats are solid; seats range in width from 19 to 23 in. (48 to 58 cm) to conform to "three-row vision" sequence. **Proscenium curtain:** Gold patterned silk damask, tableau drape, 9,000 ft² (836 m²). **Proscenium opening:** 54 ft² (16.5 m²).

ARCHITECT: Wallace K. Harrison of Harrison and Abramovitz. ACOUSTICAL CONSULTANTS: V. Lassen Jordan and Cyril M. Harris. PHOTOGRAPHS: Metropolitan Opera Archives and United Press International Photo. REFERENCES: Plans and details, John Pennino, archivist office of the Metropolitan Opera Association.

TECHNICAL DETAILS

Opera

V = 873,000 ft³ (24,724 m³)	S_a = 20,600 ft² (1,914 m²)	S_A = 24,350 ft² (2,262 m²)
S_o (pit) = 1,420 ft² (132 m²)	S_T = 25,770 ft² (2,394 m²)	N = 3,816 plus ~200 standees
H = 82 ft (25 m)	W = 110 ft (33.5 m)	L = 130 ft (39.6 m)
D = 184 ft (56.1 m)	V/S_T = 33.8 ft (10.3 m)	V/S_A = 35.8 ft (10.9 m)
V/N = 229 ft³ (6.48 m³)	S_A/N = 6.38 ft² (0.59 m²)	S_a/N = 5.40 ft² (0.50 m²)
H/W = 0.745	L/W = 1.18	ITDG = 34 msec

NOTE: The terminology is explained in Appendix 1.

17

Philadelphia

Academy of Music

*E*xcept for the Teatro Di San Carlo in Naples and the Teatro alla Scala in Milan, after which it was modeled, the Academy of Music is the oldest hall in this book. In the middle of the nineteenth century, there were no symphony orchestras to speak of in this country, but grand opera was immensely popular and enjoyed unique prestige both in Europe and in New York. The Academy was built for the express purpose of bringing this art to Philadelphia. Acoustically, it is unquestionably the finest opera house in the United States, and the most beautiful.

Venerated for nearly a century and a half of concerts and operas, beautiful and intimate, this hall and the world renowned Philadelphia Orchestra mean to America the best in music.

With a cubic volume only 40 percent greater than that of the Vienna Staatsoper, the Academy nevertheless holds 70 percent more people. Each of its 2,827 (for opera) seats is allotted only 5.5 ft² (0.5 m²), including aisles, as compared to 7.5 ft² (0.7 m²) in Vienna in the 1955 reconstruction. For symphonic music, the biggest criticism of the Academy of Music is its low reverberation time. Measured in 1992, fully occupied, at mid-frequencies, it is 1.2 sec, comparable to La Scala's 1.2 and Vienna's 1.3, but well below the 1.85 sec of Boston's Symphony Hall.

The orchestra enclosure is composed of twelve rolling "towers" (see the drawing), four on each side and four at the rear. Each has three convex sections, designed to diffuse the sound on the stage for better inter-player communication. The ceiling is made of three flat, sloped sections, as shown in the drawing.

Every conductor and music critic interviewed in 1960 said that the Academy was excellent for opera, but for symphonic music, "somewhat dry," "sound too small," "not very live." My own judgment, based on listening to concerts there many times, is that the orchestral sounds are balanced, clear, and beautiful. Basses and cellos are strong. There is a feeling of intimacy, both visually and musically. But there is no audible reverberation and the sound does not envelop one as in rectan-

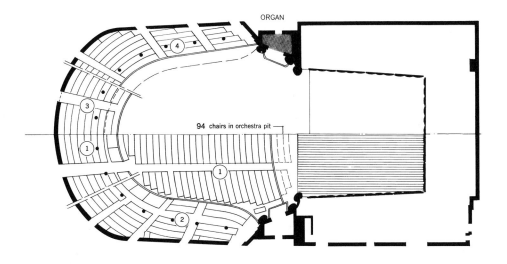

ORGAN

94 chairs in orchestra pit

SEATING CAPACITY 2827 to 2921

1 Concerts 1307; Small pit 1213; Large pit 1147

2 524

3 561

4 529

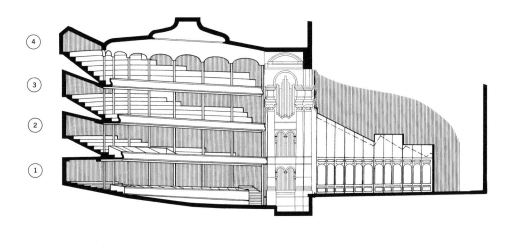

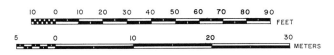

10 0 10 20 30 40 50 60 70 80 90
FEET

5 0 10 20 30
METERS

gular orchestral halls. It does not seem fair in this book to compare it with the world's great concert halls. It is really an opera house.

ARCHITECTURAL AND STRUCTURAL DETAILS

Uses: Opera, orchestra, chorus, and soloists. **Ceiling:** 0.75-in. (1.9-cm) plaster on wood lath on flat surfaces; plaster on metal wire screen on curved surfaces. **Walls:** Pine boards, 0.4 × 3 in. (1 × 7.6 cm), nailed to wooden framing. **Floors:** Two layers of 0.4-in. (1-cm) boards on joists. **Carpets:** with underpad main aisles downstairs and in the first ring (see 1 and 2 in the upper drawing). **Stage floor:** Two layers of wood on joists. **Stage height:** 52 in. (132 cm). **Seating:** Seat bottoms and seat backs fully upholstered. Bench-type backs at the upper-most level with upholstered seat bottoms and with a cushion on the bench backs.

ARCHITECTS: Napoleon E. H. C. Le Brun and Gustavus Runge. CREDITS: Drawings, seating, room, and stage enclosure details from Academy of Music. PHOTOGRAPHS: Ed Wheeler.

TECHNICAL DETAILS

Opera

V = 533,000 ft³ (15,100 m³)	S_P = 2,401 ft² (223 m²)	S_A = 15,700 ft² (1,460 m²)
S_o (pit) = 640 ft² (59 m²)	S_T = 18,740 ft² (1,740 m²)	N = 2,827
V/S_T = 28.4 ft (8.7 m)	V/S_A = 33.9 ft (10.4 m)	S_A/N = 5.55 ft² (0.52 m²)
H = 64 ft (19.5 m)	W = 58 ft (17.7 m)	L = 102 ft (31.1 m)
D = 118 ft (36 m)	H/W = 1.1	L/W = 1.76

Concerts

V = 555,000 ft³ (15,700 m³)	S_a = 14,000 ft² (1,300 m²)	S_A = 16,700 ft² (1,550 m²)
S_o = 2,350 ft² (218 m²)	S_T = 18,640 ft² (1,730 m²)	N = 2,921
V/S_T = 29.8 ft (9.07 m)	V/S_A = 33.2 ft (10.13 m)	V/N = 190 ft³ (5.38 m³)
S_A/N = 5.72 ft² (0.531 m²)	S_a/N = 4.79 ft² (0.445 m²)	ITDG = 19 msec

Note: $S_T = S_A$ + 1,940 ft² (180 m²); see Appendix 1 for definition of S_T.

NOTE: The terminology is explained in Appendix 1.

18

Philadelphia

Verizon Hall in the Kimmel Center for the Performing Arts

*P*hiladelphia's world-renown symphony orchestra has a new home, which formally opened on December 14, 2001. The orchestra had performed for 101 years in the oldest hall in America, the Academy of Music, which was built as an opera house. The change in the acoustics is enormous. Efforts to build a "symphony hall" date back to 1908, and even as recently as 1995 it seemed that the solution to a hall with a greater reverberation time might be to modify the Academy along the lines of the renovations to Orchestra Hall in Chicago. Instead, the former mayor of Philadelphia embraced the idea of the Kimmel Center as the centerpiece of his "arts-as-renovation" project. Verizon Hall, seating an audience of 2,298 plus 247 bench seats for chorus or audience, is only one part of this complex. It is adjacent to the Perelman Theater, seating 651, and a black-box experimental theater, seating 100. All are only a short distance from the Academy of Music.

The Verizon Hall is a modified "shoebox," with some resemblance in shape to a cello. Its unusual acoustical feature is means for varying the reverberation time. Up to one hundred doors at all levels can be opened into acoustic (reverberation) chambers. Preliminary data show a reverberation time (measured from -5 to -35 dB on the decay curve, using the Schroeder method of analysis) of 1.6 seconds with all doors closed, increasing to 1.7 seconds with all doors 45 degrees open, about 0.1 second below Carnegie Hall in New York, and well above the 1.3 seconds in the Academy of Music. To reduce the reverberation time for conferences, velour curtains can be drawn over the walls.

There is a canopy over the orchestra that is in three sections, each of which can be adjusted in height. Preliminary heights for the front part (facing the audience) is about 50 feet (15 meters) and the back part is about 44 feet (13.4 meters).

This hall is still undergoing adjustments and an attendee at a concert does not know whether the best acoustical condition is being presented that night. Press music reviews have been mixed, partly for this reason. The sound levels on the main floor under present conditions are weaker than those in Boston's Symphony Hall

SOUND ABSORBING
BANNER

ACOUSTICS CHAMBER

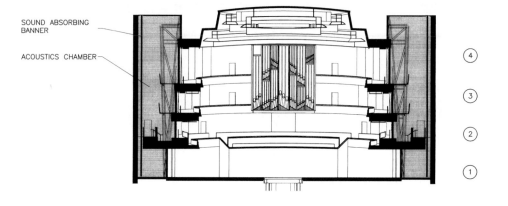

④

③

②

①

SEATING CAPACITY 2519

①   1068

②   611

③   374

④   466

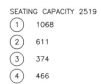

ORGAN

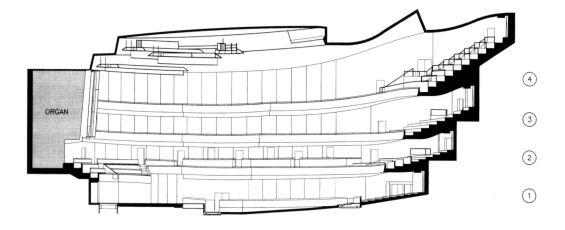

④

③

②

①

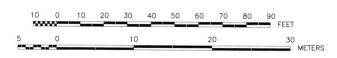

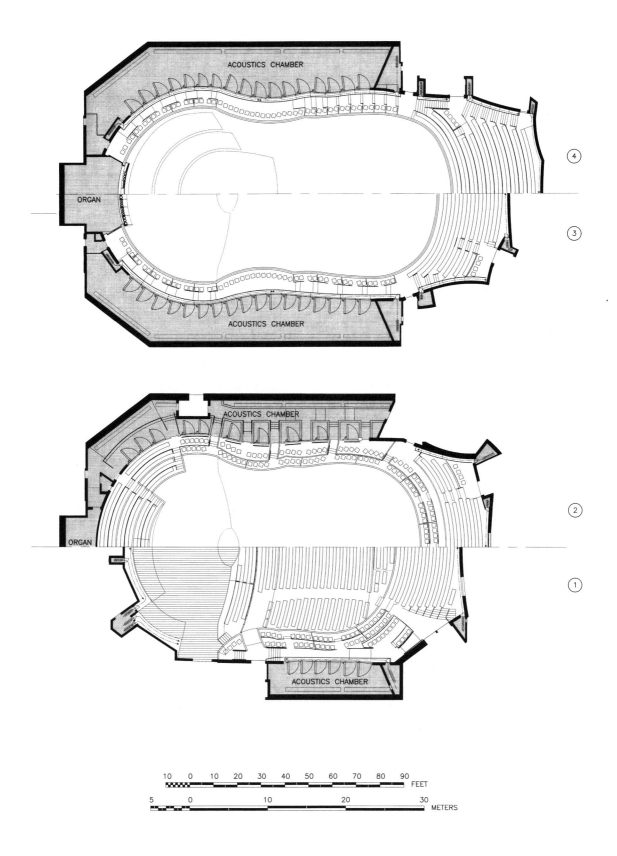

ACOUSTICS CHAMBER

ORGAN

ACOUSTICS CHAMBER

ACOUSTICS CHAMBER

ORGAN

ACOUSTICS CHAMBER

10 0 10 20 30 40 50 60 70 80 90 FEET

5 0 10 20 30 METERS

and New York's Carnegie Hall. In the balconies the sound is fuller. The bass/treble balance and the balance among orchestral sections is excellent. One appreciates, as always, the high quality of the magnificent Philadelphia Orchestra, one of America's best.

ARCHITECTURAL AND STRUCTURAL DETAILS

Uses: Concert music, pops, and organ recitals. **Ceiling:** Cement 3 in. (7.5 cm) thick with an applied layer of 0.62-in. (1.6-cm) fiberboard. **Canopy:** Multilayer. **Walls:** 12-in. (30.5-cm) block with 0.75-in. (1.9-cm) plywood and finish layer 0.5-in. (1.25-cm) fiberboard. **Sound diffusers:** Portions of balcony fronts, walls and chamber doors are textured. **Variable absorption:** 12,238 ft² (1,337 m²) sound-absorbing banners. **Floors:** 0.5-in. (1.25-cm) tongue-and-groove wooden boards, 0.5-in. plywood substrate, cemented to concrete. **Variable elements:** One hundred 4.2-in. (10.7-cm)-thick operable doors that open into cement chambers with a total volume of 262,000 ft³ (7,420 m³) and banners that drop from the ceilings of the reverberation chambers. **Walls of stage enclosure:** Portions are fixed, of poured concrete with a transparent screen in front. The movable portions are 2 to 4 layers of 0.62-in. (1.6-cm) cement board. **Stage floor:** 1-in. (2.5-cm) tongue-and-grove board over 2 in. (5 cm) thick wood laid on neoprene pads on concrete. **Stage height:** 42 in. (1.07 m). **Seats:** Seat backs and bottoms 1.25-in. (3.2-cm) padding. Choir/audience balcony seating is bench style, unupholstered.

ARCHITECT: Rafael Viñoly Architects PC. ACOUSTICAL CONSULTANT: ARTEC Consultants, Inc. PHOTOGRAPHS: Jeff Goldberg, Esto.

TECHNICAL DETAILS

V = 830,240 ft³ (23,520 m³)*	S_a = 13,224 ft²(1,229 m²)	S_A = 17,926 ft² (1,666 m²)
S_o = 2,948 ft² (274 m²)	S_T = 19,863 ft² (1,846 m²)	N = 2,519
H = 75 ft (22.9 m)	W = 84 ft (25.6 m)	L = 90 ft (27.4 m)
D = 140 ft (42.7 m)	V/S_T = 41.8 ft (12.7 m)	V/S_A = 46.3 ft (14.1 m)
V/N = 330 ft³ (9.34 m³)	S_A/N = 7.12 ft² (0.661 m²)	S_a/N = 5.25 ft² (0.49 m²)
H/W = 0.89	L/W = 1.2	ITDG = 28 msec
Note: S_T = S_A + 1,940 ft² (180 m²); see Appendix 1 for definition of S_T.		

NOTE: The terminology is explained in Appendix 1.
* Not including the 262,000-ft³ (7,420-m³) acoustic control chambers.

19

Rochester, New York

Eastman Theatre

*N*ew theaters in America no longer exhibit the expensive handiwork that went into the Eastman Theatre. Built in 1923, it embodies fabrics, metals, and decorations of a quality found in the royal palaces of old. It contains 3,347 large, comfortable seats. One is aware immediately on entering the hall of the high volume of nearly 900,000 cubic feet (25,488 m³).

The high ceiling and wide, splayed side walls direct sound primarily to the rear of the hall. Before a major renovation was made in 1972, regular concertgoers complained about a lack of clarity of the music, particularly on the main floor, which resulted from a very long initial-time-delay gap and a complete lack of early lateral sound reflections. This condition did and does not exist in the rear part of the upper balcony because lateral reflections naturally occur there, along with a shorter initial-time-delay gap.

The renovation was directed at improving the acoustics of the hall for concert music, which is presented regularly by the Minneapolis Symphony and the Eastman School orchestras. It primarily consisted of a new stage enclosure (shell) with a large storable canopy. The orchestra enclosure and canopy were designed to reflect and project more sound outward through the proscenium. The canopy is shaped and located to send a higher percentage of acoustic energy from the instruments in the front third of the orchestra to the audience, particularly onto the previously neglected main floor area. The result is an increase in the ratio of the energy of the direct sound to that of the reverberant sound and the canopy provides a short initial-time-delay gap in that area. This change spells "clarity." The reverberation times were not altered. Other objectives were restoration and preservation of the decorative features.

The theatre is used for opera. With a full 85-piece pit orchestra, half of the musicians sit under the stage. This makes it difficult for them to hear what is happening on stage and to maintain proper balance among themselves. To a listener, the instruments back under the stage sound as though they were in another room.

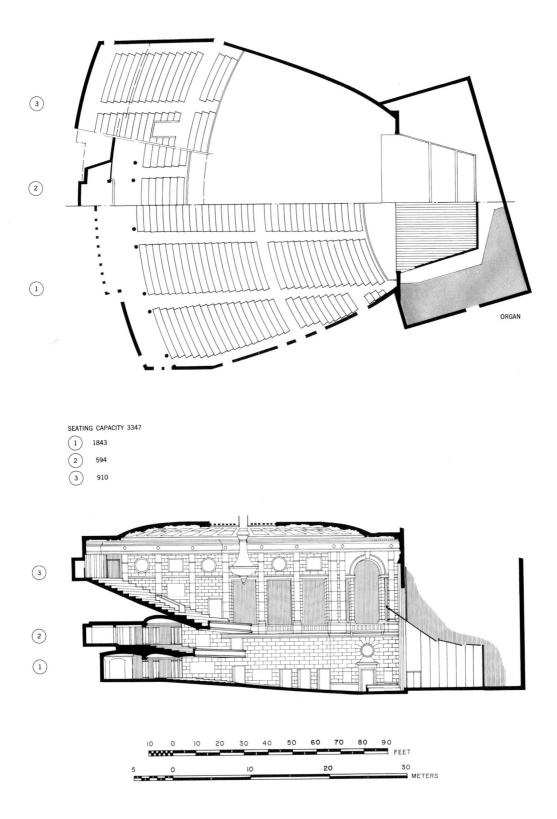

ORGAN

SEATING CAPACITY 3347

(1) 1843

(2) 594

(3) 910

10 0 10 20 30 40 50 60 70 80 90
FEET

5 0 10 20 30
METERS

The recessed pit provides better balance between the louder instruments and singers, who may be less powerful than the singers at the NY Metropolitan Opera.

⌖ ARCHITECTURAL AND STRUCTURAL DETAILS

Uses: General purpose. **Ceiling:** Plaster. **Side walls:** Plaster, approximately 50% covered with sound-absorbing materials that have been heavily painted. **Rear walls and balcony fronts:** Covered with sound-absorbing materials, heavily painted. **Floors:** Concrete. **Stage enclosure:** The sidewalls and ceiling are mostly curved plywood panels, 0.38 in. (1 cm) thick with random bracing and randomly spaced transverse ribs. **Stage floor:** Wood on joists. **Stage height:** 42 in. (107 cm). **Carpets:** On all floors. **Seating:** Fully upholstered (both sides) backrests; underseat is solid. **Pit:** Half-recessed under stage; the floor is of wood; the walls are concrete; the railing is asbestos board.

ARCHITECT: Gordon E. Kaelber. ASSOCIATE ARCHITECT: McKim, Mead and White. ARCHITECT FOR 1972 RENOVATION: Thomas Ellerbe Associates. STAGE ENCLOSURE DESIGN (1972): Paul S. Veneklasen. CREDITS: Original drawings from the *American Architect*, February 1923. PHOTOGRAPHS: Lower photograph courtesy of Ansel Adams, Eastman Kodak Company (1962); upper photograph courtesy of Paul Veneklasen.

TECHNICAL DETAILS

Concerts

V = 900,000 ft³ (25,500 m³)	S_a = 17,000 ft² (1,580 m²)	S_A = 21,750 ft² (2,021 m²)
S_o = 2,200 ft² (204 m²)	S_T = 23,690 ft² (2,201 m²)	N = 3,347
H = 67 ft (20.4 m)	W = 120 ft (36.6 m)	L = 117 ft (35.7 m)
D = 142 ft (43.3 m)	V/S_T = 38.0 ft (11.58 m)	V/S_A = 41.3 ft (12.6 m)
V/N = 269 ft³ (7.62 m³)	S_A/N = 6.5 ft² (0.60 m²)	S_a/N = 5.08 ft² (0.47 m²)
H/W = 0.56	L/W = 0.97	ITDG = 22 msec

Opera

V = 846,500 ft³ (23,970 m³)	S_P = 2,752 ft² (256 m²)	S_A = 21,750 ft² (1,907 m²)
S_T = 25,270 ft² (2,348 m²)	S_o (pit) = 770 ft² (71.5 m²)	S_{OF} (pit) = 1,750 ft² (162.5 m²)
N = 3,347	V/S_T = 33.6 ft (10.2 m)	V/S_A = 38.9 ft (11.9 m)

NOTE: The terminology is explained in Appendix 1.

20

Salt Lake City

Abravanel Symphony Hall

*S*alt Lake City is proud of its Symphony Hall, home of the Utah Symphony Orchestra. Dedicated in 1979 with special honors to retiring Maestro Maurice Abravanel, its music director for 32 years, the hall is well situated. At the entrance to the site, on the left, is a well attended garden. Just behind it is an outdoor amphitheater and sculpture court, both associated with Salt Lake's Community Art Center. The lobbies give concertgoers an uplifting experience and smiles everywhere give one a feeling of warmth and welcome.

The specification for the acoustics of the hall was that they should approach those of the highly acclaimed halls of the world. The architect, working closely with the acoustical consultant, decided on the classic "shoebox" shape, which has proven its quality in many venues. One must remark that the hall's capacity—2,812 seats, with 1,838 on the main floor—expresses optimism as to the possible number of subscribers in the Salt Lake City/Ogden, Utah, metropolitan area with a population of only 1.2 million. The metropolitan areas of Chicago, population 8.5 million; Washington, D.C., 6.8 million; and Boston, 5.5 million, support only one hall each and they seat between 2,759 and 2,582, with the smallest being the Chicago hall. The width of Salt Lake's Symphony Hall is 90 ft (27 m), which contrasts with Boston's 75 ft (22.8 m), and, along with continental seating, tends to give a visitor a feeling of entering a vast room.

The hall has the proper features for providing good acoustics. The side and rear walls are made of randomly dimensioned oak panels and the plaster ceiling closely approximates the sound-diffusing properties of the coffered ceilings of Boston and Vienna's halls. The rectangular shape provides early, lateral sound reflections which give a feeling of intimacy. Its reverberation time, with full audience, is about 1.7 sec, good for music of the Classical period and for a hall that must serve some other purposes than symphonic concerts. Compared to Boston, there are a few decibels' weakness in the bass. There is no tonal harshness and the string and woodwind sounds are excellent. The bass/treble balance and the sectional balance

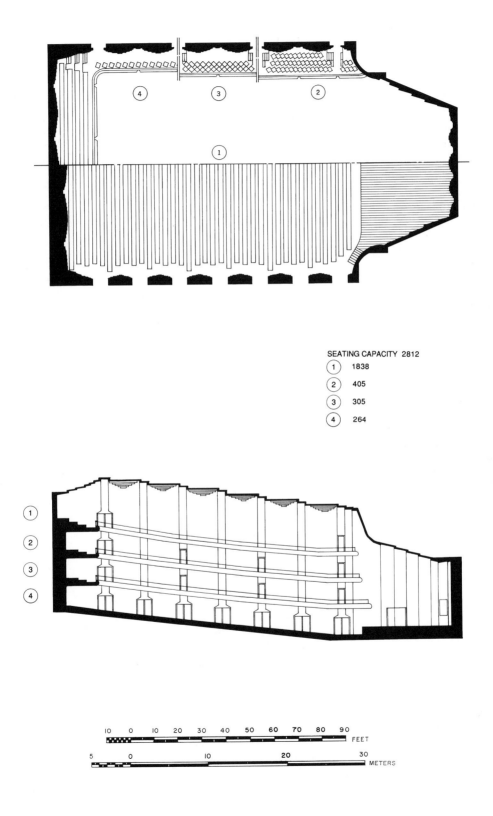

SEATING CAPACITY 2812

① 1838
② 405
③ 305
④ 264

are very good. The stage is large, 2,350 ft² (218 m²), compared to Boston's 1,600 ft² (149 m²), which may make ensemble playing more difficult.

ARCHITECTURAL AND STRUCTURAL DETAILS

Uses: Music, 90%; Speech, 10%. **Ceiling:** 1.5-in. (3.8-cm) hardrock plaster—all openings sealed with gaskets. **Walls:** Side and rear walls poured-in-place concrete with 0.75-in. (1.9-cm) plywood finish surface mounted on 0.75-in. furring strips, space between filled with fiberglass; balcony faces 1.5-in. hardrock plaster on furring strips on concrete structure. **Floors:** Main and balcony floors: 0.75-in. tongue-and-groove oak wood on furring strips on main floor and on frames in balconies. **Carpet:** None. **Stage enclosure:** See Walls above. **Stage floor:** 0.75-in. tongue-and-groove oak boards on 0.75-in. plywood subfloor on sleepers. *Stage height:* 43 in. (109 cm). **Sound-absorbing materials:** None. **Seating:** Wooden seat back; front of seat back and seat bottom upholstered with sprayed, impervious fabric over polyvinyl foam; underseat, perforated metal; arms, wooden.

ARCHITECTS: Fowler, Ferguson Kingston Ruben. ACOUSTICAL CONSULTANT: Cyril M. Harris. PHOTOGRAPHS: Schoenfeld.

TECHNICAL DETAILS

V = 688,500 ft³ (19,500 m³)	S_a = 16,000 ft² (1,486 m²)	S_A = 17,965 ft² (1,669 m²)
S_o = 2,350 ft² (218 m²)	S_T = 19,900 ft² (1,850 m²)	N = 2,812
H = 54 ft (16.5 m)	W = 96 ft (29.3 m)	L = 124 ft (37.8 m)
D = 134 ft (40.85 m)	V/S_T = 34.6 ft (10.54 m)	V/S_A = 38.3 ft (11.7 m)
V/N = 245 ft³ (6.93 m³)	S_A/N = 6.39 ft² (0.59 m²)	S_a/N = 5.69 ft² (0.528 m²)
H/W = 0.56	L/W = 1.29	ITDG = 30 msec
Note: $S_T = S_A$ + 1,940 ft² (180 m²); see Appendix 1 for definition of S_T.		

NOTE: The terminology is explained in Appendix 1.

21

San Francisco

Louise M. Davies Symphony Hall

*D*avies Symphony Hall, when it opened in 1980, was approximately circular in plan and had a maximum width on the main floor of 100 ft (30 m), about 20 ft in excess of the best of the well-liked large, rectangular concert halls. Lower wall surfaces produced useful lateral reflections, but also generated echoes back to the stage. In addition, there was a rear wall echo. The hall's flat ceiling, 68 ft (17.7 m) above platform level created an immense overhead cubic volume. The circular acrylic panels suspended above the players were sparse and helped their ensemble little.

A major remodeling of Davies Hall was undertaken during the summers of 1991 and 1992. The volume in the upper front part of the hall was reduced by building new side walls above the chorus seating—they are now 24 ft (7.3 m) closer to each other and shaped to direct sound energy to the main floor (see the photograph). A computer-controlled array of fifty-nine 6-ft (1.83-m)-square, convex panels was hung at a height of 30–35 ft (9–10 m) above the stage and the first four rows of seats. These panels improve communication onstage and provide strong early reflections to the main floor and first balcony.

The walls around the platform are 1.5 ft (0.46 m) higher than before, although the exposed wall area is less due to the height of the orchestra risers—2 ft (0.61 m) high at the sides and 3 ft (0.92 m) high at the rear. The entire rear wall of the main floor is covered with quadratic residue diffusers (QRD's) to eliminate echo. To reduce the width of the main floor to 84 ft at the front, the side walls were moved in and reshaped. Ten raised boxes were built on either side to improve sightlines and to provide early lateral reflections. The hall now seats an audience of 2,743.

The changes have worked. The sound is more intimate and clear and the bass response is greatly improved. These modifications reflect the changes in acoustical design since 1975. At that date the importance of early lateral reflections was not yet proven; the theory of multiple hanging panels was in its infancy; and QRD's were not yet invented.

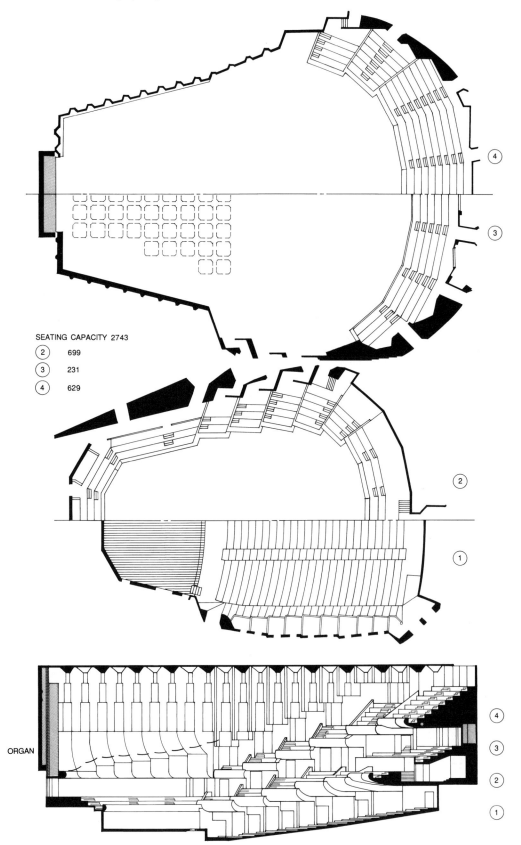

SEATING CAPACITY 2743

(2)   699
(3)   231
(4)   629

ORGAN

ARCHITECTURAL AND STRUCTURAL DETAILS

Uses: Orchestral music and recitals. **Ceiling:** 6-in. (15-cm) concrete **Walls:** Front, side, and rear walls are multi-faceted pre-cast concrete 4 in. (10 cm) thick. Above the sides of the stage, the new reflecting walls are glass fiber shells containing sand-packed steel tubes—the total weighing 40 lb/ft^2 (195 kg/m^2); balcony fronts and soffits are 2-in. (5-cm) plaster; box fronts, plaster on gypsum lath. **Floors:** Parquet over wood over small airspace and concrete base. **Carpet:** In aisles only. **Stage enclosure:** (see the text). **Stage height:** 39 in. (99 cm). **Sound-absorbing materials:** None, except QRD diffusers, which have an absorption coefficient of 0.1 to 0.25 over a wide band of frequencies. **Seating:** Molded wooden seat back; lightly upholstered front of seat back; relatively thin seat cushion with porous fabric covering; underseat, unperforated wood—all designed to reduce low-frequency sound absorption.

ARCHITECT: Skidmore Owings and Merrill. ORIGINAL ACOUSTICAL CONSULTANT: Bolt Beranek and Newman. ACOUSTICAL CONSULTANT FOR REVISIONS: Kirkegaard and Associates. PHOTOGRAPHS: Dennis Gearney. REFERENCES: L. Kirkegaard, "Concert Acoustics: The Performers' Perspective." Paper presented at the 116th meeting of the Acoustical Society of America, Honolulu, 15 November, 1988.

TECHNICAL DETAILS

V = 850,000 ft^3 (24,070 m^3)	S_a = 13,070 ft^2 (1,214 m^2)	S_A = 16,800 ft^2 (1,562 m^2)
S_o = 2,155 ft^2 (200 m^2)	S_T = 18,740 ft^2 (1,742 m^2)	N = 2,743
H = 68 ft (20.7 m)	W = 92 ft (28 m)	L = 107 ft (32.6 m)
D = 127 ft (38.7 m)	V/S_T = 45.4 ft (13.8 m)	V/S_A = 50.6 ft (15.4 m)
V/N = 310 ft^3 (8.78 m^3)	S_A/N = 6.12 ft^2 (0.57 m^2)	S_a/N = 4.76 ft^2 (0.44 m^2)
H/W = 0.74	L/W = 1.16	ITDG = 12 msec

Note: S_T = S_A + 1,940 ft^2 (180 m^2); see Appendix 1 for definition of S_T.

NOTE: The terminology is explained in Appendix 1.

22

San Francisco

War Memorial Opera House

*T*he War Memorial Opera House surpasses many of the world's opera houses in the beauty of its exterior and lobbies. The hall itself is reminiscent of other theaters built in the United States in the early 1930s. Yet it has a dignity and beauty that exceed that of many contemporary halls, partly because the architect eschewed adornments and confined the design to simple lines, a high ceiling, and a majestic proscenium.

I have attended four performances, including the world premier of *Blood Moon* by Norman Dello Joio, *Nabucco* by Verdi, and *Arshsk II* by Chukhajian. The long, thin pit gave the conductor some trouble with ensemble, but the timbre was satisfactory throughout the house. The singers were easy to hear above the orchestra, although some may have forced their voices some-what, probably because to them the hall looks huge. Most pleasant was the general liveness, greater than that of La Scala, or the Staatsoper in Vienna.

I had access to seats in four parts of the house during the operas. In the front of the house (the 15th row near the left aisle of the main floor), the reverberation tended to interfere with intelligibility of opera sung in English. But the hard materials of the interior also gave warmth to the music, and a fullness of bass not found in many European houses. At the other seats the sound was very good.

The shallow dome presents no acoustical problem, because the very large chandelier beneath it effectively diffuses the sound. G. Albert Lansburgh, one of the architects for the San Francisco Opera, told me that the dome is treated with acoustic plaster. He said that there are good sight lines to all 3,252 seats.

Seven conductors who have had experience in the San Francisco Opera House were interviewed. Each rated it good from the positions of both the conductor and the listeners. Several of them remarked that the audience hears much better in the balcony than on the main floor.

As is true of all wide halls with high ceilings, the initial-time-delay gap is somewhat too long for the listeners in the forward part of the main floor. The ceiling

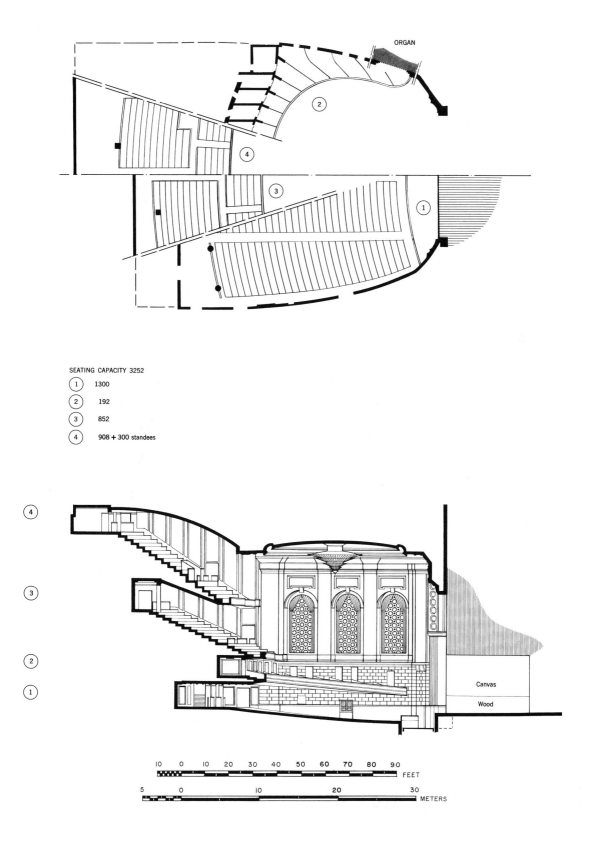

SEATING CAPACITY 3252

1 1300
2 192
3 852
4 908 + 300 standees

ORGAN

Canvas
Wood

10 0 10 20 30 40 50 60 70 80 90 FEET
5 0 10 20 30 METERS

supplies these first reflections to the fronts of the balconies and the side walls supply their first reflections to the rear of the balconies. Still, for its size and the comfort of its seating, this house is acoustically quite satisfactory.

ARCHITECTURAL AND STRUCTURAL DETAILS

Uses: Opera, general purpose. **Ceiling:** Plaster except for center domed section, which architect says is acoustic plaster. **Walls:** Plaster with some wood trim. **Floors:** Entire floor area carpeted except in upper part of top balcony where carpet is only in aisles. **Stage:** Wood over air space. **Pit:** Wooden floor on elevator; wooden rear and front walls with 1 ft (30 cm) of velvet covering over an open railing. **Stage height:** 40 in. (102 cm) above floor level at first row of seats. **Added absorptive materials:** Velvet draperies, about 1 ft (2.5 cm) high above rails of cross aisles. **Seating:** Main floor, fully upholstered with four holes 1 in. (2.5 cm) in diameter in underseat; balcony, but with hard backs. **Orchestra enclosure:** Canvas throughout except that 1/4-in. plywood covers the bottom part of the walls extending upward 10 ft (3 m) from the floor.

ARCHITECT: Arthur Brown, Jr. COLLABORATING ARCHITECT: G. Albert Lansburgh. REFERENCES: B. J. S. Cahill, "The San Francisco War Memorial Group," *The Architect and Engineer,* **111**, 11–44, 59 (1932). PHOTOGRAPHS: Courtesy of *Musical America* and *Opera News.* Details verified by the author during a visit and by management (1993).

TECHNICAL DETAILS

Opera

V = 738,600 ft³ (20,900 m³)	S_a = 16,500 ft² (1,533 m²)	S_A = 21,240 ft² (1,973 m²)
S_o (pit) = 760 ft² (70.6 m²)	S_P = 2,500 ft² (232 m²)	S_T = 24,500 ft² (2,276 m²)
N = 3,252	H = 73 ft (22.2 m)	W = 104 ft (31.7 m)
L = 120 ft (36.6 m)	D = 122 ft (37.2 m)	V/S_T = 30.1 ft (9.19 m)
V/S_A = 34.8 ft (10.6 m)	V/N = 227 ft³ (6.43 m³)	S_A/N = 6.53 ft² (0.61 m²)
H/W = 0.70	L/W = 1.15	ITDG = 51 msec

NOTE: The terminology is explained in Appendix 1.

23

Seattle

Benaroya Hall

enaroya Hall graces Seattle with a performance space designed exclusively for symphonic music. Seating 2,500, and opened in 1998, it sits in Seattle's downtown area. A component of urban renewal, the building is owned by the city, but is operated by the Seattle Symphony. The chosen site was particularly challenging. The hall sits directly over a tunnel for a major railway line, adjacent to which is an underground bus and light-rail tunnel. Threats of vibrations and noise also come from air traffic above and busy city streets to the sides.

The street approach to the hall gives little indication of its contents. A block-long, largely glass exterior wall forms one side of a high-ceilinged interior gallery that contains, on the side opposite the glass wall, a row of shops, the box office, and the entrance doors to the hall's lobby. At each end of the gallery hangs a massive, intricate chandelier crafted by the famous glass artist Dale Chihuly. Beyond the entrance doors from the gallery is a large, semicircular, four-story structure that houses the lobby. Enclosed in glass, it affords outstanding views of Puget Sound from its several levels. Above the main staircase in the lobby are two large panels by Robert Rauschenberg.

To eliminate vibrations and noise, the hall is a box floated within a box. Two levels of garage separate the railroad from the two boxes, each of the boxes is floated on rubber bearings. In the hall, outside noises are simply not audible.

The concert hall has the shoebox shape of the famous halls of Vienna, Amsterdam, and Boston, with three side and rear balconies stepped to provide good vision to their occupants. The rich dark color of the African-makoré, wooden-faced sidewalls contrasts with the ivory-toned facia and dark soffits of the balconies. Everywhere are irregularities—coffered ceilings, bulging sidewalls, erratic balcony fronts, all designed to create diffusion of reflected sound—which assure a mellow reverberation. The majority of the reviews speak favorably of the acoustics of the hall, but *USA TODAY* comments that "Critics haven't been entirely complimentary upon hearing the orchestra in its new hall . . ." The principal criticism is weakness

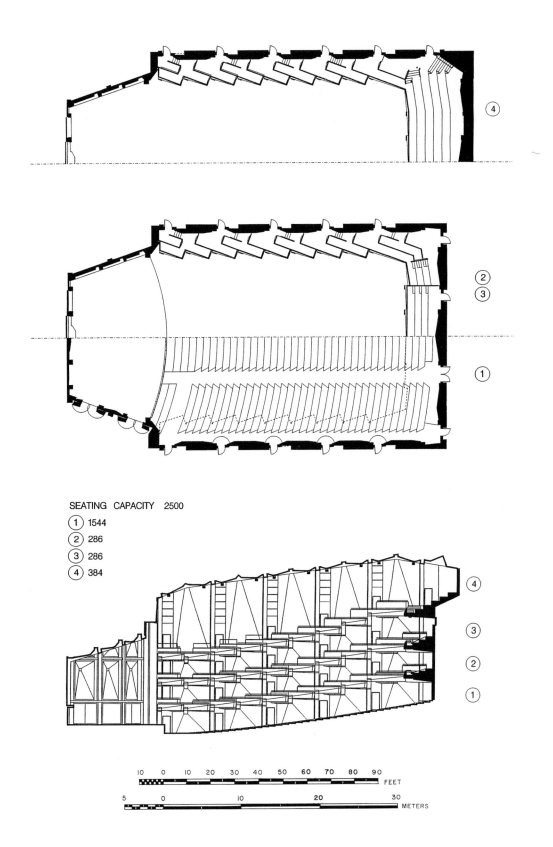

SEATING CAPACITY 2500
① 1544
② 286
③ 286
④ 384

of bass, particularly in the balcony. There are conflicting reports of the mid-frequency, fully occupied revaluation times, taken from stop chords obtained during regular concerts. One observer reports 1.95 sec (one concert) and the other 1.75 sec (three concerts). The pipe organ was installed between the two measurements.

ARCHITECTURAL AND STRUCTURAL DETAILS

Uses: Symphonic, chamber, and soloists' music. **Ceiling:** 1.5-in. (3.8-cm) plaster, with inverted polyhedron in each coffer; **Walls:** Wood particle-board panels, with cherry-mahogany veneer, shaped into irregular polyhedrons in variety of sizes, 0.75 in. (1.9 cm) thick, fastened to wood battens fixed to concrete wall—randomly spaced wood blocking behind panels; **Balconies:** Facia formed in a variety of rhomboidal shapes and sizes, and soffits, flat—all 0.75-in. plaster on metal lath. **Floors:** 0.75-in. oak tongue-and-groove strips, over 0.75-in. plywood base, supported by wood joists on concrete; **Carpets:** On aisles only, short tight pile with no underpad. **Stage enclosure:** Lower side walls are like those on auditorium side walls, while the upper side walls are polyhedrons made of plaster on metal lath, 0.75 in. thick. **Stage floor:** Same as audience floor, except sleepers instead of wood joists that rest on elastic pads isolated from a concrete stage foundation. **Stage height:** 42 in. (1.1 m) above auditorium floor. **Seating:** Back sides of seat backs are bare and upholstery has a short nap, cushion thickness chosen to give approximately same sound absorption unoccupied as when occupied.

ARCHITECTS: LMN Architects. ACOUSTICAL CONSULTANT: Cyril Harris. GROUND VIBRATION ISOLATION SYSTEM: Wilson, Ihrig & Associates. MECHANICAL SYSTEM NOISE CONTROL: Hoover & Keith, Inc. PHOTOGRAPHS: Jeff Goldberg/Esta. REFERENCE: *The Seattle Times,* June 21, 1998.

TECHNICAL DETAILS

V = 680,000 ft³ (19,263 m³)	S_a = 12,675 ft² (1,178 m²)	S_A = 15,624 ft² (1,452 m²)
S_o = 2,324 ft² (216 m²)	S_T = 17,564 ft² (1,632 m²)	N = 2500
H = 60 ft (18.3 m)	W = 85 ft (25.9 m)	L = 117 ft (35.7 m)
D = 134 ft (40.8 m)	V/S_T = 38.7 ft (11.8 m)	V/S_A = 43.5 ft (13.2 m)
V/N = 272 m³ (7.7 m³)	S_A/N = 6.25 ft² (0.58 m²)	S_a/N = 5.07 ft² (0.47 m²)
H/W = 0.71	L/W = 1.38	ITDG = 28 msec

Note: $S_T = S_A + 180$ m²; see the definition of S_T in Appendix 1.

NOTE: The terminology is explained in Appendix 1.

24

John F. Kennedy Center
for the Performing Arts, Concert Hall

The Kennedy Center for the Performing Arts, which opened in 1971, contains an opera house, a concert hall, and a drama theater. The opera and drama halls have been adequately successful, but the concert hall was never judged acoustically to be in the same league as Carnegie Hall in New York or Symphony Hall in Boston. In addition, the hall never rose to inspire a ticket holder visually—its mien was a sort of a grayish-white monotone. Finally, after a quarter century, the United States Congress appropriated $13 million of the $14 million that was spent to renovate the concert hall.

The hall was re-opened in 1997, seating 2,448. R. W. Apple, in *The New York Times,* wrote, "In the old days, members of the orchestra complained, musicians on one side of the stage could not hear those on the other; so they played louder hoping to overcome the acoustics. Violins sounded weedy and strident in the upper register, and the overall tone was sluggish and lifeless." Another leading music critic wrote, ". . . there was near-unanimity . . . that the acoustics are [now] bright, vibrant, spacious and responsive—a vast improvement on what was there before," and yet another, "Unquestionably, the hall's acoustics . . . are improved over all . . . this is now a hall positively dripping with sound."

On entering the hall, five architectural changes are immediately apparent. The stage has been modified to include audience seating in four boxes on either side at the first tier level. Choral seating is provided at the same level on the back wall of the stage beneath a new pipe organ. A set of sound-reflecting panels now hangs above the performers that are adjustable both in height and in tilt. To liven the front end of the hall acoustically, the forward ends of the third balcony have been cut back. The four dark surfaces on each upper sidewall of the stage open into hidden acoustical reverberation chambers that start behind the boxes and extend upward to the underside of the structural roof. Doors can close the chambers to reduce the reverberation, when desired, to suit the music being performed.

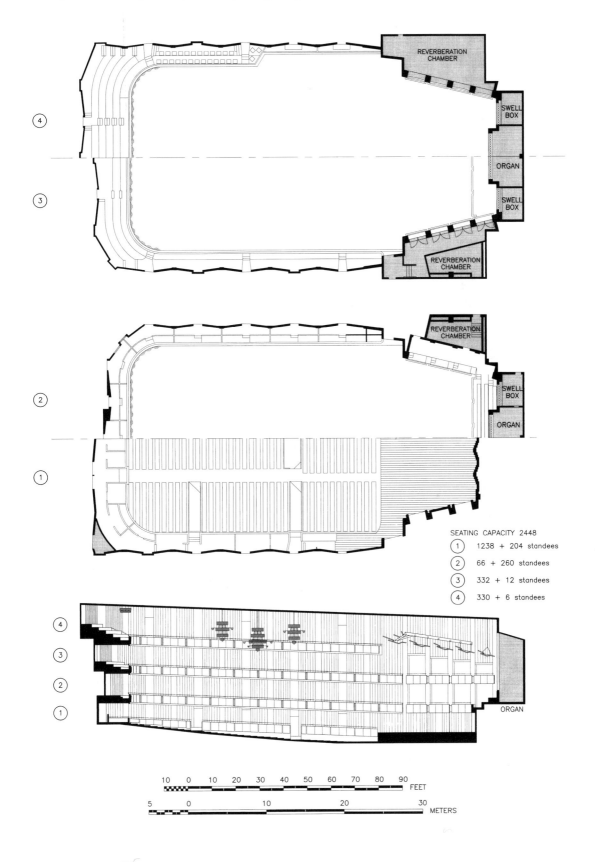

SEATING CAPACITY 2448

1 1238 + 204 standees
2 66 + 260 standees
3 332 + 12 standees
4 330 + 6 standees

10 0 10 20 30 40 50 60 70 80 90 FEET

5 0 10 20 30 METERS

The basic rectangular shape, which is excellent, has been preserved so that acoustical intimacy and spaciousness are unaffected. The changes have dramatically increased the reverberation time at low frequencies, occupied, from 1.8 to 2.0 sec, with little difference at the mid-frequencies but with greater fullness of tone. The sound-reflecting panels add brilliance, strength, and clarity to the upper string tones and should solve the musicians' ensemble problems.

ARCHITECTURAL AND STRUCTURAL DETAILS

Uses: Symphonic and chamber music, solos, and conferences. **Ceiling:** 1-in. (2.5-cm)-thick acoustical plaster. **Side walls:** 0.75-in. (1.9-cm)-thick wood panels on studs. Rear wall on the second tier level is covered with acoustical panels. **Walls and ceiling of the stage enclosure:** Combination of layers of gypsum board, concrete block and precast concrete. The ceiling is plaster (thickness varies). Acoustic canopy is 0.75-in. plywood. **Main floor:** 0.75-in. (1.9-cm) tongue-and-groove wood on top of 0.75-in. plywood over a concrete slab. **Floor on upper levels:** On the box level, there is carpet on pad; first and second tiers have the same construction as the main floor. **Stage floor:** 0.75-in. (1.9-cm) tongue-and-groove wood over 0.75-in. plywood on sleepers laid on neoprene isolators. **Height of the stage above the audience level:** 40 in. (1 m). **Carpet:** On the aisles over a pad, 0.5-in. (1.25-cm) total thickness. **Seats:** Front of backrest and top of seat bottom upholstered.

ARCHITECT: Original: Edward Durrell Stone. Renovation: Quinn Evens, Washington, D.C. ACOUSTICAL CONSULTANT: Original: C. Harris. Renovation: Jaffe Holden Acoustics. PHOTOGRAPHS: Peter Aaron/Esto.

TECHNICAL DETAILS

$V = 788{,}000$ ft³ (22,300 m³)	$S_a = 11{,}567$ ft² (1,075 m²)	$S_A = 15{,}333$ ft² (1,425 m²)
$S_o = 2{,}453$ ft² (228 m²)	$S_T = 17{,}270$ ft² (1,605 m²)	$N = 2448$
$H = 52$ ft (15.85 m)	$W = 93$ ft (28.4 m)	$L = 120$ ft (36.6 m)
$D = 131$ ft (40 m)	$V/S_T = 45.6$ ft (13.9 m)	$V/S_A = 51.4$ ft (15.6 m)
$V/N = 321.9$ ft³ (9.11 m³)	$S_A/N = 6.26$ ft² (0.582 m²)	$S_a/N = 4.73$ ft² (0.439 m²)
$H/W = 0.56$	$L/W = 1.29$	ITDG $= 40$ msec

Note: $S_T = S_A + 1{,}940$ ft² (180 m²); see Appendix 1 for definition of S_T.

NOTE: The terminology is explained in Appendix 1.

John F. Kennedy Center
for the Performing Arts, Opera House

The John F. Kennedy Center for the Performing Arts was opened in 1971 at a site along the Potomac River upstream from the Lincoln Memorial. It fills a cultural need in this city of politicians, diplomats, civil servants, companies doing business with the Federal government, and lobbyists. The center incorporates into a single building three auditoriums, the Eisenhower Theater, the Concert Hall, and the Opera House.

The size is unusual for the United States in that the house seats only 2,142 with the orchestra pit in use—56 percent of the capacity of New York's Metropolitan Opera House and 66 percent of San Francisco's Opera House. For opera lovers and singers this is good. Because the roof height of the overall building was restricted, the conventional multi-ringed, horseshoe style had to be eschewed. Instead the house has a box tier and two balconies. The audience is as close to the stage as possible.

It was expected from the beginning that grand opera would not be a dominant use. Included are musicals, soloists, speaking functions, and ballet. As a house that emphasizes voice, the reverberation time is optimum, even a little higher than that of many of the famous opera houses in Europe, measuring 1.5 sec at mid-frequencies when the house is fully occupied (taken from recordings made during the performance of an opera), compared to 1.1–1.45-sec range.

The rear and side walls of the Opera House are formed by a series of contiguous convex-shaped cylindrical surfaces that run from floor to ceiling. Not being identical in size, they provide diffusion of sound over a wider frequency range. Their average radius of curvature is 15 ft (4.6 m) and the cord length is 13 ft (4 m) and they vary in width. A short recess or doors separate one from another. The facias of the tiers are subdivided into a series of convex panels which are bowed outward, each about 5.5 ft (1.67 m) across. The overall result is good sound diffusion at all frequencies, which helps the quality of the reverberation. The center of the ceiling is a recessed circular area in which a sound-diffusing crystal chandelier hangs.

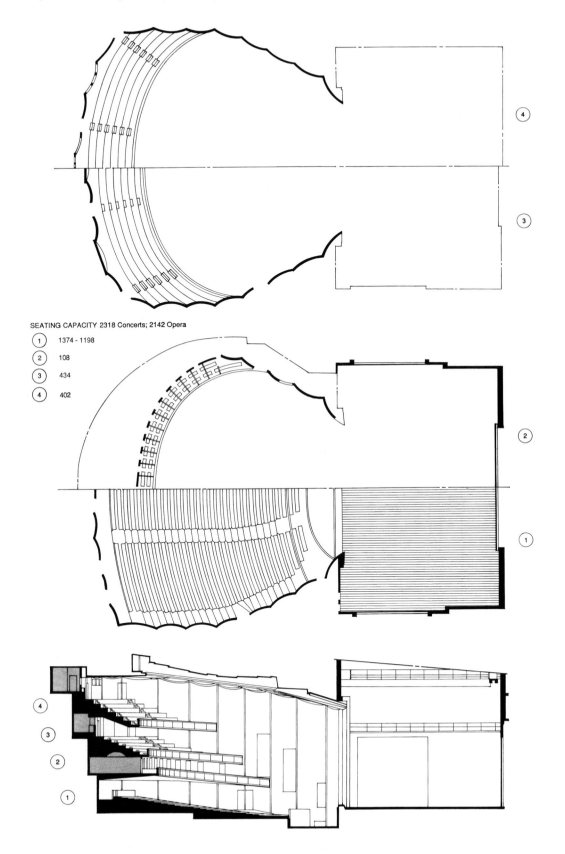

SEATING CAPACITY 2318 Concerts; 2142 Opera

① 1374 - 1198
② 108
③ 434
④ 402

No acoustical adjustments had to be made in the hall after completion. Music critics have compared the house favorably with the Staatsoper in Vienna.

ARCHITECTURAL AND STRUCTURAL DETAILS

Uses: Opera, musicals, ballet, and speaking functions. **Ceiling:** 1-in. (2.5-cm) (minimum) plaster on metal lath; underbalcony soffits 1-in. plaster. **Walls:** 1-in. plaster (see the text) on metal lath separated by 1-in. airspace from 6-in. (15 cm) solid block; rear walls are molded 0.75-in. wood paneling, fixed, with 1-in. airspace to solid block; balcony and box tier facias, 0.75-in. (1.9-cm) plaster on metal lath. **Floors:** Dense concrete. **Carpets:** Floors fully covered with thin carpet. **Stage height:** 40 in. (102 cm) above floor level at first row of seats. **Seating:** Backrest, seat bottom, and armrests fully upholstered.

ARCHITECT: Edward Durell Stone. ACOUSTICAL CONSULTANT: Cyril M. Harris. PHOTOGRAPHS: Carol Pratt. REFERENCE: Cyril M. Harris, "Acoustical design of the John F. Kennedy Center for the Performing Arts," *J. the Acoust. Soc. Am.* **51**, 1113–1126 (1972).

TECHNICAL DETAILS

$V = 460{,}000$ ft³ (13,027 m³)	$S_a = 12{,}196$ ft² (1,133 m²)	$S_A = 13{,}875$ ft² (1,289 m²)
$S_o = 1{,}173$ ft² (109 m²)	$S_P = 2{,}120$ ft² (197 m²)	$S_T = 17{,}168$ ft² (1,595 m²)
N_P (Opera) $= 2{,}142$	N_{NP} (no pit) $= 2{,}318$	$H = 56$ ft (17.1 m)
$W = 104$ ft (31.7 m)	$L = 105$ ft (32 m)	$D = 115$ ft (35.1 m)
$V/S_T = 26.8$ ft (8.17 m)	$V/S_A = 33.2$ ft (10.1 m)	$V/N_P = 215$ ft³ (6.08 m³)
$V/N_{NP} = 198$ ft³ (5.62 m³)	$S_A/N_P = 6.48$ ft² (0.602 m²)	$S_a/N = 5.69$ ft² (0.529 m²)
$H/W = 0.54$	$L/W = 1.0$	ITDG $= 33$ msec

NOTE: The terminology is explained in Appendix 1.

26

Worcester

Mechanics Hall

Opened in 1857, Mechanics Hall became known as one of the finest halls (as distinct from theaters or opera houses) in the United States. The *Worcester Daily Spy* reported, ". . . the Hall is a perfect success, . . . both for music and for speaking." In its heyday it attracted musical greats (Caruso, Paderewski, Rachmaninoff, Rubinstein) and distinguished lecturers (Thoreau, Emerson, Dickens, Mark Twain). After the great depression of the 1930s, the hall fell into disuse.

In 1977, Mechanics Hall opened again restored to its original beauty and excellent acoustics by the Worcester County Mechanics Association under the watchful eye of the Society for the Preservation of New England Antiquities. Some further renovations followed in 1990. Seating 1,343 with a 1,080-ft² (100-m²) stage, or 1,277 with a 1,400-ft² (130-m²) stage, it is closest in all dimensions to the Stadt-Casino in Basel, Switzerland, which was built 19 years later. Today, it retains its original purpose as a multi-use auditorium for concerts, conventions, large receptions, and dinners. Artists who have performed there recently include Yo Yo Ma, Itzhak Perlman, Jessye Norman, and many others. The hall boasts two concert series that bring in symphony orchestras and chamber groups from all over the world. It is used extensively for recording. The hall's organ, built in 1864, was returned to its original condition by organ-builder Fritz Noack and was rededicated in 1982.

The hall provides the intimacy and early lateral reflections common to a rectangular hall, being 80 ft (24.4 m) in width and 58 ft (17.7 m) between balcony faces. The distance from the farthest balcony seat to the stage is 98 ft (30 m), making it visually intimate. Surfaces for the diffusion of reflected sound are evident everywhere, from the coffered ceiling to the niches and pilasters on the walls. There are no sound-absorbing materials, only thin carpets in the aisles. The reverberation time, measured during intermission at a concert with full audience, but no orchestra on stage, is 1.6 sec at mid-frequencies, excellent for a multipurpose hall that normally does not present large orchestral groups. Its principal deficiency is the flat

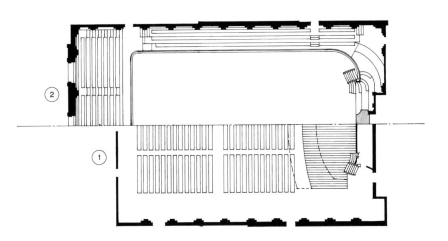

SEATING CAPACITY		SMALL GROUPS	SMALL ORCHESTRA	LARGE ORCHESTRA
Main floor	(1)	891	825	759
Balcony	(2)	518	518	518
		1409	1343	1277

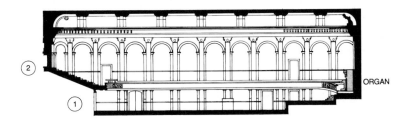

ORGAN

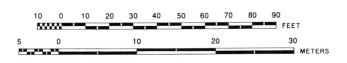

main floor, which makes neck stretching necessary to view performances. Also, organ recitalists would prefer longer reverberation times, particularly at low frequencies.

ARCHITECTURAL AND STRUCTURAL DETAILS

Uses: Music, speech, and social events. **Ceiling:** 1-in. (2.5-cm) plaster on wood lath; balcony soffits 0.6-in. (1.5-cm) gypsum board. **Walls:** 1-in. plaster on wood lath, balcony fronts wood, about 50% open. **Floors:** The finish is 0.6-in. (1.5-cm) red oak over 0.5-in. (1.25 cm) tongue-and-groove oak over two layers of 0.75-in. (1.9-cm) pine subflooring. **Carpets:** On aisles, with no underpad. **Stage floor:** 0.75-in. maple over two layers 0.6-in. wood chipboard over 1-in. wood boards over large airspace. **Stage height:** 31 in. (79 cm). **Seating:** Main floor, movable upholstered seat bottom and back on metal frame; balcony, molded plywood, upholstered seat top and front of seat back; all armrests wood.

ARCHITECT: Original hall, Elbridge Boyden; 1970s renovation, Anderson Notter Finegold, Inc.; 1990s renovation, Lamoureux Pagano & Associates, Inc. ACOUSTICAL CONSULTANT: Both renovations, Cavanaugh Tocci Associates, Inc. SOUND CONSULTANT: David H. Kaye. REFERENCES: W. J. Cavanaugh, "Preserving the acoustics of Mechanics Hall: A restoration without compromising acoustical integrity," *Technology & Conservation* (Fall 1980).

TECHNICAL DETAILS

Concert stage, small orchestra

$V = 380{,}000$ ft³ (10,760 m³)	$S_a = 5{,}823$ ft² (541 m²)	$S_A = 7{,}546$ ft² (701 m²)
$S_o = 1{,}658$ ft² (154 m²)	$S_T = 9{,}204$ ft² (855 m²)	$N = 1{,}343$
$H = 41$ ft (12.5 m)	$W = 81$ ft (24.7 m)	$L = 89$ ft (27.1 m)
$D = 99$ ft (30.2 m)	$V/S_T = 41.3$ ft (12.5 m)	$V/S_A = 50.3$ ft (15.4 m)
$V/N = 283$ ft³ (8.01 m³)	$S_A/N = 5.61$ ft² (0.52 m²)	$S_a/N = 4.34$ ft² (0.40 m²)
$H/W_l = 0.51$	$L/W = 1.1$	ITDG $= 28$ msec

NOTE: Terminology is explained in Appendix 1.

27

Buenos Aires

Teatro Colón

The Teatro Colón is one of the beautiful large opera houses of the world. With 2,487 seats plus standees, it is larger than most of the famous opera houses in Europe but smaller than the principal houses of the United States of America. It has been very successful since its dedication in 1908.

A study was made of 23 opera houses in Europe, Japan, and the Americas in 1997–99. Questionnaires were mailed to a number of important opera conductors. Twenty-one usable responses were received. Ratings of the acoustics were requested, as heard both in the audience and in the pit. The Teatro Colón was judged the best of the 23, acoustically, by a significant amount. The closest quality ratings were for the Dresden-Semperoper, Milan-Teatro alla Scalla, Tokyo-New National Theater, and Naples-Teatro di San Carlo. Acoustical tests were made in these halls relatively recently, and in every respect the measurements in the Colón confirmed the excellence of its acoustics.

One aspect of the house that pleases soloists is the well-defined reflections of adequate intensity back to the stage from the ceiling and the balcony faces, thus giving the performers a feeling of support from the house.

Regular opera-goers in Buenos Aires say that the best seats are in the top two galleries. It is only in those galleries that the violin tone from the pit comes through with its full quality, which is directly attributable to reflections from the main ceiling.

Depending on the depth of the pit below the edge of the pit railing, the sound of the pit orchestra on the main floor is somewhat muffled. When I attended several operas in the Colón, the depth of the pit was as much as 10 ft (3m).

As a concert hall, the Teatro Colón is surprisingly satisfactory. To accommodate the full orchestra the pit is generally closed over and a rather shallow concert enclosure is used, as shown in the drawings. The front side-surfaces of the boxes to either side of the stage and the fronts of boxes in the first ring reflect lateral energy to most of the main floor and, selectively, to the rings above. In the highest rings, the ceiling helps produce early reflections. The reverberation time is about 1.6 sec

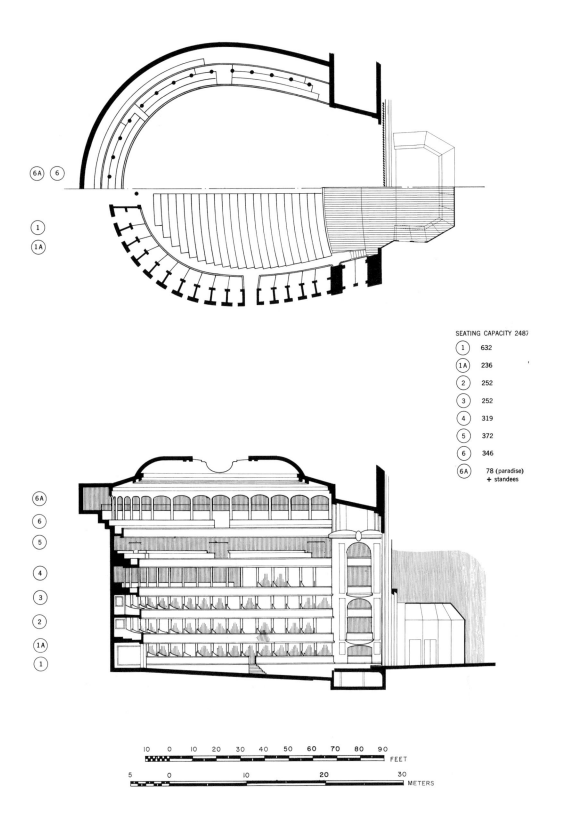

SEATING CAPACITY 2487

①	632
①A	236
②	252
③	252
④	319
⑤	372
⑥	346
⑥A	78 (paradise) + standees

10 0 10 20 30 40 50 60 70 80 90 FEET

5 0 10 20 30 METERS

for opera at mid-frequencies with full audience. For concerts it is also about 1.6 sec, less than the 1.8 sec for Carnegie Hall in New York.

ARCHITECTURAL AND STRUCTURAL DETAILS

Uses: Opera, concerts, recitals, and conferences. **Ceiling:** 1.5-in. (3.7-cm) plaster on metal lath. **Walls:** 1-in. (2.5-cm) plaster on wire lath, including balcony fronts. **Floors:** Wood. **Carpets:** On all aisles except in upper two rings. **Stage floor:** Wood over air-space; surface inclines at rate of 1 unit for each 100 units. **Orchestra pit:** Floor and sidewalls made of wood. **Stage enclosure for concerts:** 0.5-in. (2.5-cm)-thick molded fiberglass. **Seating:** Fully upholstered chairs, including rear of backrest and armrests.

ARCHITECT: Victor Meano. PHOTOGRAPHS: Burri, Magnum. ARCHITECT'S DRAWINGS: Made from the building in 1930, courtesy opera administration.

TECHNICAL DETAILS

Opera

$V = 726{,}300$ ft³ (20,570 m³)	$S_a = 15{,}200$ ft² (1,410 m²)	$S_A = 19{,}000$ ft² (1765 m²)
S_o (pit) $= 675$ ft² (63 m²)	$S_P = 3{,}402$ ft² (316 m²)	S (pit floor) $= 2{,}050$ ft² (190 m²)
$S_T = 23{,}077$ ft² (2,144 m²)	$N_{ST} = 2{,}787$	$N = 2{,}487$
$H = 87$ ft (26.5 m)	$W = 80$ ft (24.4 m)	$L = 113$ ft (34.4 m)
$D = 141$ ft (43 m)	$V/S_T = 31.5$ ft (9.6 m)	$V/S_A = 38.2$ ft (11.6 m)
$V/N_{st} = 261$ ft³ (7.39 m³)	$S_A/N_{st} = 6.8$ ft² (0.63 m²)	$S_a/N_{st} = 5.45$ ft² (0.506 m²)
$H/W = 1.09$	$L/W = 1.4$	ITDG $= 18$ msec

Concerts

$V = 760{,}000$ ft³ (21,524 m³)	$S_a = 16{,}522$ ft² (1,535 m²)	$S_A = 19{,}000$ ft² (1,765 m²)
$S_o = 2{,}470$ ft² (230 m²)	$S_T = 20{,}940$ ft² (1,945 m²)	$N = 2{,}487$
$V/S_T = 36.3$ ft (11.07 m)	$V/S_A = 40$ ft (12.2 m)	$V/N = 306$ ft³ (8.67 m³)
$S_A/N = 7.64$ ft² (0.71 m²)	$S_a/N = 6.64$ ft² (0.617 m²)	ITDG $= 18$ msec

Note: $S_T = S_A + 1{,}940$ ft² (180 m²); see Appendix 1 for definition of S_T.

NOTE: The terminology is explained in Appendix 1.

28

Sydney

Concert Hall of the
Sydney Opera House

The Concert Hall of the Sydney Opera House was opened in 1973. Contained in one of the most spectacular buildings in the world, it seats 2,679 and is the home of Sydney Symphony Orchestra. The enormous circular ceiling, which rises up to 82 ft (25 m) above the stage and radiates out and down to form about two-thirds of the walls, is paneled with white birch plywood. The lower wall, boxes, and stage are paneled with hard brown wood, brush box. Both woods are Australian. Suspended from this center point are 21 giant acrylic rings, acoustic reflectors installed to give acoustic feedback to the orchestra and some early sound reflections to the audience immediately surrounding the stage.

The orchestra platform is placed in the front fourth of the hall, with 410 seats of the 2,696 total located to its rear and 158 seats in two boxes at the sides of the stage. The ten boxes which take the place of side balconies are unusual in that they are steeply sloped and seat from 50 to 79 each. The main audience area is steeply raked compared to the main floor of classical shoe-box halls. Behind this block are two elevated seating areas, even more steeply raked, which take the place of balconies—without overhang.

The large side walls that comprise the fronts of the boxes were designed to give early sound reflections to at least half of this center area of seating. The upper seating receives early reflections from the bottoms and edges of the white birch ceiling.

The grand organ, designed and built by an Australian, Ronald Sharp, with 127 stops comprising 10,500 pipes, is said to be the largest tracker action organ in the world.

I attended a symphonic concert in the Concert Hall and was seated in the fourth side box from the rear. At that location there was not an abundance of early sound reflections. I had hoped to experience the acoustic conditions on the main floor, where the early reflections from the fronts of the side boxes certainly contribute much toward making the hall sound spacious and intimate. I was quite aware

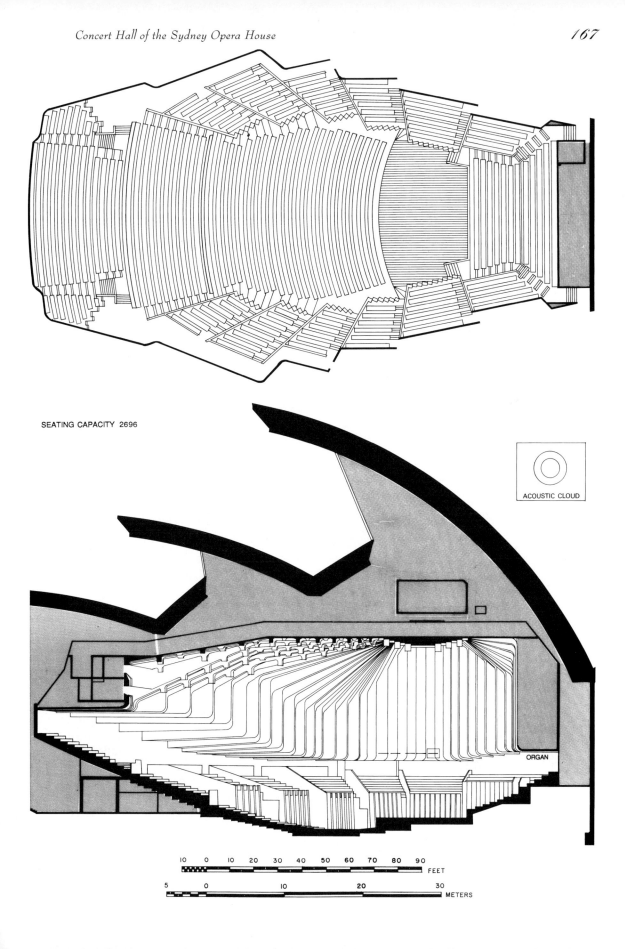

SEATING CAPACITY 2696

ACOUSTIC CLOUD

ORGAN

10 0   10   20   30   40   50   60   70   80   90
                                              FEET

5   0   10   20   30
                                              METERS

of the 2.0-sec reverberation time, which was well suited to that night's symphonic compositions. Sydney can be proud of the Concert Hall because of its beauty and acoustics.

ARCHITECTURAL AND STRUCTURAL DETAILS

[Note: This hall was originally planned as an opera house, which accounts both for its name and the steeply raked seating.]

Uses: Orchestra, chamber music, and soloists, 43%; drama and speech, 22%; school concerts, 16%; popular music, 9%. **Ceiling:** 0.5-in. (1.25-cm) plywood on 1-in. plasterboard, constituting 66% of side wall area. **Walls:** Lower 33% sidewall is 0.75-in. (1.9-cm) laminated wood. **Balcony faces:** Same as lower walls. **Floors:** 1.25-in. (3.2-cm) laminated wood over 0.75-in. air return space. **Carpets:** None. **Stage enclosure:** Same as walls. **Stage floor:** Same as floors elsewhere. **Stage height:** 50-in. at the apron. **Seating:** Backrest of 0.75-in. molded plywood; back cushion, wool-upholstered contoured polyurethane; seat bottom similar to the backrest. **Floating canopy:** 21 circular reflectors of clear acrylic and toroidal in section (see the sketch). Settings range from 27 to 35 ft (8.2 to 10.7 m) above stage level, with latter most common.

INITIAL ARCHITECT: Joern Utzon, exterior design. PRINCIPAL ARCHITECT: Peter Hall, interior design. ACOUSTICAL CONSULTANT: V. L. and N. V. Jordan. REFERENCES: V. L. Jordan, "Acoustical design considerations of the Sydney Opera House," pp. 33–53, and Peter Hall, "The design of the Concert Hall of the Sydney Opera House," pp. 54–69, *Journal and Proceedings, Royal Society of New South Wales*, **106**, 1973.

TECHNICAL DETAILS

$V = 868,600$ ft³ (24,600 m³)	$S_a = 14,666$ ft² (1,362 m²)	$S_A = 16,824$ ft² (1,563 m²)
$S_o = 1,945$ ft² (180.7 m²)	$S_T = 18,769$ ft² (1,744 m²)	$N = 2,679$
$H = 55$ ft (16.8 m)	$W = 109$ ft (33.2 m)	$L = 104$ ft (31.7 m)
$D = 146$ ft (44.5 m)	$V/S_T = 46.3$ ft (14.0 m)	$V/S_A = 51.6$ ft (15.7 m)
$V/N = 324$ ft³ (9.18 m³)	$S_A/N = 6.24$ ft² (0.58 m²)	$S_a/N = 5.47$ ft² (0.51 m²)
$H/W = 0.50$	$L/W = 0.95$	ITDG $= 36$ msec

NOTE: The terminology is explained in Appendix 1.

29

Salzburg

Festspielhaus

The Salzburg Festspielhaus, with 2,158 seats, was opened July 26, 1960. In keeping with Salzburg tradition, it was both a musical and social occasion, attracting diplomats, industrialists, and artists. The Festspielhaus has a width of nearly 112 ft between the faces of the side balcony boxes. On the main floor the width approaches 124 ft! This hugeness, both visual and acoustical, must concern the patrons who regularly attend operas and concerts in Europe's small halls.

For opera, the sound in the balcony has generally been praised. On the main floor, some seats are not as good because of the outward sloping walls, which means that lateral sound reflections are not extensive in the front part of the hall and sound reflected from the ceiling travels primarily to the rear of the hall and the balcony. The sound from the pit orchestra is also projected by the ceiling to the rear half of the main floor and by the side walls to the rear half of the main floor and to the boxes.

In 1979, to improve the sound for symphonic concerts, a new stage enclosure was installed (see the middle photo, not shown on the drawings). For aesthetic reasons, it was replaced by a second enclosure in 1993 (see the lower photo). The musicians can hear each other better and the sound is projected to the front parts of the main floor with good results. Satisfaction with the acoustics has increased greatly.

The materials of the hall are excellent from an acoustical standpoint. The walls and ceiling are of vibration-damped plaster, on reeds. Since the plaster and wood panels on the side walls are over 1.5 in. (3.8 cm) thick, they do not absorb the lower registers of the orchestra. In the balcony the sound is warm, intimate, clear, and brilliant. The reverberation time at mid-frequencies, fully occupied, is 1.5 sec, ideal for most opera, but somewhat low for symphonic music of the Romantic period.

The Festspielhaus takes its place among the more important halls of Europe because of the high quality of its musical performances. Great opera is produced by

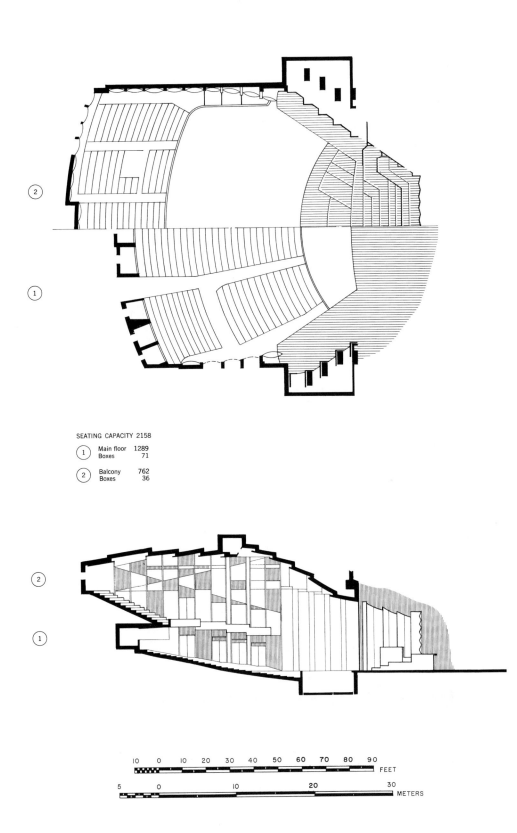

SEATING CAPACITY 2158

① Main floor 1289
Boxes 71

② Balcony 762
Boxes 36

```
10   0   10   20   30   40   50   60   70   80   90
                                                    FEET
5    0         10              20              30
                                                    METERS
```

the Vienna Opera Company and orchestral music comes from the best European and overseas traveling orchestras.

ARCHITECTURAL AND STRUCTURAL DETAILS

Uses: Opera, concerts, and drama. **Ceiling:** 1.5–2.0-in. (3.8–5-cm) plaster on reeds. **Sidewalls:** Concave-curved portions made of plaster on reeds with a thin wood layer; in front of the concave portions are convex curved panels, about 1.5 in. (3.8 cm) thick, of wood-fiber sheets. **Rear walls:** Convex wooden panels. **Floors:** Wood panels, supported above concrete by standards; seat rows and aisle steps have cemented cork linoleum covering. **Carpet:** On all aisles and in boxes, except in balcony. **Stage floor:** Wood. **Stage height:** 35.5 in. (90 cm). **Orchestra pit:** 2-in. (5-cm)-thick wood floor; walls are wooden wainscoting. **Seating:** Molded plywood; top of the seat bottom and a portion of the armrest are upholstered.

ARCHITECT: Clemens Holzmeister. ACOUSTICAL CONSULTANT: G. A. Schwaiger. PHOTOGRAPHS: upper, courtesy of the architect (1962); middle, Jens Rindel (1982); lower, Anrather Photo (1995).

TECHNICAL DETAILS

Concerts

V = 547,500 ft³ (15,500 m³)	S_a = 11,400 ft² (1,058 m²)	S_A = 14,800 ft² (1,375 m²)
S_o = 2,100 ft² (195 m²)	S_T = 16,740 ft² (1,555 m²)	N = 2158
H = 47 ft (14.3 m)	W = 108 ft (32.9 m)	L = 97 ft (29.6 m)
D = 95 ft (29 m)	V/S_T = 32.7 ft (9.97 m)	V/S_A = 37 ft (11.3 m)
V/N = 254 ft³ (7.18 m³)	S_A/N = 6.96 ft² (0.637 m²)	S_a/N = 5.28 ft² (0.490 m²)
H/W = 0.44	L/W = 0.9	ITDG = 27 msec

Opera

V = 495,000 ft³ (14,020 m³)	S_a = 10,850 ft² (1,008 m²)	S_A = 14,100 ft² (1,310 m²)
S_o (pit opening) = 800 ft² (74.3 m²)	S_o (pit floor) = 950 ft² (88.2 m²)	N = 2,158
S_T = 17,000 ft² (1,580 m²)	S_P = 2,100 ft² (195 m²)	V/S_T = 29.1 ft (8.87 m)
V/S_A = 35.1 ft (10.7 m)	V/N = 229 ft³ (6.5 m³)	S_A/N = 6.53 ft² (0.61 m²)

Note: $S_T = S_A$ + 1,940 ft² (180 m²); see Appendix 1 for definition of S_T.

NOTE: The terminology is explained in Appendix 1.

30

Grosser Musikvereinssaal

The "Grosse Saal der Gesellschaft der Musikfreunde in Wien" opened in 1870. Without doubt, the pulse of any orchestra conductor quickens when he first conducts in this renowned hall. The Vienna Philharmonic, the parade of famous conductors, and the fine music played there make this the Mecca of the old halls of Europe.

The side walls are made irregular by over forty high windows, twenty doors above the balcony, and thirty-two tall, gilded buxom female statues beneath the balcony. Everywhere are gilt, ornamentation, and statuettes. Less than 15% of the interior surfaces is made of wood. Wood is used only for the doors, for some paneling around the stage, and for trim. The other surfaces are plaster on brick or, on the ceiling and balcony fronts, plaster on wood lath.

The superior acoustics of the hall are due to its rectangular shape, its relatively small size (volume 530,000 ft³ (15,000 m³) and seats 1,680), its high ceiling with resulting long reverberation time (2.0 sec, fully occupied), the irregular interior surfaces, and the plaster interior. Any hall built with these characteristics would be an excellent hall, especially for symphonic music of the Romantic and Classical periods.

Nearly every conductor echoes Bruno Walter, "This is certainly the finest hall in the world. It has beauty of sound and power. The first time I conducted here was an unforgettable experience. I had not realized that music could be so beautiful." Herbert von Karajan, added, "The sound in this hall is very full. It is rich in bass and good for high strings. One shortcoming is that successive notes tend to merge into each other. There is too much difference in the sound for rehearsing and the sound with audience."

The Grosser Musikvereinssaal is similar acoustically to Symphony Hall in Boston. Most critical listeners agree that the clarity or definition is better than in the Amsterdam Concertgebouw. The sound in this hall is much louder than in Boston, and some feel that this is a disadvantage for a touring orchestra which may

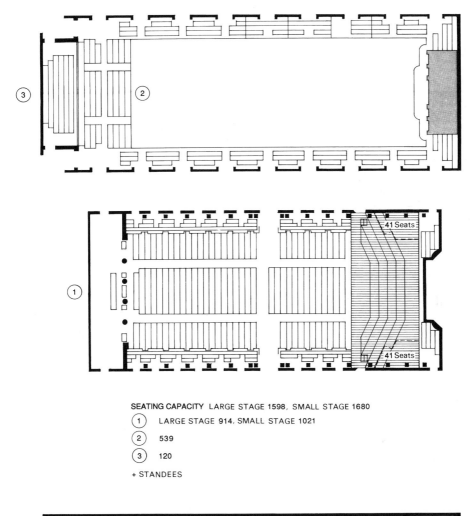

SEATING CAPACITY LARGE STAGE 1598, SMALL STAGE 1680

(1) LARGE STAGE 914, SMALL STAGE 1021

(2) 539

(3) 120

+ STANDEES

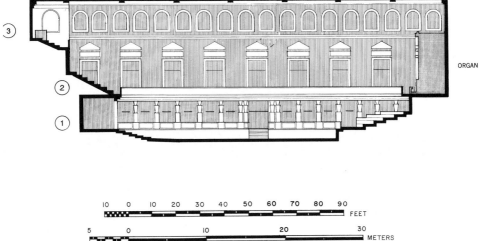

ORGAN

not be in the habit of restraining itself. Also, it is overly easy for the brass and percussion to dominate the strings. The string and woodwind tone are delicious and the sound is uniform throughout the hall.

ARCHITECTURAL AND STRUCTURAL DETAILS

Use: Orchestra and soloists. **Ceiling**: Plaster on spruce wood. **Side and rear walls**: Plaster on brick, except around the stage, where walls are of wood; doors are of wood; balcony fronts are plaster on wood. **Floors**: Wood. **Carpets**: None. **Stage floor**: Wood risers over wood stage. **Stage height**: 39 in. (1 m) above floor level. **Added absorbing material**: 200 ft² (18.6 m²) of draperies over front railings on side loges. **Seating**: Wood structure on main floor and side balconies, except that tops of seat bottoms are upholstered with 4 in. (10 cm) of cushion covered by porous cloth; rear balcony seats, plywood.

ARCHITECT: Theophil Ritter von Hansen. PHOTOGRAPHS: Courtesy of Sekretariat, Gesellschaft der Musikfreunde in Wien.

TECHNICAL DETAILS

V = 530,000 ft³ (15,000 m³)	S_a = 7,427 ft² (690 m²)	S_A = 10,280 ft² (955 m²)
S_o = 1,754 ft² (163 m²)	S_T = 12,030 ft² (1,118 m²)	N = 1,680
H = 57 ft (17.4 m)	W = 65 ft (19.8 m)	L = 117 ft (35.7 m)
D = 132 ft (40.2 m)	V/S_T = 44 ft (13.4 m)	V/S_A = 51.5 ft (15.7 m)
V/N = 315 ft³ (8.93 m³)	S_A/N = 6.1 ft² (0.57 m²)	S_a/N = 4.42 ft² (0.41 m²)
H/W = 0.88	L/W = 1.8	ITDG = 12 msec

NOTE: The terminology is explained in Appendix 1.

31

Vienna

Konzerthaus

$\mathscr{A}$t the opening of the Konzerthaus in 1913 it was claimed *to be a place for the cultivation of superior music, a focus of artistic endeavors, a building for music and a building for Vienna.* This assertion has been vigorously borne out throughout the hall's history. Containing four halls in a range of sizes, it boasts today of approximately 650 presentations and more than 30 subscription series per year. Its programs include not only the entire classical repertoire of orchestral and choir music, operas in concert, chamber music and recitals, but also jazz and folk music events.

Emperor Franz Joseph I attended the opening of the Konzerthaus. Performed were Beethoven's *Ninth Symphony* and a premier presentation of Richard Strauss's *Festliches Präludium*. The Konzerthaus is listed as an important national historical monument, representing a special example of Viennese *Art Nouveau*. Seating 1865, the main hall's gorgeous splendor was restored in 2000, along with new facilities for performers, patrons, and occupants. Together, the Konzerthaus, the Musikvereinssaal, and the Staatsoper are responsible for Vienna's worldwide reputation as a leading center for music.

One of the goals of the rehabilitation of the main hall was to make such improvements in the acoustics of the Konzerthaus as the state of the art might suggest. Although it had the sound of a typical rectangular concert hall, there had been complaints that the sound was too powerful, too reverberant—some even using the phrase "suggesting a barn." The renovation adjustments called for reducing the reverberance of the hall, particularly at low frequencies, and reducing the strength of the sound. Adding sound-absorbing materials at critical places in the hall and making some non-observable structural changes accomplished these goals. The reverberation time at 125 Hz, fully occupied, has dropped from 2.7 to 2.3 sec and that at 500 Hz from 2.0 to 1.95 sec. The binaural quality index measure has increased from 0.57 to 0.66. Musicians and conductors have commented favorably on these changes.

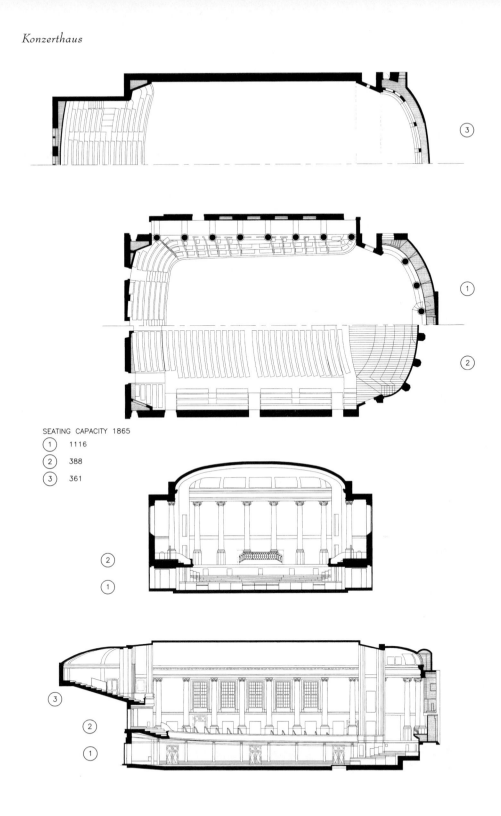

SEATING CAPACITY 1865

① 1116
② 388
③ 361

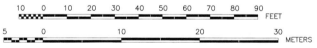

ARCHITECTURAL AND STRUCTURAL DETAILS

Uses: Orchestra, soloists, chamber music. **Ceiling:** Reed-reinforced gypsum plaster on wood framing. **Sidewalls:** 2.8-in. (7-cm) gypsum plaster on wood with 3.2-in. (8-cm) airspace over concrete wall. **Rear walls:** Fabric on perforated metal–air space–brick wall. **Stage walls:** Wood, partly removable. **Balcony fronts:** Plaster on wood. **Floors:** 0.9-in. (2.2 cm) parquet on 1.4-in. (3.5 cm) plywood; 500,000 holes for air intake. **Carpet:** 538 ft² (50 m²) on aisle behind the open side loges. **Seating:** Wooden structure, thin upholstering on the front of backrests and top of seat bottoms.

ORIGINAL ARCHITECT: Ludwig Baumann, Hermann Helmer, and Ferdinand Fellner. ARCHITECT FOR RECONSTRUCTION (1997–2000): Hans Puchhammer. ACOUSTICAL CONSULTANT (1997–2000): Karlheinz Müller, Müller-BBM. PHOTOGRAPHS: Courtesy Wiener Konzerthaus.

TECHNICAL DETAILS

Concerts (with orchestra shell):

V = 585,980 ft³ (16,600 m³)	S_a = 7,016 ft² (652 m²)	S_A = 9,480 ft² (881 m²)
S_o = 1,474 ft² (137 m²)	S_T = 10,954 ft² (1,018 m²)	N = 1,865
H = 46 ft (14 m)	W = 93 ft (28.4 m)	L = 95 ft (29 m)
D = 117 ft (35.7 m)	V/S_T = 53.5 ft (16.3 m)	V/S_A = 61.8 ft (18.8 m)
V/N = 314 ft³ (8.90 m³)	S_A/N = 5.08 ft² (0.472 m²)	S_a/N = 3.76 ft² (0.35 m²)
H/W = 0.49	L/W = 1.02	ITDG = 23 msec

NOTE: The terminology is explained in Appendix 1.

32

Vienna

Staatsoper

The main hall of the Staatsoper is modern in design. Bombed during the war and rededicated in 1955, it has the same architectural shape as the original house of 1869 but without baroque decoration and gilded statues. There are three tiers of boxes, a central box for dignitaries, and two upper galleries. The interior is dominated by red velvet and red damask, and the decoration is mainly white with gold. There is a large torus-shaped crystal chandelier against the ceiling.

The conductors who know the Staatsoper are unanimous in their opinion that the house has good sound. Bruno Walter said (1960), "The Staatsoper is the most alive of all opera halls. It is much better than New York and better than La Scala. The orchestra does not overpower the singers."

The auditorium has 1,709 seats, compared to the New York Metropolitan Opera House's 3,816 seats and San Francisco's 3,252 seats—indeed, it is a miniature. One year I attended four consecutive performances with the privilege of sitting in any part of the house. On the last evening I sat in the orchestra pit through an entire act. The sound is louder than in its American counterparts, largely because of the smaller cubic volume. On the main floor the sound is more intimate, beautifully clear and brilliant. Since it is only 64 ft (19.5 m) wide, the early lateral reflections are strong and the initial-time-delay gap is short, which accounts in large part for the superior sound.

The auditorium itself is not very live. The liveness heard during the performances I attended came primarily from the stagehouse, which is large, and quite reverberant. The canvas backdrop, the "cyclorama," is thick, heavily painted, and, hence, does not absorb high-frequency sounds.

At the ends of the pit, behind the brass (left) and percussion (right), there are small entrance chambers, open to the pit, the doors at the back of which are simply covered with red velvet. The double basses are lined up against the hard wall of the pit on the stage side, which increases their loudness.

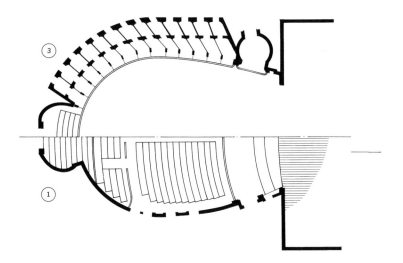

SEATING CAPACITY 1709

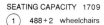

1 488 + 2 wheelchairs

2 148

3 220

4 166

5 309

6 342

7 567 standees (all levels)

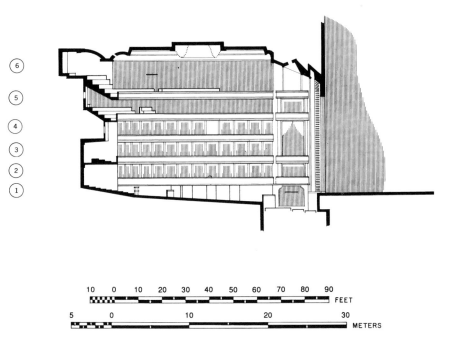

```
     10   0   10   20   30   40   50   60   70   80   90
                                                          FEET
     5    0            10             20            30
                                                          METERS
```

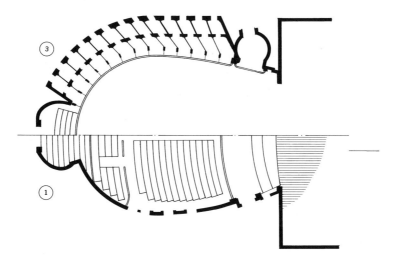

SEATING CAPACITY 1709

(1) 488+2 wheelchairs

(2) 148

(3) 220

(4) 166

(5) 309

(6) 342

(7) 567 standees (all levels)

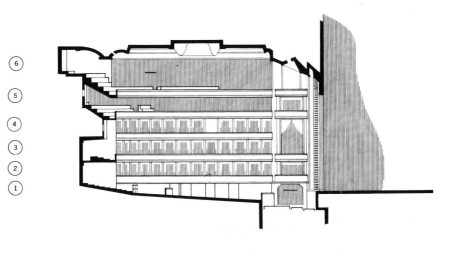

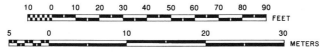

I judge the sound of the Staatsoper excellent, and I rate the main floor much better than that of La Scala. In the boxes, both houses are excellent. The live stage house of the Staatsoper also makes its sound more interesting than the dry sound of La Scala.

ARCHITECTURAL AND STRUCTURAL DETAILS

Uses: Opera. **Ceiling:** Plaster. **Walls:** In the top galleries, the walls are covered with damask behind Plexiglas to prevent damage by standees; faces of the rings are of wood; in the balcony beneath the gallery, the walls are of damask over plywood with 0.25 in. (0.64 cm) airspace between them; in the boxes the walls are of tightly stretched damask on wood. On main floor, walls are 0.5 in. (1.3 cm) wood. **Floors:** In the gallery and balcony, the floors are PVC linoleum over concrete; the main floor is of wood with carpet in the aisles; the main-floor standing room is of wood. **Pit:** On the pit floor are 3-in. (7.6-cm) wooden planks; the pit walls are of 0.75-in. (1.9-cm) wood; there are small rooms at the end of each pit, with velvet on the doors behind. **Stage height:** 41 in. (104 cm) above floor level at the first row of seats. **Seating:** the boxes have simple chairs with upholstered seats; on the main floor the seats are of solid wood except that the front of the backrests and the top of the seat bottoms are covered with 1-in. (2.5-cm) upholstering.

ARCHITECT: August Siccard von Siccardsburg and Eduard van der Nuell. AR-CHITECT FOR RECONSTRUCTION: Erich Boltenstern. ACOUSTICAL CONSULTANT FOR RECONSTUCTION: G. Schwaiger. REFERENCES: Plans, details, and photographs courtesy of the management of the Staatsoper.

TECHNICAL DETAILS

Opera

V = 376,600 ft³ (10,665 m³)	S_a = 10,000 ft² (930 m²)	S_A = 12,850 ft² (1,194 m²)
S_o (pit) = 1,150 ft² (106.8 m²)	S_P = 1,720 ft²	S_T = 15,720 ft² (1,460 m²)
N = 1,709	H = 62 ft (18.9 m)	W = 60 ft (18.3 m)
L = 98 ft (29.9 m)	D = 111 ft (33.8 m)	V/S_T = 24 ft (7.3 m)
V/S_A = 9.3 ft (8.9 m)	V/N = 220 ft³ (6.24 m³)	S_A/N = 7.75 ft² (0.72 m²)
H/W = 1.03	L/W = 1.63	ITDG = 15 msec

NOTE: The terminology is explained in Appendix 1.

33

Brussels

Palais des Beaux-Arts

nown officially as the Salle Henri Leboeuf of the Palais des Beaux-Arts, this hall has a mixed history. Opened in 1929, it was rated highly for many years. My surveys of conductors and music critics in the early 1960s established it as a world-class venue for symphonic music. I reported that a German acoustician, F. Winkel (Der Monat, February 1957, p. 77) wrote to a number of European conductors asking them to name their favorite hall anywhere in the world. The answers mentioned the Palais about as often as the Grosser Musikvereinssaal in Vienna or Symphony Hall in Boston. The reverberation time in 1961 measured, at mid-frequencies, fully occupied, about 1.7 sec, slightly lower than the halls just mentioned, but with the hall's satisfactory bass response and a plethora of early sound reflections, it was generally accepted as a superior place to perform. It contains 2,150 seats.

But over the years, preceding 1999, various renovations had taken place, some of which affected the acoustics negatively. Various openings were made in the ceiling, oversized air-conditioning outlets were installed, the stage was rebuilt with a heavy concrete surface beneath for fire safety reasons, and heavily upholstered seats were installed. The reverberation time dropped to 1.5 sec and the bass response suffered substantially. The hall's reputation dropped. To restore the hall to as near its original state as feasible architectural renovations were undertaken in 1999. The changes described above were reversed, including installing seats more favorable to bass response. As a result, the reverberation time at mid-frequencies has increased to 1.6 sec, and the bass response has been restored. Musicians using the hall in the 2000–2001 season report a significant difference and some say that it sounds nearly as it did during its earlier incarnation.

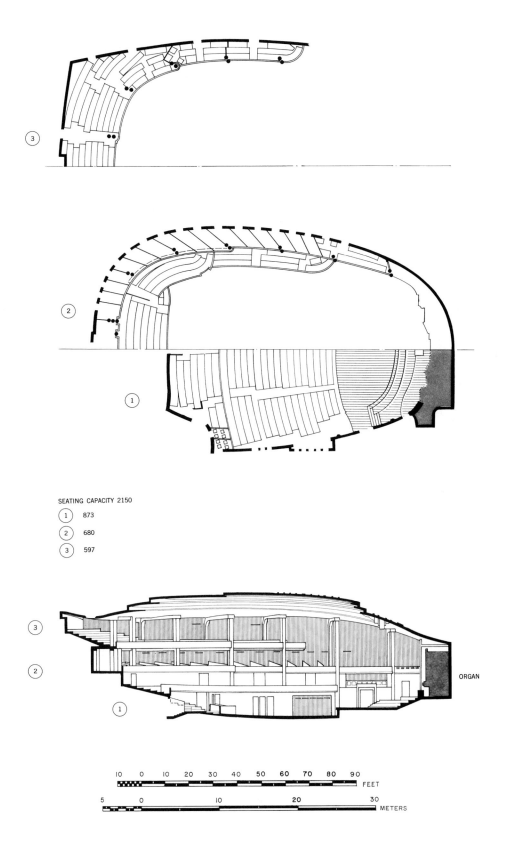

SEATING CAPACITY 2150

① 873

② 680

③ 597

ORGAN

10 0 10 20 30 40 50 60 70 80 90
                                                    FEET

5   0   10   20   30
                                                    METERS

ARCHITECTURAL AND STRUCTURAL DETAILS

Uses: Orchestra, chamber music, theater, ballet, and conferences. **Ceiling:** 75% plaster on metal lath; 20% thick glass on heavy metal frames; 5% lighting fixtures. **Rear walls and sidewalls:** Plaster on brick. **Floors:** On main floor, 0.32-in. (8-mm) wood on double layer of plywood over 3-ft (90-cm) plenum; in balconies, 0.32-in. (8-mm) wood floor on double layer of 0.32-in. (8-mm) plywood on concrete or sleepers. **Stage floor:** 0.32-in. (8-mm) oak on 1.325-in. (35-mm) on pine substructure. **Stage height:** 36 in. (92 cm). **Carpeting:** None. **Seats:** Modest upholstering.

ARCHITECTS: Original, Baron Victor Horta; year 2000 reconstruction, George Baines. ACOUSTICAL CONSULTANT FOR RECONSTRUCTION: Daniel Commins. CREDITS: Drawings, plans, photographs, and details courtesy of Director General of the hall. REFERENCES: D. Commins, "Successful acoustic design by architect Victor Horta and the Palais des Beaux-Arts concert hall, Brussels," *Proceedings of the Institute of Acoustics* **19**, pt. 3, pp. 213–220 (1997).

TECHNICAL DETAILS

V = 442,000 ft³ (12,520 m³)	S_a = 11,000 ft² (1,020 m²)	S_A = 14,000 ft² (1,300 m²)
S_o = 2,000 ft² (186 m²)	S_T = 16,000 ft² (1,486 m²)	N = 2,150
V/S_T = 27.6 ft (8.42 m)	V/S_A = 31.6 ft (9.6 m)	V/N = 206 ft³ (5.83 m³)
S_A/N = 6.51 ft² (0.60 m²)	H = 96 ft (29.3 m)	W = 76 ft (23.2 m)
L = 102 ft (31.1 m)	D = 117 ft (35.7 m)	H/W = 1.26
L/W = 1.34	S_a/N = 5.12 ft² (0.474 m²)	ITDG = 23 msec

NOTE: The terminology is explained in Appendix 1.

34
São Paulo

Sala São Paulo

The Sala São Paulo is located next to old active railway tracks. The ground floor of this building was an open-air waiting room bounded by open walkways and decorative columns, complete with palm trees. This open space was used for many years as an alternate summer location for orchestral and other musical events. In 1996, the São Paulo Symphony Orchestra and the State of São Paulo joined to investigate the feasibility of a better space for musical performances. The theater and acoustical consultant noted that the depot venue was about equal in width to that of Boston Symphony Hall and was of about the same length and height as the Vienna Musikvereinssaal. Its rectangular plan corresponded to that of both famous halls. In addition, the side galleries of the upper floors could be used for added reverberance.

The Brazilian National Institute of Historic and Artistic Heritage stipulated that the columns of the courtyard remain completely visible inside the hall. More challenging were the active railroad tracks located within a hundred feet of the facility. The solution to the railroad noise and vibration problem was to excavate the main floor and substitute a floating concrete slab that also supported the new sidewall structures, that in turn were insulated vibrationally from the existing columns. The noise level in the completed hall, with trains passing, measures NC-15.

A ceiling comprising 15 individually movable panels of wood separates the lower cubic volume from that above. When all are set at their lowest normally used level, the lower space has a volume of 635,400 ft³ (18,000 m³) and the upper 458,000 ft³ (13,000 m³). The panels can, individually, be set at any distance between 31 ft (9.5 m), thus permitting variation in the mid-frequency, occupied reverberation times from 1.7 sec up to 2.4 sec. It is thereby possible to select a degree of openness that suits the music being played. To make the hall suitable for convocations or theater, demountable seating is provided in the center, and absorbing cloth banners above the moving ceiling can be lowered to reduce the reverberation time to about 1.5 sec. The Sala São Paulo, seating 1620, opened to Brazilian

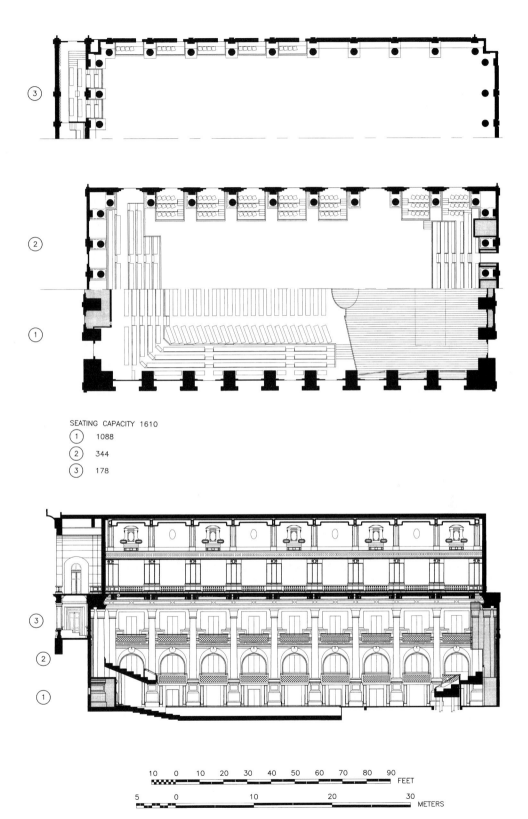

SEATING CAPACITY 1610
1 1088
2 344
3 178

10 0 10 20 30 40 50 60 70 80 90 FEET
5 0 10 20 30 METERS

acclaim in July 1999 with the Mahler *Second Symphony*. While acoustical data are limited, it seems logical to conclude that with the reverberation times reported, its modest size, and its architecture, the hall is assuredly a success.

ARCHITECTURAL AND STRUCTURAL DETAILS

Uses: Orchestras, recitals, soloists, and fashion shows. **Ceiling:** Composed of 15 adjustable panels each weighing 7.5 tons and supported by 16 steel cables, each divided into three coffered sub-modules with a range of 13–72 ft (4–22 m) above the main floor. **Canopy:** Ceiling panels can form a canopy. **Sidewalls:** Either masonry or panels of formed cement plaster on 2-in. (5-cm)-thick wood frame. *Balconies:* Supported on vibration isolated structures, 6-in. (15-cm) pre-cast concrete covered with pre-molded wood panels. **Sound diffusers:** Crowns of the columns, molded masonry figures and textured wood patterns on the balcony fronts and ceiling. **Variable sound absorption:** Acoustic banners that can be lowered into the adjustable volume above the ceiling. **Floors:** 1-in. (2.5-cm) wood over 1-in. plywood supported by a metal frame resting on a floating concrete slab. **Stage floor:** Level 2-in. (5-cm) tongue-and-groove wood mounted on metal support structure with 9.8 ft (3 m) airspace. **Stage height:** 47 in. (1.2-m). **Seats:** Plywood with high-density foam cushion covered with synthetic fabric.

ARCHITECT: Nelson Dupré, Dupré Arquitetura. ACOUSTIC AND THEATER CONSULTANT: ARTEC Consultants. PHOTOGRAPHS: Luiz Carlos Felizardo.

TECHNICAL DETAILS

V = 706,000 ft³ (20,000 m³)	S_a = 8,027 ft² (746 m²)	S_A = 11,223 ft² (1043 m²)
S_o = 4,261 ft² (396 m²)	S_T = 13,160 ft² (1223 m²)	N = 1,610
H = 60 ft (18.3 m)	W = 76 ft (23.2 m)	L = 98 ft (29.9 m)
D = 100 ft (30.5 m)	V/S_T = 53.6 ft (16.3 m)	V/S_A = 62.9 ft (19.18 m)
V/N = 438 ft³ (12.4 m³)	S_A/N = 6.97 ft² (0.648 m²)	S_a/N = 4.98 ft² (0.463 m²)
H/W = 0.79	L/W = 1.29	ITDG = 34 msec

Note: $S_T = S_A + 1{,}940$ ft² (180 m²); see Appendix 1 for definition of S_T.

NOTE: The terminology is explained in Appendix 1.

Salle Wilfrid-Pelletier

ocated in Place des Arts, Canada's largest center for the arts, the Wilfrid-Pelletier Hall is the jewel in the city's crown. Opened in 1963 with a gala concert by the Montreal Symphony Orchestra, it has served the city well since. It underwent some renovation in 1993. Seating 2,982, with a pit for 70 musicians created by removing 78 seats, the W-P Hall accommodates concerts, ballet, opera, musical comedies, and variety shows. More down-to-earth than its cousin, the Roy Thompson Hall in Toronto, it has a proscenium and a stage house of large proportions. For concerts, a stage enclosure is erected on stage, shown in the drawings. A sound system is used for non-orchestral events.

As a concert hall, it has not been without criticism. The principal objections are that the orchestral sound is not sufficiently loud on the main floor and the hall lacks intimacy. Studies have been made by several acoustical consultants with some preliminary comments: First, the ceiling of the shell is too high and the rear wall is too wide. Too much sound is trapped on the stage and the music lacks clarity. Second, there are no surfaces at or near the sides of the procenium to reflect lateral sound energy onto the main floor. Laterally reflected sound energy would give the audience a feeling of intimacy and of being enveloped by the sound. Third, the semi-open ceiling does not reflect enough sound energy to the main floor seating sections.

One acoustician reports that an experiment showed that when the back wall of the orchestra enclosure was brought forward toward the audience so that the front half of the orchestra sat over the pit, the music sounded louder, fuller, and more intimate on the main floor. If this shift were accompanied by adding reflecting surfaces on either side of the pit location and if others of proper shape were located in the space below or above the ceiling, the loudness of the orchestral sound in the audience areas would increase further and the listeners would feel more enveloped by the sound. The musicians would also sense better the acoustics of the hall. Finally, if modifications to the lower side walls and the doors were made, lateral reflections could be increased further and thus add to the feeling of spaciousness

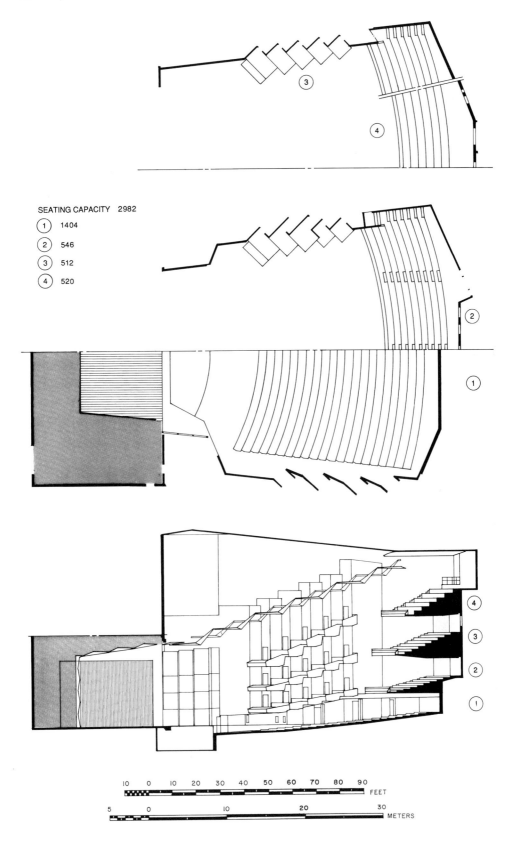

SEATING CAPACITY 2982

(1) 1404

(2) 546

(3) 512

(4) 520

FEET
10 0 10 20 30 40 50 60 70 80 90

METERS
5 0 10 20 30

and envelopment. These *suggestions* would need to be subjected to investigation in a model both for refinement and to make certain of a suitable solution.

ARCHITECTURAL AND STRUCTURAL DETAILS

Uses: Multipurpose, including classical music. **Ceiling:** Heavy plaster beneath which is hung precast plaster lattice. **Walls:** Plaster over concrete block; balcony faces molded plaster. **Floors:** Parquet on concrete, except all boxes are carpeted. **Stage enclosure:** 2-in. (5-cm) panels, each about 4 ft by 6.5 ft (1.22 by 2 m) made into irregular, off-center, shallow pyramids supported on frames. **Stage floor:** Tongue-and-groove wood over timber. **Stage height:** 36 in. (91 cm) above floor level. **Seating:** Upholstered top of seat bottom and front of seat back; underseat molded plywood; rear of seat back molded plywood; arms upholstered.

ARCHITECT: Affleck, Desbarats, Dimakopoulos, Lebensold, Michaud and Sise. ACOUSTICAL CONSULTANT: Bolt Beranek and Newman. PHOTOGRAPHS: Studio Lausanne and Panda Associates.

TECHNICAL DETAILS

V = 936,000 ft³ (26,500 m³)	S_a = 16,684 ft² (1,550 m²)	S_A = 19,020 ft² (1,767 m²)
S_o = 1,850 ft² (172 m²)	S_T = 20,870 ft² (1,939 m²)	N = 2,982
H = 77 ft (23.5 m)	W = 108 ft (32.9 m)	L = 123 ft (37.5 m)
D = 135 ft (41.2 m)	V/S_T = 44.8 ft (13.7 m)	V/S_A = 49.2 ft (15 m)
V/N = 314 ft³ (8.9 m³)	S_A/N = 6.38 ft² (0.59 m²)	S_a/N = 5.59 ft² (0.52 m²)
H/W = 0.71	L/W = 1.14	ITDG = 20 msec

NOTE: The terminology is explained in Appendix 1.

36

Toronto

Roy Thompson Hall

$\mathcal{R}$oy Thompson Hall is a semi-surround, non-proscenium hall, with only a small percentage of the audience to the sides and behind the orchestra. Seating 2,500 plus 113 seats in the choir lift, it ranks as a fairly large concert hall, but the distance from the stage to the farthest seat is only 127 ft (38.7 m).

When the Roy Thompson Hall opened in 1982 it was a striking visual success. It still appears to be a circular hall, but originally a large circular wheel was suspended overhead, surrounded by a forest of stalactites, called banners, all brilliantly colored and illuminated. Those banners were supposed to enhance the acoustics, but they seemed more to absorb the sound. The visual interior was a striking gray on gray, mostly created by paint on concrete.

The new design (see lower sketch) was dedicated in September 2002. The former interior has now been replaced by 12,000 ft² (1,100 m²) of white Canadian maple as a finish over 23 semi-cylindrical, very heavy, multilayer, "bulkheads" that surround the circular space above the upper seating tiers. Their installation has reduced the cubic volume of the hall by 13.5%.

The wheel, banners, and a number of clear acrylic discs that were suspended above the stage, have been replaced by a massive two-part canopy, weighing 38 tons (25,000 kg), that hangs overhead and serves both to return sound to the players and to distribute sound better to the main-floor audience. The canopy can be raised or lowered between 30 and 50 feet above the stage level. Larger than the size shown in the drawing, the stage has been extended 3 feet in the center, tapering off to its previous position at the sides. New two-sided panels have been installed around the stage; one side is hard, the other sound absorbing. They can be rotated to achieve a tailored orchestral balance.

Are the changes effective? One can only judge by the opening-night music critics' remarks. All agreed that there has been substantial improvement, mentioning the increased warmth of the bass and cello tones and the increased impact of the overall orchestral sound. Some mentioned, "more fibre in the woodwinds, and

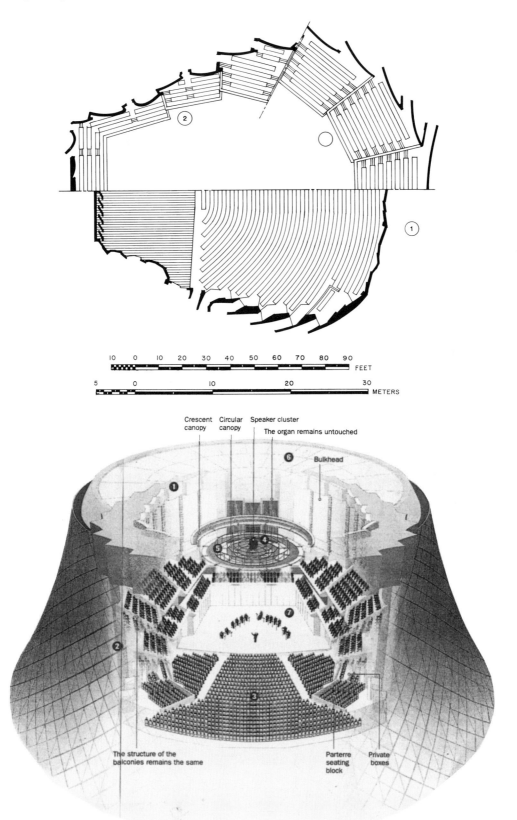

10 0 10 20 30 40 50 60 70 80 90
FEET

5   0   10   20   30
METERS

Crescent   Circular   Speaker cluster
canopy   canopy
                        The organ remains untouched

Bulkhead

The structure of the   Parterre   Private
balconies remains the same   seating   boxes
                                        block

a brighter glow in the brass." The players are particilarly pleased, now able to hear themselves and one another better. All seemed to agree that the hall, probably because of its circular construction, does not have the rich, encompassing reverberation of Vienna's Vereinssaal, or Boston's Symphony Hall. Nevertheless, the improvement is great and music in it is beautiful and distinct.

ARCHITECTURAL AND STRUCTURAL DETAILS

Uses: Primarily concert, recitals, choral events, and organ. **Ceiling:** Plaster. **Main sidewalls:** Above seating levels, 5 in. (2.5 cm) thick, nine layers of dense wood chip board; Below the seating levels, 6 in. (15 cm) poured in place concrete. **Sound absorption:** Provided on a variable basis by double-layered velour draperies. **Audience floor:** Wood on concrete. **Seating:** Upholstered seat bottom and part of seat back.

ORIGINAL ARCHITECT: Arthur Erickson and Mathers & Haldenby Associated Architects. RENOVATION ARCHITECT (2002): Kuwabara Payne McKenna Blumberg. ORIGINAL ACOUSTICAL CONSULTANT: T. J. Schultz of Bolt Beranek and Newman. RENOVATION ACOUSTICAL CONSULTANT (2002): ARTEC Consultants, Inc.

TECHNICAL DETAILS

$V = 865{,}000$ ft³ (24,500 m³)	$S_a = 14{,}000$ ft² (1,300 m²)	$S_A = 16{,}800$ ft² (1,560 m²)
$S_o = 2{,}500$ ft2 (232 m2)	$S_T = 18{,}740$ ft² (1,740 m²)	$N = 2{,}613$
$H = 76$ ft (23.2 m)	$W = 102$ ft (31.1 m)	$L = 89$ ft (27.1 m)
$D = 108$ ft (32.9 m)	$V/S_T = 46.2$ ft (14.1 m)	$V/S_A = 51.4$ ft (15.7 m)
$V/N = 331$ ft³ (9.4 m³)	$S_A/N = 6.43$ ft² (0.60 m²)	$S_a/N = 5.36$ ft² (0.498 m²)
$H/W = 0.74$	$L/W = 0.87$	ITDG $= 35$ msec

Note: $S_T = S_A + 1{,}940$ ft² (180 m²); see Appendix 1 for definition of S_T.

NOTE: The terminology is explained in Appendix 1.

37

Hong Kong

Cultural Center, Concert Hall

he Hong Kong Center for the Arts, completed in 1989, was obviously meant to bring to the city an architectural triumph like that achieved by Australia's Sydney Opera House. The impression from a neighboring skyscraper is that the motif is a combination of the United Nation's General Assembly building in New York and the tail fin of a Boeing 707 airliner. An opening article (unidentified reprint supplied by Hong Kong Cultural Centre) reads, "Hong Kong has added theatre sails to its Kowloon skyline. The curves of Hong Kong Cultural Centre may be gentler than those of Sydney Opera House but no less dramatic when viewed in the context of a skyline where all lines are straight and most are vertical. Just take in the view from the world's number one ferry trip."

Opened in November 1989 and seating 2,019, the concert hall interior resembles New Zealand's Christchurch Town Hall (Concert Hall) and the Costa Mesa (USA) Segerstrom Hall because the same acoustical consultant was responsible. The design has two features. First the upper part of the hall is not vertical as in the "shoebox" halls, but rather is lined with sloping panels that reflect the incident sound into the audience areas, while diffusing it sideways during reflections. Second, the sound is not able to reflect back and forth, as it does with parallel walls, so that energy is drained from the reverberant sound. This design creates a new acoustic, it emphasizes the early sound, making the music clear and true. It de-emphasizes the reverberant sound, both from standpoint of the energy in the reverberant sound and the length of the reverberation time. In one hall, the Glasgow Royal Concert Hall, originally of similar design, the upper reflecting panels were recently removed in order to allow more energy to remain in the reverberant sound field and to lengthen the reverberation time.

During the 1997 pre-opening musical performances, it was found that the acoustics were improved when the stage was extended and its shape and construction modified. Exactly how the sound field was modified has not been published. The acoustical consultant reports that the noise from the air-conditioning equipment,

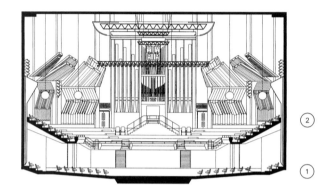

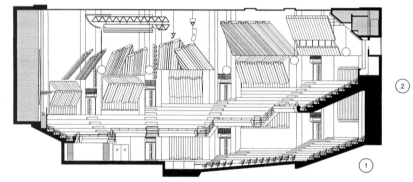

ORGAN

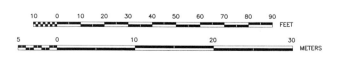

10   0   10   20   30   40   50   60   70   80   90   FEET

5   0   10   20   30   METERS

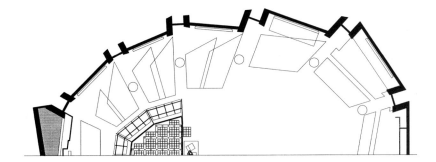

SEATING CAPACITY 2019

(1) 1045

(2) 974

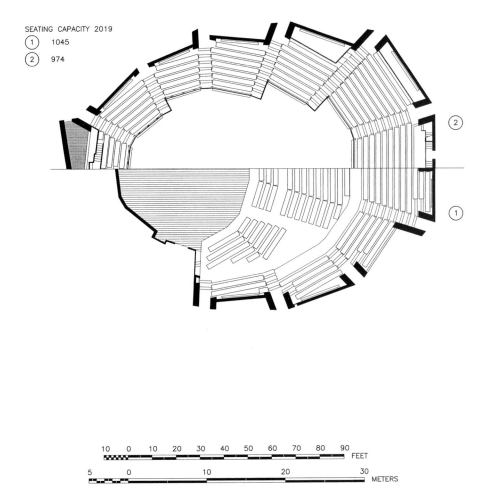

10 0 10 20 30 40 50 60 70 80 90 FEET				
5 0 10 20 30 METERS				

not his responsibility, exceeds customary standards for concert halls and infringes on the enjoyment of a performance.

ARCHITECTURAL AND STRUCTURAL DETAILS

Uses: Primarily orchestra and soloists. **Ceiling:** 1.0-in. (2.5-cm) "glasscrete" coffers suspended below plate girders. The top of the horizontal area of the coffers (about 75%) was plastered with 1.4-in. (3.5-cm) cement plaster. **Walls:** Plaster on concrete. **Internal reflectors:** QRD panels constructed of composite plasterboard/plywood and oak veneer, total 1.4 in. (3.5 cm) thick. The lower reflectors serve the lower level seating, the upper ones serve the balconies. **Floors:** Oak on concrete at front, oak on framing at rear. **Carpets:** None. **Stage enclosure:** The photographs show QRD diffusing panels around the stage. The overhead sound reflector is 60% QRD reflectors, surrounded by 0.7-in. (1.7-cm) Plexiglas curved panels. Its normal concert height is 25 ft (7.5 m). For organ concerts it is raised to 49 ft (15 m). **Stage floor:** Wooden strips over small airspace on sleepers on concrete. **Stage height:** 42 in. (1.07 m) above floor level at first row of seats. **Seating:** Fully upholstered, including armrests.

ARCHITECT: Jose Lei. ACOUSTICAL CONSULTANT: Harold Marshall. PHOTO-GRAPHS: Courtesy Leisure and Cultural Services Department of Hong Kong Government.

TECHNICAL DETAILS

V = 750,000 ft³ (21,250 m³)	S_a = 9,264 ft² (861 m²)	S_A = 11,954 ft² (1,111m²)
S_o = 2,668 ft² (248 m²)	S_T = 13,890 ft² (1,291 m²)	N = 2,019
H = 65.6 ft (20 m)	W = 106.6 ft (32.5 m)	L = 82 ft (25 m)
D = 80.36 ft (24.5m)	V/S_T = 54 ft (16.5 m)	V/S_A = 62.7 ft (19.1m)
S_A/N = 5.92 ft² (0.55 m²)	S_a/N = 4.59 ft² (0.426 m²)	V/N = 371 ft³ (10.52 m³)
H/W = 0.615	L/W = 0.769	ITDG = 27 msec

Note: $S_T = S_A + 1,940$ ft² (180 m²); see Appendix 1 for definition of S_T.

NOTE: The terminology is explained in Appendix 1.

38

Shanghai

Grand Theatre

*A*s part of the modernization of Shanghai, the government built the Shanghai Grand Theatre as a major step forward toward recognition of the value of performing arts. The theatre both gives Shanghai's native inhabitants a chance to share in the cultural beauty of the world and satisfies the cultural preferences of the recent huge influx of international companies and their associated personnel.

The Grand Theatre, seating 1,895, opened in August 1998 with a performance of Swan Lake by the National (Beijing) Ballet company. It is located at the west side of the Shanghai People's Plaza, facing the Shanghai Museum, with underground parking. As yet, the theatre is not associated with any musical organization and there is no resident opera company. A month after its opening, the Florence, Italy, Opera gave a memorable performance of *Aida*. Between then and the end of 2000, the performances were ballet, 103; orchestral concerts, 68; Western opera, 31; and a large number of Chinese shows, including Beijing opera, local opera, and drama.

In 1999, the Hamburg Ballet, the Suisse Ballet, the British Royal Ballet, a British opera, and a Dusseldorf opera gave performances. In 2000, a Swiss opera, the British Royal Orchestra, the German Broadcast Orchestra, the Baden-Baden Philharmonic Orchestra, and an American ballet performed there. A highlight of 2001 was the appearance of the Philadelphia Orchestra.

Orchestral concerts are accommodated by a removable orchestral enclosure on stage that is maneuvered by an "air jet cushion" system. It can be dismounted in three sections for easy storage.

It is difficult to rate the acoustics of this theatre, because there is nothing comparable in China. Guest musicians from abroad are satisfied with the sound in the hall, and, in comparison with most other halls in China, it is outstanding.

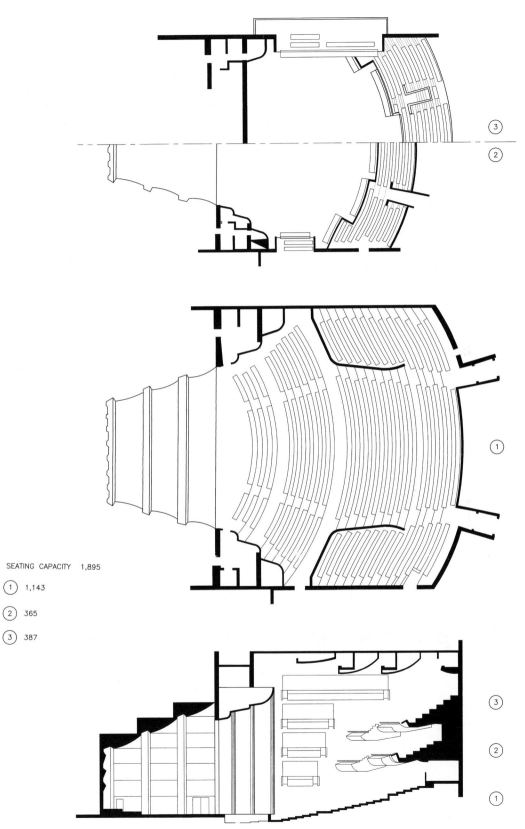

SEATING CAPACITY 1,895

① 1,143

② 365

③ 387

ARCHITECTURAL AND STRUCTURAL DETAILS

Uses: Opera, symphonic concert, Chinese opera, drama. **Ceiling:** Two layers cement gypsum board, each 0.78 in. (2 cm) thick. **Sidewalls:** Outer walls are concrete; decorative walls, lower part (below 11.5 ft, 3.5 m), granite, wooden grille with movable drapery behind for RT adjustment and decorated hardboard panel for remainder. **Rear walls:** Outer walls are concrete, remainder wooden grille with movable drapery behind. **Acoustical absorbing material:** When all exposed, 3,228 ft² (300 m²). **Carpets:** On main aisles with underpad. **Seating:** Thick upholstering on top of seat bottoms and front of seat backs. **Stage enclosure:** Removable—hard board on steel frame. **Noise level:** NR-23.

ARCHITECT: Jean-Marie Charpentier. INTERIOR: STUDIO, Architectural Design Firm. JOINT DESIGNER: East China Design and Research Institute of Architecture, Shanghai. ACOUSTICAL CONSULTANT: J. Q. Wang. ACOUSTICAL DESIGN: K. S. Zhang of above Institute, Shanghai.

TECHNICAL DETAILS

Opera

V = 458,900 ft³ (13,000 m³)	S_a = 6,983 ft² (649 m²)	S_A = 9,060 ft² (842 m²)
S_{pit} = 914.6 ft² (85 m²)	S_p = 2,666 ft² (247.8 m²)	S_T = 12,641 ft² (1,175 m²)
N = 1,676	V/S_T = 36.3 ft (11.06 m)	V/S_A = 50.6 ft (15.44 m)
V/N = 273.8 ft³ (7.76 m³)	S_A/N = 5.41 ft² (0.502 m²)	S_a/N = 4.17 ft² (0.39 m²)

Concerts

V = 529,500 (15,000 m³)	S_a = 7,381 ft² (686 m²)	S_A = 9,695 ft² (901 m²)
S_o = 1,431 ft² (133 m²)	S_T = 11,126 ft² (1,034 m²)	N = 1,895
H = 66 ft (20.1 m)	W = 116 ft (35.4 m)	L = 91 ft (27.7 m)
D = 109 ft (33.2 m)	V/S_T = 47.6 ft (14.51 m)	V/S_A = 54.6 ft (16.65 m)
V/N = 279 ft³ (7.91 m³)	S_A/N = 5.12 ft² (0.475 m²)	S_a/N = 3.89 ft² (0.362 m²)
H/W = 0.57	L/W = 0.78	ITDG = 40 msec

NOTE: The terminology is explained in Appendix 1.

39

Copenhagen

Radiohuset, Studio 1

The Radiohuset is the home of the Danish Radio Symphony Orchestra. The stage and the great organ dominate the room, and the audience area is small, seating 1,081, compared to that in a conventional concert hall. Opened in 1945, the cylindrical ceiling, low over the stage, has caused on-stage dissatisfaction among the orchestra players and has resulted in two major revisions. In the most recent renovation, completed in 1989, a canopy of acrylic panels was hung over the stage; a number of sloped large reflecting panels (see the photograph) were attached to the stage side walls; and a new floor, more favorable to bass and cellos, was installed along with hydraulic risers on stage. These recent changes have improved communication among the orchestra members and have provided more intimate sound to the audience than when the hall was without panels. The hall is also better acoustically for recordings. The reverberation time, 1.5 sec, fully occupied, is on the low side, the optimum for symphonic music in a hall of this size is about 1.7 sec.

The object of the acrylic panels is to improve communication on stage, facilitate orchestral timing, and to create better balance and blend. The size and number of panels for the Radiohuset followed studies by J. H. Rindel (1990). He found that a large number of smaller panels, each slightly curved, as compared to larger ones used elsewhere, give a more uniform reflection of the entire range of frequencies. The open area in the panel array should be about 50%. Installed in Studio 1 are 61 panels, each 5.9 × 2.62 ft (1.8 × 0.8 m), 0.3 in. (0.8 cm) thick, and slightly curved.

In the ceiling of the Radiohuset, there are low-frequency sound-absorbing Helmholtz-type resonators, 105 of which are (currently, 1994) tuned to absorb sound at 98 Hz and 45 at 64 Hz. Their purpose is to remove low-frequency standing waves between the floor and the curved ceiling. Because of the small volume of the hall, 420,000 ft³ (11,900 m³), the sound of a full symphony is loud, which helps create a feeling to the listener of being enveloped by the sound field and adds spaciousness to the orchestral music. The reports are that the 1989 changes have been

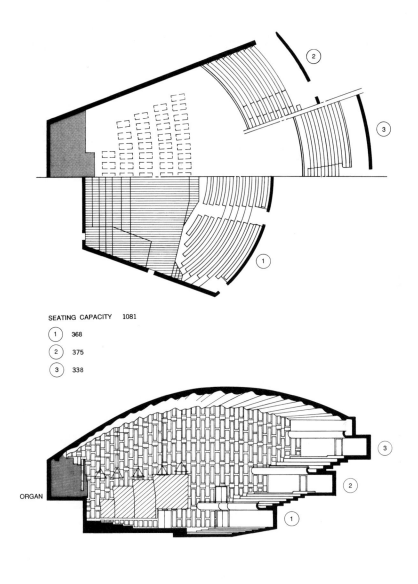

SEATING CAPACITY   1081

① 368

② 375

③ 338

ORGAN

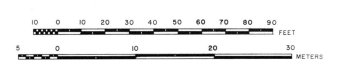

well accepted by the orchestra. Reviews by music critics are mixed, some saying that the sound is much the same as before the stage renovations.

ARCHITECTURAL AND STRUCTURAL DETAILS

Uses: All types of broadcasting. **Ceiling:** 3-in. (8-cm) concrete roof, covered by 1-in. (2.5-cm) tongue-and-groove wood strips (no airspace). **Sidewalls:** 0.83-in. (2.1-cm) wood with either 1-in.- or 4-in.- (2.5- or 10-cm)-deep airspace, about 50% each. **Rear walls:** Mineral wool, 25% perforated, covered with thin wood; above second balcony, 0.5-in. (1.2 cm) wood panels with airspace. **Floor:** Parquet over concrete. **Stage floor:** 0.9-in. (2.2 cm) parquet on joists spread apart 16 in. (40 cm), over 14 in. (36 cm) airspace; rear part has 26 hydraulic risers. **Stage height:** 27.5 in. (70 cm). **Seating:** Upholstered in leather, underseat unperforated.

ARCHITECT: Vilhelm Lauritzen. ACOUSTICAL CONSULTANT: Jordan Akustik. RESEARCH: A. C. Gade and J. H. Rindel from Technical University of Denmark. REFERENCE: N. V. Jordan, J. H. Rindel & A. C. Gade, "The new orchestra platform in the Danish Radio Concert Hall," *Proceedings of Nordic Acoustical Meeting,* Lulea, Sweden, 1990.

TECHNICAL DETAILS

V = 420,000 ft³ (11,900 m³)	S_a = 6,512 ft² (605 m²)	S_A = 7,761 ft² (721 m²)
S_o = 3,100 ft² (288 m²)	S_T = 9,701 ft² (901 m²)	N = 1,081
H = 58 ft (17.7 m)	W = 110 ft (33.5 m)	L = 61 ft (18.6 m)
D = 75 ft (22.9 m)	V/S_T = 43.29 ft (13.2 m)	V/S_A = 54.1 ft (16.5 m)
V/N = 388 ft³ (11.00 m³)	S_A/N = 7.2 ft² (0.67 m²)	S_a/N = 6.02 ft² (0.560 m²)
H/W = 0.53	L/W = 0.55	ITDG = 29 msec

Note: $S_T = S_A + $ 1,940 ft² (180 m²); see Appendix 1 for definition of S_T.

NOTE: The terminology is explained in Appendix 1.

40

Odense

Nielsen Hall in Odense Koncerthus

*D*uring the 1960s Odense underwent a major transformation—new streets, a four-lane highway, and urban renewal. Only a small part, the Hans Christian Anderson quarter, was preserved. On the east side the concert and congress building was erected, as well as an adjacent hotel. To the west, stands the Anderson museum and his house, as well as some low-rise provincial residences. To the north is a sizeable park. To avoid overpowering the other buildings, the main floor and concert stage of the Koncerthus are sunken below ground level; the front of the balcony starts at ground level.

The Carl Nielsen Hall is named after the best-known Danish composer, who was born on the island Fyn on which Odense is now the main city, the third largest in Denmark. The Nielson Hall opened in September 1982, with the Queen and Prince in attendance. The Koncerthus complex also contains a chamber music hall and rehearsal facilities for a full-size symphony orchestra. Conferences and congresses find the site ideal because it is near several hotels.

Financing for the hall came from the city, an independent committee operates it. Seating 1,320, and having the only built-in pipe organ in Denmark, excepting the Danish Radiohuset in Copenhagen, it is widely recognized as the most successful concert hall in the country. It serves, primarily, as the home for the Odense Symphony Orchestra. Each fourth year the hall becomes the venue for the International Carl Nielsen Violin Competition.

On entering the hall, one is immediately aware of its rectangularity and the sound-diffusing baffles, "birds," flying overhead, designed by the artist Gunnar Aagaard Anderson. Miniature protruding balconies on the sidewalls both create diffusion of the sound and provide surfaces for enabling lateral reflections of early sound to the seating areas. The louvered spaces above the balconies are air-conditioning outlets. The two large white surfaces to either side of the choir/audience seating are concrete panels with doors. The ceiling is a succession of deep triangles,

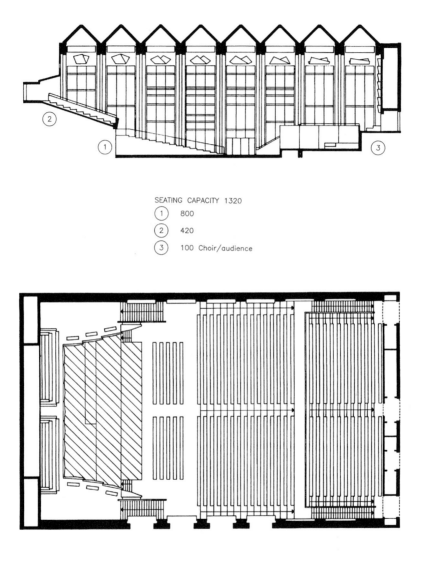

SEATING CAPACITY 1320

(1) 800

(2) 420

(3) 100 Choir/audience

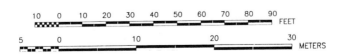

designed to create diffusion of the reverberant sound field. The users of the hall speak highly of its acoustics.

ARCHITECTURAL AND STRUCTURAL DETAILS

Uses: Symphonic concerts, 50%; chamber music, soloists, and other. **Ceiling:** Molded concrete with some absorption for reverberation control; hanging diffusers. **Sidewalls:** Plasterboard with wood veneer, 2 in. (5 cm) thick; slotted areas that are used for fine tuning. **Floor:** Wood, 1.5 in. (3.8 cm) thick on joists with air chamber beneath. **Stage floor:** Wood, 2 in. (5 cm) thick. **Height of the stage above the main floor:** 35.5 in. (0.8 m). **Carpet:** None. *Seats:* Front of backrest and top of seat upholstered; rear of the backrest is cloth covered; underside of the seat is perforated.

ARCHITECT: P. Hougaard Nielsen and C. J. Norgaard Pedersen with Lars Møller and Brite Rørbaek responsible. ACOUSTICAL CONSULTANT: V. Lassen Jordan and Niels V. Jordan. PHOTOGRAPHS: K. H. Petersen. REFERENCES: Arkitektur, DK, No. 8, 1982.

TECHNICAL DETAILS

V = 494,200 ft³ (14,000 m³)	S_a = 5,900 ft² (548 m²)	S_A = 7,000 ft² (651 m²)
S_o = 1,894 ft² (176 m²)	S_T = 8,900 ft² (827 m²)	N = 1,320
H = 42 ft (12.8 m)	W = 91 ft (27.7 m)	L = 108 ft (32.9 m)
D = 102 ft (31.1 m)	V/S_T = 55.5 ft (16.9 m)	V/S_A = 70.6 ft (21.5 m)
V/N = 374 ft³ (10.6 m³)	S_A/N = 5.30 ft² (0.493 m²)	S_a/N = 4.47 ft² (0.415 m²)
H/W = 0.46	L/W = 1.187	ITDG = 40 msec

NOTE: The terminology is explained in Appendix 1.

41

Birmingham

Symphony Hall

The City of Birmingham has reestablished itself as a major European city through its Broad Street Development with a quality convention center and Symphony Hall which opened in 1991. No less an achievement was the engagement of Simon Rattle as Principal Conductor and Artistic Advisor of the Birmingham Symphony Orchestra, who quickly raised it to world class level.

The entrance to the hall is by way of a mall, a tall, fully glazed street, which is dominated on one side by Symphony Hall, with its curved walls and four cascading foyer levels that provide a dazzling spectacle of light and color at nighttime.

On entering the hall through the doors near the rear, one is struck by the beautiful interior, with intimate warm soft red tones and generous use of wood, stainless steel, and granite. The hall is a combination of a classical rectangular hall with the rear third curved and multi-tiered like an opera house. Seating 2,211, including the choir space behind the orchestra, it provides ample wall areas, not widely separated, for early, lateral reflections necessary to an intimate sound. The strong lines of the balconies that sweep to the front of the room draw one's eyes to focus on the pipe organ.

A special acoustical feature of the hall is the reverberation chamber. It is accessed by mechanized concrete doors that surround the organ and also are located on the upper side walls. These doors can be opened in any number to create variable reverberation times that can be matched to the type of music or to other uses of the hall. To provide adjustability to the clarity of the sound a large wood canopy is provided over the orchestra and the front of the audience, which can be raised and lowered. The reverberation also can be reduced by a series of 3-in.-deep sound-absorbing panels, mounted on rails, which can be moved in and out of the hall. The normal occupied mid-frequency reverberation time is 1.85 sec.

It is difficult for a one-time auditor to describe the acoustics because of the great variability that can be provided in both the reverberation and the clarity, any

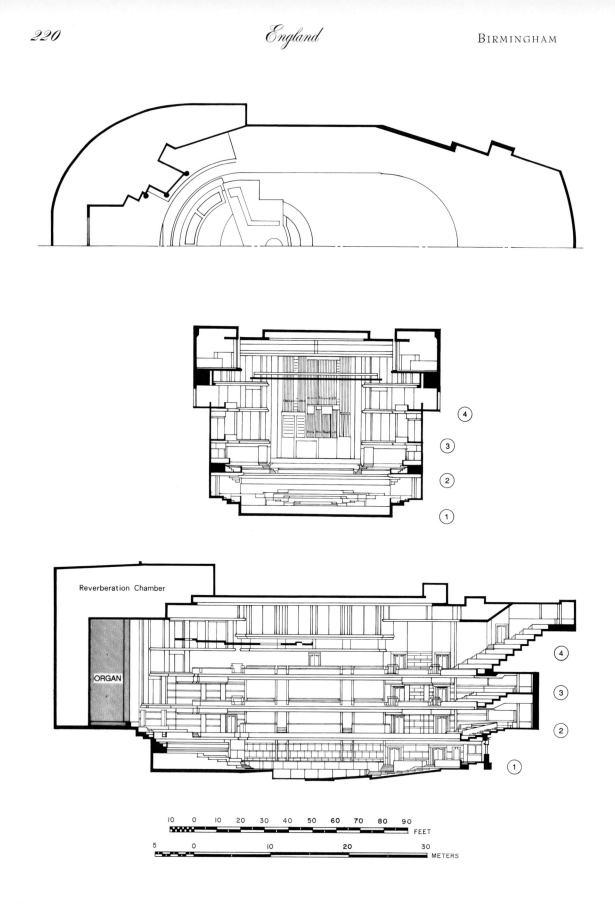

Reverberation Chamber

ORGAN

④
③
②
①

④
③
②
①

10   0   10   20   30   40   50   60   70   80   90
                                              FEET

5   0   10   20   30
                                              METERS

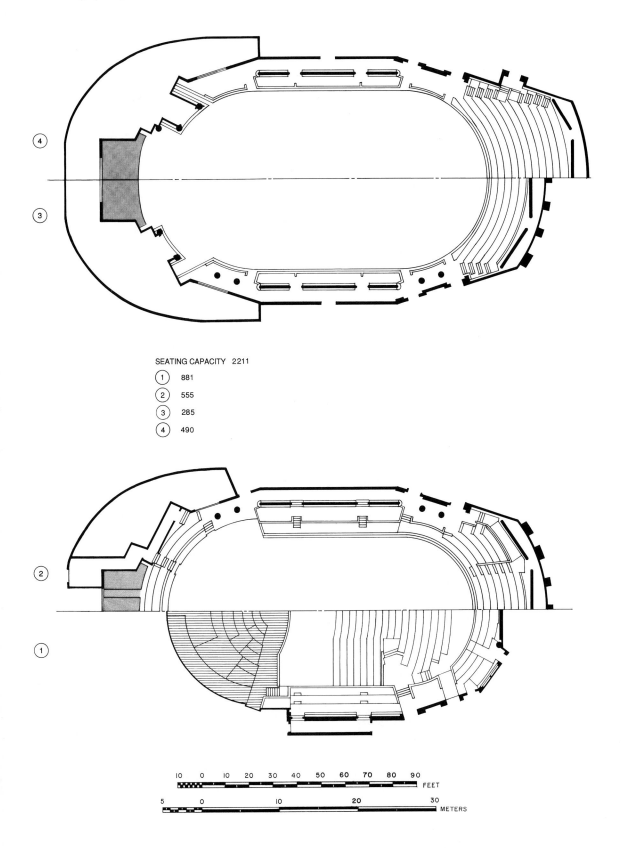

SEATING CAPACITY 2211

(1)	881
(2)	555
(3)	285
(4)	490

10 0 10 20 30 40 50 60 70 80 90
FEET

5 0 10 20 30
METERS

combination of which might be in use during that visit. Critical reviews have mostly been ecstatic. One reviewer complained of "not enough reverberance."

ARCHITECTURAL AND STRUCTURAL DETAILS

Uses: Concerts and conferences. **Ceiling:** Precast concrete with plaster cover; balcony soffits, three layers of 2-in. (5-cm) plasterboard. **Canopy:** 4.5-in. (11.4-cm) timber on steel frame, area 2,900 ft² (270 m²); lowest regularly used height is 30 ft (9 m). **Walls:** Wood veneered panels on battens infilled with sand/cement, plus large areas of plaster on structural walls. **Terrace fronts at main floor level:** 0.75–1.5-in. (2–4-cm) granite. **Variable absorption:** Motor-operated, fabric covered panels of 3-in. (7.5-cm) mineral wool, hung on wood frames, totaling 2,050 ft² (625 m²) area. **Floors:** 0.9-in. (2.2-cm) wood strip on 0.75-in. (1.9-cm) plywood on battens infilled with sand/cement screed over concrete base. **Carpet:** None. **Reverberation chambers:** 369,000 ft³ (10,300 m³) hard-surfaced concrete, containing operable two-layer velour drapes; opened by 20 remotely controlled doors of 6-in. (15-cm) concrete, totaling 2,100 ft² (195 m²). **Stage height:** 47 in. (1.2 m). **Seating:** Upholstered front of seat back and upper side seat.

ARCHITECT: Percy Thomas Partnership. ACOUSTICAL AND THEATER CONSULTANT: ARTEC Consultants, Inc. PHOTOGRAPHS: Courtesy of the architect. REFERENCE: "International Convention Centre Birmingham," *Architecture Today,* 17, April 1991.

TECHNICAL DETAILS

V = 883,000 ft³ (25,000 m³)	S_a = 11,100 ft² (1,031 m²)	S_A = 14,210 ft² (1,320 m²)
S_o = 3,000 ft² (279 m²)	S_T = 16,150 ft² (1,500 m²)	N = 2,211
H = 75 ft (22.9 m)	W = 90 ft (27.4 m)	L = 104 ft (31.7 m)
D = 132 ft (40.2 m)	V/S_T = 54.7 ft (16.67 m)	V/S_A = 62.1 ft (18.9 m)
V/N = 400 ft³ (11.3 m³)	S_A/N = 6.42 ft² (0.60 m²)	S_a/N = 5.02 ft² (0.466 m²)
H/W = 0.83	L/W = 1.15	ITDG = 27 msec
Note: $S_T = S_A$ + 1,940 ft² (180 m²); see Appendix 1 for definition of S_T.		

NOTE: The terminology is explained in Appendix 1.

42

Sussex

Glyndebourne Opera House

"*G*lyndebourne's new opera house marks the start of a fresh chapter in the life of one of England's most venerable summer institutions. The exquisitely crafted building is an impressive technical achievement which not only combines a sensuous materiality with an austere formalism, but also contains enough studied eccentricity to ensure the survival of Glyndebourne's indomitable spirit and cultural tradition." Thus starts a special edition of *The Architectural Review* (Summer 1994).

The Glyndebourne Summer Festival lasts for 13 weeks, followed in the autumn by a 3-week Sussex Season. The opera house seats 1,243 and was opened in May 1994. The exterior building is a large oval providing acoustical isolation from over-flying aircraft noise. Besides the usual foyers and spaces inside for musicians and staff, it surrounds a cylindrical audience hall, capped by its own protruding conical roof, and a rectangular fly (stage) tower that extends well above the main roof.

The circular space of the audience chamber presented a major challenge to the acoustical consultants. The advantage is that the farthest listener is close to the front of the stage, here about 100 ft (30.5 m). The danger is that the singers' voices will be unevenly distributed throughout the seating area. The reverberation time at mid-frequencies, fully occupied, is 1.25 sec, within the optimum range for Italian opera. Particularly important is the low noise level. Adverse focusing effects and the need for uniform distribution of the music were both addressed by the ceiling design, the shapes of the balcony fronts, the forms of the rear walls of the boxes, and the addition of convex panels at several levels. There is no carpet and only a small amount of sound-absorbing materials. A major acoustical achievement is the carrying of the balcony fronts to the front of the stage so that surfaces are available to augment, by reflection, the direct sound from the singers. *The New York Times* reported (Sunday, August 15, 1999), "So it's not a uniformly perfect acoustic. But in the main floor and in the front rows of the rings it's miraculous."

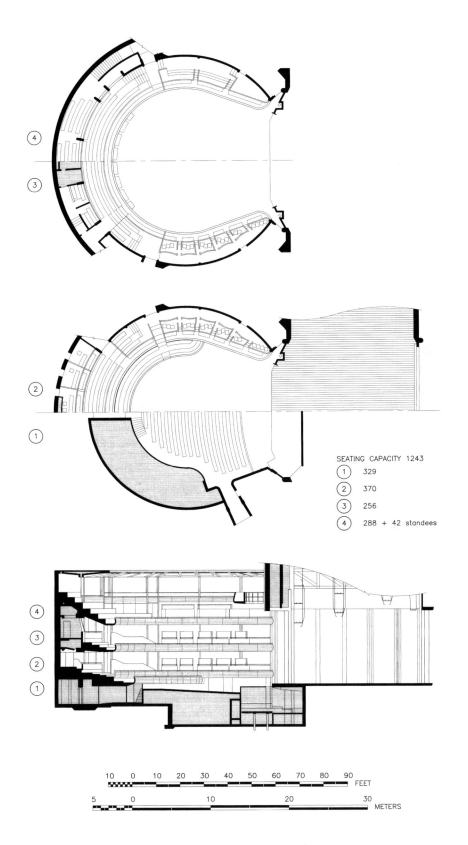

SEATING CAPACITY 1243

① 329
② 370
③ 256
④ 288 + 42 standees

10 0 10 20 30 40 50 60 70 80 90 FEET

5 0 10 20 30 METERS

ARCHITECTURAL AND STRUCTURAL DETAILS

Uses: Opera. *Center ceiling:* 6-in. (15-cm) precast concrete, height of ceiling chosen to achieve desired reverberation time; soffits of balconies also 6-in. (15-cm) precast concrete. **Sidewalls:** (separating auditorium from rest of building) two layers, 8.6-in. (22-cm) brickwork with 2-in. (5-cm) cavity; the fronts of balconies, rear walls of boxes, and upper walls are covered all or in part by 1-in. (2.5-cm) pitch pine. **Floors:** Soffits of balconies that form the floors of the balconies above are 1.8-in. (4.5-cm) wood; main floor is 2 in. (5 cm) on steel supports, perforated by ventilation holes which in turn are fed by concrete ducts into a plenum beneath the floor. **Pit:** Poured concrete sidewalls to which 1-in. (2.5-cm) wood is affixed on spacers; pit rail above floor level at first row of seats, 0.5-in. (1.25-cm) wood, made semi-perforate by a slatted treatment, with extra material provided if an unperforated pit rail is desired. **Carpets:** None. *Seating:* Molded plywood with high backs and slotted seat bottom; upholstering covers ca 95% of the front of seat back and is 1.2 in. (3 cm) thick; that covering the seat bottom is 1.8 in. (4.5 cm) thick; arms not upholstered; ventilation from holes in floor is expelled through a perforated cylinder adjacent to the pedestal support. **Noise level:** PNC-15

ARCHITECT: Michael, Hopkins & Partners. ACOUSTICAL CONSULTANT: Arup Acoustics. REFERENCES: *The Architectural Review,* Special edition, Summer 1994. R. Harris, "The acoustic design of the Glyndebourne Opera House," *Proc. IOA International Auditorium Conference,* England (1995) (www.ioa.org.uk).

TECHNICAL DETAILS

V = 275,000 ft³ (7,790 m³)	S_a = 6,004 ft² (558 m²)	S_A = 7,543 ft² (701 m²)
S_{pit} = 1,173 ft² (109 m²)	S_{proc} = 1,612 ft² (150 m²)	S_T = 10,328 ft² (960 m²)
N = 1243	H = 62 ft (18.9 m)	W = 25 ft (7.6 m)
L = 89 ft (27.1 m)	D = 96 ft (29.3 m)	V/S_T = 26.6 ft (8.11 m)
V/S_A = 36.4 ft (11.1 m)	V/N = 221 ft³ (6.27 m²)	S_A/N = 6.07 ft² (0.564)
S_a/N = 4.83 ft² (0.449 m²)	H/W = 2.48 L/W = 3.56	ITDG = 20 msec

NOTE: The terminology is explained in Appendix I.

43

Liverpool

Philharmonic Hall

*P*hilharmonic Hall, now seating about 1,800, is pleasant, warm, and comfortable. The sidewalls and ceiling form a shell inside the basic construction composed of a series of sections, each about 15 ft (4.6 m) wide, that, in the back part of the hall, extend down to the balcony floor and in the front part to a level equal to that of the soffit of the balcony overhang. These sections are arranged in echelon, with the front edge of each section rolled back to leave a recess for lighting. An audience of over a hundred partially surrounds the orchestra.

Opened in 1939, Philharmonic Hall recently has undergone extensive renovations, without noticeably changing its visual appearance. The renovations include enlargement of the stage, replacement of the plaster sidewalls and ceiling surfaces with concrete, addition of a front row to the balcony, and the installation of quadratic residue diffusers QRD's (see Chapter 4) over the entire rear wall. The QRD's replace sound-absorbent, rock-wool batts that reduced the reverberation time, and they provide useful lateral reflections to those in the upper balcony. The new concrete interior reflects the bass sounds which were formerly absorbed by the thin plaster. Carpet was removed from under the seats and replaced with wood on concrete. New chairs have been installed that reduce the audience absorption. Although occupied reverberation time data are not available, it is estimated that the changes increase it about 0.1 sec at mid-frequencies.

In 1960, Gerald McDonald, then General Manager of the Royal Liverpool Philharmonic Society, said, "From the point of view of artists performing in the hall, the fact that there are only 613 [then] seats enclosed by a ring of most attractive boxes seating over a hundred, with a further audience of 943 [then] in the balcony, is unusual and stimulating. Its relatively low reverberation time and beautiful well-defined tone make it uniquely favorable for baroque and Classical music starting with Bach, for modern music, and for any music that is cerebral in nature or requires refinement of definition." The improved reverberation time should, today, make the

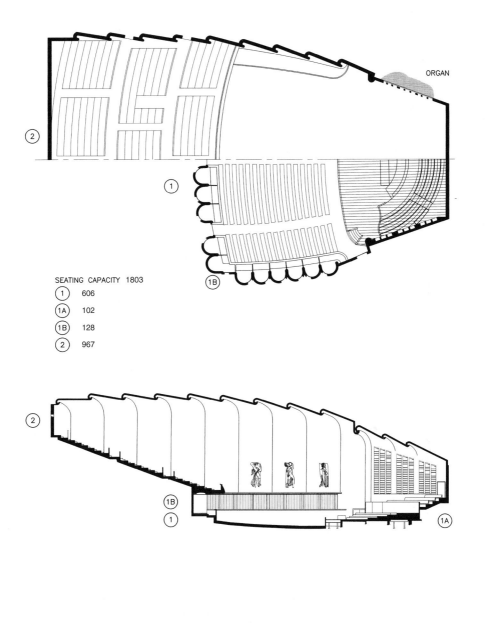

ORGAN

SEATING CAPACITY 1803

①   606

①A   102

①B   128

②   967

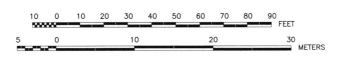

10   0   10   20   30   40   50   60   70   80   90
FEET

5   0   10   20   30
METERS

hall excellent for music of the late Classical period with greater fullness of tone and richness of bass.

[Note: The bottom photo is from 1960 and does not show the enlarged stage and the decreased seating on the main floor and behind the stage. The stage photo is current.]

ARCHITECTURAL AND STRUCTURAL DETAILS

Uses: Symphonic music, cinema, and speech. **Ceiling and sidewalls:** Concrete, divided above the boxes by lighting coves into 15-ft (4.6-m) sections; boxes are of wood. **Rear upper wall:** Partially covered by quadratic residue diffusers (QRDs). **Floors:** Wood over concrete. **Stage enclosure:** One side wall is occupied by a pipe-organ grille; the remaining surfaces are concrete. **Stage floor:** Boards on joists over airspace. **Stage height:** 30 in. (76 cm). **Seating:** Wooden construction with upholstering only on the seats and front of the backrests.

ORIGINAL ARCHITECT: Herbert J. Rowse. ORIGINAL ACOUSTICAL CONSULTANT: Hope Bagenal. RENOVATION ARCHITECT: Brock Carmichael Co. RENOVATION ACOUSTICAL CONSULTANT: Kirkegaard and Associates. PHOTOGRAPHS: Main hall, courtesy of the City of Liverpool; stage, Jonathan Keenan. REFERENCE: M. Barron, *Auditorium Acoustics and Architectural Design,* E & FN SPON, Chapman & Hall, New York and London (1993).

TECHNICAL DETAILS

V = 479,000 ft³ (13,570 m³)	S_a = 10,560 ft² (981 m²)	S_A = 13,720 ft² (1,275 m²)
S_o = 1,720 ft² (160 m²)	S_T = 15,440 ft² (1,435 m²)	N = 1,803
H = 46 ft (14 m)	W = 98 ft (30 m)	L = 94 ft (28.6 m)
D = 119 ft (36.3 m)	V/S_T = 31 ft (9.46 m)	V/S_A = 34.9 ft (10.64 m)
V/N = 266 ft³ (7.54 m³)	S_A/N = 7.62 ft² (0.71 m²)	S_a/N = 5.87 ft² (0.545 m²)
H/W = 0.47	L/W = 0.96	ITDG = 25 msec

NOTE: The terminology is explained in Appendix 1.

44

London

Barbican Concert Hall

The Barbican Concert Hall was originally planned as a conference facility with a design that boasted audience proximity to the stage and a relatively low reverberation time of 1.4 seconds. When it was decided to use it for concert music, it was too late to raise the ceiling to give it a higher reverberation time and to decrease the width and increase the length to make it closer to a shoebox shape. An architectural feature that has been a disaster acoustically is two 12-ft (3.7-m)-deep ceiling beams, necessary to support the roof and the outdoor plaza above, that run the full width of the hall. Until recently, these beams have prevented most of the ceiling from sending reflections of the sound to the upper tiers of the audience area. Listeners there complained of a lack of excitement and a feeling that they were isolated from the performance. Throughout the hall, the most serious deficiencies were lack of bass caused by thin wood paneling and echoes caused by delayed reflections from the side and angled rear walls.

The recent renovations have significantly improved the acoustics without appreciably increasing the reverberation time. The principal changes in 1994 and 2001 involve the ceilings over the stage and the audience chamber and the elimination of thin wood paneling wherever possible to strengthen the bass. Also, the stage has been enlarged. The canopy over the stage and a series of reflecting panels over the audience area are the most visible changes in 2001. The canopy and ceiling reflectors, integrated in design, comprise 35 burgundy-colored stainless-steel reflectors, the largest of which is nearly 66 ft (20 m) in length, each made up of between 4 and 17 panels. These reflecting panels can be adjusted to different heights and angles so as to distribute the sound equally to the players and to the forward part of the audience. Placed around and between the large trusses, new reflecting panels were designed to improve the sound for those seated farther back. The hall in its usual configuration seats 1,924.

The Guardian wrote, "The refurbished auditorium is a revelation . . . the new sense of immediacy and detail is compelling, and the physicality of the sound genuinely exciting."

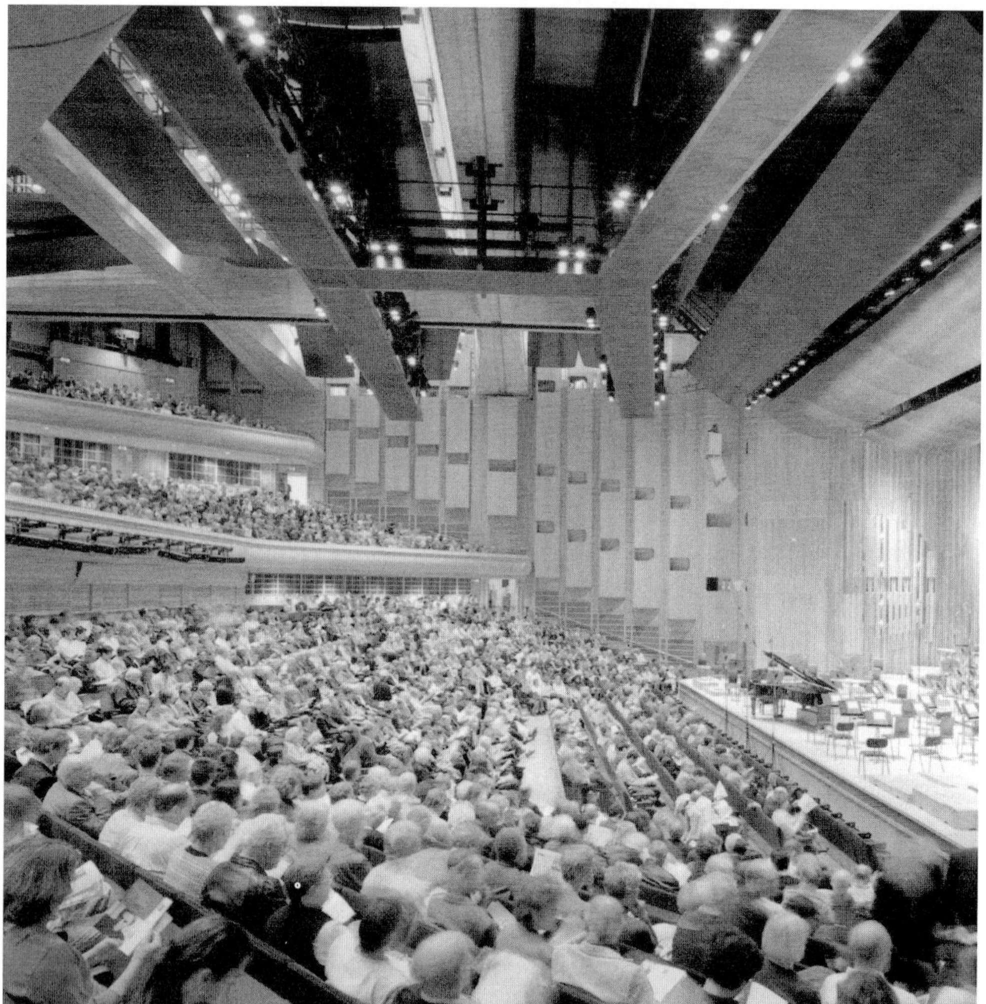

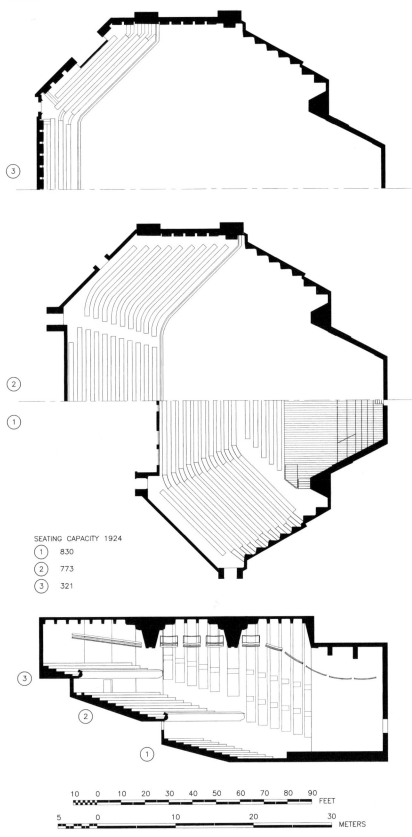

SEATING CAPACITY 1924

① 830

② 773

③ 321

FEET 10 0 10 20 30 40 50 60 70 80 90

METERS 5 0 10 20 30

For London, these improvements are vital, not only for concerts of the London Symphony Orchestra but also for concerts of many visiting performing groups.

ARCHITECTURAL AND STRUCTURAL DETAILS

Uses: Symphonic music, chamber music, recitals, and some conferences. **Ceiling:** Solid concrete, below which hang a number of large burgundy-colored stainless-steel reflecting panels that are adjustable in angle. **Sidewalls:** Upper sidewalls 2.83-in. (7.2-cm) single-layer, braced plywood panels on frames; lower sidewalls are the same but are selectively covered with felt 0.125 in. (0.32 cm) thick; rear walls are covered with diffusers like those in New York's Carnegie Hall. **Carpets:** None. **Stage enclosure:** Timber with sculptured surface over concrete. **Stage canopy:** A large portion of the 35 burgundy-colored stainless-steel panels in the room are located here, each of which is made up of 4 to 17 panels, and are placed in steeply ascending order above the stage. **Stage floor:** Partly wooden over airspace, and four elevators. **Stage height:** 3 ft (0.91 m). **Seating:** Top of seat bottom, front of backrest, and armrests are upholstered.

ORIGINAL CONSULTANT ARCHITECTS (1982): Chamberlain, Powell & Bonn. ORIGINAL ACOUSTICAL CONSULTANT: Hugh Creighton. ARCHITECT FOR RENOVATION (2001): Caruso St John. ACOUSTICAL CONSULTANT FOR RENOVATIONS (1994, 2001): Kirkegaard & Associates. PHOTOGRAPHS: Robert Moore, London.

TECHNICAL DETAILS

$V = 600,000$ ft³ (17,000 m³)	$S_a = 11,820$ ft² (1,100 m²)	$S_A = 13,610$ ft² (1,265 m²)
$S_o = 2,250$ ft² (209 m²)	$S_T = 15,550$ ft² (1,445 m²)	$N = 1,924$
$H = 47$ ft (14.3 m)	$W = 129$ ft (39.3 m)	$L = 90$ ft (27.4 m)
$D = 107$ ft (32.6 m)	$V/S_T = 38.6$ ft (11.76 m)	$V/S_A = 44.1$ ft (13.43 m)
$V/N = 312$ ft³ (8.84 m³)	$S_A/N = 7.07$ ft² (0.66 m²)	$S_a/N = 6.14$ ft² (0.57 m²)
$H/W = 0.364$	$L/W = 0.70$	ITDG = 25 msec

Note: $S_T = S_A + 1,940$ ft² (180 m²); see Appendix 1 for definition of S_T.

NOTE: The terminology is explained in Appendix 1.

45

London

Royal Albert Hall

On Wednesday, March 29, 1871, Queen Victoria, on one of her rare public appearances after the death of the Prince Consort, appeared in the royal box of the newly completed Royal Albert Hall of Arts and Sciences, surrounded by members of the Royal Family, to participate in the official opening. The official record (see References, Clark, p. 58) relates: The "Prince of Wales . . . began to read a welcoming address . . . speaking distinctly in a clear voice that could be heard in all parts of the building; in many parts it could be heard twice, a curious echo bringing a repetition of one sentence as the next was begun." Thus began 100 years of experiments in an attempt to transform into a concert hall a space that is much too large ever to be fully successful for this purpose.

The Royal Albert Hall holds 5,222 persons seated plus about 500 standees. It is nearly elliptical in shape with a cubic volume of about 3,060,000 ft³ (86,650 m³), four times as great as the largest regular well-liked concert hall in the world. Its uses are many—pageants, lectures, exhibitions, choral presentations, symphony concerts, proms (pops) concerts, solo recitals, even athletic events. One of the buildings most famous features is the great pipe organ, installed in 1871, and rebuilt for a second official presentation in 1934. The Royal Albert Hall fulfills London's need for a hall with a large seating capacity.

The principal acoustical problems of the Royal Albert Hall, particularly before the 1970 additions, arise from its size. They are (1) a decrease in loudness of the direct and early sound emanating from the performers as it travels such large distances in the hall (consider a voice trying to fill nearly six times more cubic volume of air than there is in a typical European concert hall); and (2) echoes, owing to the return of very-long-delayed reflections from distant surfaces back to the front of the hall.

To alleviate the echoes, but further weakening the loudness of the sound, a large velarium of cloth weighing 1.25 tons (1,136 kg) was hung and remained in place until 1949. This velarium was raised and lowered from time to time in an

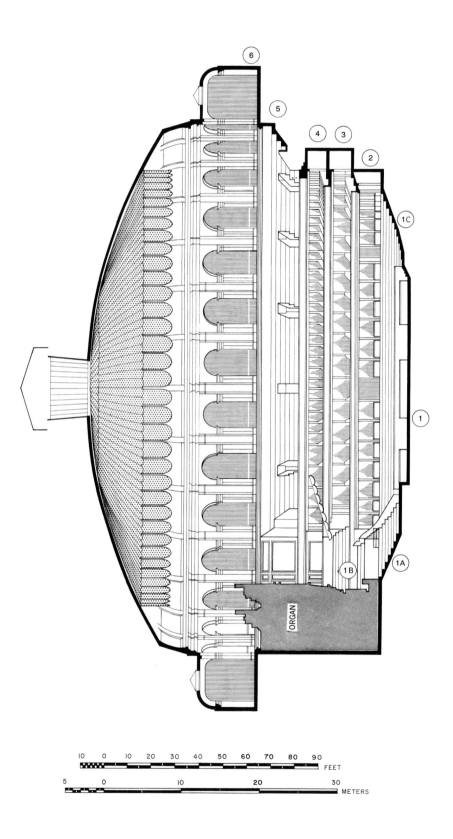

ORGAN

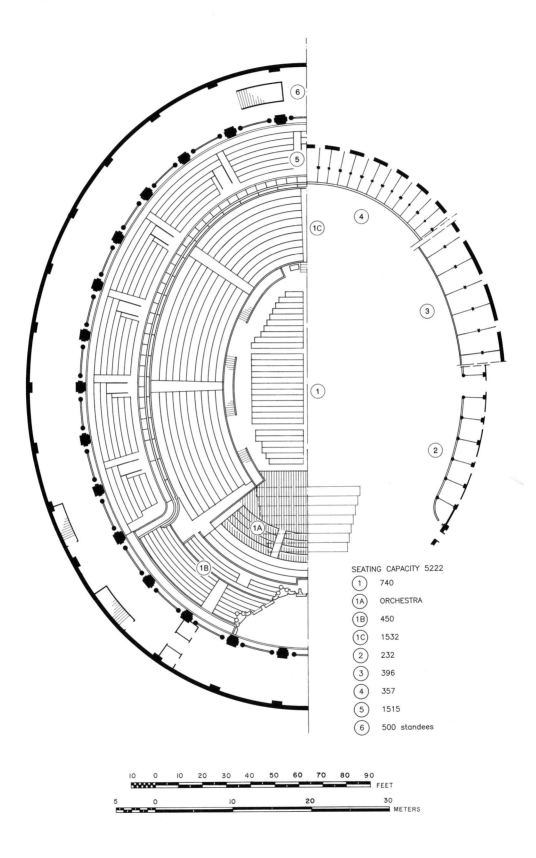

SEATING CAPACITY 5222

(1)	740
(1A)	ORCHESTRA
(1B)	450
(1C)	1532
(2)	232
(3)	396
(4)	357
(5)	1515
(6)	500 standees

```
10   0   10  20  30  40  50  60  70  80  90
                                              FEET

5    0        10              20         30
                                              METERS
```

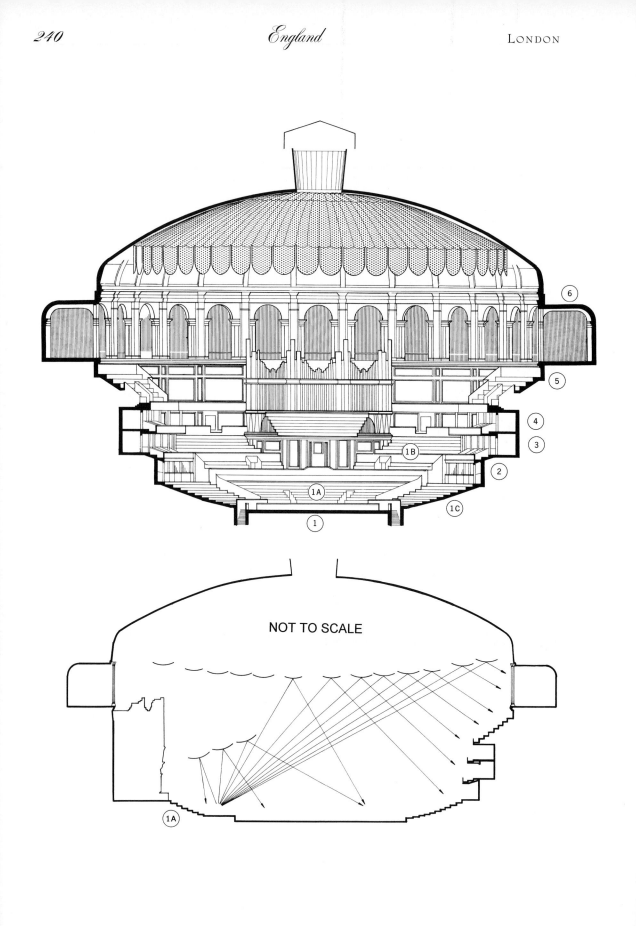

NOT TO SCALE

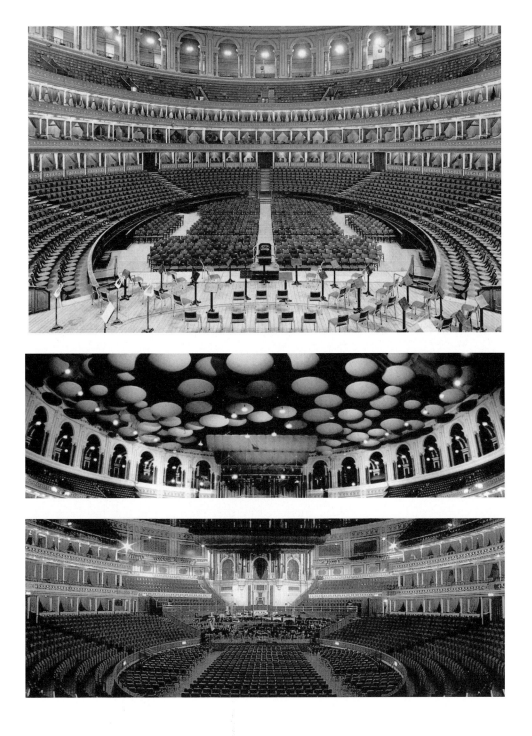

effort to improve the acoustics. In 1934, we read, "The echo persisted and there was even some suggestion that it had grown worse."

I take the following information on the Royal Albert Hall from M. Barron (see References, 1993, pp. 117–125):

> It was only in 1968–1970 that the echo problem was finally suppressed by suspending 134 "flying saucer" diffusers at the level of the Gallery ceiling [see the photograph and sketch] . . . Two aspects were also dealt with during this period. . . . The reverberation time [that] was particularly long at mid-frequencies . . . was rendered more uniform with frequency . . . by placing absorbent which was most efficient at these frequencies on the upper side of the saucers . . . This location for absorbent is particularly apt because it increases the effect of the saucers by absorbing some remaining sound reflected from the dome. Another problem with a large space is maintaining sufficient early reflections when some seats are inevitably remote from useful surfaces. The reflections from the suspended saucers are very valuable at the upper seating levels. A reflector above the stage, as first used in 1941, has been retained to provide [early sound] reflections particularly to the [main floor] and [lower side] seating. [The middle photo shows the saucers. Recent revisions have changed the front, as shown in the lower photo.]

The measured occupied reverberation time (measurements by M. Barron in 1982) is 2.4 sec at mid-frequencies. This is not disastrous in a room this size. The more serious problem is the weakness of orchestral sound, partly because it is absorbed to some extent by the upper surfaces of the saucers, and partly because it has to be distributed over 5,080–6,000 heads, compared to 1,680 in Vienna's Grosser Musikvereinssaal. Perhaps, as electronic sound reinforcement becomes more common, the level can be increased by that means. The hall has been very successful in presenting popular (pops or proms) concerts—audiences of over 4,000 are common.

I heard Tchaikovsky's *1812 Overture* there. The great organ sounded out at the climax and the "cannon" boomed forth. The chimes clanged loudly through their part and, finally, in combination with the military fanfare theme, the *Russian Imperial Anthem* thundered forth. The general reverberation swelled with the increased vigor of the composition. Above all the great organ sounded like the voice of Jupiter. The audience was left breathless and tingling. It is for these moments of ecstasy that Albert Hall continues to exist.

ARCHITECTURAL AND STRUCTURAL DETAILS

Uses: General purpose. **Ceiling:** Roof over balcony is of plaster directly on the structure; fluted inner dome with perforated inner skin backed by mineral wool, plus 134 hanging sound-diffusing reflectors—"flying saucers" which are glass reinforced plastic dishes with 1.5-in. (3.8-cm) glass fiber blanket on top. **Walls:** Generally plastered, but the back of the boxes are of thin wood. **Floors:** Arena floor is 1 in. (2.5 cm) on wooden joists over large airspace; balcony is of concrete; box floors are of wood. **Stage enclosure:** Glass-reinforced canopy over orchestra. **Stage floor:** Thick wood over airspace. **Stage height:** 40 in. (102 cm). **Carpets:** Often used on arena floor. **Seating:** Arena and main floor surrounding, tops of seats and fronts of backrests are upholstered with cloth covering—all other surfaces are of wood; boxes, loose occasional chairs upholstered the same as for arena; balcony, theater-type tip-up chairs upholstered in cloth; gallery, 300 wooden chairs for use in the standing areas.

DESIGNERS: Capt. Francis Fowke (who died during design in 1865) and Lt. Col. H. Y. D. Scott (who succeeded Fowke and continued as Director of Works under the supervision of a "Committee on Advice"). PHOTOGRAPHS: Courtesy of M. Barron and Royal Albert Hall. REFERENCES: Ronald W. Clark, *The Royal Albert Hall,* Hamish Hamilton, London, 1958; M. Barron, *Auditorium Acoustics and Architectural Design,* FN SPON, Chapman & Hall, New York and London (1993).

TECHNICAL DETAILS

V = 3,060,000 ft³ (86,650 m³)	S_a = 29,000 ft² (2,700 m²)	S_A = 37,800 ft² (3,512 m²)
S_o = 1,900 ft² (176 m²)	S_T = 39,700 ft² (3,688 m²)	N = 5,222 + 500 standees
H = 118 ft (36 m)	W = 155 ft (47 m)	L = 146 ft (44.5 m)
D = 140 ft (42.7 m)	V/S_T = 77.1 ft (23.5 m)	V/S_A = 81 ft (24.7 m)
V/N = 586 ft³ (16.6 m³)	S_A/N = 7.24 ft² (0.672 m²)	S_a/N = 5.55 ft² (0.517 m²)
H/W = 0.76	L/W = 0.94	ITDG = 15 msec

NOTE: The terminology is explained in Appendix 1.

46

London

Royal Festival Hall

The Royal Festival Hall, opened in 1951, is said to host more important musical events annually than any other large concert hall. The architecture is a triumph, a fascinating mosaic of stairways, carpets, lighting, and interior vistas. The approach up the stairs and through foyers is so cleverly made and so beautiful that it promises an evening of excitement. The designers have achieved beauty through proportion and color and admirably contrived lighting. Dominant features in the hall itself are a large three-part canopy over the stage, projecting box fronts on the sidewalls, a large pipe organ, and a steeply raked main floor.

From 1900 to 1948 only a handful of large halls had been built in the world, and most of them before modern acoustical measuring equipment was available. Thus Festival Hall, seating 2,901, was the site for most of the experiments in concert hall acoustics in the first half of the twentieth century.

The general consensus by all types of listeners is that the hall is "too dry," not reverberant enough, particularly at low frequencies, and that the bass tone is weak. The primary cause was the lack of technical information on how much sound an audience absorbs when seated in modern theater chairs. The consultants used an absorbing indicator of 0.33 per person, compared to today's accepted 0.57 per person. The result: a reverberation time at 500 Hz of 1.5 sec, compared to the design goal of 2.2 sec. At the low frequencies the reduced reverberation time is also exacerbated by the open area of the boxes, the lightweight ceiling, and other sound-absorbing surfaces.

My analysis, backed by interviews, is that the definition is excellent and the hall is very good for piano, chamber music, and modern music. The hall is not as effective for music of the late Classical and Romantic periods. The lack of bass is the most serious problem.

Several well-known consultants and architects have been engaged to make recommendations for major revisions in Festival Hall. It appears that the reverber-

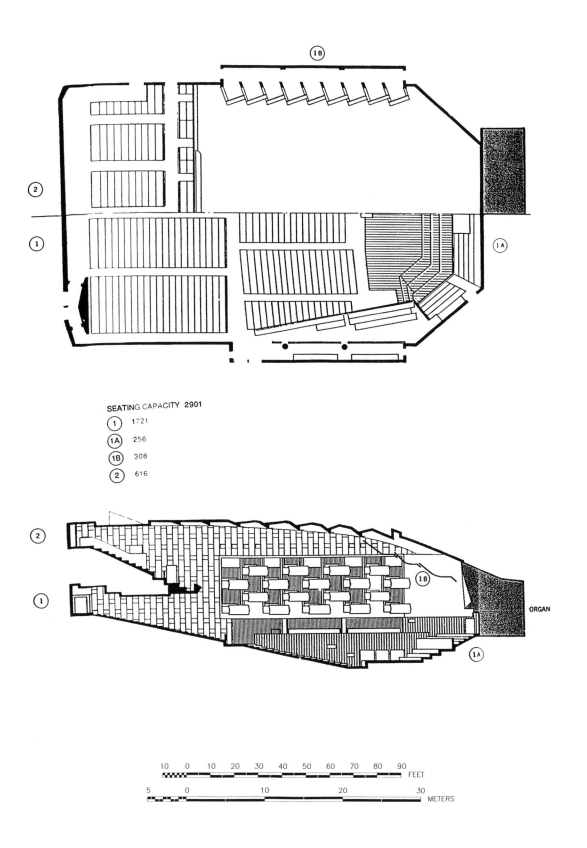

SEATING CAPACITY 2901

(1)	1721
(1A)	256
(1B)	308
(2)	616

ORGAN

```
10   0   10   20   30   40   50   60   70   80   90
                                                    FEET

5    0              10              20              30
                                                    METERS
```

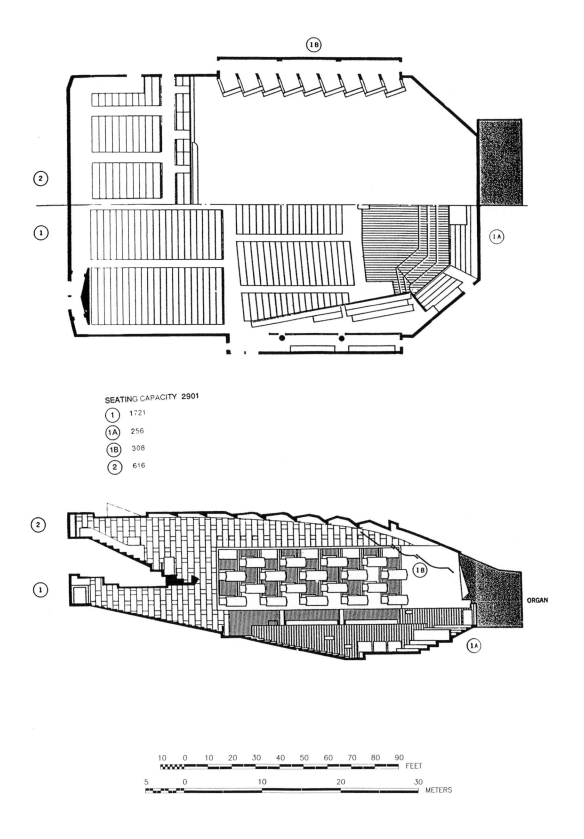

SEATING CAPACITY 2901

1   1721
1A   256
1B   308
2   616

ORGAN

| 10 | 0 | 10 | 20 | 30 | 40 | 50 | 60 | 70 | 80 | 90 | FEET |
| 5 | 0 | | 10 | | 20 | | 30 | METERS |

ation time can be increased somewhat and that the bass can be strengthened. As this is written, it is said that work can begin when funding becomes available.

ARCHITECTURAL AND STRUCTURAL DETAILS

Uses: Classical concerts, plus summer ballet season. **Ceiling:** 2–3-in. (5–7-cm) plaster, prefabricated with wood lath and rags for binding. **Side walls and rear walls:** 0.4-in. (1-cm) elm plywood with 3–4-in. (7–10-cm) airspace behind; the doors, the rears of the side balconies, and the rear walls are covered with leather, backed by rock wool with several inches of airspace behind; exposed wood wool, 2 in. (5 cm) thick is used in 33% of the cornice regions. **Floors:** Principally thin, hard, compressed cork on concrete. **Carpets:** Transverse aisles carpeted. **Stage floor:** Wooden boards over airspace. **Stage height:** 30 in. (76 cm). **Sound-absorbing materials:** See walls and ceiling. **Seating:** Molded wooden seat back, upholstered front of seat back and top of seat bottom with porous fabric; underseat, perforated wood; choir seats partly upholstered in leather; armrests upholstered in leather.

ARCHITECT: London County Council Architects' Department, R. H. Mathews, J. L. Martin with E. Williams and P. M. Moro. ACOUSTICAL CONSULTANT: Hope Bagenal with Building Research Station (H. R. Humphreys, P. H. Parkin, and W. A. Allen). PHOTOGRAPHS: Courtesy of Royal Festival Hall. REFERENCES: P. H. Parkin, W. A. Allen, H. J. Purkis, and W. E. Scholes, *Acustica* **3**, 1–21 (1953); E. Meyer and R. Thiele, *Acustica,* **6**, 425–444 (1956); M. Barron, *Auditorium Acoustics and Architectural Design,* E & FN Spon, Chapman and Hall, New York and London (1993).

TECHNICAL DETAILS

$V = 775{,}000$ ft³ (21,950 m³)	$S_a = 16{,}500$ ft² (1,540 m²)	$S_A = 21{,}230$ ft² (1,972 m²)
$S_o = 1{,}860$ ft² (173 m²)	$S_T = 23{,}090$ ft² (2,145 m²)	$N = 2{,}901$ (with box seats)
$H = 50$ ft (15.2 m)	$W = 106$ ft (32.3 m)	$L = 121$ ft (36.8 m)
$D = 126$ ft (38.4 m)	$V/S_T = 33.6$ ft (10.2 m)	$V/S_A = 36.5$ ft (11.1 m)
$V/N = 267$ ft³ (7.56 m³)	$S_A/N = 7.32$ ft² (0.68 m²)	$S_a/N = 5.69$ ft² (0.53 m²)
$H/W = 0.47$	$L/W = 1.14$	ITDG $= 34$ msec

NOTE: The terminology is explained in Appendix 1.

47

London

Royal Opera House

*T*he Royal Opera House, rehabilitated in 2000 and seating 2,157 plus 108 standees, is one of the world's most important centers for opera. The site was originally a convent garden owned by the Abbey of Westminster. In 1630–33, the garden was laid out as a residential square according to the design of the English landscape architect Inigo Jones. In 1632 a theater for plays was opened where the ROH now stands. In 1734–1737 George Frideric Handel emphasized opera. In 1847 it became the Royal Italian Opera House. After several fires, the theater was rebuilt for the third time in 1858 and is now the Royal Opera House.

London profits from the Royal Opera House's long heritage and interest in producing all that is internationally good in opera, whether Italian, French, Austrian, German, English, Slavic, or other. It has presented the world's finest singers and has boasted of celebrated music directors—Frederich Gye, Augustus Harris, Sir Thomas Beecham, Bruno Walter, Rafael Kubelik, Georg Solti, Colin Davis, and Bernard Haitink. The Royal Opera House is a boon to artists because London appreciates opera and the house is small enough to be intimate and truly enjoyable.

Also home to the Royal Ballet, the House is in use nearly every day and evening the year around. The orchestra of the Royal Opera House, with musicians of the highest quality, not only accompanies the opera and ballet performances, but it performs on concert stages. During the recent two years when the House was closed for rehab, the orchestra appeared at a number of the world's leading concert venues.

The acoustical quality of the house has improved markedly as a result of the renovations. The bass, which formerly was diminished by thin wood, is now in balance. Nearly all carpet has been eliminated and the chairs of modest, but comfortable, upholstering also act to preserve the bass. As is true in other houses of similar design, singers are not as loud when they move upstage from the proscenium line. Special reflecting surfaces on the audience side of the proscenium would help this problem, but they would have to be part of the basic architectural design. All in all, I rate this house among the top ten of the world's prominent opera houses.

48

Manchester

Bridgewater Hall

The motivations for constructing the Bridgewater Hall were to provide a new home for the Manchester Hallé orchestra and to upgrade a lame site made available by the Central Manchester Development Corporation. To one side of the hall lies the historic center of the city, another side faces a vast exhibition hall, and the remaining sides front on an area that rapidly has become a venue for offices, restaurants, and chic bars. A seven-year planning and construction period followed its conception, resulting in a design process that produced a unique architectural entity. Approaching from the city center, one comes upon a dramatic glass prow, sparkling when illuminated.

The foyers beyond the entrance provide a grand introduction to the auditorium, whose interior shape both suggests the rectangularity of a classic shoebox hall and the vineyard design of the Berlin Philharmonie. One's eyes are drawn toward four features. First, we see the impressive organ with its 5,000 pipes. In front of these, over the stage, are convex transparent sound reflectors through which the platform receives its illumination. One next sees the tall, spindly columns that support the ceiling. And, finally, one observes the unusual general hall lighting composed of ropes of low-intensity lamps. The hall opened in 1996.

The seating capacity is 2,357; no person is more than 123 ft (37.4 m) from the front of the stage. Owing to the steep slope of the vineyard seating, spectators have a full view of the orchestra, in line with the architect's wish that every instrument be visible; although there are a few side seats that do not see 100 percent of stage. The front of each tier provides necessary early lateral sound reflections toward various parts of the audience. The high ceiling provides the cubic volume needed to produce a mid-frequency reverberation tine of 2.0 sec, with the hall fully occupied. To promote diffusion of the reflected sound, irregularities are provided on the ceiling. The balcony fronts and the vertical walls are relatively smooth, even though the acoustical consultant recommended sound-diffusing irregularities.

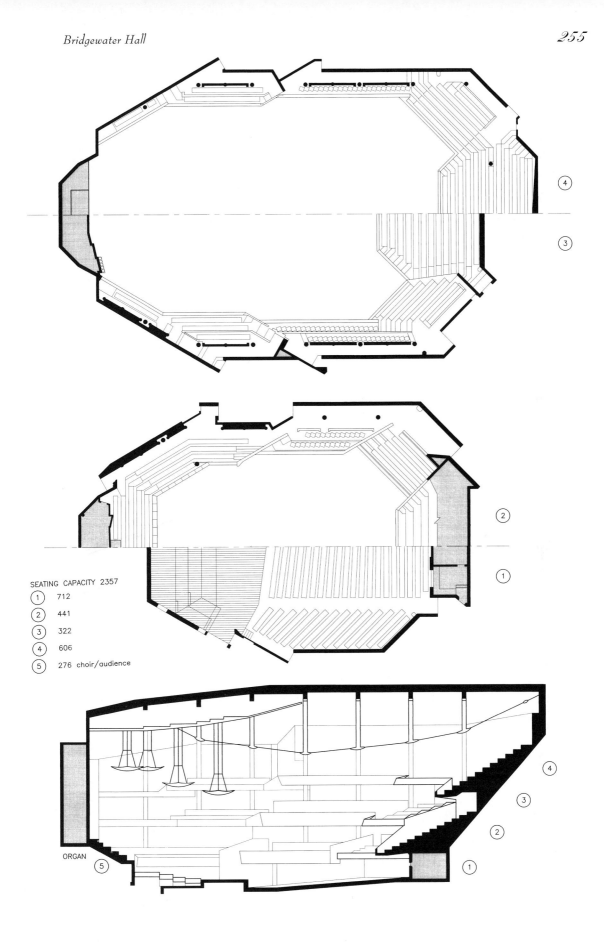

SEATING CAPACITY 2357

1 712
2 441
3 322
4 606
5 276 choir/audience

ORGAN

ARCHITECTURAL AND STRUCTURAL DETAILS

Uses: Primarily symphonic, choral, and organ music, but also chamber music and solos, and a host of amplified sound events. **Ceiling:** Pre-cast, sound-scattering concrete panels. **Side walls:** Inner wall plaster; balcony fronts, and reflectors, 1.2-in. (3-cm) glass-reinforced gypsum (veneer or paint). **Stage surround:** Stage enclosure formed by surround choir and terrace (vineyard) seating. **Floors:** Red oak wood imbedded in concrete. **Suspended canopy:** Array of curved glass panels (convex downward) over stage and choir seating, adjustable in height in sections. **Stage floor:** Tongue-and-groove boards over airspace with grille openings in wall just above the stage surface. **Stage height:** 35 in. (90 cm) above front-floor surface. **Carpet:** None. **Seats:** Plywood on metal frames; upholstering on front of seat back and top of seat bottom; arm rests wooden; underside of seats perforated. **Vibration isolation:** To isolate the hall from an adjacent light railway, the hall, weighing 36,000 tons, is supported on 3.5-Hz springs. **Noise level:** PNC 15.

ARCHITECT: Renton Howard Wood Levin. ACOUSTICAL CONSULTANT: Arup Acoustics. PHOTOGRAPHS: Dennis Gilbert. REFERENCES: *Proceedings of the Institute of Acoustics*, **19**, 129–133 (1997). *RIBA Profile* **103**, 1–29 (1996).

TECHNICAL DETAILS

V = 882,500 ft³ (25,000 m³)	S_a = 14,558 ft² (1,353 m²)	S_A = 17,334 ft² (1,611 m²)
S_o = 2,970 ft² (276 m²)	S_T = 19,271 ft² (1791 m²)	N = 2357
H = 77 ft (23.5 m)	W = 86 ft (26.2 m)	L = 97 ft (29.6 m)
D = 130 ft (39.6 m)	V/S_T = 45.8 ft (14 m)	V/S_A = 50.9 ft (15.5 m)
V/N = 374 ft³ (10.6 m³)	S_A/N = 7.35 ft² (0.683 m²)	S_a/N = 6.18 ft² (0.57 m²)
H/W = 0.90	L/W = 1.13	ITDG = 39 msec
Note: $S_T = S_A$ + 180 m²; see definition of S_T in Appendix 1.		

NOTE: The terminology is explained in Appendix 1.

Sibelius/talo

The Sibelius Hall in Lahti, Finland, is unique. It is the only concert hall built with wood as a basis for its construction. Even before an architect was engaged, the thought was motivated by Finland's celebration of the "Year of Wood in 1996." The Town of Lahti, the government, and the Finnish forestry industry funded the hall.

Even the process by which the architect was chosen was unique. The acoustical consultant was first engaged by the town with the charter that the hall had to be constructed strictly according to his instructions. The vote to proceed with the construction was taken by the Town Council in May 1998. The 1996 campaign was followed by the "Era of Wood in 1997–2000," whose patron was the Finnish government. With the Sibelius Hall as the centerpiece of timber engineering, wood is now said by the town "to have regained its position as an 'officially' recognized building material." The site is along beautiful Lake Vesijärvi, remote from the noise of a downtown area.

The Sibelius Hall is the home of Sinfonia Lahti and is used for congresses suited to a hall seating 1,250 (of which 150 seats are for audience or choir). The hall is basically rectangular, but interesting curves in the balcony fronts and the main floor seating section give it an inviting appearance. Ground-floor-to-roof and full front-to-aft reverberation chambers are located at the sides of the hall with 188 controlled doors coupling them to the hall at all levels. The visible exterior is all glass with no floor levels showing. Inside this are the outer sidewalls of the reverberation chambers. These walls are made of two thick layers of wood separated by a sand-filled space 7 in. (18 cm) thick. The wooden panels of this sandwich are made from a type of laminated-veneer lumber called "Kerto-Wood." An adjustable-height, three-part "Kerto-Wood" canopy hangs above the stage.

Being a small hall with a proven shape and every effort made to preserve the bass, it is not surprising that the opening night reviews by a number of European music critics were favorable.

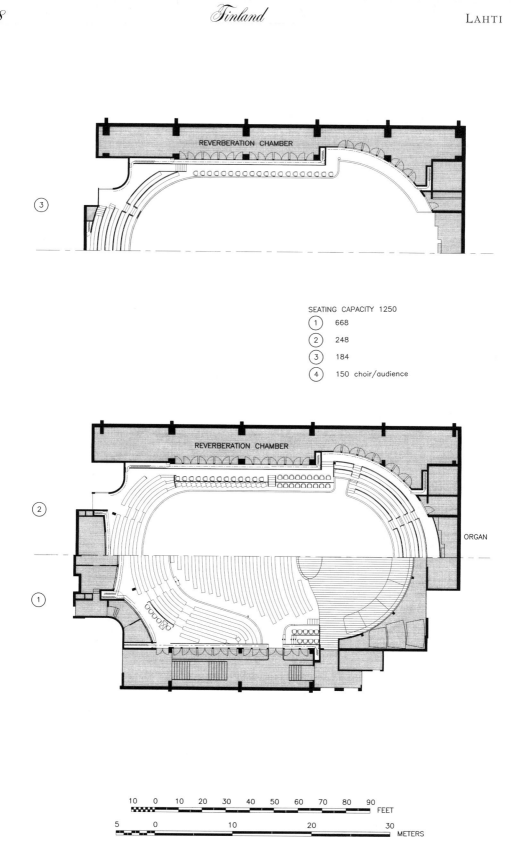

SEATING CAPACITY 1250

① 668
② 248
③ 184
④ 150 choir/audience

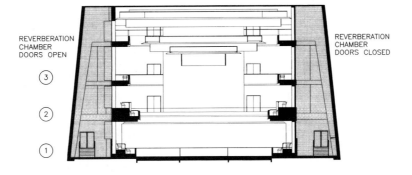

REVERBERATION
CHAMBER
DOORS OPEN

REVERBERATION
CHAMBER
DOORS CLOSED

③

②

①

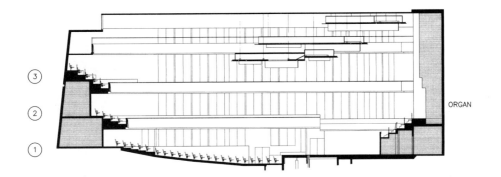

③

②

①

ORGAN

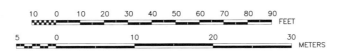

ARCHITECTURAL AND STRUCTURAL DETAILS

Uses: Symphonic and chamber music, soloists, recordings, conferences. **Ceiling:** Laminated veneer wood ("Kerto-Wood") 2.7 in. (6.9 cm) thick, over 30-cm-deep airspace, over another laminated wood layer 2 in. (5.1 cm) thick. **Hanging canopies:** "Kerto-Wood" 2.7 in. (6.9 cm) thick with applied layer of 0.75-in. (1.9-cm)-thick plywood. Range of travel between 39 ft (12 m) and 49 ft (15 m) above stage floor. **Balcony undersides:** Gypsum wallboard 0.5 in. (1.2 cm) thick over airspace. **Sidewalls:** Outer sidewalls of reverberation chambers made from a sandwich of 2.7 in. (6.9 cm) "Kerto-Wood," layer of sand 7 in. (18 cm) thick and 2 in. (5.1 cm) "Kerto-Wood." The 188 doors to chambers also of same thick veneer wood. Most other interior surfaces same. **Balcony faces:** Veneer plywood 0.35 in. (0.9 cm) supported by wood members 15 in. (50 cm) on center. **Stage walls:** Semi-circular arrangement of reversible panels—reflective side 1-in. (2.5-cm) plywood, and, absorptive side 2-in. (5-cm)-thick glass fiber boards covered with perforated metal. **Main floor:** 0.5-in. (1.2-cm) parquet bonded to 0.25-in. (0.6-cm) cork bonded to concrete. **Balcony floors:** Same, except 2.7-in. (6.9-cm) "Kerto-Wood" instead of concrete. **Stage floor:** 1-in. (2.5-cm) tongue-and-groove flooring over 1-in. (2.5-cm) substrate, supported by 2 × 6-in. (5 × 15-cm) timbers on edge, on neoprene pads, on concrete. **Height of the stage above the audience level:** 39 in. (1 m). **Seating:** Molded plywood back and seat with 1.5-in. (3.7-cm) padding covered with porous fabric; wooden armrests.

ARCHITECT: Hannu Tikka and Kimmo Lintula of the firm Arkkitehtityöhuone, APRT ACOUSTICAL CONSULTANT: ARTEC Consultants Inc. THEATER PLANNING: ARTEC Consultants, Inc. PHOTOGRAPHS: 2000 Voitto Niemelä.

TECHNICAL DETAILS

V = 547,150 ft³ (15,500 m³)*	S_a = 6,100ft² (567 m²)	S_A = 8,156 ft² (758 m²)
S_o = 1,950 ft² (181 m²)	S_T = 10,110 ft² (940 m²)	N = 1250
H = 62 ft (19 m)	W = 75 ft (doors closed)	L = 138 ft (42 m)
D = 98 ft (30 m)	V/S_T = 54.2 ft (16.5 m)	V/S_A = 67.1 ft (20.4 m)
V/N = 438 ft³ (12.4 m³)	S_A/N = 6.52 ft² (0.61 m²)	S_a/N = 4.88 ft² (0.45 m²)
H/W = 0.83 (closed)	L/W = 1.84 (closed)	ITDG = 20 msec

*Note: Not including 247,000 ft³ (7000 m³) periphery reverberation chamber.

NOTE: The terminology is explained in Appendix 1.

50

Paris

Opéra Bastille

The Opéra Bastille is larger than all but a few opera houses in the world, three of the biggest being those in Milan (2,289), Buenos Aires (2,487), and New York (3,816). The Bastille seats 2,700 and it is unique in Europe because of the government's requirement that all spectators have a full frontal view of the stage. To meet this demand, the architect had to reject the multi-ringed house made so popular by the Teatro Alla Scala of Milan, inaugurated in 1778, and instead chose a two-balcony design, with moderate overhangs. The Opéra Bastille opened in July 1989 to some troublesome reviews.

The acoustical consultants designed the means to provide the various parts of the audience with the necessary early lateral sound reflections, including two-dimensional sound-diffusing surfaces on the sides and above the proscenium, and on the front sidewall, and curved box fronts that spread the incident sound over large areas. The reverberation time, fully occupied, at mid-frequencies is 1.55 sec, about the same as that for the highly regarded Buenos Aires Opera Colón, and well above the 1.3-sec average for the best known European houses and for the Paris Opéra Garnier (with a RT of 1.1 sec).

As usual with large halls, the sound is better in the balcony, particularly the rear part of the second balcony, than on the main floor center because of the higher density of early lateral reflections farther back. The reverberation is heard in all parts of the house, but more in the rear part. The measured early sound energy compared to the later reverberant energy is nearly optimum for opera promoting clarity of the singing voice.

One early review speaks of the unevenness of the sound in the audience as a singer moves back from the proscenium line. Some say the bass sound is weaker than desired. The musicians indicate that communication among the orchestral sections in the pit is not ideal and that ensemble playing is difficult. Various experiments to improve the acoustics in the pit have been performed by rearranging

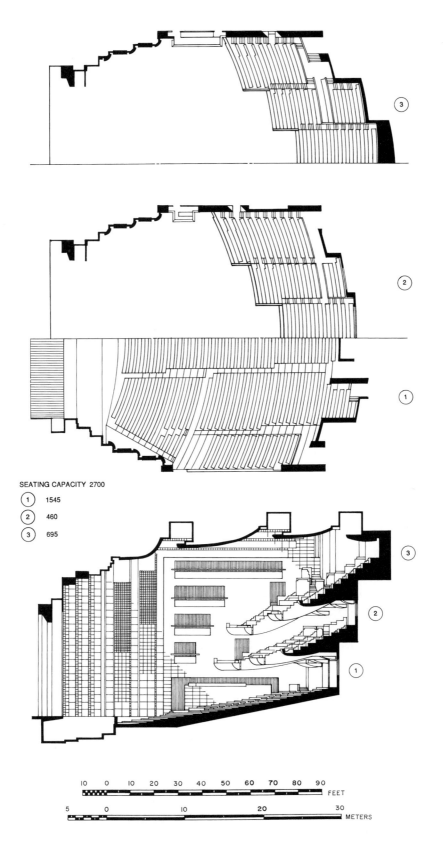

SEATING CAPACITY 2700

1 1545
2 460
3 695

the orchestra and making use of the full flexibility of the pit, whose width, depth, and overhangs can be changed.

The original plan was to have all opera in the Bastille and all ballet in Garnier. For some time now, the houses share nearly 50/50 in both. Certainly, the Opéra Garnier with its small size, plush interior, gold trimmings, and great heritage will always be loved by Parisians and will remain a formidable competitor.

ARCHITECTURAL AND STRUCTURAL DETAILS

Uses: Opera and ballet, 90%; Symphonic concerts, 10%. **Ceiling:** Clamped glass tiles, 0.25 in (6 mm) thick. **Walls:** Side and rear walls, granite 1.2 in. (3 cm) thick, fastened to concrete wall by hooks, over an air gap filled with mineral wool 1–2 in. (2.5–5 cm) deep; two-dimensional QRD diffuser (see Chapter 4) used for fixed part of the vertical proscenium area and on the front side walls. **Movable proscenium:** Diffusing wood structures hidden behind highly transparent perforated metal sheet. **Floors:** Wood cemented to concrete below seats; stone tiles in the aisles of the main floor; carpet (no underpad) in the aisles of the balconies. **Sound-absorbing materials:** None except glass ceiling planned for low-frequency absorption. **Pit:** Size is variable, all surfaces are wood—0.8 in. (2 cm) for pit walls, 2 in. (5 cm) air slots in floor. **Seating:** Rear of seat back is wood; upperside seat bottom is upholstered with polyurethane foam and covered with porous velvet fabric; underside, perforated wood.

ARCHITECT: Carlos Ott. ACOUSTICAL CONSULTANTS: Mueller-BBM. ASSOCIATED ACOUSTICAL CONSULTANTS: Centre Scientifique et Technique des Salles. PHOTOGRAPHS: Courtesy of Opéra Bastille. REFERENCE: Bruno Suner, "Ott. L'Opéra Bastille," L'Architecture D'Aujourd'Hu.

TECHNICAL DETAILS

V = 741,500 ft³ (21,000 m³)	S_a = 13,650 ft² (1,268 m²)	S_A = 16,382 ft² (1,522 m²)
S_o (pit) = 2,000 ft² (186 m²)	S_P = 2,600 (242)	S_T = 21,000 ft² (1,951)
N = 2,700	H = 70 ft (21.3 m)	W = 106 ft (16.2 m)
L = 123 ft (31.1 m)	D = 148 ft (39 m)	V/S_T = 35.3 ft (10.8 m)
V/S_A = 45.3 ft (13.8 m)	S_A/N = 6.1 ft² (0.56 m²)	V/N = 275 ft³ (7.8 m³)
H/W = 0.66	L/W = 1.16	ITDG = 41 msec

NOTE: The terminology is explained in Appendix 1.

51

Opéra Garnier

This famous opera house, which opened in 1875, was designed by the architect Charles Garnier. Garnier pursued diligently the elusive factors of good acoustics, but in his book, *The Grand Opera in Paris,* he confesses that he finally trusted to luck, "like the acrobat who closes his eyes and clings to the ropes of an ascending balloon." "Eh bien!" he concludes, "Je suis arrivé!"

The Opéra Garnier has very good acoustics—not the finest but very good. Garnier said, "The credit is not mine. I merely wear the marks of honor." He continues,

> It is not my fault that acoustics and I can never come to an understanding. I gave myself great pains to master this bizarre science, but after fifteen years of labor, I found myself hardly in advance of where I stood on the first day. . . . I had read diligently in my books, and conferred industriously with philosophers—nowhere did I find a positive rule of action to guide me; on the contrary, nothing but contradictory statements. For long months, I studied, questioned everything, but after this travail, finally I made this discovery. A room to have good acoustics must be either long or broad, high or low, of wood or stone, round or square, and so forth.

The baroque interior of the Opéra Garnier combines tan-gold plaster "carvings" and gold trim with burgundy-red plush upholstery and box linings. The recent paintings on the ceiling (not in the photo), beneath which hangs a grand chandelier, place this theater in an age of elegance far removed from the frugal reconstruction of La Scala or the almost chaste interior of the Vienna Staatsoper. On the main floor, there are only 21 closely spaced rows, the last 8 of which are elevated above those in front. The architectural effect of this raised area is to foreshorten the main floor and give the hall an intimate appearance.

The reason for the good acoustics is the room's small size. The audience area is 11,000 ft² (1,020 m²) with 2,131 seats. Eighty-six of the seats fold down into

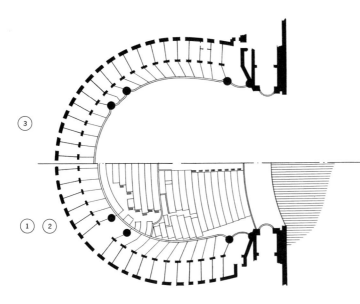

SEATING CAPACITY 2131

(1)	787
(2)	246
(3)	242
(4)	268
(5)	588 + standees

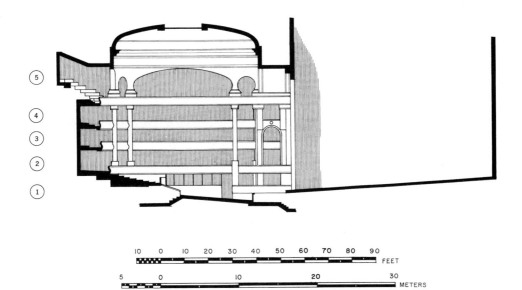

```
   10   0   10   20   30   40   50   60   70   80   90
                                                        FEET

5    0          10              20              30
                                                  METERS
```

the aisles during performances. The reverberation time is very low, the orchestra is said to dominate the singers, and the best seats in the house are in the front rows of the upper two rings, opposite the stage. Herbert von Karajan said (1960), "The acoustics of the Paris Opera House are wonderful."

ARCHITECTURAL AND STRUCTURAL DETAILS

Uses: Opera and ballet. **Ceiling:** Dome is 0.10-in. (2.5-mm) steel; above the top level (amphitheater) the ceiling is plaster; ceilings of the boxes are stretched damask cloth 1 in. away from the plaster. **Walls:** All visible surfaces (faces of the rings, columns, capitals, and so forth) are plaster; the dividers between boxes are 0.5-in. (1.25-cm) wood covered with damask; inside and top of the balcony railing are covered with plush; rear of boxes are separated from cloak room by velour curtains; walls at the rear of the amphitheater are of plaster. The front of the rail dividing the main floor is of wood; rear is lined with carpet. **Floors:** Wood throughout the house. **Carpets:** Main floor and all boxes fully carpeted; no carpet in amphitheater. Note: Cyclorama on stage is of aluminum with canvas facing. **Orchestra pit:** Floor is of wood over airspace; side walls are of solid wood. **Stage height:** 26 in. (66 cm) above floor level at first row of seats. **Seating:** Main floor seats are fully upholstered on all surfaces except for wood trim; boxes have upholstered chairs; in amphitheater the seat bottoms are fully upholstered, backrests are of open wood.

ARCHITECT: Charles Garnier. REFERENCES: Plans, details, and photographs courtesy of the management of the Opéra Garnier.

TECHNICAL DETAILS

Opera

V = 352,000 ft³ (10,000 m³)	S_a = 9,700 ft² (900 m²)	S_A = 12,120 ft² (1,126 m²)
S_o (pit) = 840 ft² (78 m²)	S_P = 2,632 ft² (244 m²)	S_T = 15,600 ft² (1,448 m²)
N = 2,131	N_T = 2,231	H = 68 ft (20.7 m)
W = 62 ft (18.9 m)	L = 91 ft (27.7 m)	D = 106 ft (32.3 m)
V/S_T = 22.7 ft (6.9 m)	V/S_A = 29 ft (8.87 m)	V/N = 165 ft³ (4.68 m³)
S_A/N = 5.75 ft² (0.53 m²)	H/W = 1.1	L/W = 1.47
ITDG = 17 msec		

NOTE: The terminology is explained in Appendix 1.

52

Paris

Salle Pléyel

or half a century, Salle Pléyel was Paris's principal concert hall. It opened in 1927. The architect believed that once-reflected sound largely sets acoustical quality. But the parabolic shape that he chose primarily reflected sound to the rear of the hall, although he tilted the side walls inward slightly to bring some sound to the audience areas. No attention was paid to reverberation time. A fire in 1928 did major damage. The redesign seemed influenced by the American consultant, F. R. Watson, whose credo was "Design the auditorium so that for listening the sound will be comparable to outdoor conditions." Upholstered seats were installed and sound-absorbing materials were applied to all surfaces surrounding the balconies. The entire floor was carpeted. Some changes were made in 1981, mainly stage extension, the addition of stage ceiling reflectors, more stage ventilation, and removal of a considerable amount of sound-absorbing material.

In 1994, extensive changes were made to increase the energy in the lateral sound reflections and to distribute them more evenly over the floor by the use of large MLS sound-diffusing reflectors spaced out from the side walls. Changes were made in the stage ceiling to improve communication among the musicians. The balcony fronts were made more irregular and a small amount of sound-absorbing material was added to the rear wall above the second balcony to reduce echoes heard primarily on stage. These changes are on the accompanying drawings, but not in the main-hall photographs. Following new ownership, the hall is currently not used by a major orchestra.

The hall seats 2,386 and the unoccupied reverberation time at mid-frequencies is 2.0 sec. Fully occupied it is 1.6 sec, below the optimum of about 1.9 sec for the usual symphonic repetoire. Except for 860 ft^2 (80 m^2) near the proscenium of the stage, the ceiling is covered with a hanging array of wooden strips, 2 in. (5 cm) wide, separated by 3 in. of space and located about 6.5 ft (2 m) (average) below the hard ceiling. Orchestral balance is adjusted by mechanical platforms, adjustable ceiling reflectors, and tilting back-stage prismatic reflectors.

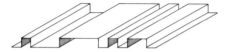

reflector, diffuser panel

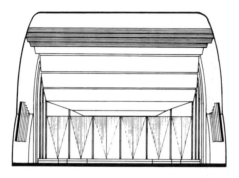

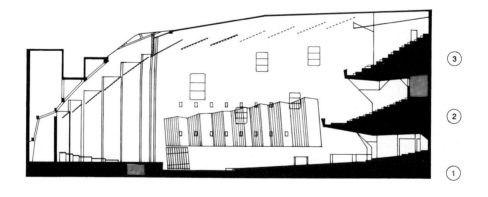

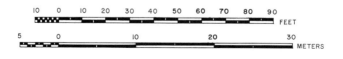

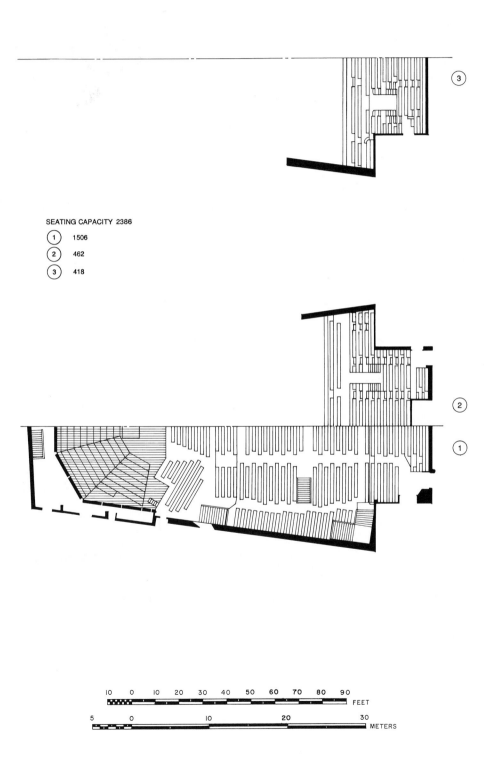

SEATING CAPACITY 2386

(1) 1506

(2) 462

(3) 418

Paris music critics, in 1994, generally stated that the string sound is clearer and the orchestral balance is better; the woodwinds are softened and clearer; and, overall, the orchestral tone is richer and less "frontal."

ARCHITECTURAL AND STRUCTURAL DETAILS

Uses: Music and recitals, 80%; speech and others, 20%. **Ceiling:** Brick, lightly plastered, beneath which, except for 860 ft² (80 m²) at the front, is hung a sawtooth wooden-slat structure, each slat 2 in. (5 cm) wide, spaced 3 in. (7.5 cm) apart. **Walls:** Sidewalls are heavy brick, plastered and painted, to which are attached wide-band MLS acoustic diffusers, 13–16 ft (4–5 m) high, tilted at the same angle as the sidewalls; 50% of rear walls and walls under the balconies are lightweight gypsum boards with air spaces of variable thickness. **Floors:** Wooden, including balconies. **Stage enclosure:** Movable wooden panels, varnished, 1 in. (2.5 cm) thick. Panels forming the stage back wall are large and prism-shaped (see the photograph). **Stage height:** 3 ft (1 m). **Sound-absorbing materials:** Some above rear of orchestra; also above second balcony on rear wall; gypsum panels on the rear walls below the two balconies absorb low frequencies. **Seating:** Molded wooden seat back, upholstered front of seat back and top of seat bottom.

ORIGINAL ARCHITECT AND ENGINEER: Gustave Lyon. RENOVATION ARCHITECT, 1994: M. De Portzampare. ACOUSTICAL CONSULTANT, 1994: Albert Yaying Xu. PHOTOGRAPHS: Paul Maurer.

TECHNICAL DETAILS

V = 547,000 ft³ (15,500 m³)	S_a = 8,396 ft² (780 m²)	S_A = 11,390 ft² (1,058 m²)
S_o = 2,605 ft² (242 m²)	S_T = 13,330 ft² (1,238 m²)	N = 2,386
H = 61 ft (18.6 m)	W = 84 ft (25.6 m)	L = 100 ft (30.5 m)
D = 120 ft (36.6 m)	V/S_T = 41.0 ft (12.52 m)	V/S_A = 48 ft (14.6 m)
V/N = 229 ft³ (6.5 m³)	S_A/N = 4.77 ft² (0.44 m²)	S_a/N = 3.52 ft² (0.327 m²)
H/W = 0.73	L/W = 1.19	ITDG = 35 msec

Note: $S_T = S_A + 1,940$ ft² (180 m²); see Appendix 1 for definition of S_T.

NOTE: The terminology is explained in Appendix 1.

53

Baden-Baden

Festspielhaus

The first large opera house in Europe to be financed by private money, actually a real-estate fund, is the Festspielhaus in Baden-Baden, built on land donated by the city. The historical railway station next door serves as its foyer. In Europe, all opera houses receive subsidies except the Festspielhaus Baden-Baden, and that has posed a severe challenge for its sponsors. Opened in 1998, and off to a rocky financial start, the Festspielhaus is presently a huge success.

The Festspielhaus is of contemporary design, however, the entrance through the historical railway station, which has wonderful paintings on both the ceiling and the sidewalls—even the ticket windows are attractive—give this opera house a special flavor. The size of the house is also unique, with 2,392 seats, second only (in Europe) to the Opéra de la Bastille in Paris.

From the acoustical standpoint, the primary goal was to direct the sound strength to the audience areas, and away from the ceiling, floor, the sides of the proscenium, and the side walls. Sound-absorbing material was kept to a minimum. Because the height of the room is great, 72 ft (22 m), the ceiling above the orchestra pit is stepped downward toward the singers. This also serves as part of the ceiling for the concert stage. These steps provide communication among the orchestra players and provide early reflections to the audience on the main floor. Additional sound reflectors, suspended from the ceiling, provide desirable reflections to the balconies. Music critics have generally spoken favorably about the acoustics, although the reverberation time for opera (normal semi-closed set) is high, 1.9 sec at mid-frequencies, fully occupied.

The media have vigorously discussed the entire project, including private ownership and the architectural design. It appears that the Festspielhaus Baden-Baden still must become accepted into the family of state-supported opera houses of Europe. The owners believe that they have made a great beginning and they speak confidently for the future.

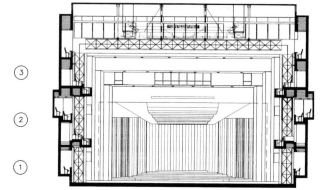

③

②

①

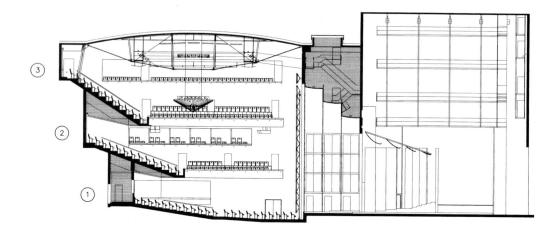

③

②

①

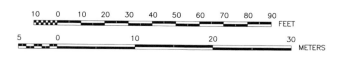

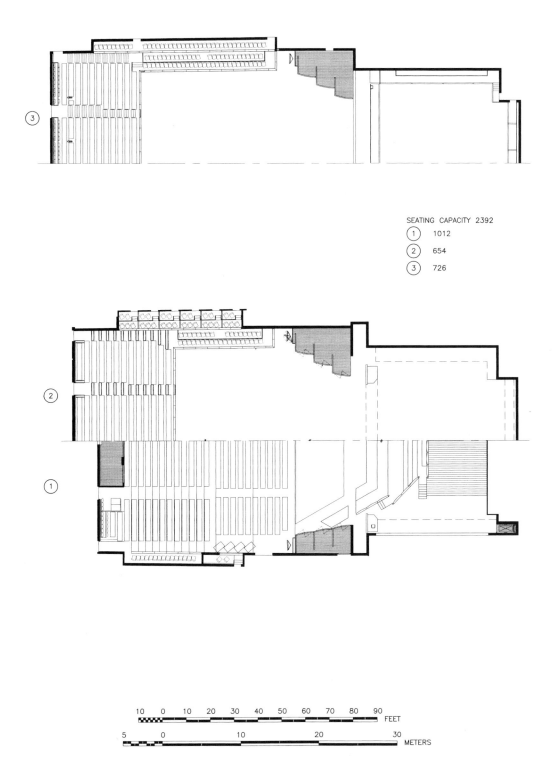

SEATING CAPACITY 2392

① 1012

② 654

③ 726

10 0 10 20 30 40 50 60 70 80 90
FEET

5 0 10 20 30
METERS

Architectural and Structural Details

Uses: Orchestra, opera, soloists, chamber music, chorus, and ballet. **Ceiling:** Upward-curved, concrete slab; two lighting bridges, hanging 11.5 ft (3.5 m) below the ceiling, with convex glass, sound-reflectors attached directly to their undersurfaces. **Side walls:** Plaster directly on brick or concrete walls; the balcony fronts are plaster, irregular to provide sound diffusion. **Rear walls:** Wooden panels, varying in thickness from 0.3–1.4 in (0.8–3.6 cm) with airspaces behind to provide sound absorption over a range of low frequencies. **Stage enclosure:** Wooden. **Floors:** Concrete or wooden panels supported by standards above concrete—all covered with cork linoleum. **Stage floor:** Pine wood. **Stage height:** 3.3 ft (1 m). **Carpet:** None. **Seats:** Plywood, perforated on the seat bottom, covered with thick upholstering wherever people come into contact while seated. **Orchestra pit:** Height of pit floor variable, with adjustable side-wall panels for changing the area of the pit floor; some panels can be rotated to provide either sound-reflecting or -absorbing surfaces.

ARCHITECT: Atelier Wilhelm Holzbauer. ACOUSTICAL CONSULTANT: Karlheinz Müller, Müller-BBM. PHOTOGRAPHS: Courtesy Festspielhaus Baden-Baden.

TECHNICAL DETAILS

Concerts (with orchestra shell)

V = 709,500 ft³ (20,100 m³)	S_a = 12,320 ft² (1,145 m²)	S_A = 15,290 ft² (1,421 m²)
S_o = 3,040 ft² (282 m²)	S_T = 17,227 ft² (1,601 m²)	N = 2,300 (+92 standees)
H = 72 ft (22 m)	W = 98 ft (30 m)	L = 92 ft (28 m)
D = 115 ft (35m)	V/S_T = 41.2 ft (12.6 m)	V/S_A = 46.4 ft (14.1 m)
V/N = 308 ft³ (8.74 m³)	S_A/N = 6.65 ft² (0.62 m²)	S_a/N = 5.36 ft² (0.50 m²)
H/W = 0.735	L/W = 0.94	ITDG = 38 msec

Opera (curtain lowered)

V = 692,000 ft³ (19,600 m³)	S_a = 12,320 ft² (1,145 m²)	S_A = 15,290 ft² (1,421 m²)
S_{pit} = 1,173 ft² (109 m²)	Sp = 2,322 ft² (216 m²)	S_T = 18,785 ft² (1,746 m²)
N = 2,300 (+92 standees)	H and W, as above	L = 117 ft (35.7 m)
D = 131 ft (40 m)	V/S_T = 36.8 ft (11.2 m)	V/S_A = 45.2 ft (13.8 m)
V/N = 308 ft² (8.74 m²)	S_A/N = 6.65 ft² (0.62 m²)	S_a/N = 5.36 ft² (0.50 m²)
H/W = 0.735	L/W = 1.19	ITDG = 26 msec

NOTE: The terminology is explained in Appendix 1.

54

Bayreuth

Festspielhaus

The Festspielhaus in Bayreuth is probably the world's most unusual opera house. Its design is unique; it was conceived by the composer Richard Wagner to satisfy his own image of how an opera house should look and sound; and it responds well only to the music of its master and best to *Der Ring Des Nibelungen* and *Parsifal*.

In 1876, Wagner's dream of a building for his operas solidified into reality. Not only does the sunken pit (shown in the drawings) project under the stage, but the conductor and the strings are under a solid wooden cover. Their music radiates outward from a slot over the middle of the orchestra which extends from one end of the pit to the other. The whole orchestra is out of sight and the musicians can play, unseen, in their shirtsleeves!

The seating area of the audience is fan-shaped, but the side walls of the theater are parallel. To fill in the progressively wider space between the walls and the seats toward the front of the hall, a series of seven piers is employed, each one penetrating deeper into the hall than the one behind it and each capped by a column that extends to capitals just beneath the ceiling. Although the ceiling is flat (I walked on it), one has the impression that it rises from the back toward the front, like a great awning stretched over a Greek amphitheater.

The average height of the ceiling above the sloping floor of the Festspielhaus is great, with the result that the reverberation time is long. It is 1.55 sec at middle frequencies when the theater is fully occupied, which is especially favorable to Romantic Wagnerian music.

The sunken-pit design is the center of endless controversy. One purpose of the sunken pit is to give greater balance between the singers and the orchestra. But this was not the only feature Wagner had in mind. He desired the unusual and dramatic effect of a "mystical abyss." He expected to create acoustically a mysterious sound, emanating from an invisible orchestra, with a modified, somewhat uncanny

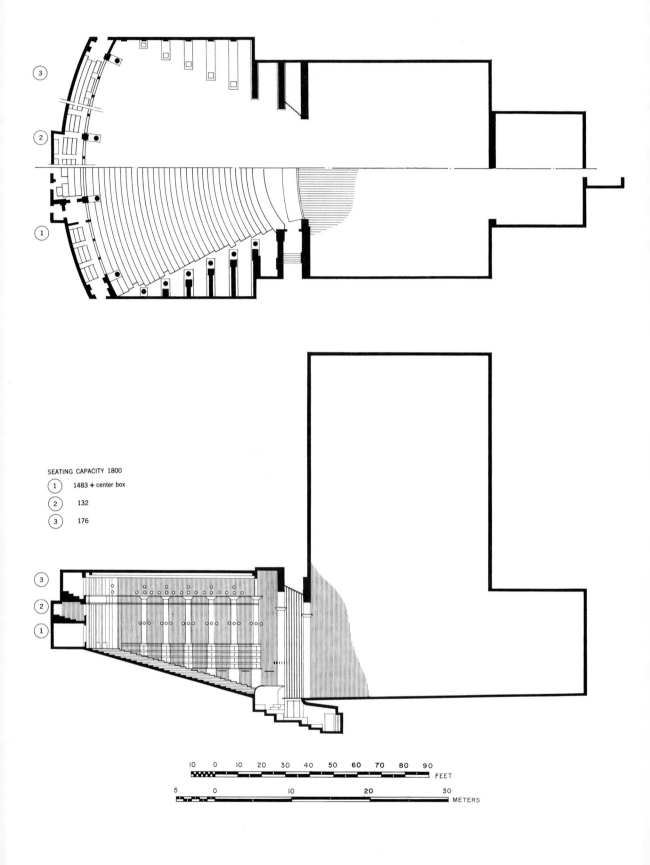

SEATING CAPACITY 1800

(1)   1483 + center box

(2)   132

(3)   176

10   0   10   20   30   40   50   60   70   80   90
                                              FEET

5   0   10   20   30
                                          METERS

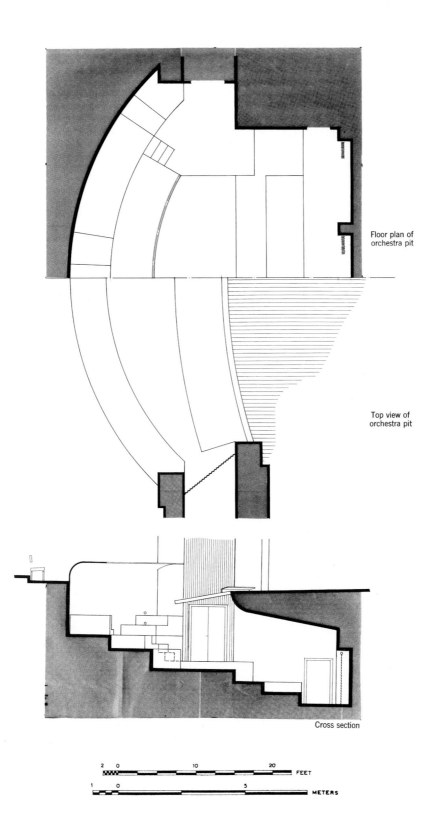

Floor plan of
orchestra pit

Top view of
orchestra pit

Cross section

timbre. One can easily conclude that the overriding purpose of the Bayreuth pit was to emphasize the drama, rather than to preserve the vocal-orchestral balance.

Most conductors love to perform in the Bayreuth pit. Every note can be heard at the podium. The orchestra can "let out the throttle" and not drown out the singers. They feel that, because the blending of the sound takes place in the pit, it merges in the form intended by the composer.

Joseph Wechsberg's description of his experience in the August 18, 1956, issue of *The New Yorker* is the most colorful I have read:

> Then there was silence, and out of the darkness came a sustained E flat—so low that I couldn't distinguish exactly when the silence ended and sound began. Nor could I be sure where the sound came from; it might have come from the sides of the auditorium, or the rear, or the ceiling. It was just there. Slowly the orchestra began to play melodic passages, barely audible at first and gradually increasing in volume until the auditorium was filled with music—the music of the water of the Rhine. When the curtain parted, the whole stage seemed filled with water—blue-green waves, ebbing and flowing in precise synchronization with the music. The Rhine Maidens appeared from behind a rock, pretty, slim and wearing golden one-piece bathing suits, the music rose and fell back to the pianissimo; and Woglinde started her "Weia! Waga!" It took me a moment to realize that there were no props and no stage set; the whole scene—rock and all—were created by means of projected film and light. The music, the singing, the waters, and the lights blended perfectly. Although the brass dominated, it did not sound brassy, as it often does in the large orchestra Wagner calls for. By accident or design, the strings, particularly the first violins, were somewhat subdued. . . . I was under a spell.

Some conductors have suggested that the cover over the violins be removed. Herbert von Karajan told me that he tried this experiment. He removed the cover, but he found that, with the rest of the orchestra buried beneath the stage the result was not satisfactory and the cover was replaced. At another time a perforated hood was tried. It was said to have given good results, but when objections were raised by one of the conductors, the old hood was re-installed.

In some productions, almost no teaser curtains are used and there is a minimum of hanging scenery. Also, a heavy, sound-reflecting cyclorama canvas ordinarily surrounds the acting area. As a result, the stage-house reverberation can be long, although the singers' voices project well into the house when the full undraped height of the proscenium is employed. The stage is lower than the audience, and thus the forestage is important as a sound reflector.

The interior finish is largely of plaster, either on brick or on wood lath. The ceiling combines wood and plaster. The horizontal ceiling contributes short-time-delay reflections at most seats. The projecting wings on the sides of the hall give a desirable mixing of the sound in the house. The Festspielhaus is small—it seats only 1,800 persons—and thus its size alone favors its acoustical quality.

ARCHITECTURAL AND STRUCTURAL DETAILS

Uses: Wagner's opera. **Ceiling:** 0.5-in. (1.25-cm) plaster on reeds over 0.5-in. wood; wooden carvings used as decorations. **Rear and side walls:** Plaster on brick or wood lath; the round columns and part of their capitals are of thick wood; the wing nearest the stage is closed off with corrugated asbestos sheet. **Seating:** Seats wood with cane bottoms.

ARCHITECT: Otto Brueckwald. CREDITS: Drawings from Edwin O. Sachs, *Modern Opera Houses and Theatres,* Vol I., London, B. T. Batsford, 1896–1897. Seating count from a box-office plan and other details, courtesy of Secretary. PHOTOGRAPHS: Lauterwasser, Uberlingen/Bodensee. OTHER: Pit drawings were developed from measurements by the author during the visit.

TECHNICAL DETAILS

$V = 364{,}000$ ft³ (10,308 m³)	$S_a = 8{,}125$ ft² (755 m²)	$S_A = 9{,}100$ ft² (845 m²)
S_o (pit) $= 371$ ft² (34.5 m²)	$S_P = 1{,}640$ ft² (152 m²)	$S_T = 11{,}111$ ft² (1,032 m²)
$N = 1{,}800$	$V/S_T = 32.8$ ft (10 m)	$V/S_A = 40$ ft (12.2 m)
$V/N = 202$ ft³ (5.72 m³)	$S_A/N = 5.1$ ft² (0.47 m²)	$H = 42$ ft (12.8 m)
$W = 109$ ft (33.2 m)	$L = 106$ ft (32.3 m)	$D = 111$ ft (33.8 m)
$H/W = 0.385$	$L/W = 0.97$	ITDG $= 14$ msec

NOTE: The terminology is explained in Appendix 1.

Kammermusiksaal der Philharmonie

The Kammermusiksaal follows the concept of "Music in the Center" that was the design premise of the neighboring Philharmonie Konzertsaal. Seating about 1,138, it was completed in 1987 and embodies features consistent with the new multimedia age. The reverberation time is about 1.8 sec at mid-frequencies, occupied. The stage in the center of the hall is surrounded by two tiers of circular seating areas. The lower tier, seating 419 listeners, plus 40 on stage, is symmetrical and to that audience it constitutes an intimate hall suited to the traditional musical repertoire of soloists or groups up to the size of a chamber orchestra. The side walls around the stage are tilted to provide cross reflections that help the performers hear each other. The parapet surrounding the first tier was made larger than normal to attain greater strength of early lateral sound reflections for the audience. Above the stage at a height of about 30 ft (9 m) hang nine multi-sided sound reflectors in a circular arrangement, that provide early reflections to the players and to those seated in the lower tier.

The upper tier with 619 seats is asymmetrical and the architect describes the relation of the symmetry to asymmetry in the design of the room as corresponding to a dialog between the traditional and the future. The convex, tent-shaped ceiling provides early reflections to the listeners in the upper tier. Again, sound diffusion is provided by 95 pyramidal diffusing elements in the ceiling which do not also serve, as in the main hall, as sound-absorbing resonators. All walls are shaped to diffuse sound and are positioned to prevent echoes.

Three elevated sections, containing 101 seats, can be used for special musical presentations, such as those required by the music of Giovanni Gabrieli and Claudio Monteverdi, or by modern works. Special rooms, with operable windows, are provided around the periphery for multimedia and broadcasting controls.

Loudspeakers are located in the center and around the edges of the convex tent ceiling and at various places outdoors, thus permitting unusual sound effects associated with such music as that of composer Luigi Nono.

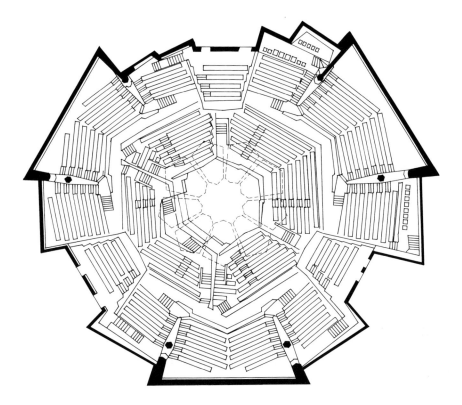

SEATING CAPACITY 1138

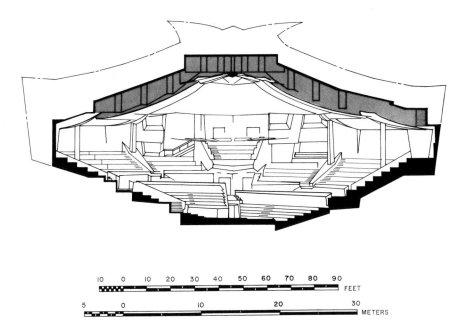

```
10   0   10  20  30  40  50  60  70  80  90
                                              FEET

5      0           10              20        30
                                              METERS
```

ARCHITECTURAL AND STRUCTURAL DETAILS

Uses: Chamber groups, soloists, and dance. **Ceiling:** Three layers—the roof, an intermediate plate, and the actual ceiling, which is 1.2-in. (3-cm) plaster on metal lath. **Suspended over-stage panels:** 9 multi-sided polyester panels, each about 90 ft² (8.4 m²) in area, with equal airspace between and variable in height, 26–40 ft (8–12 m) above the stage. **Sound-diffusing units:** 95 pyramidal-shaped, sound-diffusing units are located in the ceiling. **Sidewalls:** Part of the side walls are thin wood over airspace to control low-frequency reverberation. The parapets are faced with Jurassic limestone plaster. **Stage floor:** Wooden floor on planks over airspace. **Audience floor:** Oak parquet cemented to pre-cast slabs. **Seating:** Seat back, molded veneered plywood, the upper part bent vertical so that maximum sound reflection occurs when occupied; cushion on front of seat back does not extend to top; seat bottom is upholstered on top and the underside is covered with cloth only and is perforated; armrests are upholstered.

ARCHITECT: Edgar Wisniewski. ACOUSTICAL CONSULTANT: Lothar Cremer with Thomas Fuetterer. PHOTOGRAPHS: Courtesy of the architect and M. Barron. REFERENCE: E. Wisniewski, *Die Berliner Philharmonie und ihr Kammermusiksaal Der Konzertsaal als Zentralraum*, Gebr. Mann Verlag (1993).

TECHNICAL DETAILS

V = 388,400 ft³ (11,000 m³)	S_a = 6,650 ft² (618 m²)	S_A = 8,720 ft² (810 m²)
S_o = 840 ft² (78.2 m²)	S_c = 204 ft² (19 m²)	S_T = 9,764 ft² (907 m²)
N = 1,138	H = 37 ft (11.3 m)	W = 159 ft (48.5 m)
L = 60 ft (18.3 m)	D = 62 ft (18.9 m)	V/S_T = 39.8 ft (12.1 m)
V/S_A = 44.5 ft (13.6 m)	V/N = 341 ft³ (9.66 m³)	S_A/N = 7.7 ft² (0.71 m²)
H/W = 0.23	L/W = 0.38	ITDG = 20 msec

NOTE: The terminology is explained in Appendix 1.

56

Berlin

Konzerthaus Berlin
(formerly, Schauspielhaus)

The old Schauspielhaus, opened in 1821, was totally destroyed in World War II. Restored to the original design of architect Karl Friedrich Schinkel and recently renamed Konzerthaus Berlin, it takes its place among the most architecturally distinguished halls in the world. Every detail inside—the gorgeous parquetry floor, the pictorial coffered ceiling, the white walls with illuminated sculptures, the balcony fronts with gold ornamentation, the stainless steel organ pipes, and the crystal chandeliers—presage an extraordinary experience.

The hall, including the choir rows, seats between 1,507 and 1,677, depending on the size of stage. Sound-diffusing surfaces are everywhere. To promote clarity-enhancing early reflections to the audience, the side walls of the stage were made as high as possible. To enable the players to hear each other better, the upper part of the wall, beneath the balustrade and molding around the stage, are tilted inward.

The reverberation time, with audience, is now 2.0 sec at mid-frequencies, rising to 2.2 sec at low frequencies. Although, before opening, only about 20 percent of a judgment group thought the reverberation too large, afterwards there were reservations. The reverberation was overwhelming, rising to 2.7 sec at low frequencies. Corrective steps took the form of a large number of sound-absorbing resonators placed in the ceiling and tuned to reduce the low-frequency reverberation. Even now, for a hall this size, the optimum reverberation time at all frequencies is usually considered to be 0.1 to 0.2 sec lower than in this hall.

Seated on the main floor, the listener is immersed in the reverberant sound from all directions. Although there are no special surfaces directing the early sound reflections onto the audience, the ratio of early reflected sound energy to later reverberant energy is about the same, i.e., the same clarity, as in other rectangular halls (e.g., Vienna and Boston). The tonal and loudness balances among instruments are excellent. The hall is loud both because of its smaller size (about 60 percent of the number of seats in the Boston hall) and the high reverberation times.

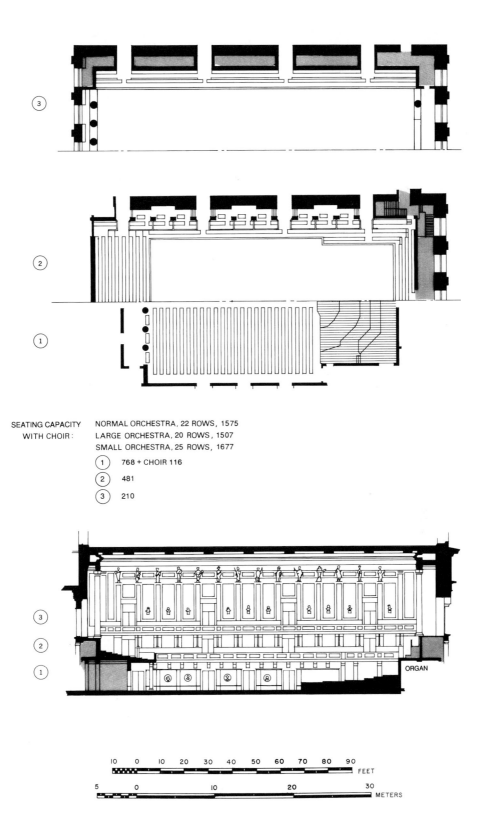

SEATING CAPACITY
WITH CHOIR:

NORMAL ORCHESTRA, 22 ROWS, 1575
LARGE ORCHESTRA, 20 ROWS, 1507
SMALL ORCHESTRA, 25 ROWS, 1677

① 768 + CHOIR 116
② 481
③ 210

ORGAN

10 0 10 20 30 40 50 60 70 80 90 FEET

5 0 10 20 30 METERS

For music of the Classical and Romantic periods the acoustics equal the best in the world.

ARCHITECTURAL AND STRUCTURAL DETAILS

Uses: Classical music. **Ceiling:** Plaster. **Walls:** Plaster over concrete block; balcony faces, plaster. **Resonators:** Helmholtz-type, tuned from 130 Hz to 300 Hz, installed in the ceiling at the slot just above the upper side wall molding. **Floors:** Parquet on concrete. **Stage enclosure:** Sidewalls, plaster. **Stage floor:** Tongue-and-groove wood over timber 3 in. (8 cm) thick; risers adding 51–59 in. (1.3–1.5 m). **Stage height:** Variable from 31 in. (0.8 m) to 95 in. (2.5 m). **Seating:** Wood chairs, upholstered seat bottom and backrest.

ARCHITECT: Architekten-kollektivs: Prasser. ACOUSTICAL CONSULTANTS: Wolfgang Fasold, Ulrich Lehmann, Hans-Peter Tennhardt, Helgo Winkler PHOTOGRAPHS: Jens Huebner. REFERENCES: W. Fasold, U. Lehmann, H. Tennhardt and H. Winkler, "Akustische Massnahmen im Schauspielhaus, Berlin," *Bauforschung, Baupraxis,* **181,** 1–20 (1986); W. Fasold and U. Stephenson, "Gute Akustik von Auditorien, Planung Mittels Rechnersimulation und Modellmesstechnik," *Bauphysik* **15,** 20–49 (1993).

TECHNICAL DETAILS

$V = 530,000$ ft³ (15,000 m³)	$S_a = 8,440$ ft² (784 m²)	$S_A = 10,150$ ft² (943 m²)
$S_o = 1,700$ ft² (158 m²)	$S_T = 11,850$ ft² (1,101 m²)	$N = 1,575$
$H = 58$ ft (17.7 m)	$W = 68$ ft (20.7 m)	$L = 79$ ft (24.1 m)
$D = 84$ ft (25.6 m)	$V/S_T = 44.7$ ft (13.6 m)	$V/S_A = 52.2$ ft (15.9 m)
$V/N = 336$ ft³ (9.53 m³)	$S_A/N = 6.4$ ft² (0.60 m²)	$S_a/N = 5.36$ ft² (0.50 m²)
$H/W = 0.85$	$L/W = 1.16$	ITDG $= 2.5$ msec

NOTE: The terminology is explained in Appendix 1.

57

Berlin

Berlin Philharmonie

"Music in the Center" was the overriding postulate of architect Hans Scharoun. He felt that the normal placement of the orchestra at one end of a hall prevents audience and musicians from communicating freely and intensely. The result is a most dramatic room, dedicated in 1963, with 250 of the 2,218 seats directly behind the orchestra and about 300 on either side. In addition, there are about 120 places on stage and spaces for 44 handicapped listeners. No listener is more than 100 ft (30 m) away from the stage, compared to 133 ft in Boston Symphony Hall (seating 2,612).

Philharmonie Hall has become one of the models of successful acoustical designs, pioneering the concept of the "vineyard" style hall. The acoustical consultant agreed on the advantage of breaking the audience into blocks, so that the first row in each block receives unimpeded direct sound. The seats in many of the blocks receive early lateral reflections from the side walls that surround them, including the wall behind. The fronts of the terraced blocks provide early reflections for both the musicians and the audience seated in the middle of the hall. Additional early reflections are provided to the orchestra and audience by ten large suspended panels hung above the stage. The seats in the upper blocks receive early reflections from the convex, tent-shaped ceiling.

In the audience sections in front of the orchestra, the sound is beautiful, clear, balanced, and with a liveness that completely surrounds one. The principal disadvantage is that those seated to the rear, or near rear, of the stage hear a different sound: the trumpets and trombones radiate forward, and the French horns backward. The sound from piano and singers is also troubling, a large part of the upper registers are projected forward. Fortunately, the visual impression of viewing the conductor face-on favorably shapes one's judgment of the acoustics. The mid-frequency reverberation time, fully occupied, is 1.9 sec. The bass is controllable by adjusting the 136 pyramid-shaped low-frequency resonators in the ceiling.

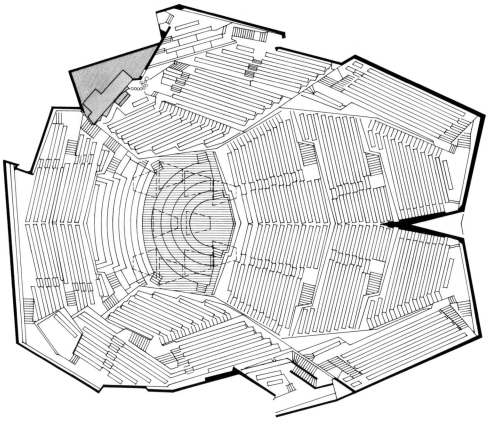

SEATING CAPACITY 2218 + 120 chorus

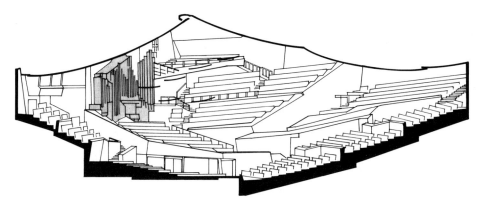

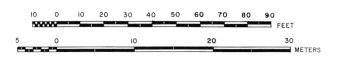

ARCHITECTURAL AND STRUCTURAL DETAILS

Uses: Concerts, primarily. **Ceiling:** Suspended (1.2 in. [3 cm, up to 4 cm at center]), chalk-gypsum plaster on expanded metal. **Suspended stage panels:** 10 trapezoidal polyester panels, each 81 ft² (7.5 m²) in area, 50% open space between, variable in height 32–40 ft (10–12 m) above the stage. **Ceiling sound-absorbing units:** 136 pyramidal-shaped, combination sound-diffusing, low-frequency Helmholtz-resonator-type absorbing boxes. **Sidewalls:** Part of sidewalls are thin wood over airspace. The parapets are faced with Jurassic limestone plaster. **Stage side walls:** Arranged to reflect sound back to the musicians. **Stage floor:** Wooden floor on planks over airspace. **Stage height:** 30 in. (76 cm). **Audience floor:** Oak parquet in asphalt base over precast slabs. **Seating:** Seat back, molded veneered plywood, the upper part bent vertical (see the photograph) so that maximum sound reflection occurs when occupied; cushion on front of seat back does not extend to top; seat bottom is upholstered on top and the underside is covered with cloth only and is perforated; armrests are wooden.

ARCHITECT: Hans Scharoun. ACOUSTICAL CONSULTANT: Lothar Cremer with Joachim Nutsch. PHOTOGRAPHS: Foto R. Friedrich. REFERENCES: L. Cremer, "Die raum- und bauakustischen Massnahmen beim Wiederaufbau der Berliner Philharmonie," *Die Schalltechnik* **57**, 1–11 (1964); R. S. Lanier, "Acoustics in-the-round at the Berlin Philharmonic," *Architectural Forum* **120**, 99–105 (1964).

TECHNICAL DETAILS

V = 741,300 ft³ (21,000 m³)	S_a = 11,380 ft² (1,057 m²)	S_A = 14,900 ft² (1,385 m²)
S_o = 1857 ft² (1,72.5 m²)	S_T = 16,765 ft² (1,558 m²)	N = 2,218 + 120 on stage
H = 42 ft (12.8 m)	W = 140 ft (42.7 m)	L = 95 ft (29 m)
D = 98 ft (30 m)	V/S_T = 42.2 ft (13.5 m)	V/S_A = 49.8 ft (15.2 m)
V/N = 317 ft³ (9 m³)	S_A/N = 6.72 ft² (0.62 m²)	S_a/N = 4.87 ft² (0.455 m²)
H/W = 0.3	L/W = 0.68	ITDG = 21 msec

NOTE: The terminology is explained in Appendix 1.

Beethovenhalle

The Beethovenhalle project in Bonn is one of the impressive cultural centers that have been built in Germany since World War II. Separated from the Rhine by a narrow mall on one side, and backed by ample parking space at the rear, the hall is well situated. Viewed from an airplane, its striped molded roof suggests a lively whale at play. Its lobbies are typical of contemporary architecture in Germany, a little aseptic but relieved by colorful murals. The Beethoven Halle opened in 1959 and after a 1983 fire was rebuilt identically.

The architecture of the interior takes the visitor by surprise. The main floor, which contains 1,030 seats, is flat, and the hall may be used for banquets, dances, or exhibitions. An asymmetrical balcony seats 377 listeners. On entering, one's eyes are immediately carried to the ceiling, which looks like a many-tufted yellow bedspread, held by an upward pull of gravity against a domed ceiling. From the ceiling down, the interior of the hall suggests the classic struggle between architecture and acoustics; the architect wants a majestic domed ceiling, and the acoustician, faced with the difficult acoustics usual to domes, must find a means of scattering the sound.

The 1,760 acoustical elements on the ceiling are designed not only to scatter the sound impinging on them but also to absorb sound in the region of 125 Hz. Low-frequency sound absorption is always debatable. Musicians and music critics express a preference for a rich, strong bass—such as results from a room finished in heavy plaster. Yet the possible focusing of the dome, especially at the lower frequencies, had to be combated. The Beethoven Hall has a satisfactory bass response, although not as strong as before the fire.

After the fire the upholstering on the seats was replaced with a thicker polyvinyl padding, which has reduced the reverberation time by a small amount, particularly the bass.

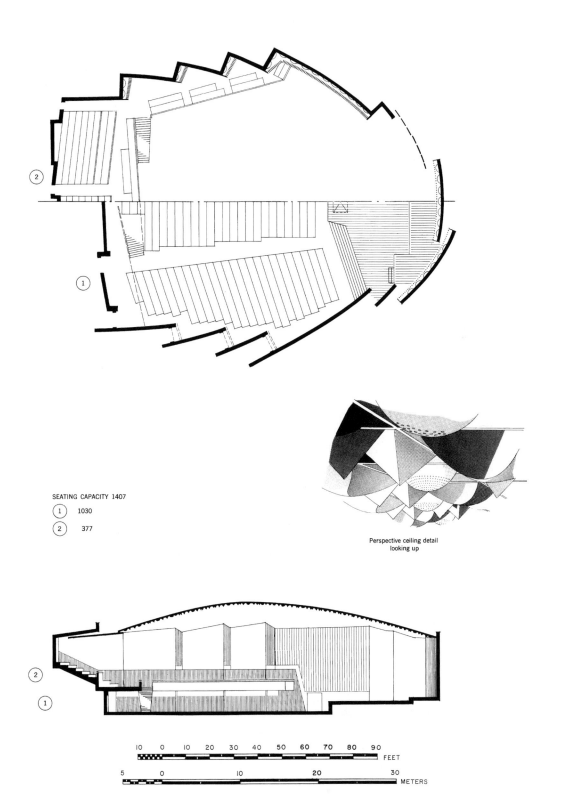

②

①

SEATING CAPACITY 1407

①   1030

②   377

Perspective ceiling detail
looking up

②

①

10   0   10   20   30   40   50   60   70   80   90
FEET

5   0   10   20   30
METERS

ARCHITECTURAL AND STRUCTURAL DETAILS

Uses: Concerts, meetings, and social events. **Ceiling:** Beneath the concrete roof of the hall is hung a rhombic (egg crate) grid of 1-in. (2.5-cm) reinforced-gypsum sheets. Inside each unit a "sound scatterer" is placed, made from a gypsum-vermiculite mixture, 0.3 in. (0.8 cm) thick; each of these 1,760 scatterers, with a height of 1 ft (30 cm) [the total area covered being 12,000 ft² (1,115 m²)], is either a double pyramid, a spherical segment, or a cylindrical segment obliquely cut on both sides with the curvature facing downward. **Sidewalls:** An acoustically transparent grid of vertical slats covers some vertically oriented cylindrical diffusers, each with a chord of 3 ft (1 m) separated from each other by 1 or 2 ft. Some damped resonators tuned to 250 Hz are located between each pair of the cylinders; the four sections of wall above the balcony on either side of the hall that face the audience are 0.75-in. (1.9-cm) wooden panels over airspaces that vary between 0.4 in. (1 cm) and 3.5 in. (9 cm) deep; behind half the panels there are 2-in. (5-cm) glass fiber blankets; the lower walls are of plaster. **Floor:** Oak parquet. **Stage floor:** Wood over airspace. **Stage height:** 43.5 in. (110 cm). **Seats:** The tops of the seat bottoms and the fronts of the seat backs are thickly upholstered.

ARCHITECT: Siegfried Wolske. ACOUSTICAL CONSULTANTS: Erwin Meyer and Heinrich Kuttruff. PHOTOGRAPHS: Courtesy of the architect. REFERENCES: *Deutsche Bauzeitung* **65**, 59–75 (1960); E. Meyer and H. Kuttruff, in a letter to the editor of *Acustica* **9**, 465–468 (1959).

TECHNICAL DETAILS

Concerts

V = 555,340 ft² (15,728 m²)	S_a = 9,300 ft² (864 m²)	S_A = 12,000 ft² (1,115 m²)
S_o = 2,200 ft² (204 m²)	S_T = 13,940 ft² (1,295 m²)	N = 1,407
H = 40 ft (12.2 m)	W = 120 ft (36.6 m)	L = 114 ft (34.8 m)
D = 125 ft (38.1 m)	V/S_T = 39.7 ft (12.13 m)	V/S_A = 46.3 ft (14.1 m)
V/N = 395 ft³ (11.2 m³)	S_A/N = 8.5 ft² (0.79 m²)	S_a/N = 6.61 ft² (0.614 m²)
H/W = 0.33	L/W = 0.95	ITDG = 27 msec

Note: $S_T = S_A$ + 1,940 ft² (180 m²); see Appendix 1 for definition of S_T.

NOTE: The terminology is explained in Appendix 1.

59

Dresden

Semperoper

One of Europe's most elegant opera houses, the Semperoper has suffered a troubled history. Created by the architect Gottfried Semper in 1838–41, it was destroyed by fire in 1869. With Gottfried Semper again as architect, it opened anew in 1879, this time built of stone and marble, only to be destroyed in 1945 by Allied bombing. The theater was rebuilt exactly as before. The cornerstone for the reconstructed building was laid June 24, 1977, and the House opened on February 13, 1985, with a production of Wolfgang Amadeus Mozart's *Così fan tutte*.

The Semperoper is a commanding building on the west bank of the Elbe River. A tour through the entire house is breathtaking. Starting with the splendid foyer, one admires the richly decorated staircase and enters the auditorium that is dominated by a lavish stage curtain and gorgeous balcony fronts. The quality of workmanship in all the public spaces is beyond belief, including stunning baroque ceilings, "marbelization" of columns as they were done originally, and a multitude of chandeliers, sconces, and paintings everywhere. Seating only 1,300, it is one of the most intimate of opera houses. In the four rings above the main floor, there are no dividing walls, so that all listeners hear the performance equally.

Lighting for the stage is provided from the "boxes" nearest the proscenium, from the space above the fourth ring, and through small openings in the ceiling. The rear wall of the concert shell is 26 ft (8 m) behind the proscenium opening.

The rating of the acoustical quality of the Semperoper, made by 21 well-known opera conductors, is next to that of the highest, the Opera Colón in Buenos Aires, and equal to that of La Scala in Milan (Hidaka and Beranek, 2000). Its mid-frequency reverberation time with normal scenery in the stage house, occupied, is an optimum 1.6 sec. Its binaural quality index measures higher than that of any other house and the strength of its bass is equal to that of the two houses just mentioned. When the house is used with stage enclosure for concerts, the measured reverberation time, occupied, at mid-frequency, is 1.9 sec, also optimum.

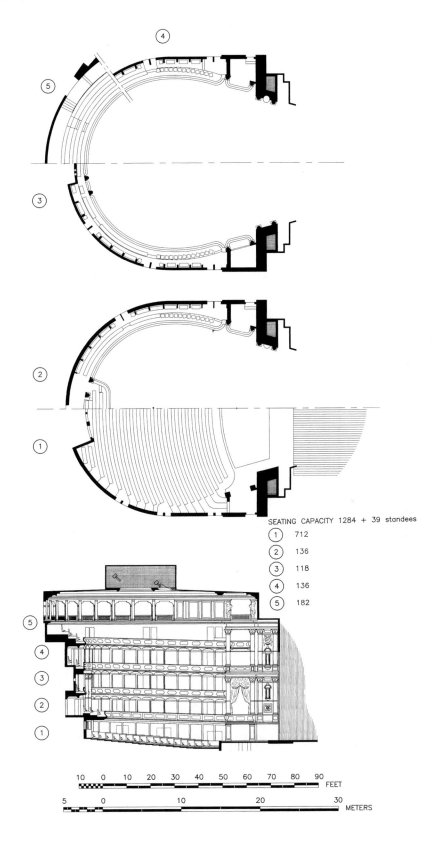

SEATING CAPACITY 1284 + 39 standees

① 712
② 136
③ 118
④ 136
⑤ 182

10 0 10 20 30 40 50 60 70 80 90
FEET

5 0 10 20 30
METERS

ARCHITECTURAL AND STRUCTURAL DETAILS

Uses: Principally opera, with concerts and occasional ballet. **Ceiling:** Suspended gypsum (stucco) varying in thickness from 1.5 to 2.8 in. (4 to 7 cm). **Walls:** Brick or reinforced concrete approximately 1 ft (30 cm) thick. **Floors:** 0.8-in. (2-cm) parquet wood directly on concrete base. **Pit:** Usual size about 20×66 ft (6×20 m), but can be extended 3.5 ft (1 m) in length. **Seats:** Heavily upholstered, back, front, and armrests, with air conditioning in back. **Concert enclosure:** 0.4-in. (1-cm) plywood on wooden framing. **Pipe organ:** Located on one of the movable side stages, for use in concerts and musical theater.

CHIEF ARCHITECT FOR THE RECONSTRUCTION: Wolfgang Hänsch. ACOUSTICAL CONSULTANT: Walter Reichardt and (for model measurements) Wolfgang Schmidt. REFERENCES: "Die neue Semperoper in Dresden," *Kulturbauten,* Heft 2 (1985), with acoustics paper, Wolfgang Schmidt, "Die Raumakustik im Zuschauerraum"; Wolfgang Schmidt, "Der Wieferaufbau der Semper-Oper.-Raumakustik im Zuschauerraum," *Bühnentechnische Rundschau,* Heft 6, pp. 15ff. (1979).

TECHNICAL DETAILS

Opera

$V = 441,200$ ft³ (12,500 m³)	$S_a = 6,790$ ft² (631 m²)	$S_A = 9,318$ ft² (866 m²)
$S_{pit} = 1,291$ ft² (120 m²)	$S_p = 1,800$ ft² (167.2 m²)	$S_T = 12,409$ ft² (1,153 m²)
$N = 1,284$ ($+39$ standees)	$H = 62.3$ ft (19 m)	$W = 88.2$ ft (26.9 m)
$L = 84$ ft (25.6 m)	$D = 98$ ft (29.9 m)	$V/S_T = 35.55$ ft (10.84 m)
$V/S_A = 47.3$ ft (14.4 m)	$V/N = 344$ ft³ (9.74 m³)	$V/S_A = 47.3$ ft (14.43 m)
$V/S_a = 65$ ft (18.8 m)	$H/W = 0.706$	$L/W = 0.95$

Concerts

$V = 457,500$ ft³ (12,960 m³)	$S_a = 6,790$ ft² (631 m²)	$S_A = 9,318$ ft² (866 m²)
S_o(Extra) $= 490$ ft² (45.6 m²)	$S_T = 11,100$ ft² (1,032 m²)	$N = 1,284$ ($+39$ standees)
$H = 62.3$ ft (19 m)	$W = 88.2$ ft (26.9 m)	$L = 65$ ft (19.8 m)
$D = 80$ ft (24.4 m)	$V/S_T = 41.2$ ft (12.56 m)	$V/S_A = 49.1$ ft (15 m)
$V/N = 356.3$ ft³ (10.1 m³)	$S_A/N = 7.26$ ft² (0.67 m²)	$S_a/N = 5.29$ ft² (0.49 m²)
$H/W = 0.706$	$L/W = 0.736$	ITDG $= 35$ msec

NOTE: The terminology is explained in Appendix 1.

60

Leipzig

Gewandhaus

The predecessor to the Gewandhaus was the "Neues" Gewandhaus, completed in 1884 and destroyed in World War II. It was preceded by the "Altes" Gewandhaus, which spanned the period from 1781 to 1894. Those older Gewandhauses were both famous, the 1781 one because of its excellent orchestra and small size (400 seats) and the "Neues" because of its great orchestra and famed acoustics (1,560 seats). Notwithstanding the repute of those previous "shoebox"-type halls, the architect had courage to strike out in a contemporary direction.

Opened in 1981, the Gewandhaus is a striking building, located in the city center on the Augustus-Platz. At night the lights of the splendid foyer glisten and enliven the grand plaza. The concert hall's interior is dramatic in a very different sense than that of Berlin's Konzerthaus. The art that shows everywhere in the Konzerthaus is missing completely in the Gewandhaus. One is impressed instead by the white balcony fronts that contrast with the red upholstering, the assemblage of acoustical sound-diffusing boxes on the darkened side walls and the majestic pipe organ in the front of the hall. With the audience size of 1,900, the architect choose to make the hall fan/rhombic-shaped and to locate the platform about one-fourth of the room length from the front wall. The audience is seated on all sides of the stage. With this plan, the farthest auditor is only about 110 ft (33.5 m) from the front of the stage, compared to Boston Symphony Hall's 133 ft (40.5 m).

Much of the audience in a typical fan-shaped hall will receive too little early reflected sound energy unless special acoustical additions are made to the wall surfaces. An extreme example of such treatment are the very large sloping panels found in the Town Hall in Christchurch, New Zealand. In the Gewandhaus the solutions were to employ a large number of sound-reflecting tilted surfaces on the side walls (see the photograph and drawing) and to tilt the balcony fronts at various angles in the hall to direct early sound energy to the listeners. Around the stage the balcony fronts that form the upper wall are tilted inward to provide cross-reflections for the musicians, which make ensemble playing easier. The shaped ceiling supplies early

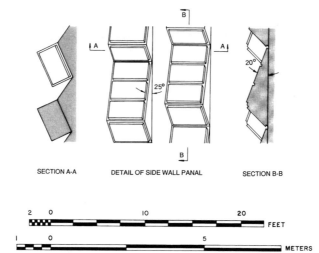

SECTION A-A   DETAIL OF SIDE WALL PANAL   SECTION B-B

SEATING CAPACITY 1900

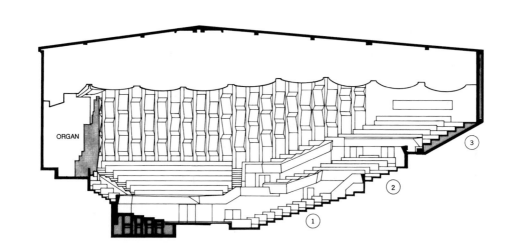

ORGAN

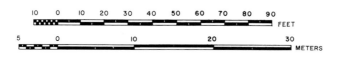

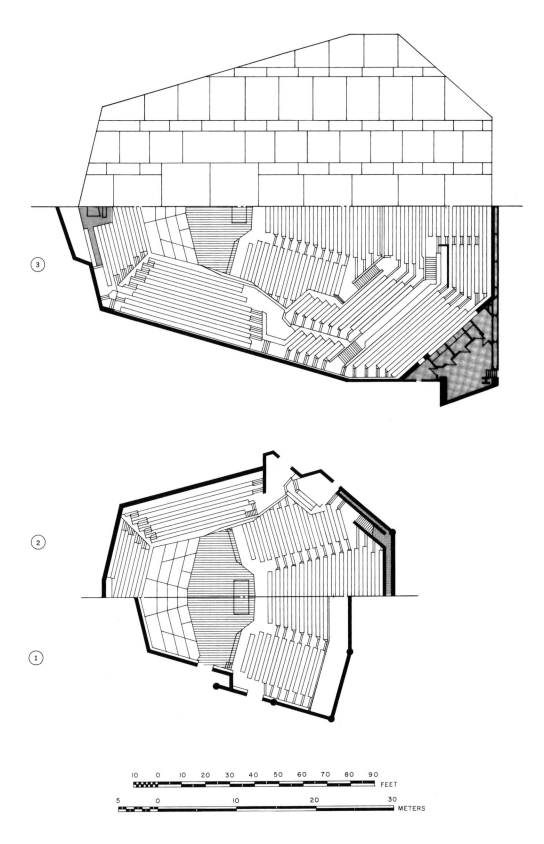

FEET

METERS

sound reflections to most of the audience in hall. The curved panels over the stage return early reflected sound energy to the musicians and to the audience sitting near the stage. The reverberation time is about 2.0 sec at mid-frequencies with full occupancy, and has almost the same value at all frequencies below 1,400 Hz. The low-frequency absorption that reduces the reverberation time somewhat more than expected may be caused by the large number of openings in the ceiling for air conditioning, lights, loudspeakers, and lines. Also, the heavily upholstered seats absorb low frequencies.

With full audience, this listener was surprised to sense that the early sound energy predominated over the later reverberant sound energy. Only in fortissimo passages and after musical stops did the reverberation seem to take an active part in the music. The high walls surrounding the stage seem to shield the side walls from receiving a significant part of the sound energy, and the stage enclosure as a whole seems to direct a large part of the energy to the audience in the rear half of the hall. These factors would explain why more energy is not sent into the upper part of the hall where the reverberant sound is developed. On the side walls themselves, about 50% of the surfaces are directed to the upper part of the hall in an effort to strengthen the reverberation.

The tone quality is excellent and the orchestral balance is good, except in the seats on either side of the orchestra where the sections nearest the listener are louder.

A group of 50 subjects, chosen in equal number from music professionals, concertgoers, and acousticians, were asked to judge the hall's quality in the months before the hall opened. They listened to a range of symphonic compositions in five parts of the hall. All judged the loudness and clarity to be very good. About 20% felt that the room did not take an active enough part in shaping the sound and that the reverberation time was a little short. Strangely, about 5% felt that the reverberation time was a little long, which indicates the range of subjective judgments even from a selected group. About 40% said that the overall impression was less than perfect. But the unanimous response of the group as a whole was "sehr gut" or "gut" (very good or good). The general conclusion was that the acoustics of the hall had received high marks.

ARCHITECTURAL AND STRUCTURAL DETAILS

Uses: Classical music. **Ceiling:** At least 1.4 in. (3.5 cm) of gypsum board (8.2 lb/ft², 40 kg/m²) formed as cylindrical sections, of several different lengths and widths (see the reflected ceiling plan), with a different radii of curvature of about 26 ft (8 m), oriented to direct reflections of sound energy to the various seating areas; larger panels are introduced over the podium to reflect more energy back to the musicians and to the seating areas near the stage; an unusually complex roof structure was used to eliminate noise. **Walls:** The walls around the platform and the sidewalls are two thicknesses of plywood, each 0.7 in. (1.8 cm) thick with a steel plate 0.12 in. (0.3 cm) thick cemented to the back; in the back parts of the side walls and the rear wall, the steel plates were eliminated, so that those areas resonate at a frequency of 63 to 110 Hz; a complex structure was used in the outer walls to eliminate noise. **Balcony fronts:** 1.1-in. (3 cm) artificial stone. **Floors:** parquet cemented to concrete base. **Stage floor:** Two layers of wood, each 1.6 in. (4 cm) thick. **Stage height:** 27.5 in. (70 cm). **Seating:** Rear of seat back molded plywood; upholstering chosen to give same absorption unoccupied as occupied; the wood part of the seatbacks is higher, therefore larger, in the steeper raked audience sections.

ARCHITECT: Leipziger Architekten Gemeinschaft, Dr. Skoda & Partner. ACOUSTICAL CONSULTANTS: Same as References. PHOTOGRAPHS: Foto Gert Mothes. REFERENCES: W. Fasold, E. Kuestner, H. Tennhardt, and H. Winkler, "Akustische Massnahmen im Neuen Gewandhaus, Leipzig," *Bauforschung, Baupraxis* **117**, 1–33 (1982); W. Fasold and U. Stephenson, *Gute Akustik von Auditorien*, Ernst & Sohn, Verlag, fur Architektur und technische Wissenschaften, Berlin (1993).

TECHNICAL DETAILS

V = 742,000 ft³ (21,000 m³)	S_a = 11,150 ft² (1,036 m²)	S_A = 12,880 ft² (1,197 m²)
S_o = 1,945 ft² (181 m²)	S_T = 14,825 ft² (1,378 m²)	N = 1,900
H = 65 ft (19.8 m)	W = 118 ft (36 m)	L = 106 ft (32.3 m)
D = 108 ft (32.9 m)	V/S_T = 50 ft (15.2 m)	V/S_A = 57.6 ft (17.6 m)
V/N = 390 ft³ (11 m³)	S_A/N = 6.78 ft² (0.63 m²)	S_a/N = 5.87 ft² (0.545 m²)
H/W = 0.55	L/W = 0.90	ITDG = 27 msec

NOTE: The terminology is explained in Appendix 1.

61

Munich

Herkulessaal

The Herkulessaal, located in the Residenz Museum of Munich, is an important venue for the rich musical life of the city. It was constructed in 1953 with the same proportions and dimensions as the old throne room of the Royal Palace, which was destroyed during World War II. The hall is rectangular in shape, seating 1,287 listeners on a flat floor. Along the sides and rear of the main floor are many rather large columns that partially conceal the access space at the ends of the rows of continental seating. None of the main floor seating is overhung by the single balcony. The ceiling is coffered and constructed from thin plywood, backed by sound-absorbing material (Beranek, 1996, p. 434). As part of the construction, the architect requested that old Flemish Gobelin tapestries, each with an area of 300 ft^2 (28 m^2), be hung on the upper side walls. Ten in number, these were removed some 15 years ago and put in museum storage. The tapestries have been replaced by exact photographic copies produced on cotton cloth using a large-format, digital-printing technique. The cloth is backed by a relatively thick mat and is spaced from the walls about 1 in. (2.5 cm).

Because of its relatively small size, the acoustical defects of the hall are largely disregarded. The deficiencies: The large columns shield the side walls from the direct sound of the performing entity, so that the room is deficient in lateral reflections. The thin wood of the ceiling absorbs the bass tones (100–400 Hz), although this is compensated to some degree by the "tapestries," which absorb the higher frequencies. The lack of upper side wall irregularities and so little hard surface area destroy the magnificent reverberation heard in halls like Boston's Symphony Hall and Vienna's Musikvereinssaal. Also, the flat floor results in a loss of direct high-frequency sound in the rear of the main floor owing to the absorption by intervening heads.

An attempt was made in the early years of the hall to overcome the deficiency in high-frequency sound at the rear of the hall by adding curved-plastic sound-

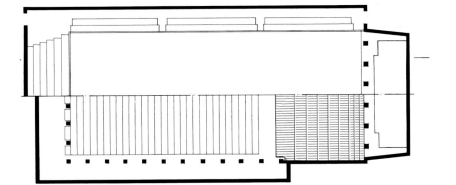

SEATING CAPACITY 1287

① 853

② 434

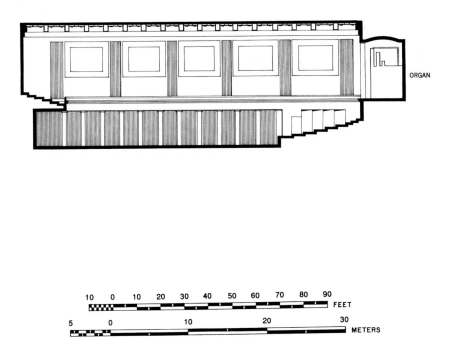

ORGAN

10 0 10 20 30 40 50 60 70 80 90 FEET

5 0 10 20 30 METERS

reflecting panels above the stage (faintly seen in the photograph). These have been removed owing to aging. Audience complaints after removal were minimal.

The reverberation time at mid-frequencies is nearly 2.0 sec with full occupancy, which is specially suited for orchestral music. Music critics say that the acoustics are "average," some adding that the sound is more distant when heard in the rear seats of the main floor and that the reverberation does not seem fully connected with the music.

ARCHITECTURAL AND STRUCTURAL DETAILS

Uses: Orchestra, 60%; chamber music, 10%; chorus, 5%; soloists, 10%; other, 15%. **Ceiling:** 0.2–0.4-in. (0.5–1-cm) plywood, backed by rock wool over 4–10-in. (10–25-cm) airspace; laboratory tests indicate that the ceiling has the following absorption coefficients: 100 Hz, 0.25; 200 Hz, 0.33; 400 Hz, 0.22; 800 Hz, 0.12, and 1,600 Hz and above, about 0.035. **Sidewalls:** Plaster on solid brick, except that 4,000 ft² (372 m²) consist of 0.25-in. (0.64-cm) sheets of gypsum board separated by 2 in. (5 cm) from the solid wall behind. **Rear wall:** Plastered brick; great doors are metal and plywood. **Added absorptive material:** Ten tapestries or heavy draperies hung on upper walls for architectural reasons. **Floors:** Tile on concrete and tile on the wood of the risers. **Stage floor:** 2-in. (5-cm) wooden planks over airspace. **Stage height:** 40 in. (102 cm). **Seating:** Upholstered on both sides of backrest and top of seat bottom. Underseat is solid.

ARCHITECT: Rudolf Esterer. ACOUSTICAL CONSULTANT: L. Cremer. PHOTOGRAPHS: Courtesy of the management. REFERENCES: L. Cremer, *Die Schalltechnik* **13**, 1–10 (1953); L. Cremer, *Gravesaner Blatter* **2/3**, 10–33 (1956); E. Meyer and R. Thiele, *Acustica*, **6**, 425–444 (1956).

TECHNICAL DETAILS

Concerts

V = 480,000 ft³ (13,590 m³)	S_a = 6,304 ft² (585.7 m²)	S_A = 7,250 ft² (674 m²)
S_o = 1,810 ft² (168 m²)	S_T = 9,060 ft² (842 m²)	N = 1,287
H = 51 ft (15.5 m)	W = 72 ft (22 m)	L = 105 ft (32 m)
D = 108 ft (32.9 m)	V/S_T = 53 ft (16.1 m)	V/S_A = 66.2 ft (20.2 m)
V/N = 373 ft³ (10.6 m³)	S_A/N = 5.63 ft² (0.52 m²)	S_a/N = 4.9 ft² (0.455 m²)
H/W = 0.71	L/W = 1.46	ITDG = 24 msec

NOTE: The terminology is explained in Appendix 1.

62

Munich

Philharmonie am Gasteig

The Gasteig is a clear statement of Munich's desire to be a major cultural center in Germany. The name is taken from "gachen footpath" and its site is high enough to command vistas of the center of the city, the river Isar, the famous Deutsches science museum, and old residential sections. In planning stages for 14 years, it has five principal components, a concert hall, multipurpose theater, music conservatory, college, and library. The Philharmonie, which opened in November 1985, presents a whole new appearance in the world of concert halls. Finished in American red oak, with red seat upholstering, it is visually striking. The multitudinous acoustic panels, free-standing from the walls and ceiling, mean that every effort was made to realize good acoustics.

Seating 2,387, plus 100 chorus, it is 287 seats larger than the Berlin Philharmonie and 587 seats larger than the Leipzig Gewandhaus. The great size, 1,050,000 ft³ (29,800 m³), necessary to achieve an occupied reverberation time of 1.95 sec at mid-frequencies, combined with the architect's concept of two wings, brought special problems to the acoustical consultant. A fan-shaped plan and a rising ceiling, without embellishment, would mean that early sound reflections would not reach listeners in the front two-thirds of the hall. To solve this problem, free-standing, reflecting panels on the side walls and a suspended sound-reflecting ceiling, diffuse in cross-section, and a stepped audience area with intermediate reflecting walls (vineyards) were introduced to provide multiple reflections within the first 100 msec to each listener.

The sound is different in various seating areas; not bad anywhere, just of varying texture. In the exact center of the seating areas the music sounds as though played in a cathedral. When seated off the center line, especially in elevated levels, the sound is well-balanced and of high quality, without giving the impression of a very large hall. The stage is very large, and often musical groups move toward the back wall for acoustical support, using the front and the elevated podium steps as reflecting surfaces. This is particularly helpful to soloists.

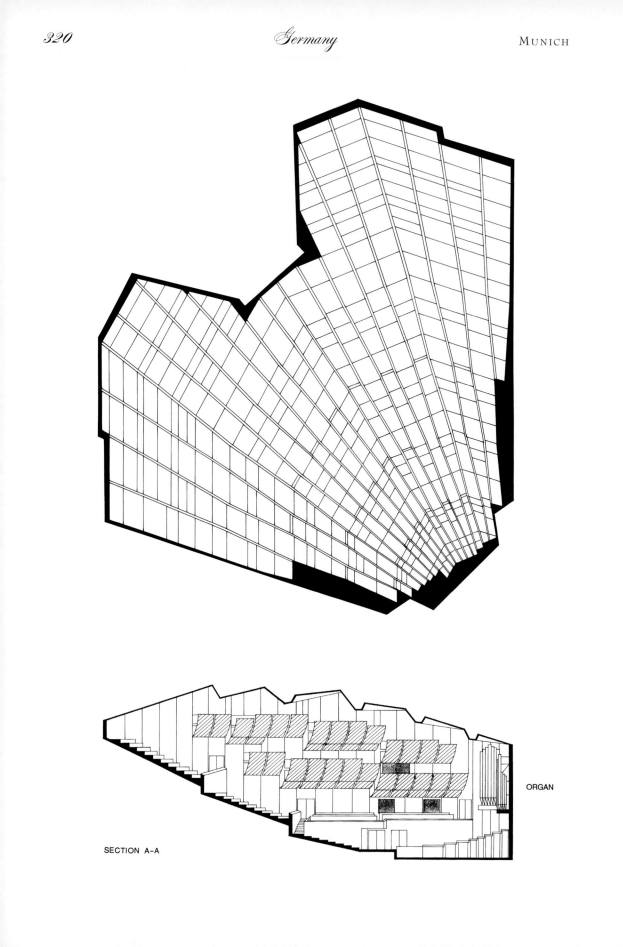

ORGAN

SECTION A-A

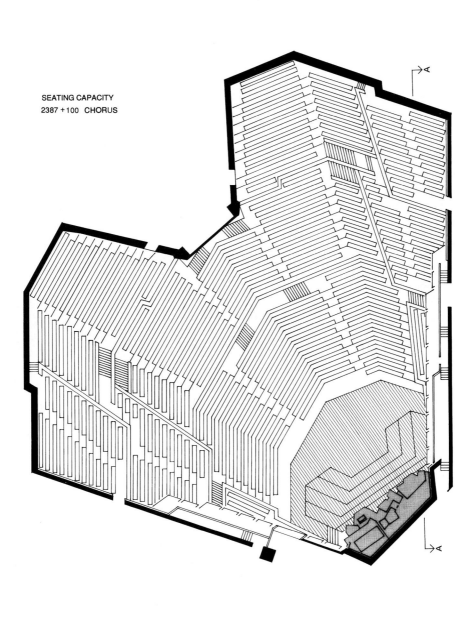

SEATING CAPACITY
2387 +100 CHORUS

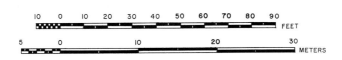

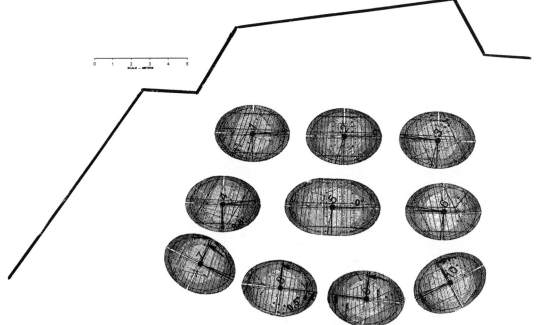

Since the hall's opening, Plexiglas reflectors have been added above the stage to improve acoustical contact within the orchestra (see photograph and drawing).

ARCHITECTURAL AND STRUCTURAL DETAILS

Uses: Primarily classical music. **Ceiling:** Wooden, with a uniform weight of 6 lb/ft² (30 kg/m²); wooden suspended panels are hung from the ceiling (see the photographs); veneer of American red oak. **Walls:** Concrete base with wooden particle-board lining weighing 2 to 10 lb/ft² (10 to 50 kg/m²); veneer of American red oak. **Floors:** Wood on prefabricated thin concrete plates. **Stage enclosure:** Reflections from rear wall, some from low side walls, more from organ and hung Plexiglas panels. **Stage floor:** Wood on wooden joists. **Stage height:** 30 in. (75 cm). **Seating:** Molded plywood backs, perforated wood underseats, upholstering on seat backs and on seat tops; arms are wood.

ARCHITECTS: C. F. Raue, E. Rollenhagen, G. Lindemann, G. Grossmann. ACOUSTICAL CONSULTANT: Mueller-BBM. PHOTOGRAPHS: Sigrid Neubert. REFERENCES: H. A. Mueller, U. Opitz, G. Volberg, "Structureborne sound transmission from the tubes of a subway into a building for a concert hall," *Proceedings of Internoise* **11**, 715–718 (1980); H. A. Mueller, U. Opitz, "Anweng der raumakustischen Modelltechniken bei der Planung des Konzertsaales der Muenchner Philharmonie," *Buildungswerk des Verbandes Deutscher Tonmeister,* 306–313 (1984); H. A. Mueller, U. Opitz, J. Reinhold, "Akustische Wirkung eines Raumes auf die ausfuehrenden Musiker," *db Deutsche Bauzeitung* **123**, 69–81 (1989).

TECHNICAL DETAILS

V = 1,050,000 ft³ (29,737 m³)	S_a = 14,305 ft² (1,329 m²)	S_A = 17,640 ft² (1,639 m²)
S_o = 2,476 ft² (230 m²)	S_T = 19,580 ft² (1,819 m²)	N = 2,387 + 100 chorus
H = 48 ft (14.6 m)	W = 168 ft (51.2 m)	L = 134 ft (40.8 m)
D = 145 ft (44.2 m)	V/S_T = 53.63 ft (16.35 m)	V/S_A = 59.5 ft (18.1 m)
V/N = 440 ft³ (12.45 m³)	S_A/N = 7.39 ft² (0.69 m²)	S_a/N = 5.99 ft² (0.557 m²)
H/W = 0.29	L/W = 0.80	ITDG = 29 msec

Note: S_T = S_A + 1,940 ft² (180 m²); see Appendix 1 for definition of S_T.

NOTE: The terminology is explained in Appendix 1.

63

Stuttgart

Liederhalle, Beethovensaal

*S*ituated on one of Stuttgart's busiest streets, the Liederhalle contains four halls, two of which seat 2,000 persons—one general purpose and the other for concerts—and the other two seat 750 and 350 persons. Of these, the 750-seat hall is the most spectacular from the outside; the Beethovensaal, from the inside. This concert hall has an unusual shape, like a grand piano, chosen for architectural and not acoustical reasons. The striking balcony rises like a grand staircase from the left side of the main floor, and with graceful lines soars over the main floor seats and sweeps around the rear of the hall to the wall on the right-hand side.

A large, convex concrete wall, on which there is a mosaic of painted wooden pieces and gold threads, connects the left side of the stage to the rising portion of the balcony. The right-hand wall of the hall, which is finished in teak, is irregular with projecting boxes and control booths for radio and television broadcasting. All the other walls are of teak. The main floor is flat. The ceiling is interestingly contoured to provide sound diffusion and to give desirable short-time-delay sound reflections. Reflecting panels, hanging over the orchestra, direct the sound of the strings to the audience at the rear of the main floor to overcome the disadvantages of intervening heads between listeners and the stage.

The Beethovensaal was renovated in 1992. Measurements by the acoustical consultants show that the modernizing has not affected the acoustical conditions. The hall has about the same reverberation time (1.6 sec) as Severance Hall in Cleveland (USA) but it is less reverberant than Symphony Hall in Boston (1.9 sec) and the Grosser Musikvereinssaal in Vienna (2.0 sec).

In general, visiting conductors have enjoyed this hall, saying that on the podium one can hear everything, even the smallest error in the performance. Some listeners speak of the hall as acoustically excellent. Others say it is very good, but that it falls below the quality of the Vienna hall. In my one concert experience, I found the sound quite satisfactory. The sound in the balcony is excellent. In the

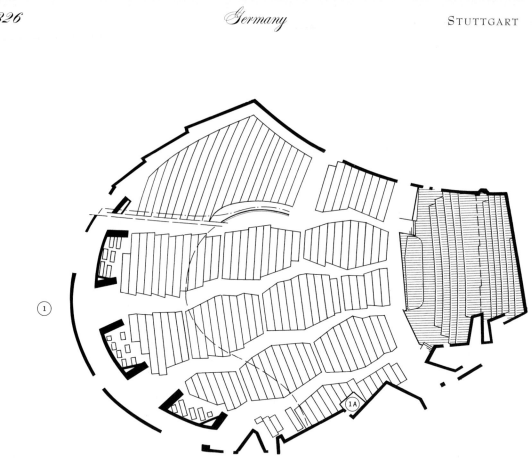

SEATING CAPACITY 2000

(1) 1175

(1A) 25

(2) 800

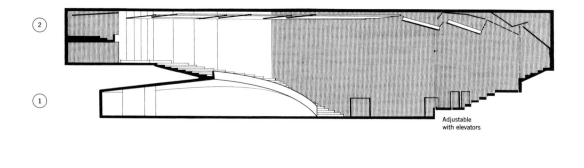

Adjustable
with elevators

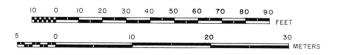

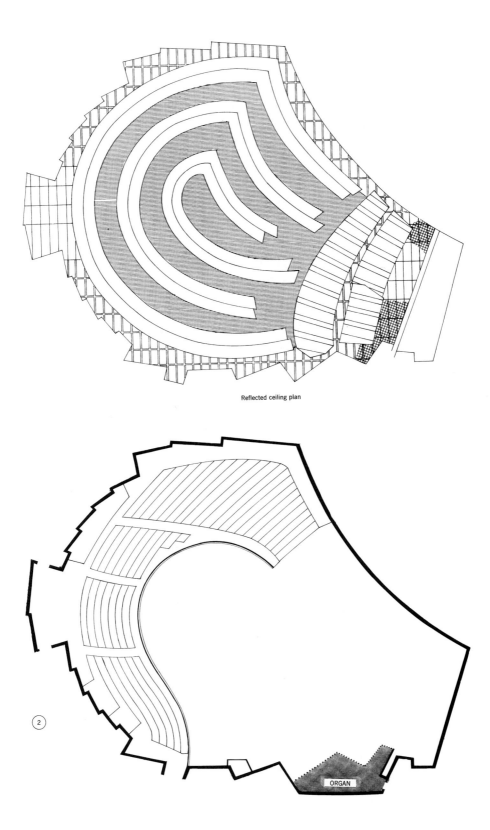

Reflected ceiling plan

ORGAN

rear half of the main floor, not under the balcony, I especially enjoyed the intimate, brilliant sound. I have not listened in the front half of the main floor.

ARCHITECTURAL AND STRUCTURAL DETAILS

Uses: Orchestra, organ, chamber, soloists, and general. **Ceiling:** Gypsum plaster on metal lath except for a strip about 10 ft (3 m) near the walls, one-third of which is 0.62-in. (1.6-cm) fiberboard over airspace and two-thirds slotted fiberboard backed by a layer of fiberglass. **Walls:** The left convex wall is concrete; other walls are plywood that varies from 0.38 in. (9.5 mm) to 0.88 in. (2.2 cm) thick over an airspace 1–5 in. (2.5–12.7 cm) deep; some panels are slotted with fiberglass behind; the slotted panels and the thin panels are backed with a fiberboard egg-crate structure; these variations yield walls that absorb sound over a wide range of low frequencies. **Floors:** Wood parquet; no carpet. **Stage enclosure:** Largely 0.88-in. plywood; the organ is located behind closeable "jalousies" on the right-hand side of the stage; splayed 0.88-in. (2.2-cm) plywood reflectors are hung over the orchestra and the front two rows of seating. **Stage floor:** 2-in. (5-cm) boards on elevators. **Stage height:** 49 in. (1.24 m). **Seating:** The front of the backrest and the top of the seat are upholstered; the under-seats are solid; the armrests are upholstered in leather.

ARCHITECT: A. Abel and R. Gutbrod. ACOUSTICAL CONSULTANTS: Lothar Cremer, Helmut Mueller and L. Keidel. PHOTOGRAPHS: Courtesy of the management. REFERENCES: L. Cremer, L. Keidel, and H. Mueller, *Acustica* **6**, 466–474 (1956); "Konserthaus Stuttgarter Liederhalle," Dr. Pollert Verlag, Stuttgart (1956).

TECHNICAL DETAILS

$V = 565,000$ ft³ (16,000 m³)	$S_a = 10,800$ ft² (1,000 m²)	$S_A = 14,000$ ft² (1,300 m²)
$S_o = 1,900$ ft² (176 m²)	$S_T = 16,500$ ft² (1,533 m²)	$N = 2,000$
$H = 44$ ft (13.4 m)	$W = 119$ ft (36.2 m)	$L = 137$ ft (41.8 m)
$D = 134$ ft (40.8 m)	$V/S_T = 34.2$ ft (10.4 m)	$V/S_A = 40.36$ ft (12.3 m)
$V/N = 282.5$ ft³ (8.0 m³)	$S_A/N = 7.0$ ft² (0.65 m²)	$S_a/N = 5.4$ ft² (0.5 m²)
$H/W = 0.37$	$L/W = 1.15$	ITDG $= 29$ msec

NOTE: The terminology is explained in Appendix 1.

64

Athens

Megaron, the Athens Concert Hall

𝒯he Athens Concert Hall Megaron encompasses two facilities, one dubbed "The Friends of Music," seating 1962 plus 30 spaces for wheelchairs, and the other "The Dimitris Mitropoulos Hall," seating 494. In daily reference, the large hall described here is usually called the "main hall" of the Megaron. The Megaron was to be completed in the early 1970s and the concrete-shell structure was erected at that time. For financial reasons, completion of the hall was delayed until 1991. With a new architect and acoustical consultant, a drastic change was made to the interior space in order to obtain greater acoustical intimacy, spaciousness, and clarity. The balcony was converted into an assemblage of parallel boxes, with fascia that provide the early lateral reflections to the audience areas that are necessary for good sound. The walls behind these boxes were brought in to reduce the acoustical width of the hall. The distance between opposite balcony faces is now about 88 ft (27 m) compared to 137 ft (42 m) in the original design. Irregularities on the sidewalls and ceiling provide necessary diffusion of the sound field as well as a better distribution of early sound reflections.

An unique feature is provision of means for transforming the concert hall into an opera house, although with limited stage depth. The six balcony boxes at the front sides of the stage can be retracted and four or five rows of seats removed to form a pit. Two towers imbedded several meters back from the front of the stage floor rise to form a proscenium. Space above the stage is then opened to reveal a fly (stage) tower from which a main curtain and other scenery can be dropped. About 2 m of the forestage is adjustable in height and can be lowered to create a full-size orchestra pit. For the large pit the seat count is reduced to 1700. With proscenium and pit available, the hall is also used for ballet and theater.

I attended a performance of Mahler's *Second Symphony,* with the Boston Symphony Orchestra on stage. I was seated in the rearmost box on the right side of the hall. At that location, the acoustics of the hall were entirely satisfactory. I was told that in some seats, in particular, those nearest the sides of the stage, the

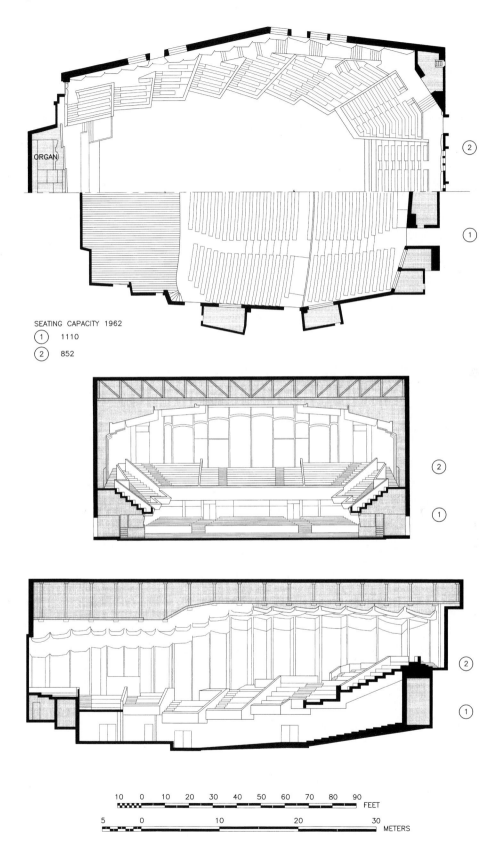

ORGAN

SEATING CAPACITY 1962
(1)   1110
(2)   852

(2)

(1)

(2)

(1)

(2)

(1)

10 0 10 20 30 40 50 60 70 80 90
FEET

5   0   10   20   30
METERS

sound is not as good. All in all the Megaron has made a significant contribution to the cultural facilities of Athens.

ARCHITECTURAL AND STRUCTURAL DETAILS

Uses: Symphonic concerts, chamber music, soloists, opera, and theater. **Ceiling:** Thick wood, with weight of 6 lb/ft² (30 kg/m²). **Walls:** Concrete with wood cladding. **Floors:** Parquet, glued on chipboard on a steel substructure. **Carpet:** None. **Stage height:** 30 in. (98 cm) above floor level at first row of seats. **Seating:** Plywood frame, seat bottoms upholstered, armrests wooden.

ARCHITECT: 1971, building frame, Heinrich Keilholz; 1991, Elias Scrubelos. ACOUSTICAL CONSULTANT: Müller-BBM.

TECHNICAL DETAILS

Concerts

V = 674,200 ft³ (19,100 m³)	S_a = 10,050 ft² (934 m²)	S_A = 12,729 ft² (1,183 m²)
S_o = 3,088 ft² (287 m²)	S_T = 14,656 ft² (1363 m²)	N = 1,962
V/S_T = 46 ft (14 m)	V/S_A = 53 ft (16.1 m)	V/N = 344 ft³ (9.73 m³)
S_A/N = 6.49 ft² (0.603 m²)	S_a/N = 5.12 ft² (0.476 m²)	H = 56 ft (17.1 m)
W = 117 ft (35.7 m)	L = 104 ft (31.8 m)	D = 106 ft (32.3m)
H/W = 0.48	L/W = 0.89	ITDG = 38 msec

Operas

V = 529,500 ft³ (15,000 m³)	S_a = 9,270 ft² (862 m²)	S_A = 11,735 ft² (1,090 m²)
S_o = 1,237 ft² (115 m²)	S_p = 3,440 ft² (320 m²)	S_T = 16,412 ft² (1,525 m²)
V/S_T = 32.3 ft (9.84 m)	V/S_A = 45.1 ft (13.8 m)	N = 1700
V/N = 311 ft³ (8.82 m³)	S_A/N = 6.9 ft² (0.64 m²)	S_a/N = 5.45 ft² (0.51 m²)
H and W, as above	L = 111 ft (33.8 m)	D = 113 ft (34.3 m)
H/W = 0.48	L/W = 0.95	ITDG = 40 msec

NOTE: The terminology is explained in Appendix 1.

65

Budapest

Magyar Állami Operahaz

The Budapest State Opera House has a long and distinguished history. The house is located in the center of the city, on the Pest side. Hungary, a country with a population equal to that of New York City, has produced an extraordinary number of important composers, directors, and musicians. Music is taught in the schools, and attendance at musical events is at a high level. Budapest, the size of Greater Boston, possesses three venues for its musical performances, the Large Hall in the Academy of Music, seating 1,170 persons; the Erkel Theatre, rebuilt in 1951, seating 2,340 (for both concerts and opera); and the State Opera House, seating 1,277.

The State Opera House, Magyar Állami Operahaz, opened on September 27, 1884, with the Emperor Franz Ferdinand I in attendance. Performed were excerpts from two of Ferenc Erkel's operas and one act from Wagner's *Lohengrin*. Gustav Mahler was music director of the House between 1888 and 1891. Some years after World War II, the house was declared unsafe. A big reconstruction took place between 1981 and 1984. The auditorium and surrounding spaces were rebuilt exactly as they were before. The stage area underwent a major expansion and is now equipped with the finest of stage machinery. On September 27, 1984, exactly a hundred years to the day after its founding, the house re-opened.

The gold and marble grand stairway and foyers with their exquisite chandeliers provide a fitting entrance to the auditorium. The colors of gold and red dominate the house's all-wooden interior. The reverberation time, fully occupied and at middle frequencies, is a little over 1.2 sec, and the bass strength is about the same as those of Milan's La Scala. The small size, 314,000 ft³ (8,900 m³), makes it one of the most intimate grand opera houses in the world. The low reverberation time prevents it from becoming overly loud. Needless to say, its acoustics are judged excellent for opera, although they are less favorable for symphonic music.

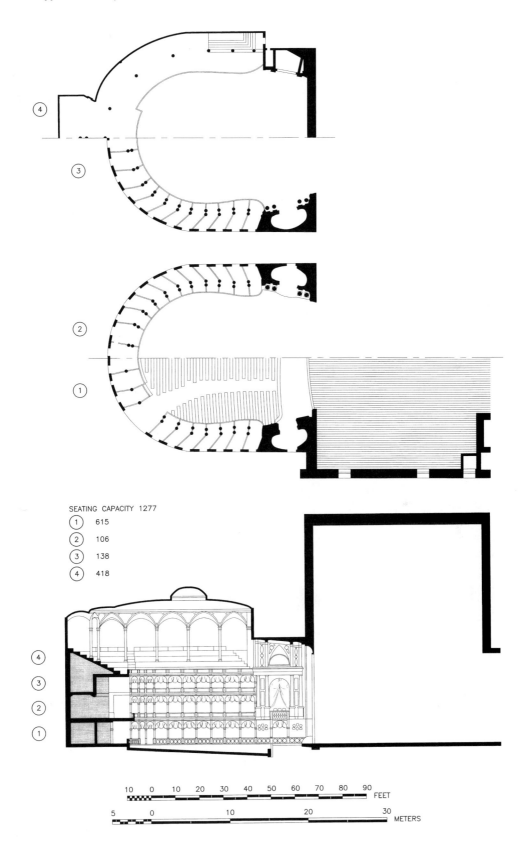

SEATING CAPACITY 1277

- (1) 615
- (2) 106
- (3) 138
- (4) 418

ARCHITECTURAL AND STRUCTURAL DETAILS

Uses: Opera principally, concerts occasionally. **Interior materials:** All wooden. **Concerts:** The pit floor is raised to the stage level and the orchestra sits in front of the proscenium curtain, below the large state boxes on either side. **Carpets:** On all aisles and on floors of boxes. **Seats:** Fully upholstered except for arms, which are wooden.

ARCHITECT: Built from plans of the tender, Miklos Ybl. The original construction took nine years and used a totally new stage hydraulic system, first in the world, which served without problems for 84 years. RECONSTRUCTION ARCHITECT: Maria Siklos.

TECHNICAL DETAILS

Opera

$V = 314{,}170$ ft^3 (8,900 m^3)	$S_a = 5{,}574$ ft^2 (518 m^2)	$S_A = 6{,}929$ ft^2 (644 m^2)
$S_{pit} = 624$ ft^2 (58 m^2)	$S_p = 1{,}710$ ft^2 (159 m^2)	$S_T = 9{,}263$ ft^2 (861 m^2)
$N = 1{,}277$	$H = 61$ ft (18.6 m)	$W = 50$ ft (15.2 m)
$L = 82$ ft (25 m)	$D = 103$ ft (31.4 m)	$V/S_T = 33.9$ ft (10.3 m)
$V/S_A = 45.3$ ft (13.8 m)	$V/N = 246$ ft^3 (6.97 m^3)	$S_A/N = 5.42$ ft^2 (0.50 m^2)
$S_a/N = 4.36$ ft^2 (0.40 m^2)	$H/W = 1.22$ $L/W = 1.64$	ITDG $= 23$ msec

NOTE: The terminology is explained in Appendix 1.

Pátria Hall,
Budapest Convention Centre

The Pátria Hall of the Convention Centre opened in 1985. A wide-fan-shaped room, it posed difficult acoustical problems. Nevertheless, with the seating capacity of 1,750 and a mid-frequency reverberation time of 1.6 sec, fully occupied, the hall has been selected as a concert venue by touring orchestras.

The acoustician's solution to the wide-fan plan was to design a system of reflecting plates for the side walls that directed sound to the upper rear part of the auditorium so that it would be reflected across to the other side of the hall and back. This raised the reverberation time from a computed low of 1.3 to 1.6 sec.

The necessary lateral reflections to the main floor of the hall are provided by the side walls and ceiling of the stage and additional early reflections by the suspended reflectors in the fore-part of the room. These stage reflectors, both on the sides and at the front edge, resemble fingers, with different slopes, that reflect sound energy both to the musicians and to the middle of the main-floor seating. As is the case in many halls, the best sound is to be heard in the front center of the balcony.

The stage has an area of 1,680 ft² (156 m²) and can be extended 3.9 ft (1.2 m) to accommodate a large orchestra and large chorus. For recitals or chamber music concerts chairs are placed on the stage for audience seating. Although the balcony overhang is large, the opening is also large and the front edge is shaped to reflect sound to the rear seats. Much of the floor is flat to accommodate convention and banquet activities.

The reflecting structures on the side walls and ceiling are 1.6-in.-thick layered plates that do not absorb the low frequencies appreciably and the chairs on the main floor are not heavily upholstered. Therefore, the bass ratio (ratio of low to middle frequency reverberation times) and the low-frequency level is high enough to contribute to the feeling of spaciousness in the hall.

One attendance at an all-Bartok concert in this hall by the Chicago Symphony Orchestra under Maestro George Solti was an experience to be remembered.

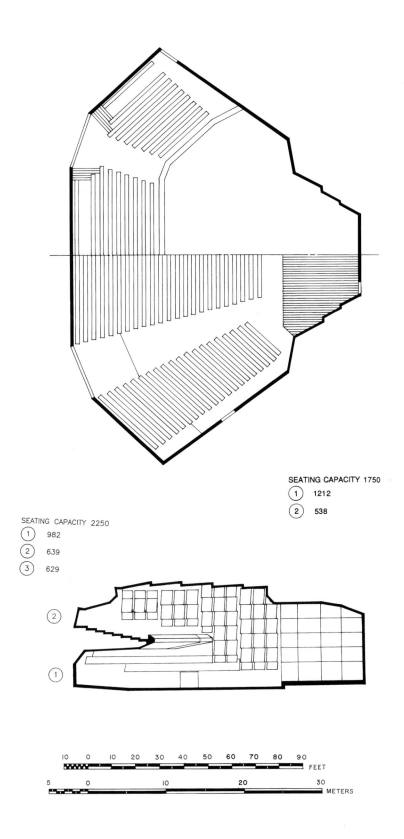

SEATING CAPACITY 1750

① 1212

② 538

SEATING CAPACITY 2250

① 982

② 639

③ 629

②

①

```
10   0   10   20   30   40   50   60   70   80   90
                                                      FEET

5        0              10              20            30
                                                      METERS
```

ARCHITECTURAL AND STRUCTURAL DETAILS

Uses: Concerts, chamber music, and recitals. **Ceiling:** Thick plaster; hung panels are gypsum board, 1 in. (2.5 cm) thick. **Walls:** Concrete covered by sound-reflecting panels, 1.6 in. (4 cm) thick, suspended from steel tubing; the lowest part of the sidewalls are splayed concrete; rear walls behind seating, splayed plyboards, with low-frequency resonant absorbers. **Floors:** Parquet cemented to solid concrete. **Carpet:** None. **Stage height:** 35 in. (0.90 m). **Seating:** Plywood seat back, covered with stretched fabric; plywood underseat with polyvinyl cushion; backrest covered by porous fabric; armrests wooden.

ARCHITECT: J. Finta. ACOUSTICAL CONSULTANTS: W. Fasold, B. Marx, H. Tennhardt, and H. Winkler. REFERENCE: Fasold, Marx, Tennhardt, and Winkler, "Raumakustische Massnahmen im Budapester Kongresszentrum," *Bauakademie der Deutschen Demokratischen Republik,* Berlin (1986).

TECHNICAL DETAILS

V = 473,150 ft³ (13,400 m³)	S_a = 12,270 ft² (1,140 m²)	S_A = 13,840 ft² (1,286 m²)
S_o = 1,680 ft² (156 m²)	S_T = 15,520 ft² (1,442 m²)	N = 1,750
H = 43 ft (13.1 m)	W = 138 ft (42.1 m)	L = 86 ft (26.2 m)
D = 87 ft (26.5 m)	V/S_T = 30.5 ft (9.3 m)	V/S_A = 34.2 ft (10.4 m)
V/N = 270 ft³ (7.66 m³)	S_A/N = 7.9 ft² (0.73 m²)	S_a/N = 7.0 ft² (0.65 m²)
H/W = 0.31	L/W = 0.62	ITDG = 44 msec

NOTE: The terminology is explained in Appendix 1.

67

Belfast

Waterfront Hall

Located on the River Lagan, the dramatic Waterfront Hall building resembles a massive alien spacecraft. In form, it is a five-story cylinder with two appendages, one containing a small auditorium and the other an outdoor dining terrace. Heavily glazed, two-thirds of the periphery contains foyers wrapped around the hall that are open to the public.

Entering the hall itself is a dramatic experience—one encounters terraced seating running the length of the sidewalls and also behind the stage, and an unusual, circular, sound-reflecting "ceiling" hung from the geometrical center of the flat ceiling that covers the whole space beneath the dome. The hall seats 2,250, but with the main-floor seats removed and the floor converted to standing space for promenade (pops) concerts, the audience size increases to 3,000. The percentage of the audience seated to the sides and rear of the stage is only about 12 percent.

The hall is multipurpose, with primary emphasis on orchestral use. A very high reverberation time, 2.0 sec at mid-frequencies, fully occupied, was chosen. To make allowance for the amplified music and speech uses, retractable absorbent fabric drapes hang behind each of the two lighting bridges. However, even with the drapes, the reverberation time drops only to 1.7 sec, fully occupied. About a dozen large circular panels hang at a height of about 38 ft (11.6 m) above the stage, to enable the musicians to hear each other and to provide mixing of the sound.

For theater or opera, a high portion of the ceiling directly above the stage contains a fly grid from which scenery can be hung. Such scenery will reduce the reverberation time further. The distance between the farthest listener and the front of the stage is 119 ft (36 m).

Sound diffusion is provided by the hung central "ceiling," by various panels below the ceiling, by the fronts of the simultaneous-interpretation and broadcasting rooms, the organ platform, galleries around the stage area, sloping balcony fronts and lighting bridges. It is not apparent to what extent lateral early sound reflections are provided.

ARCHITECTURAL AND STRUCTURAL DETAILS

Uses: Symphonic and choral concerts, recitals, conferences, and sports. **Ceiling:** Underside of concrete roof, pre-cast concrete blocks; Decorative ceiling, acoustically transparent, timber slat areas, and fixed concentric rings of acoustic reflectors composed of convex curved plywood on steel frames; underside of seating tiers, thick plasterboard with airspace above. **Orchestra canopy:** Circular (different sizes) convex DRG panels, adjustable in height and inclination. **Walls:** Plaster on concrete blocks; sides and rear of orchestra seating and sides of main floor seating—thick veneered wooden paneling. **Floor:** Heavy wood, covered with thin pile nylon carpet. **Stage floor:** Heavy wood with elevators to provide orchestra risers or to join the main floor for arena events. **Choir area:** Also used for orchestra seating; Can extend down to orchestra level, by use of seating wagons and stage risers, together with a removable section of the choir railing. **Adjustable sound-absorbing areas:** Motorized drapes in ceiling void and manually pulled drapes on sidewalls. **Seats:** Molded plywood: Lower part of seatback and upper side of seat bottom upholstered; choir seated on upholstered benches; majority of main-floor seats mounted on wagons, so that, when removed, provide a flat arena surface.

ARCHITECT: Robinson & McAllwaine. ACOUSTICAL CONSULTANT: Sandy Brown Associates. THEATER CONSULTANT: Carr and Angier.

TECHNICAL DETAILS

$V = 1{,}088{,}000$ ft^3 (30,800 m^3)	$S_a = 11{,}513$ ft^2 (1,070 m^2)	$S_A = 14{,}000$ ft^2 (1,301 m^2)
$S_o = 2{,}150$ ft^2 (200 m^2)	$S_T = 15{,}935$ ft^2 (1,481 m^2)	$N = 2{,}250$
$H = 63.6$ ft (19.4 m)	$W = 98.4$ ft (30 m)	$L = 98.4$ ft (30 m)
$D = 125$ ft (38 m)	$V/S_T = 68.2$ ft (20.8 m)	$V/S_A = 77.7$ ft (23.7 m)
$V/N = 484$ ft^3 (13.7 m^3)	$S_A/N = 6.22$ ft^2 (0.58 m^2)	$S_a/N = 5.12$ ft^2 (0.48)
$H/W = 0.64$	$L/W = 1.0$	ITDG $= 34$ msec

Note: $S_T = S_A + 1{,}940$ ft^2 (180 m^2); see Appendix 1 for definition of S_T.

NOTE: The terminology is explained in Appendix 1.

67

Belfast

Waterfront Hall

$\mathcal{L}$ocated on the River Lagan, the dramatic Waterfront Hall building resembles a massive alien spacecraft. In form, it is a five-story cylinder with two appendages, one containing a small auditorium and the other an outdoor dining terrace. Heavily glazed, two-thirds of the periphery contains foyers wrapped around the hall that are open to the public.

Entering the hall itself is a dramatic experience—one encounters terraced seating running the length of the sidewalls and also behind the stage, and an unusual, circular, sound-reflecting "ceiling" hung from the geometrical center of the flat ceiling that covers the whole space beneath the dome. The hall seats 2,250, but with the main-floor seats removed and the floor converted to standing space for promenade (pops) concerts, the audience size increases to 3,000. The percentage of the audience seated to the sides and rear of the stage is only about 12 percent.

The hall is multipurpose, with primary emphasis on orchestral use. A very high reverberation time, 2.0 sec at mid-frequencies, fully occupied, was chosen. To make allowance for the amplified music and speech uses, retractable absorbent fabric drapes hang behind each of the two lighting bridges. However, even with the drapes, the reverberation time drops only to 1.7 sec, fully occupied. About a dozen large circular panels hang at a height of about 38 ft (11.6 m) above the stage, to enable the musicians to hear each other and to provide mixing of the sound.

For theater or opera, a high portion of the ceiling directly above the stage contains a fly grid from which scenery can be hung. Such scenery will reduce the reverberation time further. The distance between the farthest listener and the front of the stage is 119 ft (36 m).

Sound diffusion is provided by the hung central "ceiling," by various panels below the ceiling, by the fronts of the simultaneous-interpretation and broadcasting rooms, the organ platform, galleries around the stage area, sloping balcony fronts and lighting bridges. It is not apparent to what extent lateral early sound reflections are provided.

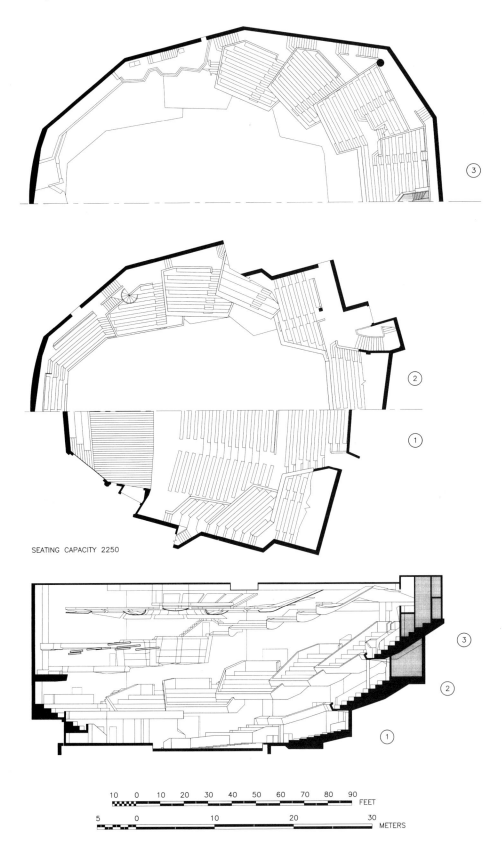

SEATING CAPACITY 2250

10   0   10   20   30   40   50   60   70   80   90   FEET

5   0   10   20   30   METERS

ARCHITECTURAL AND STRUCTURAL DETAILS

Uses: Symphonic and choral concerts, recitals, conferences, and sports. **Ceiling:** Underside of concrete roof, pre-cast concrete blocks; Decorative ceiling, acoustically transparent, timber slat areas, and fixed concentric rings of acoustic reflectors composed of convex curved plywood on steel frames; underside of seating tiers, thick plasterboard with airspace above. **Orchestra canopy:** Circular (different sizes) convex DRG panels, adjustable in height and inclination. **Walls:** Plaster on concrete blocks; sides and rear of orchestra seating and sides of main floor seating—thick veneered wooden paneling. **Floor:** Heavy wood, covered with thin pile nylon carpet. **Stage floor:** Heavy wood with elevators to provide orchestra risers or to join the main floor for arena events. **Choir area:** Also used for orchestra seating; Can extend down to orchestra level, by use of seating wagons and stage risers, together with a removable section of the choir railing. **Adjustable sound-absorbing areas:** Motorized drapes in ceiling void and manually pulled drapes on sidewalls. **Seats:** Molded plywood: Lower part of seatback and upper side of seat bottom upholstered; choir seated on upholstered benches; majority of mainfloor seats mounted on wagons, so that, when removed, provide a flat arena surface.

ARCHITECT: Robinson & McAllwaine. ACOUSTICAL CONSULTANT: Sandy Brown Associates. THEATER CONSULTANT: Carr and Angier.

TECHNICAL DETAILS

V = 1,088,000 ft³ (30,800 m³)	S_a = 11,513 ft² (1,070 m²)	S_A = 14,000 ft² (1,301 m²)
S_o = 2,150 ft² (200 m²)	S_T = 15,935 ft² (1,481 m²)	N = 2,250
H = 63.6 ft (19.4 m)	W = 98.4 ft (30 m)	L = 98.4 ft (30 m)
D = 125 ft (38 m)	V/S_T = 68.2 ft (20.8 m)	V/S_A = 77.7 ft (23.7 m)
V/N = 484 ft³ (13.7 m³)	S_A/N = 6.22 ft² (0.58 m²)	S_a/N = 5.12 ft² (0.48)
H/W = 0.64	L/W = 1.0	ITDG = 34 msec
Note: $S_T = S_A$ + 1,940 ft² (180 m²); see Appendix 1 for definition of S_T.		

NOTE: The terminology is explained in Appendix 1.

Binyanei Ha'Oomah

The Jerusalem Congress Hall, containing 3,142 seats and opened in 1960, was designed initially for conventions. The shape of the hall, the balcony around the stage, and the inward sloping walls were fixed by that purpose. Construction was started in 1950, suspended for several years, and resumed with the purpose changed to include musical performances. Such modifications as could be made were then recommended by the acoustical consultant. The reverberation time was increased to 1.75 sec at mid-frequencies with full audience. The hall is constructed of materials that preserve the warmth and liveness of the sound.

The overall architectural effect is one of simplicity and nobility. The rich wood interior is especially designed not to absorb the bass. When flooded with light, the white coffers overhead act to diffuse the light and provide uniform illumination to the seating areas.

The Binyanei Ha'Oomah has received favorable comments from the conductors who have used it. Israeli Philharmonic Orchestra members prefer it to the Mann Auditorium, primarily because of its longer reverberation time.

A typical review of a performance by the Netherlands Kamerorkest:

> But as soon as the [chamber music] concert started, beauty of tone and the complete union of the 20 musicians into a single body made this concert one of the finest heard for some time. The sensitive acoustics allowed one to enjoy every phrase and every shade of dynamics employed in a highly polished presentation.

Though it is not as good for concert performances as smaller rectangular concert halls (for example, the Grosser Musikvereinssaal in Vienna), Binyanei Ha'Oomah is superior for symphonic music to most of the larger multipurpose halls built in its time. Its acoustics are satisfactory for lectures and conferences, many of which can be heard without an amplifying system, but a state-of-the-art sound system is incorporated.

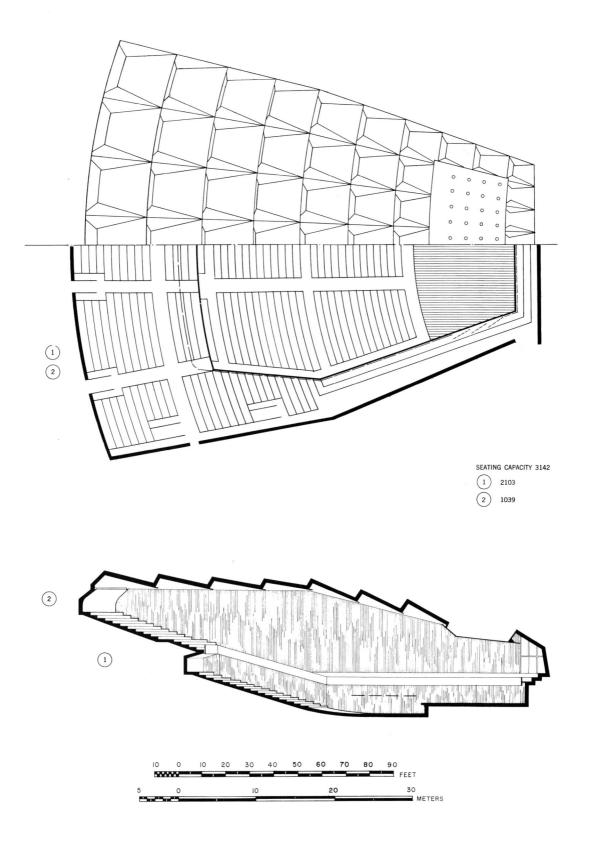

SEATING CAPACITY 3142

① 2103

② 1039

ARCHITECTURAL AND STRUCTURAL DETAILS

Uses: Orchestra, general music, and meetings. **Ceiling:** Prefabricated from 2-in. (5-cm) gypsum plasterboard. **Sidewalls:** 0.5-in. (1.25-cm) plywood affixed to 2-in. plaster. **Rear walls:** Upper, 0.5-in. plywood over airspace; lower, 0.5-in. plywood on masonry. **Floors:** Asphalt tiles on concrete. **Carpet:** None. *Stage enclosure:* Ceiling, 0.5-in. plywood, fabricated into a movable canopy; back wall: 0.5-in. plywood on masonry. *Side walls:* Large plywood wooden doors, adjustable into sawtooth splays. **Stage floor:** Heavy wood over air space. **Stage height:** 44 in. (112 cm). **Seating:** Tops of seat bottoms and fronts of backrests upholstered; underseats perforated with glass fiber blanket inside.

ARCHITECT: Rechter, Zarhy, Rechter. ACOUSTICAL CONSULTANT: Bolt Beranek and Newman. PHOTOGRAPHS: Courtesy of the architects. REFERENCES: L. L. Beranek and D. L. Klepper, "The acoustics of the Binyanei Ha'Oomah Jerusalem Congress Hall," *J. Acoust. Soc. Am.* **33**, 1690–1698 (1961).

TECHNICAL DETAILS

V = 873,000 ft³ (24,700 m³)	S_a = 18,000 ft² (1,672 m²)	S_A = 23,000 ft² (2,137 m²)
S_o = 2,800 ft² (260 m²)	S_T = 24,940 ft² (2,317 m²)	N = 3,142
H = 45 ft (13.7 m)	W = 100 ft (30.5 m)	L = 122 ft (37.2 m)
D = 145 ft (44.2 m)	V/S_T = 35.0 ft (10.66 m)	V/S_A = 38 ft (11.6 m)
V/N = 278 ft³ (7.9 m³)	S_A/N = 7.32 ft² (0.68 m²)	S_a/N = 5.72 ft² (0.532 m²)
H/W = 0.45	L/W = 1.22	ITDG = 26 msec

Note: $S_T = S_A$ + 1,940 ft² (180 m²); see Appendix 1 for definition of S_T.

NOTE: The terminology is explained in Appendix 1.

69

Tel Aviv

Fredric R. Mann Auditorium

The Fredric R. Mann Auditorium is a pleasant contemporary concert hall. The unusual arrangement of the seating reduces the apparent size of the hall, even for the viewer in the balcony. The pattern and lighting of the ceiling and the warm interior of wood give this hall a handsome yet relaxed appearance.

The design of the Mann Auditorium was begun in 1951 and the hall opened in 1957. On the advice of the late Serge Koussevitsky and the Israel Philharmonic Orchestra, the building committee and the architects wanted a hall that resembled Kleinhans Music Hall in Buffalo rather than one of the older rectangular halls.

It was planned that the reverberation time in the Mann hall should be about 1.9 sec, fully occupied, a number about halfway between the reverberation time in Boston Symphony Hall (then thought to be 2.3 sec, fully occupied) and that in Kleinhans Hall in Buffalo (then thought to be 1.5 sec). Unfortunately, the design preceded the author's findings in 1957 that the sound absorption of an audience must be calculated on the basis of the number of square meters that the chairs sit over and not on the number of people in them. The result, for a certain cubic volume, is that the reverberation time will be lower in the completed hall than that calculated. The reverberation time in the Mann hall measures 1.55 sec at mid-frequencies, fully occupied. The hall seats 2,715.

Because of its low reverberation time, there have been complaints by the members of the Israel Philharmonic Orchectra and music critics. The orchestra wrote, "Our main complaint is the lack of a certain sonority which we especially observe in the sound of our string body and which is particularly noticeable at the climaxes of the orchestral music."

Plans are underway for a major revision of the hall which should reduce the amount of thin wood, which absorbs the bass, and increase the height of the inner and stage roofs, which would increase the reverberation time.

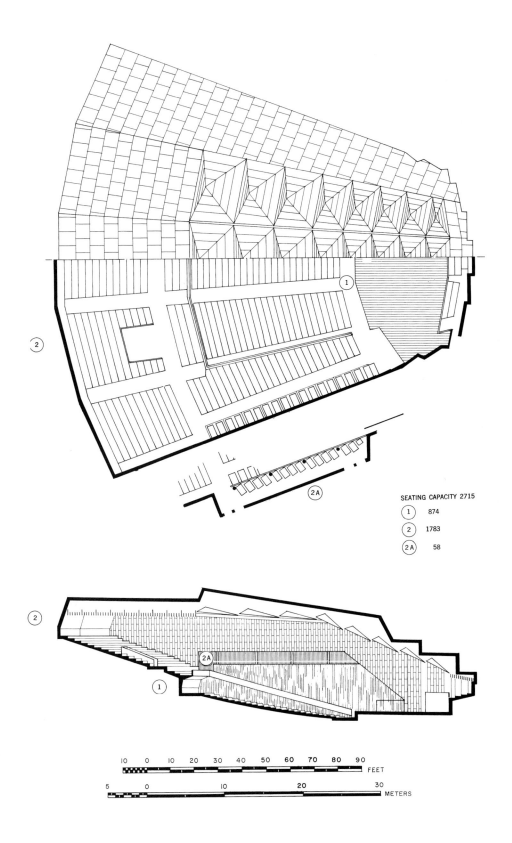

SEATING CAPACITY 2715

① 874

② 1783

②A 58

10 0 10 20 30 40 50 60 70 80 90
FEET

5   0   10   20   30
METERS

ARCHITECTURAL AND STRUCTURAL DETAILS

Uses: Orchestra, chamber music, and soloists. **Ceiling:** Solid upper ceiling, 0.75-in. (1.9-cm) plaster; hung center portion (1/3 of area) is pyramid shaped to diffuse sound, made of 0.16-in. (0.4-cm) asbestos board; visible hung portion is 6-in. (15-cm) deep wooden egg-crate sections, each 12 × 12 in. (30 × 30 cm). **Side walls and rear walls:** On the upper side walls are three types of panels, each 18 × 56 in. (46 × 142 cm). One-third are 0.25-in. (0.6-cm) ash plywood, to the back of which a 2 × 2 × 0.75-in. (5 × 5 × 1.9-cm) egg-crate structure is glued. One-third have the same egg-crate backing, but with two layers of 0.25-in. plywood. The remaining one-third are a sandwich of two 0.25-in. plywood panels, with about 0.5-in (1.25-cm) of softwood filling, making a solid panel of 1 in. (2.5 cm). The lower part of the side walls is vertical slats about 2-in. (5 cm) wide, with no airspaces behind; the rear wall is 0.25-in. plywood panels. **Carpet:** None. **Floor:** Vinyl tile on concrete. **Stage enclosure:** None. **Stage floor:** Heavy wood over airspace. **Stage height:** 30 in. (76 cm). **Seating:** Front of backrest and top of seat are upholstered.

ARCHITECT: Z. Rechter and D. Karmi. ASSOCIATE ARCHITECT: J. Rechter. ACOUSTICAL CONSULTANT: Bolt Beranek and Newman. PHOTOGRAPHS: Courtesy of the architects. REFERENCES: L. L. Beranek, "Acoustics of the F. R. Mann Concert Hall," *J. Acoust. Soc. Am.* **31**, 882–892 (1959).

TECHNICAL DETAILS

V = 750,000 ft³ (21,240 m³)	S_a = 14,500 ft² (1,350 m²)	S_A = 18,300 ft² (1,700 m²)
S_o = 2,100 ft² (195 m²)	S_T = 20,240 ft² (1,880 m²)	N = 2,715
H = 40 ft (12.2 m)	W = 77 ft (23.5 m)	L = 100 ft (30.5 m)
D = 127 ft (38.7 m)	V/S_T = 37.05 ft (11.30 m)	V/S_A = 41.0 ft (12.49 m)
V/N = 276 ft³ (7.82 m³)	S_A/N = 6.74 ft² (0.626 m²)	S_a/N = 5.34 ft² (0.497 m²)
H/W = 0.52	L/W = 1.30	ITDG = 30 msec

Note: $S_T = S_A$ + 1,940 ft² (180 m²); see Appendix 1 for definition of S_T.

NOTE: The terminology is explained in Appendix 1.

Teatro Alla Scala

*O*n August 3, 1778, La Scala opened its doors to a future of great music, glamor, and unequaled tradition. The house was built by Maria Theresa of Austria (which then ruled Milan) and its architect was Giuseppi Piermarini. The exterior of La Scala looks today very much as it did in 1778. Until World War II it experienced only minor changes—in 1883 it got electric lights. In 1943 La Scala was the victim of a bombing attack; the walls remained standing, but little else. In May 1946 it reopened, "restored to music by Arturo Toscanini," its appearance almost as before. Every original detail was preserved.

La Scala is a beautiful and engaging theater. It is horseshoe in plan with high balcony faces and a vaulted ceiling 6 ft higher at the center than at the sides. A recent study found that 22 world conductors placed it "very good," exceeded in rating only by the Teatro Colón in Buenos Aires and the Semperoper in Dresden. But acoustical measurements, discussed in Chapter 5, show that its excellent acoustical quality exits only in the boxes at all levels, not on the main floor.

On January 1, 2002, La Scala was closed to be reopened in the fall of 2004. The entire stage area and fly tower are to be completely modernized. Extensive areas of carpet will be taken up and the wooden floors left bare. New seats, with less upholstering, will be installed. A spokesman for La Scala said, "But the truth of the matter is that if you ask the musicians, they will tell you that the acoustics are not that great now. The wood floors will improve the sound."

The openings to the boxes are only 4.5 ft² (1.37 m²), so that the "wall" that they present is only about 40 percent open; that is to say, 60 percent of the incident sound is reflected back into the center space. This large reflecting area and the resulting small cubic volume of the house makes the sound more intimate. Because of the carpet and plush seats, the reverberation time is short, 1.24 sec at midfrequencies, fully occupied. The Buenos Aires and Dresden houses have reverberation times of 1.6 sec. Owing to the "wall" the sound is returned to the stage with an intensity not reached in other large opera houses. The vaulted ceiling returns

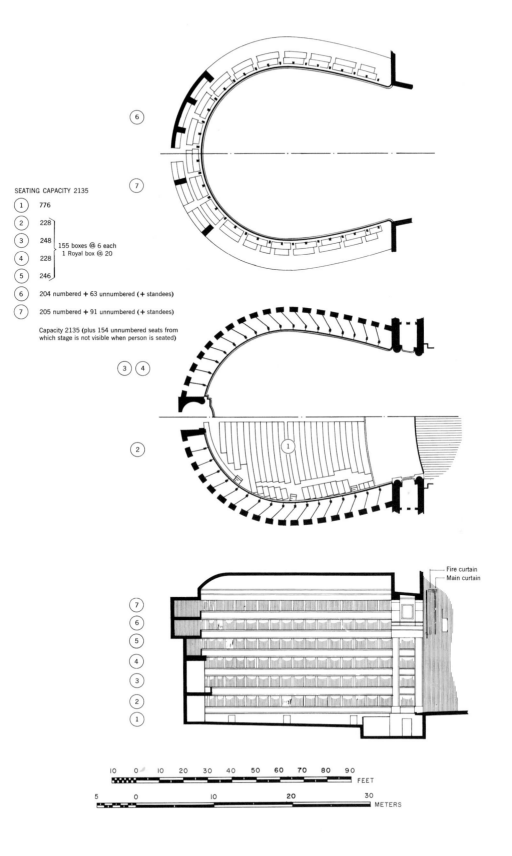

SEATING CAPACITY 2135

① 776

② 228 ⎫
③ 248 ⎬ 155 boxes @ 6 each
④ 228 ⎪ 1 Royal box @ 20
⑤ 246 ⎭

⑥ 204 numbered + 63 unnumbered (+ standees)

⑦ 205 numbered + 91 unnumbered (+ standees)

Capacity 2135 (plus 154 unnumbered seats from which stage is not visible when person is seated)

Fire curtain
Main curtain

10 0 10 20 30 40 50 60 70 80 90
FEET

5 0 10 20 30
METERS

the sound of the orchestra and the singers to the conductor, clear and loud. For these reasons, both singers and conductors are enthusiastic about the acoustics.

ARCHITECTURAL AND STRUCTURAL DETAILS

Uses: Opera and ballet. **Ceiling:** Vaulted, 6 ft (1.8 m) higher at the center than the sides; plaster on lath attached to 1 × 4-in. (2.5 × 10-cm) longitudinal boards; there are no irregularities (coffers) on the ceiling. **Walls:** Box faces are of plaster 42 in. (1.07 m) high; vertical columns at railings are plaster 6 in. (15.2 cm) wide; each box opening is 54 × 54 in. (1.37 × 1.37 m). **Floors:** Wood over 3-ft (0.91-m) airspace over concrete. **Carpets:** On all floors. **Pit floor:** Wooden floor, flat, elevator type; often located about 10 ft (3 m) below stage level. **Stage height:** 60 in. (152 cm) above floor level at first row of seats. **Seating:** Front of the backrests and the top of the seat bottoms are upholstered.

ARCHITECT: Giuseppe Piermarini. REFERENCES: Plans, details, and photographs courtesy of the management of La Scala.

TECHNICAL DETAILS

Opera

$V = 397,300$ ft³ (11,252 m³)	S_o (pit) $= 1,200$ ft² (111 m²)	S_o (pit floor) $= 1,350$ ft² (125.4 m²)
$V = 318,200$ ft³ (9,012 m³)*	$S_A = 14,000$ ft² (1,300 m²)	$S_P = 2,400$ ft² (223 m²)
$S_T = 17,600$ ft² (1,635 m²)	$S_{AMF} = 7,700$ ft² (715 m²)**	$S_{TMF} = 11,300$ ft² (1,050 m²)**
$N = 2,289$ (w/unnumbered)	$N_T = 2,489$ (N + standees)	$V/S_T = 22.6$ ft (6.9 m)
$V/S_A = 28.3$ ft (8.65 m)	$V/N = 174$ ft³ (4.92 m³)	$S_A/N = 6.1$ ft² (0.57 m²)
$H = 63$ ft (19.2 m)	$W = 66$ ft (20.1 m)	$L = 99$ ft (30.2 m)
$D = 105$ ft (32 m)	$H/W = 0.95$ $L/W = 1.5$	ITDG $= 20$ msec

*Volume in open space above main floor.

**Main floor seating only.

NOTE: The terminology is explained in Appendix 1.

71

Naples

Teatro di San Carlo

The original Teatro di San Carlo was built by King Charles of Bourbon in 1737 and destroyed by fire in 1816. Architect Antonio Niccolini was engaged by King Ferdinand to rebuild and restore the Teatro di San Carlo as it had been before. Re-opened in 1817, the stage was enlarged. The decorations were renewed and the ceiling today bears a painting of that date by Giuseppe Cammarano, which depicts Apollo introducing the greatest poets in the world to the goddess Minerva.

An orchestra pit was built in 1872 at the recommendation of Giuseppe Verdi. A new chandelier with electric lighting was introduced in 1890. Since that date a new foyer and a wing for dressing rooms has been added. To meet safety regulations, the seating capacity has been reduced to 1414. The 1737 opening date precedes that of Teatro Alla Scala in Milan by 41 years. The San Carlo and La Scala theaters are two of the largest surviving implementations of the traditional Baroque theater for opera. The first opera house was the Teatro di San Cassiano that opened in Venice in 1637. The orchestra was between the stage and the audience, although there was no pit. The form of the Italian-style house has been in vogue among architects for more than two centuries. Certainly, its acoustical characteristics have influenced the composers of Italian operas.

The author was involved recently (see References) in a survey of two-dozen well-known opera houses in ten countries. The acoustical quality of each were garnered from 21 leading opera conductors via questionnaires on which the recipients rated the houses on a linear scale with "poor" and "one of the best" on the two ends, and intervening steps of "passable," "good," and "very good." The acoustical rating of the sound in the audience areas placed San Carlo fifth ("very good"), barely behind La Scala, Dresden's Semperoper, and Tokyo's New National Theatre. For the sound as heard by the conductor in the pit, it ranked second only to Buenos Aires' Teatro Colón ("One of the best").

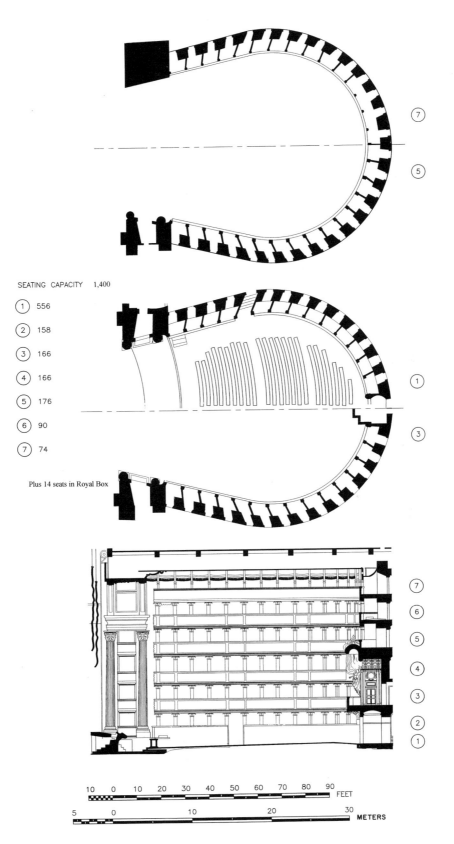

SEATING CAPACITY 1,400

1. 556
2. 158
3. 166
4. 166
5. 176
6. 90
7. 74

Plus 14 seats in Royal Box

10 0 10 20 30 40 50 60 70 80 90 FEET

5 0 10 20 30 METERS

ARCHITECTURAL AND STRUCTURAL DETAILS

Uses: Opera, symphonic music, and ballet. **Ceiling:** Canvas painted by Cammarano; upper side of the canvas is glued to flat framework composed of thin wooden ribbons in basket weave; poured gypsum above the ribbons makes 2–2.75-in. (5–7-cm) rigid layer suspended from wooden beams by tie rods. **Side walls:** Decorated plaster over masonry at main floor level. Wooden balustrades on the outer face at higher levels. **Walls on either side of proscenium:** Two half-columns on each side made of masonry covered with plaster and wood; four proscenium boxes are inserted between each half-column pair; proscenium arch made of wood decorated with stucco. **Rear wall and sidewalls in boxes:** Masonry with wooden door in rear; side walls: masonry in the rear fourth of their length; front part is double layer of wood, 2 in. (5 cm) thick; thin velvet covers the walls. **Walls in pit:** Wood panels on steel frame fixed to masonry walls. **Main floor:** Heavy wood, covered with parquet, over 28-in. (70-cm) airspace. **Floor in boxes:** Concrete, with ceramic tiles. **Floor in pit:** Wood covered with wood tiles; height changeable. **Carpet:** Thin and only in the central aisle. **Seats:** on main floor, well upholstered chairs with wooden lower seatbottom; wooden chairs with velvet cushions in boxes.

ARCHITECT: 1737, Antonio Medrano; 1817, Antonio Niccolini. ACOUSTICAL CONSULTANT: None. PHOTOGRAPHS: Press Office of the S. Carlo. REFERENCES: G. Iannace, C. Ianniello, L. Maffei and R. Romano, "Room acoustic conditions of performers in an old opera house," *J. Sound Vib.* **232**, 17–26 (2000); and "Objective measurement of the listening condition in the old Italian opera house "Teatro di San Carlo," *J. Sound Vib.*, **232**, 239–249 (2000).

TECHNICAL DETAILS

V = 483,600 ft³ (13,700 m³)	S_a = 7,941 ft² (738 m²)	S_A = 10,222 ft² (950 m²)
S_{pit} = 1,162 ft² (108 m²)	S_p = 2,900 ft² (270 m²)	S_T = 14,284 ft² (1,327 m²)
N = 1,414	H = 80 ft (24.4 m)	W = 73 ft (22.2 m)
L = 112 ft (34 m)	D = 119 ft (36.3 m)	V/ST = 33.85 ft (10.32 m)
V/S_A = 47.3 ft (14.4 m)	V/N = 345 ft³ (9.78 m³)	S_A/N = 7.30 ft² (0.68 m²)
S_a/N = 5.67 ft² (0.527 m²)	H/W = 1.10 L/W = 1.53	ITDG = 31 msec

NOTE: The terminology is explained in Appendix 1.

Concert Hall

The Kyoto City Foundation for the Promotion of Classical Music has as its motto, "Kyoto to be a culturaly free city open to the exchange of cultures of the world." The Kyoto Symphony Orchestra is owned by the City of Kyoto, the only orchestra in Japan owned by a municipality. The Concert Hall, seating 1,840, opened in October 1995, celebrating the 1,200th anniversary of the transfer of the Capital of Japan to Kyoto.

Paul Goldberger, architecture critic, *The New York Times,* wrote on July 2, 1995, "The hall is crisp and sharp and as serene as a Japanese temple. Its mood is defined in part by its materials: white oak floor, natural wood grillwork lining the bottom half of the room, white plaster at the top, side columns of gray-green Pietro Serena stone and seats of a subtle blue-gray velvet. Like all first-rate halls, Kyoto Concert Hall blends a sense of intimacy with monumentally. The room is warm enough to feel comforting, spare enough to defer to the primacy of the music."

With a reverberation time of 2.0 sec at mid-frequencies, fully occupied, the hall is well suited to today's taste in classical music. I heard organ music in the semi-occupied hall, and the sound was full and warm.

The hall has interesting acoustical features. On the main floor there are no balcony overhangs on the sides; at the rear only two rows of seats are overhung. In the balcony behind the stage there are 156 seats and to the sides there are about 140 seats. The distance from the front of the stage to the farthest listener is only about 118 ft (36 m), compared to Boston's 133 ft (40.5 m). To provide sound diffusion at the higher frequencies, the concrete sidewalls at the main floor and first balcony levels are first covered with a 0.4-in. (1-cm)-thick artificial wood cemented to concrete and a randomly spaced lattice work composed of strips 2.4 × 2.8 in. (6 × 7 cm) of artificial wood firmly cemented to the 0.4-in. board. Two-thousand obliquely cut cubes with an average height of about 1 ft (30 cm) cover the center of the ceiling and are designed to diffuse the sound.

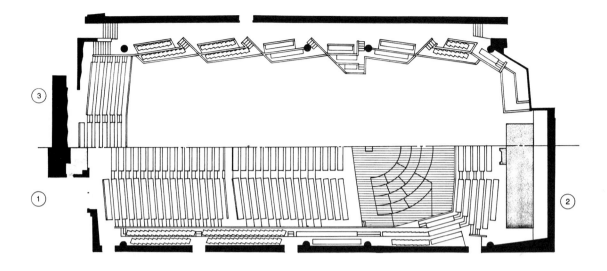

SEATING CAPACITY 1840

(1)   980

(2)   460

(3)   400

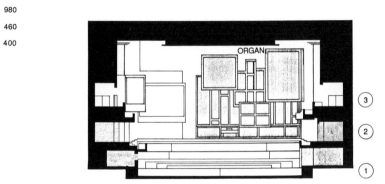

ORGAN

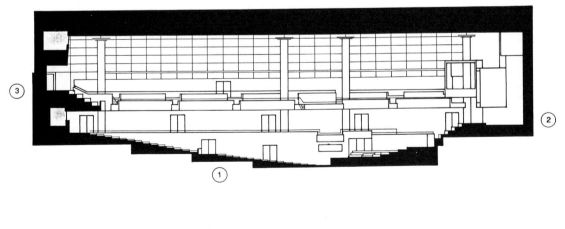

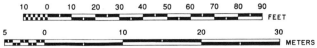

| 10 | 0 | 10 | 20 | 30 | 40 | 50 | 60 | 70 | 80 | 90 | |
FEET

| 5 | 0 | | 10 | | 20 | | 30 | |
METERS

ARCHITECTURAL AND STRUCTURAL DETAILS

Uses: Orchestra, chamber music, soloists, chorus, organ. **Ceiling** (center): Mortar on wire lath, 2.4 in. (6 cm), with 2,000 sound-diffusing, obliquely cut, cubes with heights of 6–12 in. (15–30 cm), formed from 0.4-cm steel plate, filled with 1.6-in. (4-cm) mortar. **Side ceiling:** Fiber-reinforced concrete 1.2 in. (3 cm). **Sidewalls:** Upper, mostly 1.2-in. (3-cm) fiber reinforced concrete; sidewalls at 3rd floor, artificial oak wood 0.8 in. (2 cm) cemented to solid concrete and 0.3-mm natural oak cemented to that; at 1st and 2nd balcony levels, randomly spaced lattice work of 2.4 × 2.8-in. (6 to 7-cm) artificial wood cemented to 0.4 in.(1 cm) artificial wood board, in turn cemented to solid concrete; rear wall, gypsum board, 1.2 in. (3 cm); behind organ, solid concrete; around stage, same lattice work as for side walls except separated from concrete wall by 24–45-in. airspace. **Floors:** Audience in general, 0.75-in. (1.8-cm) Japanese oak on plywood 0.6 in. (1.5 cm), on wood joists above a large airspace over a concrete slab; side balcony of 3rd floor, Japanese oak cemented to solid concrete. **Stage floor:** Tongue-and-groove pure Japanese Hinoli cypress, 1.6 in. (4 cm), on wood joists, selected by audition. **Seating:** Seat back, molded plywood; front side of seat back upholstered with polyurethane foam, 2.4 in.(6 cm); rear side polyurethane foam, 0.25 in.(0.6 cm) thick; upper side of seat bottom with polyurethane foam, 1.6 in.(4 cm); bottom side; armrests wood.

ARCHITECT: Arata Isozaki & Associates. ACOUSTICAL CONSULTANT: Nagata Acoustics, Inc. PHOTOGRAPHS: Katsuaki Furudate. REFERENCES: C. Ishiwata, K. Oguchi, Y. Toyota & M. Nagata, "Acoustical design of Kyoto Concert Hall," *Proc. 3rd Joint Meeting of ASA/ASJ, J. Acoust. Soc. Am.* **100**, 2706(A) (1996); K. Oguchi, C. Ishiwata, Y. Toyota & M. Nagata.

TECHNICAL DETAILS

V = 706,000 ft³ (20,000 m³)	S_a = 9,533 ft² (886 m²)	S_A = 12,503 ft² (1,162 m²)
S_o = 2,550 ft² (237 m²)	S_T = 14,440 ft² (1,342 m²)	N = 1,840
H = 49.2 ft (15 m)	W = 119 ft (35.3 m)	L = 115 ft (35.1 m)
D = 118 ft (36 m)	V/S_T = 48.9 ft (14.9 m)	V/S_A = 56.5 ft (17.2 m)
V/N = 384 ft³ (10.9 m³)	S_A/N = 6.80 ft² (0.632 m²)	S_a/N = 5.18 ft² (0.48 m²)
H/W = 0.42	L/W = 0.98	ITDG = 32 msec

Note: $S_T = S_A$ + 1,940 ft² (180 m²); see Appendix 1 for definition of S_T.

NOTE: The terminology is explained in Appendix 1.

73

Osaka

Symphony Hall

The Osaka Symphony Hall, which opened in 1982 on the occasion of the 30th-anniversary of the Asahi Broadcasting Corporation, was the first concert-only hall built in Japan. It has a very pleasant ambience, in large part because the most remote seat is only 98 ft (30 m) from the stage. It is used daily for classical music concerts and has a good reputation—a tribute to the care taken in acoustical design. A primary goal was to achieve an optimum reverberation time of between 1.8 and 2.0 sec, fully occupied.

For a hall that seats only 1,702 persons, it is surprisingly wide, the main floor width measuring 92 ft (28 m) and the first-balcony cross-hall width 119 ft (36.3 m). This poses some difficult acoustical problems. To produce a satisfactorily short, initial-time-delay gap, 26 suspended, convex, sound-reflecting panels are positioned over the stage. Early lateral reflections are directed to the main floor seating areas both by these suspended panels and by the downward-sloping balcony fronts. Such reflections are not provided by the lower side walls because of the presence of large-scale diffusing elements (see the photograph). Concert reverberation is provided by a high ceiling, from which are hung an array of 40 convex panels, chosen to simulate the acoustical diffusion provided in classical European halls by ceiling coffers. The stage is very wide and large in area (3,070 ft², 285 m²).

I attended a concert by an 85-piece symphonic orchestra in 1993. The violins sat 2 m from the front of the stage and 4 m from each side of the stage, so that the players were using an area equal to that of the entire stage in Boston's Symphony Hall. Where I sat in the first balcony, third row, 14 seats off center, the sound was excellent. The reverberation time was satisfactory, subjectively 2 sec at mid-frequencies. The balance among sections of the orchestra was flawless and the ensemble was perfect. As expected, the orchestra seemed very close, visually and acoustically. On the main floor the various sections of the orchestra did not blend as well, probably because of the great width of the hall and insufficient early lateral reflections from the lower side walls at mid- and high frequencies.

Osaka is justly proud of this beautiful and musically satisfactory hall.

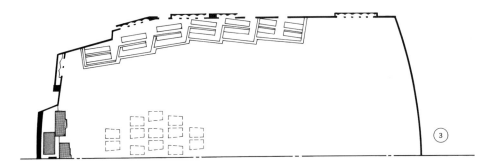

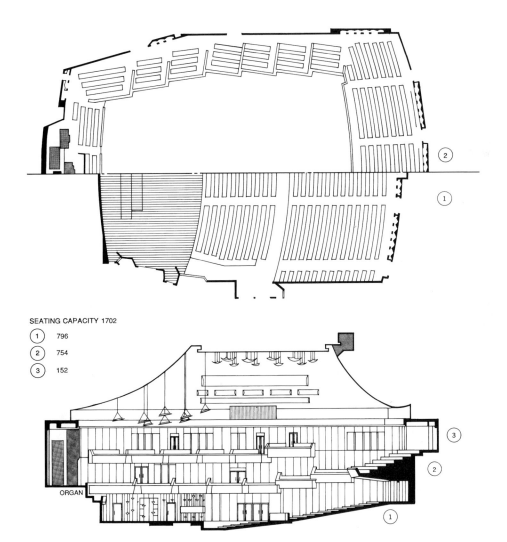

SEATING CAPACITY 1702

(1)   796
(2)   754
(3)   152

ORGAN

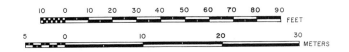

ARCHITECTURAL AND STRUCTURAL DETAILS

Uses: Classical music concerts. **Ceiling:** Two layers of gypsum or calcium board, total thickness 1.5 in. (3.8 cm), some openings. **Walls:** Plastered concrete, with a gypsum board layer to which is cemented artificial marble; balcony faces are plywood over concrete. **Floors:** Cork tile over concrete. **Carpet:** None in hall. **Stage walls:** Same as side walls. **Stage floor:** 0.75-in. tongue-and-groove hardwood on 0.75-in. plywood subfloor on sleepers. **Stage height:** 37 in. (94 cm). **Sound-absorbing materials:** none. **Seating:** Wooden seat back; front of seat back and seat bottom upholstered with porous fabric over polyvinyl foam; underseat, perforated wood; arms, wooden.

ARCHITECTS: Taisei Corporation. ACOUSTICAL CONSULTANTS: Kiyoteru Ishii and Hideki Tachibana. PHOTOGRAPHS: Courtesy of Taisei Corporation.

TECHNICAL DETAILS

V = 628,500 ft³ (17,800 m³)	S_a = 9,774 ft² (908 m²)	S_A = 13,304 ft² (1,236 m²)
S_o = 3,068 ft² (285 m²)	S_T = 15,244 ft² (1,416 m²)	N = 1,702
H = 68 ft (20.7 m)	W = 104 ft (31.7 m)	L = 93 ft (28.3 m)
D = 100 ft (30.5 m)	V/S_T = 41.2 ft (12.57 m)	V/S_A = 47.2 ft (14.4 m)
V/N = 369 ft³ (10.45 m³)	S_A/N = 7.81 ft² (0.725 m²)	S_a/N = 5.74 ft² (0.533 m²)
H/W = 0.65	L/W = 0.89	ITDG = 40 msec
Note: $S_T = S_A + 1,940$ ft² (180 m²); see Appendix 1 for definition of S_T.		

NOTE: The terminology is explained in Appendix 1.

74

Sapporo

"Kitara" Concert Hall

*S*apporo, the largest city located on Japan's northernmost island Hokkaido, has a concert hall, nicknamed "Kitara," that attracts prominent classical-music artists from within Japan and abroad. Seating 2,008, Kitara is the home of the Sapporo Symphony Orchestra and the Pacific Music Festival. On entering the hall, one sees two huge, sweeping, curved surfaces that focus on the stage, behind which stands a beautiful organ, all capped by a three-part canopy that is surrounded by an array of lights.

The concert hall opened on July 4, 1997, and is modeled after the Philharmonie in Berlin and the Suntory Hall in Tokyo. In all three, about 30 percent of the audience is seated to the sides and rear of the orchestra, thus reducing the distance of the farthest auditor from the podium. The balconies are a collection of terraces, eight in number, that not only subdivide the audience, but serve, through their fascia, to distribute much of the music's sound energy directly to the seating areas. Massive sidewall and ceiling constructions preserve the bass sounds. The multiple curved surfaces around the periphery diffuse the reverberant sound.

In the design, emphasis was placed on providing strong early lateral reflections of the orchestra's sound directly to the audience areas. This emphasis on reflecting early sound energy from the balcony fronts means that less energy than is normal in other halls is available to create reverberant energy; thus the reverberation time could be low. However, this possible deficiency is compensated for by increasing the cubic volume so that it is about 60 percent greater then that found in a classical rectangular hall with the same seating capacity. The RT at mid-frequencies with full audience is (calculated) between 1.8 and 2.0 sec.

Because of the large volume, the average strength of the sound is less than that in a smaller hall, thus making it particularly favorable to large orchestral forces such as are required to play compositions like those of Mahler. Large pipe-organ concerts are very effective.

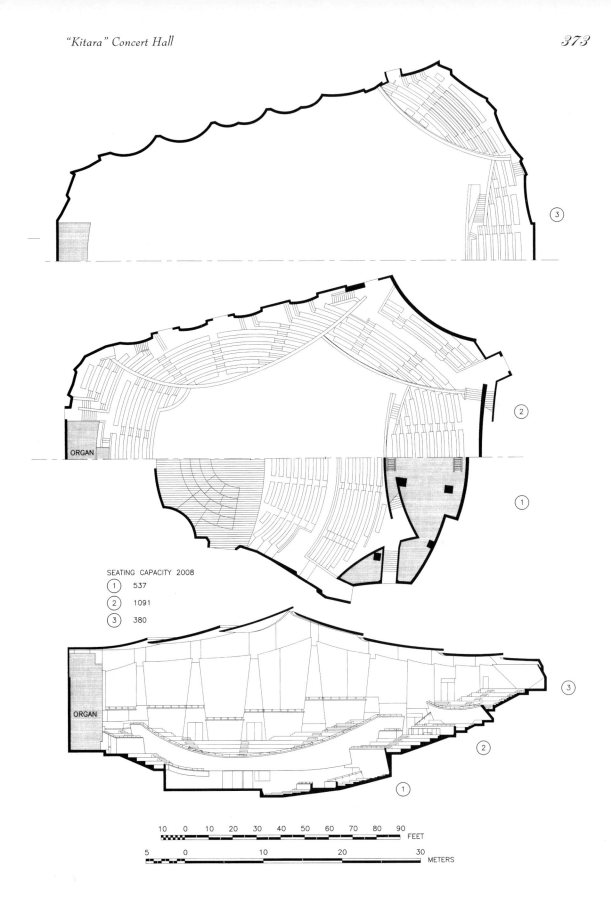

SEATING CAPACITY 2008

① 537
② 1091
③ 380

ARCHITECTURAL AND STRUCTURAL DETAILS

Uses: Symphonic and choral concerts; voice, instrumental, and organ recitals. **Ceiling:** 6-in. (15-cm) pre-cast concrete. **Stage canopy:** Adjustable-height, suspended, three-part reflectors, 2 in. (5 cm) total thickness composed of multiple layers of glass-fiber-reinforced gypsum board. **Sidewalls:** Outer walls, massive concrete; curved inner walls, one to three layers glass-fiber-reinforced gypsum board with wood veneer. **Frontal wall** (behind organ): Poured concrete with wood veneer. **Rear wall:** 50% open grille, over 2-in. (5-cm) glass-fiber blanket, over airspace. **Terrace (balcony) fronts:** Poured concrete with wood veneer. **Audience floors:** Wooden parquet on concrete. **Carpet:** None. **Stage floor:** Two layers wood (Hinoki cypress), over airspace. **Stage height:** 31 in. (80 cm). **Seats:** Seat bottom, perforated; front of back and upper part of seat upholstered; armrests, upholstered.

ARCHITECT: Hokkaido Engineering Consultants. ACOUSTICAL CONSULTANT: Nagata Acoustics. ORGAN BUILDER: Alfred Kern (French). REFERENCE: Y. Toyota, A. Ozawa, and K. Naniwa, "Acoustical design of the Sapporo concert hall," *Proceedings of 16th International Congress on Acoustics,* Seattle, WA, 2457–2458 (1998).

TECHNICAL DETAILS

V = 1,016,640 ft³ (28,800 m³)	S_a = 10,900 ft² (1,013 m²)	S_A = 15,473 ft² (1,438 m²)
S_o = 2,582 ft² (240 m²)	S_T = 17,410 ft² (1,618 m²)	N = 2,008
H = 70 ft (21.3 m)	W = 100 ft (30.5 m)	L = 72 ft (22 m)
D = 117 ft (35.7 m)	V/S_T = 58.4 ft (17.8 m)	V/S_A = 65.7 ft (20.0 m)
V/N = 506 ft³ (14.34 m³)	S_A/N = 7.70 ft² (0.716 m²)	S_a/N = 5.43 ft² (0.50 m²)
H/W = 0.7	L/W = 0.72	ITDG = 45 msec

Note: $S_T = S_A$ + 1,940 ft² (180 m²); see definition of S_T in Appendix 1.

NOTE: The terminology is explained in Appendix 1.

Bunka Kaikan

*P*rior to 1975, most Japanese halls used for the performance of music had been built as city halls or general-purpose public halls. Dramas, public speeches, and traditional Japanese opera (Kabuki) were their mainstay. All had stage houses and used a concert stage enclosure for symphonic, chamber, and recital music. The most enduring one for concerts is the 2,327 seat Bunka Kaikan in Ueno Park, located along with several art museums and a zoo.

The hall, dedicated in 1961, is diamond shaped. The width at the proscenium is 66 ft (20 m), at the rear, 60 ft (18 m), and at the center 112 ft (34 m). Two large, very irregular walls extend from the stage into the hall along the entire sides of the main floor. The orchestra enclosure is striking in appearance. In plan, it is fan shaped. Starting at a height of about 20 ft (6 m) above the rear stage riser, the stage ceiling soars upward to double the height at the proscenium. A permanent section of the ceiling extends into the hall, reaching a height of 56 ft (17 m) above stage level. This steep sound-reflecting canopy is modulated in convex, curved segments so that reflected sound spreads nearly evenly over the stage, most of the main floor, and the lower balconies. A large part of the side walls of the stage are unusual, and are not seen in any other hall. They consist largely of horizontal parallel slats ranging in widths from one to a few inches, spaced 4 in. (10 cm) apart (see the photograph). Behind these slats are vertically splayed surfaces.

With a reverberation time of only 1.5 sec, occupied, at mid-frequencies, the hall is surprisingly live. This is explained by the large number of early reflections that come from the canopy, the front sides of the stage, the irregular side walls, and the balcony fronts, giving an early sound decay time EDT of 1.85 sec (average throughout the hall). On the main floor, the overall impression is one of satisfactory bass and lovely string tone. A solo violin is very clear and resonant, aided by the 1.85 sec EDT. The brass is in good balance with the strings. The woodwinds have an optimal sound. The public gives Bunka Kaikan high marks.

74

Sapporo

"Kitara" Concert Hall

*S*apporo, the largest city located on Japan's northernmost island Hokkaido, has a concert hall, nicknamed "Kitara," that attracts prominent classical-music artists from within Japan and abroad. Seating 2,008, Kitara is the home of the Sapporo Symphony Orchestra and the Pacific Music Festival. On entering the hall, one sees two huge, sweeping, curved surfaces that focus on the stage, behind which stands a beautiful organ, all capped by a three-part canopy that is surrounded by an array of lights.

The concert hall opened on July 4, 1997, and is modeled after the Philharmonie in Berlin and the Suntory Hall in Tokyo. In all three, about 30 percent of the audience is seated to the sides and rear of the orchestra, thus reducing the distance of the farthest auditor from the podium. The balconies are a collection of terraces, eight in number, that not only subdivide the audience, but serve, through their fascia, to distribute much of the music's sound energy directly to the seating areas. Massive sidewall and ceiling constructions preserve the bass sounds. The multiple curved surfaces around the periphery diffuse the reverberant sound.

In the design, emphasis was placed on providing strong early lateral reflections of the orchestra's sound directly to the audience areas. This emphasis on reflecting early sound energy from the balcony fronts means that less energy than is normal in other halls is available to create reverberant energy; thus the reverberation time could be low. However, this possible deficiency is compensated for by increasing the cubic volume so that it is about 60 percent greater then that found in a classical rectangular hall with the same seating capacity. The RT at mid-frequencies with full audience is (calculated) between 1.8 and 2.0 sec.

Because of the large volume, the average strength of the sound is less than that in a smaller hall, thus making it particularly favorable to large orchestral forces such as are required to play compositions like those of Mahler. Large pipe-organ concerts are very effective.

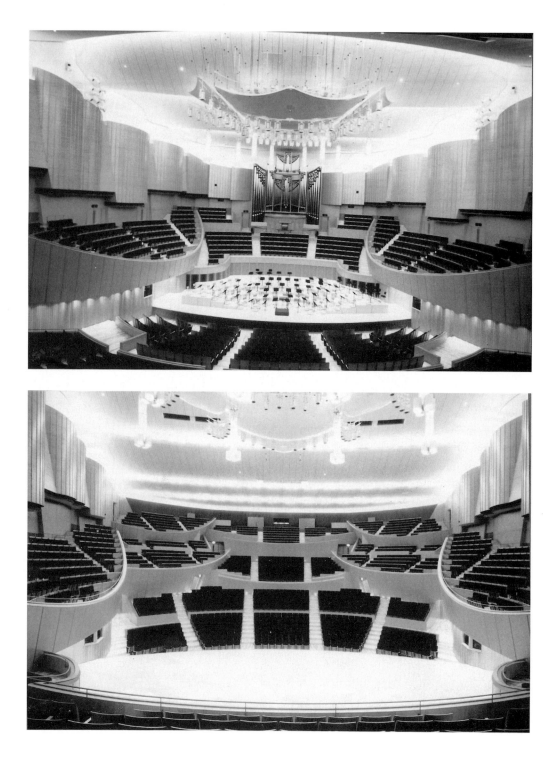

ORGAN

③

②

ORGAN

①

SEATING CAPACITY 2008
① 537
② 1091
③ 380

③

②

①

10 0 10 20 30 40 50 60 70 80 90 FEET

5 0 10 20 30 METERS

ARCHITECTURAL AND STRUCTURAL DETAILS

Uses: Symphonic and choral concerts; voice, instrumental, and organ recitals. **Ceiling:** 6-in. (15-cm) pre-cast concrete. **Stage canopy:** Adjustable-height, suspended, three-part reflectors, 2 in. (5 cm) total thickness composed of multiple layers of glass-fiber-reinforced gypsum board. **Sidewalls:** Outer walls, massive concrete; curved inner walls, one to three layers glass-fiber-reinforced gypsum board with wood veneer. **Frontal wall** (behind organ): Poured concrete with wood veneer. **Rear wall:** 50% open grille, over 2-in. (5-cm) glass-fiber blanket, over airspace. **Terrace (balcony) fronts:** Poured concrete with wood veneer. **Audience floors:** Wooden parquet on concrete. **Carpet:** None. **Stage floor:** Two layers wood (Hinoki cypress), over airspace. **Stage height:** 31 in. (80 cm). **Seats:** Seat bottom, perforated; front of back and upper part of seat upholstered; armrests, upholstered.

ARCHITECT: Hokkaido Engineering Consultants. ACOUSTICAL CONSULTANT: Nagata Acoustics. ORGAN BUILDER: Alfred Kern (French). REFERENCE: Y. Toyota, A. Ozawa, and K. Naniwa, "Acoustical design of the Sapporo concert hall," *Proceedings of 16th International Congress on Acoustics*, Seattle, WA, 2457–2458 (1998).

TECHNICAL DETAILS

$V = 1,016,640$ ft^3 (28,800 m^3)	$S_a = 10,900$ ft^2 (1,013 m^2)	$S_A = 15,473$ ft^2 (1,438 m^2)
$S_o = 2,582$ ft^2 (240 m^2)	$S_T = 17,410$ ft^2 (1,618 m^2)	$N = 2,008$
$H = 70$ ft (21.3 m)	$W = 100$ ft (30.5 m)	$L = 72$ ft (22 m)
$D = 117$ ft (35.7 m)	$V/S_T = 58.4$ ft (17.8 m)	$V/S_A = 65.7$ ft (20.0 m)
$V/N = 506$ ft^3 (14.34 m^3)	$S_A/N = 7.70$ ft^2 (0.716 m^2)	$S_a/N = 5.43$ ft^2 (0.50 m^2)
$H/W = 0.7$	$L/W = 0.72$	ITDG $= 45$ msec

Note: $S_T = S_A + 1,940$ ft^2 (180 m^2); see definition of S_T in Appendix 1.

NOTE: The terminology is explained in Appendix 1.

75

Tokyo

Bunka Kaikan

*P*rior to 1975, most Japanese halls used for the performance of music had been built as city halls or general-purpose public halls. Dramas, public speeches, and traditional Japanese opera (Kabuki) were their mainstay. All had stage houses and used a concert stage enclosure for symphonic, chamber, and recital music. The most enduring one for concerts is the 2,327 seat Bunka Kaikan in Ueno Park, located along with several art museums and a zoo.

The hall, dedicated in 1961, is diamond shaped. The width at the proscenium is 66 ft (20 m), at the rear, 60 ft (18 m), and at the center 112 ft (34 m). Two large, very irregular walls extend from the stage into the hall along the entire sides of the main floor. The orchestra enclosure is striking in appearance. In plan, it is fan shaped. Starting at a height of about 20 ft (6 m) above the rear stage riser, the stage ceiling soars upward to double the height at the proscenium. A permanent section of the ceiling extends into the hall, reaching a height of 56 ft (17 m) above stage level. This steep sound-reflecting canopy is modulated in convex, curved segments so that reflected sound spreads nearly evenly over the stage, most of the main floor, and the lower balconies. A large part of the side walls of the stage are unusual, and are not seen in any other hall. They consist largely of horizontal parallel slats ranging in widths from one to a few inches, spaced 4 in. (10 cm) apart (see the photograph). Behind these slats are vertically splayed surfaces.

With a reverberation time of only 1.5 sec, occupied, at mid-frequencies, the hall is surprisingly live. This is explained by the large number of early reflections that come from the canopy, the front sides of the stage, the irregular side walls, and the balcony fronts, giving an early sound decay time EDT of 1.85 sec (average throughout the hall). On the main floor, the overall impression is one of satisfactory bass and lovely string tone. A solo violin is very clear and resonant, aided by the 1.85 sec EDT. The brass is in good balance with the strings. The woodwinds have an optimal sound. The public gives Bunka Kaikan high marks.

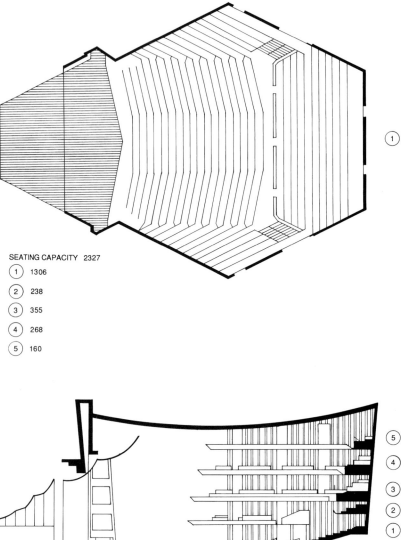

SEATING CAPACITY 2327

① 1306
② 238
③ 355
④ 268
⑤ 160

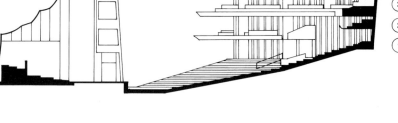

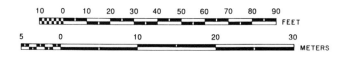

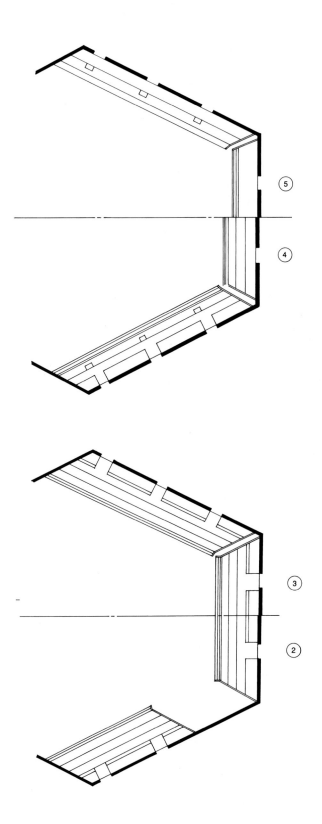

ARCHITECTURAL AND STRUCTURAL DETAILS

Uses: Orchestral music, drama, public functions, and recitals, **Ceiling:** 1-in. plaster on wire lath suspended beneath heavy structure. **Walls:** Concrete. **Floors:** Concrete. **Carpet:** In aisles only. **Stage enclosure:** Walls (see the text). **Stage height:** 19.7 in. (50 cm). **Acoustic reflectors:** The two side walls extending from the proscenium to the back of the balconies are heavily modulated to produce diffuse reflections. **Sound-absorbing materials:** On rear walls. **Seating:** Wooden seat back; front of seat back and top of seat bottom upholstered with porous fabric facing; underseat, wood, unperforated; arms not upholstered.

ARCHITECT: Kunio Maekawa. ACOUSTICAL CONSULTANT: NHK Technical Research Laboratories. PHOTOGRAPHS: Courtesy of NHK. REFERENCES: Y. Makita, M. Nagata, A. Usuba, H. Tsukuda, H. Takato, T. Yamamoto, Y. Sakamoto, T. Yamaguchi, H. Nakajima, R. Nishimura, and A. Mizoguchi, "Acoustical design of the Tokyo Metropolitan Festival Hall," *Technical Journal of NHK* **15**, No. 1–2 (1963).

TECHNICAL DETAILS

$V = 610{,}900$ ft³ (17,300 m³)	$S_a = 10{,}581$ ft² (983 m²)	$S_A = 14{,}000$ ft² (1,301 m²)
$S_o = 2{,}600$ ft² (241 m²)	$S_T = 15{,}940$ ft² (1,481 m²)	$N = 2{,}327$
$H = 57$ ft (17.4 m)	$W = 87$ ft (26.5 m)	$L = 104$ ft (31.7 m)
$D = 118$ ft (36 m)	$V/S_T = 38.3$ ft (11.68 m)	$V/S_A = 43.6$ ft (13.3 m)
$V/N = 262$ ft³ (7.42 m³)	$S_A/N = 6.02$ ft² (0.56 m²)	$S_a/N = 4.55$ ft² (0.422 m²)
$H/W = 0.66$	$L/W = 1.2$	ITDG $= 14$ msec
Note: $S_T = S_A + 1{,}940$ ft² (180 m²); see Appendix 1 for definition of S_T.		

NOTE: The terminology is explained in Appendix 1.

Dai-ichi Seimei Hall

*U*nique in several ways, the Dai-ichi Seimei Hall, opened in 2001, is the first privately owned musical-performance facility in Tokyo whose operation is closely tied to the surrounding community through a non-profit organization. Located on the manmade Harumi Island in the Tokyo Bay, which is connected to the city by two bridges and a subway, the hall is tucked in the space between three office skyscrapers. Oval in shape, with a mid-frequency reverberation time of 1.6 sec and seating 767, it is intended primarily for chamber music.

The non-profit organization, called the Triton Arts Network, initially plans to emphasize string quartets, which currently have no prominent center in Tokyo. At least once a month, on Wednesdays, quartets that have won major competitions worldwide, and those that are attracting attention in Japan, will be invited to present "award concerts." Special master classes conducted over a two-week period precede a special performance on Christmas Eve.

The oval shape of the Dai-ichi Seimei Hall was chosen to bring performers and listeners closer together. The interior is bright, featuring toned white oak. Acoustically, an oval shape for a hall is challenging. The reflection patterns of the hall were studied using a 20:1 scale model. All surfaces involved in early reflections are made irregular to give the music a rich "patina." To avoid focusing on the longitudinal centerline, the designers placed sound-absorbing materials in the rear parts of the lighting coves that surround the ceiling. Schroeder diffusers, some with absorbent materials, are located on the walls beneath the rear balcony to eliminate echoes. A "simulated audience" made by placing a suitable cloth over all seats was specially helpful in the tuning period after the hall was completed (Beranek and Hidaka, 2000). This procedure makes detailed acoustical tests for the "occupied" condition possible and eliminates the anxiety of "how will it sound on opening night?" Criticism by critics, musicians, and audiences has been favorable as detailed in a paper by Okano and Beranek (2002).

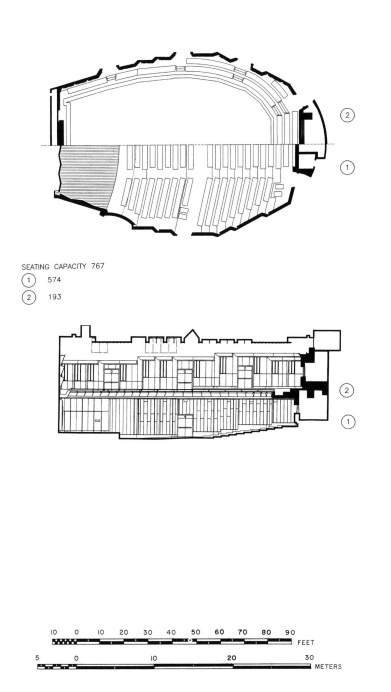

SEATING CAPACITY 767

① 574

② 193

10 0 10 20 30 40 50 60 70 80 90 FEET

5 0 10 20 30 METERS

ARCHITECTURAL AND STRUCTURAL DETAILS

Uses: Classical concerts, especially chamber music, soloists and conferences. **Ceiling:** 1.25-in. (3.2-cm) fiber-reinforced gypsum board, 10.4 lb/ft² (51 kg/m²). **Sidewalls:** 0.25-in. (0.6-cm) wood-like silicate board over 1.0 in. (2.5 cm) fiber-reinforced gypsum board, 9 lb/ft² (44 kg/m²). **Floors:** Main floor and floors at upper levels, 0.2-in (0.5-cm) coated cork tile cemented to solid concrete. **Carpet:** None. **Stage floor:** 0.6-in. (1.5-cm) laminated cypress glued over 0.6-in. (1.5-cm) plywood with joists at 18-in. (45-cm) intervals. **Seats:** Wood with cushions at the top of seat bottom and front of seat back, each 1.6 in. (4 cm) thick. **Stage height:** 30 in. (0.75 m).

ARCHITECT: Basic design: Urban Development Corporation and Nikken Sekkei Ltd; Detailed design: Takenaka Corporation (joint venture). ACOUSTICAL CONSULTANT: Takenaka Research & Development Institute and Leo L. Beranek. PHOTOGRAPHS: Technical Art Corporation. REFERENCES: T. Okano, "Overhead and near-head sound localization in an oval-shaped hall," *J. Acoust. Soc. Am.* Paper submitted, 2002. T. Okano and L. Beranek, "Subjective evaluation of a concert hall's acoustics using a free-format-type questionnaire and companion with objective measurements," *J. Acoust. Soc. Am.* Paper submitted, 2002.

TECHNICAL DETAILS

V = 240,000 ft³ (6,800 m³)	S_a = 4,110 ft² (382 m²)	S_A = 5,789 ft² (538 m²)
S_o = 1,120 ft² (104 m²)	S_T = 6,909 ft² (642 m²)	N = 767
H = 38.4 ft (11.7 m)	W = 66.9 ft (20.4 m)	L = 73 ft (22.3 m)
D = 73 ft (22.3 m)	V/S_T = 34.7 ft (10.6 m)	V/S_A = 41.45 ft (12.6 m)
V/N = 313 ft³ (8.86 m³)	S_A/N = 7.54 ft² (0.70 m²)	S_a/N = 5.36 ft² (0.50 m²)
H/W = 0.57	L/W = 1.09	ITDG = 24 msec

Note: When the stage is expanded, S_o = 1,506 ft² (140 m²); S_a = 3,900 ft² (363 m²); S_A = 5,165 ft² (480 m²); N = 714. N includes 8 wheelchairs.

NOTE: The terminology is explained in Appendix 1.

77

Tokyo

Hamarikyu Asahi Hall

The Hamarikyu Asahi Hall takes its place in Tokyo as a location for musical performance hospitable to intimate concert music. Small orchestras, chamber groups, and soloists demand a close relationship to their audiences. The solution here is a hall of intimate size, proven shape, with a properly tailored stage.

Seating 552 persons, the hall was dedicated in 1992. Tokyo's building code prevents the use of wood except on the stage floor. A marble-like material with a corrugated surface is used on the walls below the balcony. Elsewhere, all walls and the ceiling are either gypsum board or artificial wood. A thin wooden ply, for appearance, is cemented to the hard backing above the balcony and on the walls surrounding the stage. The random vertical corrugations on the marble-like facing were added to scatter the high frequencies above 1,500 Hz into the reverberant sound field, thus reducing the high-frequency energy in the early specular reflections and giving the early reflected sound a more mellow quality (see Chapter 4). The balcony fronts reflect early lateral energy to the seating areas. Similarly, the ceiling is coffered to scatter the first reflections and to diffuse the reverberant sound field. A tilted, corrugated panel is located around the top of the stage enclosure to provide good communication among the players. The stage depth was held to 23 ft (7 m) and the area to 786 ft² (73 m²) to favor small performing groups.

Measurements in the completed hall disclosed that the seats were absorbing more low-frequency energy than those tested during the design period. The result was a peak in the reverberation time of 0.2 sec between 500 and 2,000 Hz. This peak was flattened by adding a thin sheet of sound-absorbing material behind an impervious face in the "white" spaces above the balcony (see the photograph).

Performers have expressed satisfaction with the stage acoustics and the sound is uniform in quality throughout the hall. The acoustical measurements show an occupied mid-frequency reverberation time of 1.7 sec.

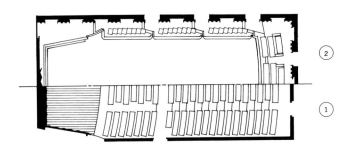

SEATING CAPACITY 552

(1) 448

(2) 104

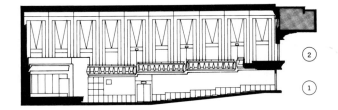

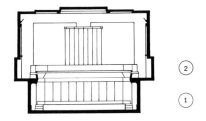

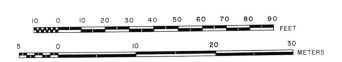

ARCHITECTURAL AND STRUCTURAL DETAILS

Uses: Chamber music and recitals. **Ceiling:** 0.83-in. (2.1-cm) gypsum board. **Over-stage reflecting panels:** Chevron shapes, artificial wood (phenol foam board mixed with glass fiber). **Walls above balcony and balcony fronts:** Thin ply of maobi wood cemented to 0.31-in. (0.8-cm) artificial wood. **Walls around main floor:** Corrugated marble-like urethane-coated silicate board cemented to 1.7-in. (4.2-cm) gypsum board. **Floors:** Parquet cemented to solid concrete. **Carpet:** None. **Stage height:** 30 in. (0.75 m). **Seating:** Upholstered both sides of seat back and seat with polyvinyl cushion covered by porous fabric; armrests wooden.

ARCHITECT: Takenaka Corporation, Tokyo. ACOUSTICAL DESIGN CONSULTANT: Leo L. Beranek. ACOUSTICAL RESEARCH LABORATORY: Takenaka Research and Development Institute. PHOTOGRAPHS: Taisuke Ogawa. REFERENCE: T. Hidaka et al., "Acoustical design and evaluation of Hamarikyu Asahi Hall," *Proc. Acoust. Soc. Jpn.*, March (1993) (in Japanese).

TECHNICAL DETAILS

V = 204,800 ft³ (5,800 m³)	S_a = 3,046 ft² (283 m²)	S_A = 4,252 ft² (395 m²)
S_o = 785 ft² (73 m²)	S_T = 5,037 ft² (468 m²)	N = 552
H = 39 ft (12 m)	W = 49 ft (15 m)	L = 81 ft (24.7 m)
D = 80 ft (24.4 m)	V/S_T = 37.3 ft (11.4 m)	V/S_A = 43.8 ft (13.3 m)
V/N = 371 ft³ (10.5 m³)	S_A/N = 8.48 ft² (0.79 m²)	S_a/N = 5.52 ft² (0.513 m²)
H/W = 0.8	L/W = 1.65	ITDG = 16 msec

NOTE: The terminology is explained in Appendix 1.

78

Metropolitan Art Space Concert Hall

*O*pened in 1990, the Tokyo Metropolitan Art Space Concert Hall, seating 2,017, is one of four large concert halls of Tokyo. Several of the Tokyo metropolitan symphony orchestras hold part of their subscription concerts here. The entrance to the Art Space is by way of a huge all-glass atrium, 92 ft (28 m) high. Audiences take a single escalator to the fifth floor, and then go up another escalator or ascend a flight of stairs.

The broad design goal was to follow the general shape of the Bunka Kaikan. A much larger reverberation time was prescribed, 2.15 sec at mid-frequencies, fully occupied, compared to 1.5 sec for the Bunka Kaikan. Early reflections from many directions following the direct sound are provided to all parts of the seating area. The measured results are in agreement with the design concept. The reverberation time rises from 2.15 sec at 500 Hz to 2.55 sec at 125 Hz. The sound level is remarkably constant on the main floor, and drops off as normally expected in the first balcony, but regains its level in the second balcony. The clarity measure is normal for good halls.

Over the stage, a large canopy, adjustable segmentally, provides communication among the players and sends early reflections to the main floor. The balcony faces are tilted to send early sound reflections to the seating areas and large sound reflecting and diffusing elements are provided on the side walls.

Listening to a concert from the main floor center, featuring Mahler's *Second Symphony*, gave me a very different impression from that in the Bunka Kaikan. The music was very much louder—one is enveloped by the sound, almost drowned in it—and the clarity is less. Because of the rise in reverberation time at low frequencies, the bass and cello sounds are very strong, primarily on the main floor. The players have found the sound on stage quite acceptable.

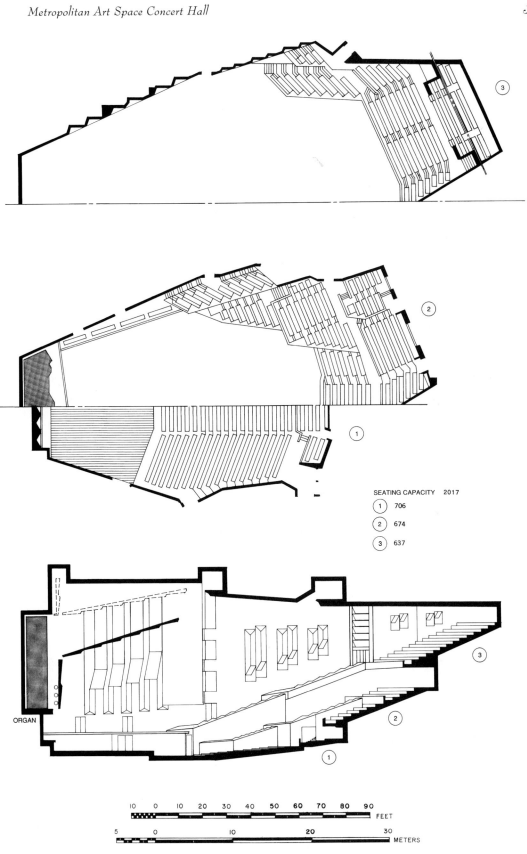

SEATING CAPACITY   2017

①	706
②	674
③	637

ORGAN

10   0   10   20   30   40   **50**   **60**   **70**   **80**   **90**
FEET

5   0   10   **20**   30
METERS

ARCHITECTURAL AND STRUCTURAL DETAILS

Uses: Music only. **Ceiling:** 1.4-in. (3.6-cm) plasterboard, above which is large airspace; cloth finish above seating; perforated board applied on the surrounding areas. **Walls:** Sloping; lower side walls concrete, covered with 0.3-mm walnut veneer, upper side walls marble veneer (see the photographs); rear wall at first balcony level, glass wool behind perforated board; balcony fronts, plaster covered with walnut veneer. **Sound diffusers:** On side walls, marble veneer. **Floors:** Japanese oak, penetrated by ventilation mushrooms. **Carpets:** None. **Stage enclosure:** Lower part, same as lower side walls; upper part marble veneer; canopy, 1.4-in. (3.6-cm) plaster board with cloth finish; canopy, adjustable for small concerts, completely concealing pipe organ; full height for organ and large concerts. **Stage floor:** 1.6-in. (4-cm) Japanese oak over 1-in. (2.4-cm) board. **Stage height:** 31.5 in. (80 cm). **Seating:** Underseats and rear backs, molded plywood; porous cloth over polyurethane foam on front of backs and top of seat; armrests, solid.

ARCHITECT: Yoshinobu Ashihara & Associates. ACOUSTICAL CONSULTANT: Nagata Acoustics, Inc. PHOTOGRAPHS: Kokyu Miwa. REFERENCE: M. Nagata, S. Ikeda, and K. Oguchi, "Acoustical design of the Tokyo Metropolitan Art and Cultural Hall," paper presented at 120th Meeting of Acoustical Society, November 1990, *J. Acoust. Soc. Am.* **88**, p. S112 (abstract).

TECHNICAL DETAILS

V = 880,000 ft^3 (25,000 m^3)	S_a = 10,000 ft^2 (929 m^2)	S_A = 14,120 ft^2 (1,312 m^2)
S_o = 2,230 ft^2 (207 m^2)	S_T = 16,060 ft^2 (1,492 m^2)	N = 2,017
H = 51 ft (15.5 m)	W = 92 ft (28 m)	L = 115 ft (35 m)
D = 153 ft (46.6 m)	V/S_T = 54.8 ft (16.75 m)	V/S_A = 62.3 ft (19.0 m)
V/N = 436 ft^3 (12.4 m^3)	S_A/N = 7.0 ft^2 (0.65 m^2)	S_a/N = 4.96 ft^2 (0.46 m^2)
H/W = 0.55	L/W = 1.25	ITDG = 27 msec

Note: $S_T = S_A$ + 1,940 ft^2 (180 m^2); see Appendix 1 for definition of S_T.

NOTE: The terminology is explained in Appendix 1.

New National Theatre, Opera House

apan's first National Performing Arts Center has two parts—the New National Theatre (NNT), owned and operated by the Japan Arts Council; and the Tokyo Opera City (TOC), owned by a consortium of industrial companies and operated by the Tokyo Opera City Cultural Foundation. The NNT is composed of the Opera House, the "Playhouse," and a small experimental theater, "The Pit." The TOC, adjacent to the NNT, consists of a concert hall, a 54-story skyscraper, an art gallery, and a galleria with shops and restaurants. The complex covers nearly 11 acres (44,090 m²) and is located in Shinjuku district.

Seating 1,810, the opera house opened October 10, 1997. The opening night audience, which included the Emperor and Empress of Japan, heard remarks by Prime Minister Ryutaro Hashimoto, who stated that with the NNT/TOC, Tokyo had become a cultural center of the world.

The architect states, "The design of the Opera House facilitates the exchange of energies between those on stage and those attending the performance. . . . The people in the audience are positioned so that they surround the stage as much as possible, and the balconies encourage empathy amongst them as they face each other across the hall."

The opera house has a main floor that first spreads and then becomes box-like. Each of the three balconies has parallel faces. The acoustical design goals were to achieve uniform projection of the singers' voices over the audience at sound levels somewhat higher than in conventional opera houses. To achieve those goals an architectural "acoustical trumpet" was devised as shown in the drawings by the seven surfaces labeled "T." The overhead sound reflector supplies the necessary clarity. The balcony-front "T"s assure optimum spaciousness and loudness. Also, the balcony faces are shaped to provide early lateral reflections, to aid in uniform sound distribution, and to increase the loudness. The reverberation time, occupied, at mid-frequencies, is 1.5 sec.

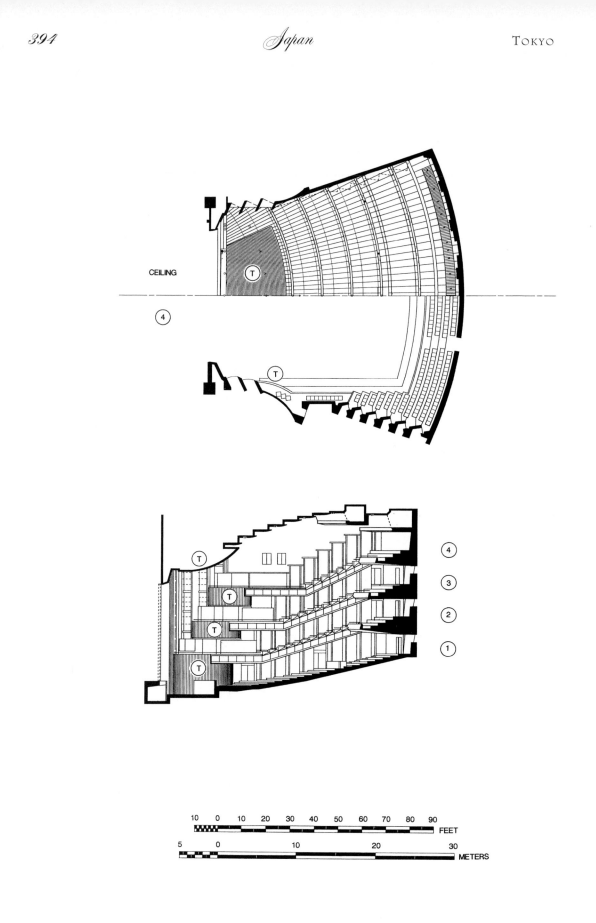

CEILING

4

T

T

T

T

T

T

4

3

2

1

10 0 10 20 30 40 50 60 70 80 90 FEET

5 0 10 20 30 METERS

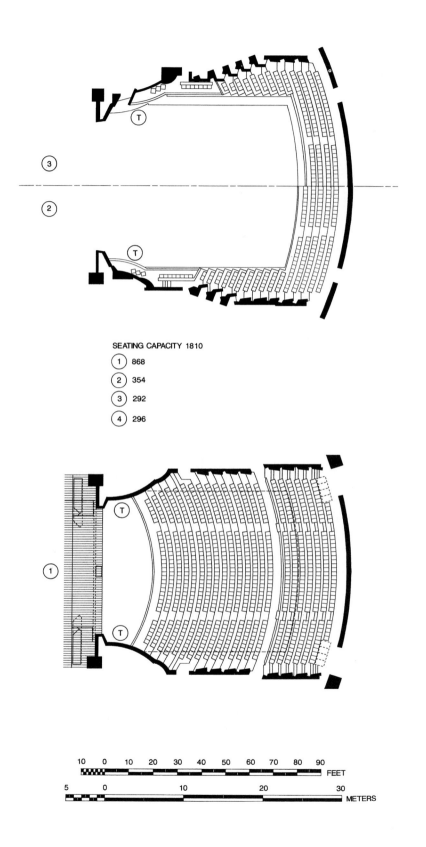

SEATING CAPACITY 1810

1 868
2 354
3 292
4 296

Music critics and performers agree: "The balances between singers and orchestra and the high and low tones are just right. The conductor can play full strength and yet, every voice and every instrument are clear—the definition is just right. . . . The hall is a success."

ARCHITECTURAL AND STRUCTURAL DETAILS

Uses: Grand and popular opera and ballet. **Ceiling** (main): Two layers plywood, each 0.5 in.(1.2 cm). **Balcony soffits:** Two layers gypsum boards, each 0.5 in. (12.5 mm). **Sidewalls** (*main*): Two layers, 1.18 in. (3 cm). **Reflectors:** Large "T," over proscenium, two layers of plywood boards, each 0.25 in. (6 mm); balcony front "T"s, wooden panel, 1.18 in. (3 cm), cemented silky oak with sound-diffusing grooves, 1 in. (2.5 cm). **Balconies:** faces, wood panel, 2 in. (5 cm); soffits, two layers gypsum board, each 0.5 in. (1.25 cm). **Absorbing materials:** Portion of balcony soffit beneath 1st balcony rear, 95 m², covered with glass fiber 1 in. (2.5 cm) to reduce echoes; rear wall, main floor, 34 m² covered with glass fiber, 1 in. (2.5 cm). **Floors:** Audience, two layers— one of tongue-and-groove wood flooring 0.5 in. (1.2 cm) and one of plywood 0.9 in. (1.5 cm) over concrete; pit floor, two layers—upper one of pine board, 1.2 in. (3.0 cm); lower of plywood, 0.5 in. (1.5 cm). **Carpet:** None. **Seating:** 65% of backrest covered with cushion, 0.8 in. (2.0 cm); seat bottom covered with cushion, 3.9 in. (10 cm); armrests wooden.

ARCHITECT: Takahiko Yanigasawa + TAK Associated Architects, Inc. ACOUSTICAL CONSULTANT: Leo Beranek and Takenaka Research and Development Institute. PHOTOGRAPHS: From audience, Osamu Murai C; from stage, Shinkenchiku-sha Co. REFERENCES: L. L. Beranek, T. Hidaka, and S. Masuda, "Acoustical design of the Opera House of the New National Theatre, Tokyo, Japan," *J. Acoust. Soc. Am.* **107**, 355–367 (2000).

TECHNICAL DETAILS

V = 512,000 ft³ (14,500 m³)	S_a = 9,590 ft² (891 m²)	S_A = 12,400 ft² (1153 m²)
S_o (pit) = 1,100 ft² (102 m²)	S(proc) = 2205 ft² (205 m²)	S_T = 15,705 ft² (1460 m²)
N = 1810	H = 67.2 ft (20.5 m)	W = 85.3 ft (26 m)
L = 101 ft (30.7 m)	D = 112.5 ft (34.3 m)	V/N = 283 ft³ (8.0 m³)
V/S_T = 32.6 ft (9.9 m)	V/S_A = 41.3 ft (12.58 m)	S_A/N = 6.9 ft² (0.64 m²)
H/W = 0.79	L/W = 1.18	ITDG = 28 msec

NOTE: The terminology is explained in Appendix 1.

80

Tokyo

NHK Hall

The Japanese Broadcasting Corporation (NHK) uses this hall, seating 3,677 persons, primarily as a concert hall, but also for opera, variety shows, and popular music. Opened in 1973, it is the home of the NHK Symphony Orchestra, which holds subscription concerts six days a month. A semi-fan plan was adopted, fan at front and parallel walls in rear half. At each level, the audience is divided into three areas. NHK Hall takes its place as one of the largest concert halls in the world used by a major orchestra. The concert stage can be expanded from 2,100 ft² (196 m²) up to 3,360 ft² (310 m²) in three steps, by removing up to 150 seats.

The audience on the main floor receives early reflections from the stage enclosure, from the impressive wedge-shaped balcony fronts, and from the succession of overhead curved reflectors that begin at the rear wall of the stage and continue as an integral part of the ceiling over most of the audience. Shown clearly in the photographs are protruding elements attached to the surfaces of the side walls of the stage and the auditorium which spread, by diffusion, early sound reflections. The greatest distance of a listener from the non-augmented stage front is 156 ft (47.5 m). The mid-frequency reverberation time, with full audience, is about 1.7 sec, a little on the low side for symphony concerts, but excellent for the many uses made of the hall.

I have attended two concerts in NHK Hall, one by the Boston Symphony Orchestra and the other by the NHK orchestra. I sat in three seats, one on the main floor, the second 15 seats off center in the second balcony, and the third on the aisle near an exit stair in the first balcony. I was surprised at the similarity of the sound at the three seats and at its strength, although it was a little weaker than than in Boston. The orchestral sections were well balanced and the sound was clear and warm. The lower reverberation time and the lack of reverberant sound energy from behind made the sound more directional than in classical, smaller rectangular halls. The Boston players expressed no dissatisfaction.

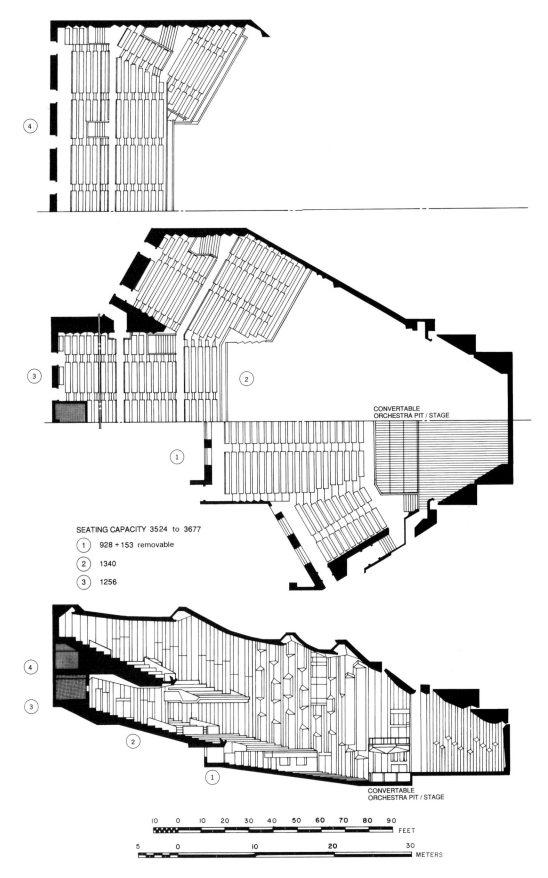

SEATING CAPACITY 3524 to 3677

① 928 + 153 removable

② 1340

③ 1256

CONVERTABLE ORCHESTRA PIT / STAGE

CONVERTABLE ORCHESTRA PIT / STAGE

10 0 10 20 30 40 50 60 70 80 90
FEET

5 0 10 20 30
METERS

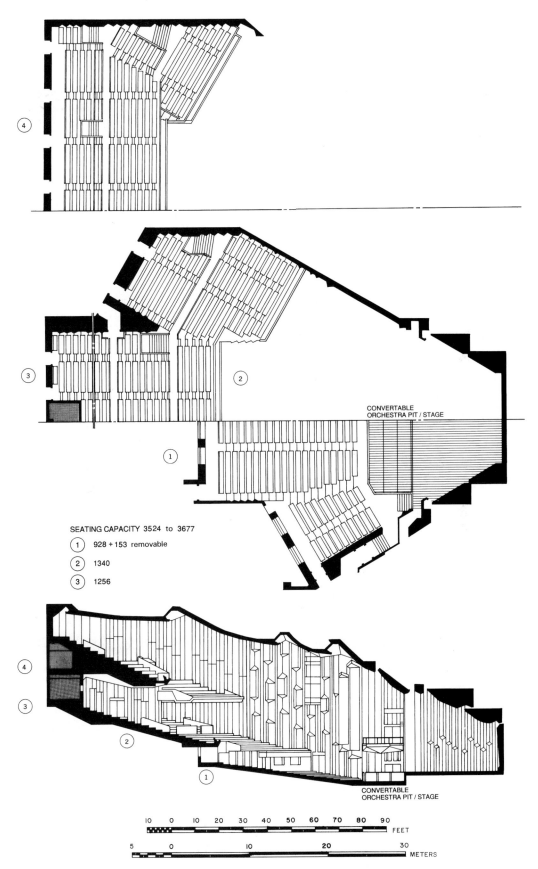

SEATING CAPACITY 3524 to 3677

(1) 928 + 153 removable

(2) 1340

(3) 1256

CONVERTABLE
ORCHESTRA PIT / STAGE

CONVERTABLE
ORCHESTRA PIT / STAGE

ARCHITECTURAL AND STRUCTURAL DETAILS

Uses: Primarily concerts, variety shows, musical comedies, and popular audience-participation TV productions. **Ceiling:** 0.2-in. (5-mm) asbestos board. **Stage ceiling and sidewalls:** 0.6-in. (1.7-cm) compound panel. **Main and balcony side walls:** 1.2-in. (3-cm) mortared, reinforced concrete and 0.8-in. (2-cm)-thick asbestos board. **Sound absorption:** In rear part of ceiling, two large areas of perforated asbestos board with 1-in. (2.5-cm) glasswool and airspace behind; same for rear walls of balconies. **Stage floor:** 1.9 in. (4.8 cm) wooden strips. **Stage height:** 35 in. (89 cm). **Audience floor:** vinyl tile on concrete. **Carpet:** Thin, on aisles. **Seating:** Tops of seats and fronts of the backrests are upholstered with cushion; armrests, wood.

ARCHITECT: Nikken Sekkei Ltd., Tokyo. ACOUSTICAL CONSULTANT: Technical Research Laboratories of NHK. PHOTOGRAPHS: Courtesy of Takashi Nishi, NHK. REFERENCES: T. Yamamoto, M. Nagata, and H. Tsukuda, "Acoustical design of the New NHK Hall," NHK Technical Monograph, No. 19, 37 pages (March 1972).

TECHNICAL DETAILS

$V = 890,000$ ft³ (25,200 m³)	$S_a = 15,700$ ft² (1,458 m²)	$S_A = 19,600$ ft² (1,821 m²)
$S_o = 2,077$ ft² (193 m²)	$S_T = 21,540$ ft² (2,001 m²)	$N = 3,677$
$H = 49$ ft (14.9 m)	$W = 109$ ft (33.2 m)	$L = 120$ ft (36.6 m)
$D = 156$ ft (47.6 m)	$V/S_T = 41.3$ ft (12.6 m)	$V/S_A = 45.4$ ft (13.8 m)
$V/N = 242$ ft³ (6.85 m³)	$S_A/N = 5.33$ ft² (0.50 m²)	$S_a/N = 4.27$ ft² (0.40 m²)
$H/W = 0.45$	$L/W = 1.10$	ITDG $= 23$ msec
Note: $S_T = S_A + 1,940$ ft² (180 m²); see Appendix 1 for definition of S_T.		

NOTE: The terminology is explained in Appendix 1.

81

Tokyo

Orchard Hall

Orchard Hall is the largest hall in a cultural complex called "Bunkamura" (cultural village), situated in a busy and flourishing subcenter of Tokyo. Orchard Hall has a stage house whose configuration can be tailored to suit symphonic concerts, opera, ballet, and smaller musical groups. The hall opened in 1989 with the Bayreuth Festival Company from Germany, featuring Wagner's music.

Rectangular in shape and seating 2,150, it closely approximates the width and length of Boston Symphony Hall. The hall is beautiful, with clean contemporary lines. The visual ceiling is transparent to sound; the acoustical ceiling is 10 ft (3 m) above. Various sizes of stage enclosure are accomplished with telescoping "shelters" that provide platform areas from 940 ft^2 (87 m^2) to 2,680 ft^2 (249 m^2). To vary reverberation time, areas of sound-absorbing material can be exposed, or for smaller performing groups, the volume can be reduced by up to 10%, with a corresponding decrease in reverberation time. The reverberation time, with medium-size shelter, which is normally used for symphonic music, is about 1.8 sec at mid-frequencies at full occupancy. Orchard hall has a clearness or transparency of sound that distinguishes it from many others, probably because it has less "baroque" ornamentation than the classical rectangular halls in which there is greater diffusion of the sound.

I have attended two symphonic concerts there, in May 1991 and June 2001. At the first concert (Tchaikovsky and Brahms), I sat in two seats, one on the main floor before intermission and in the center of the first balcony after intermission. At the second concert (Mahler's 2nd) I sat in the center of the first balcony. The sound quality was good and seemed to come mainly directly from the stage with minor support from the sidewalls. Also, the sound was weaker on the main floor. The sections of the orchestra were well balanced, and the orchestra did not overpower the voices in the Mahler.

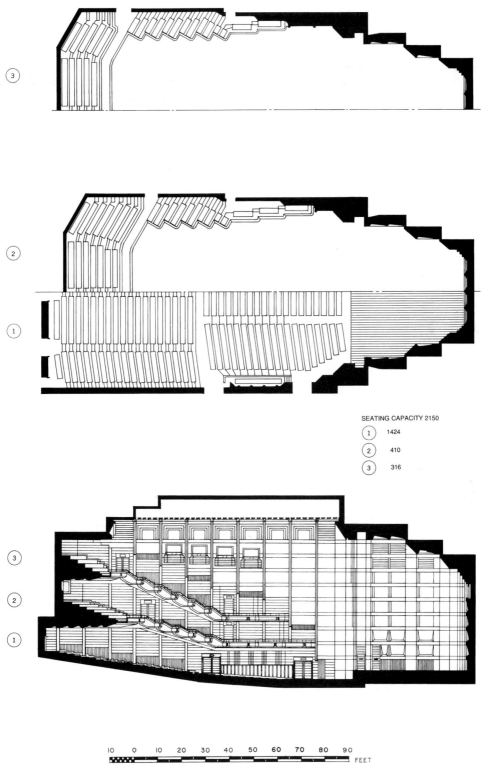

SEATING CAPACITY 2150

(1)	1424
(2)	410
(3)	316

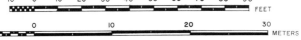

10   0   10   20   30   40   50   **60 70 80**   90
FEET

5   0   10   **20**   30
METERS

ARCHITECTURAL AND STRUCTURAL DETAILS

Uses: Concerts, 70%; small groups, 10%; opera and ballet, 20%. **Ceiling:** Two layers of wood chip board, each 0.7 in. (1.8 cm). **Visual ceiling:** Acoustically transparent screen. **Sidewalls:** Same as ceiling, with sound-diffusing elements. **Balcony facia:** Pre-cast gypsum affixed to concrete. **Sound-absorbing surfaces:** Rear wall under first balcony, absorbs low frequencies, composed of hard, thin covering with slits and sound-absorbing material behind. **Floors:** 0.6-in. (1.5-cm) wood impregnated in acrylic plastic, on 0.35-in. (0.9-cm) plywood, cemented to concrete. **Carpet:** None. **Stage enclosure:** Built in three pieces, which telescope into each other—for largest stage with chorus extension, the stage volume is 141,000 ft² (4,000 m³); without extension, 121,800 ft³ (3,450 m³); for medium stage, 94,600 ft³ (2,680 m³); for smallest stage, 50,850 ft³ (1,440 m³). **Stage floor:** Resilient wood, 1.2 in. (3 cm) thick, over two plywood sheets 1 in. (2.4 cm) total, over large airspace. **Stage height:** 43 in. (1.1 m). **Seating:** Front of seat back and top of seat bottom upholstered.

ARCHITECT: Yuzo Mikami and MIDI Architects. ACOUSTICAL CONSULTANT: Kiyoteru Ishii and Kimura Laboratory of Nihon University. PHOTOGRAPHS: Osamu Murai. REFERENCES: Yuzo Mikami and Kiyoteru Ishii, "Design philosophy of the Orchard Hall and its development," *Architectural Acoustics and Noise Control* **19**, 61–68 (1990) (in Japanese).

TECHNICAL DETAILS

V = 723,850 ft³ (20,500 m³)	S_a = 10,760 ft² (1,000 m²)	S_A = 14,140 ft² (1,314 m²)
S_o = 2,337 ft² (217 m²)	S_T = 16,080 ft² (1,494 m²)	N = 2,150
H = 75 ft (23 m)	W = 80 ft (24.4 m)	L = 126 ft (38.4 m)
D = 131 ft (40 m)	V/S_T = 45.0 ft (13.7 m)	V/S_A = 51.2 ft (15.6 m)
V/N = 337 ft³ (9.53 m³)	S_A/N = 6.58 ft² (0.61 m²)	S_a/N = 5.00 ft² (0.465 m²)
H/W = 0.94	L/W = 1.57	ITDG = 26 msec

Note: $S_T = S_A$ + 1,940 ft² (180 m²); see Appendix 1 for definition of S_T.

NOTE: The terminology is explained in Appendix 1.

82

Tokyo

Suntory Hall

*I*n 1986, Suntory Hall opened on the occasion of the 60th anniversary of Suntory, Ltd., Japan's largest distillery. Located in the busiest section of Tokyo's Akasaka, it has an enclosed courtyard and gracious lobby. Entering the hall, one is awed by the beautiful light-colored wood interior and the "vineyard" style in which the audience is arranged in blocks of various heights. Side surfaces of the individual terraces, made of smooth white marble, provide early reflections to most of the seats, simulating those that are found in the traditional rectangular concert hall. The roomy and comfortable seats, numbering 2,006, were demanded by Suntory's Chairman, Keizo Saji, a large man by Japanese standards.

The reverberation time for Suntory, fully occupied, is 2.0 sec at mid-frequencies, equal to that of the best-liked concert halls and just right for the baroque repertory especially enjoyed by the Japanese. Diffusion of the reverberant sound field is accomplished by splayed side wall surfaces and stepped sections in the convex ceiling. The stage area is 2,530 ft² (235 m²) and the musicians receive early sound reflections from ten convex fibreglass panels, each with an area of 43 ft² (4 m²) at a height of 39 ft (12 m). With a small orchestra, the double basses are likely to be far from a wall, thus losing the boost that their sound would otherwise receive.

The acoustical data show that some parts of Suntory Hall receive more early reflections (less than 40 msec), which is desirable, than others. The best seats, in this regard, are in the rear half of the hall.

This hall is one of the most popular in Tokyo, in large part because the world's best orchestras on tour tend to go to Suntory. It is also used heavily by five of Japan's symphony orchestras. Another reason for its reputation is that touring orchestras generally praise those halls with high reverberation times, whereas for concentrated rehearsals, necessary at home, a lower RT is often preferable. In all, Suntory Hall is beautiful, the acoustics are very good, and it serves the musical public very well.

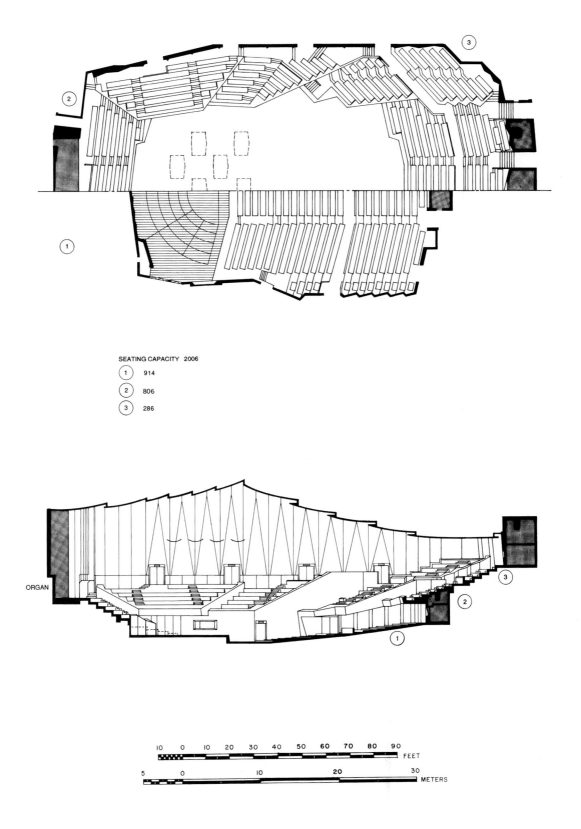

SEATING CAPACITY 2006

(1) 914

(2) 806

(3) 286

ORGAN

10 0 10 20 30 40 50 60 70 80 90
FEET

5 0 10 20 30
METERS

ARCHITECTURAL AND STRUCTURAL DETAILS

Uses: Primarily classical music. **Ceiling:** Two layers of 0.5-in. (1.25-cm) gypsum board beneath airspace. **Sidewalls:** Two principal portions, flat and splayed; flat portions are triangles with base at dado line, two layers 2-in. woodchip board, 0.3-in. (0.8-cm) plaster board and thin wood ply; splayed portions resemble pillars, with point at dado line, widening and deepening until widest portions touch just at the ceiling line and then continue under the ceiling; these splayed portions are one or two layers of 0.7-in. (1.8-cm) chip board, plus plaster board and wood ply; walls on sides of the organ, one layer of chip board; back walls splayed (see the drawing) with combination of solid and perforated board; several small areas 2-in. fiberglass blanket behind grille; sides of terraces are marble held by mortar to concrete. **Floor:** 0.5-in. hardwood, on 0.5-in. plywood, on concrete. **Stage enclosure:** Marble (see the photograph). **Stage floor:** 1.6-in. (4-cm) wood on 0.5-in. plywood, on 2.5 × 2.5-in. (6-cm) sleepers, 1 ft (30 cm) on centers. **Stage height:** 31 in. (80 cm) high. **Seating:** Front of backs and seat tops, upholstered.

ARCHITECT: Yasui Architects. ACOUSTICAL CONSULTANT: Nagata Acoustics, Inc. PHOTOGRAPHS: courtesy of the architect.

TECHNICAL DETAILS

V = 740,000 ft³ (21,000 m³)	S_a = 11,220 ft² (1,042 m²)	S_A = 14,680 ft² (1,364 m²)
S_o = 2,530 ft² (235 m²)	S_T = 16,620 ft² (1,544 m²)	N = 2,006
H = 54 ft (16.5 m)	W = 102 ft (31.1 m)	L = 100 ft (30.5 m)
D = 118 ft (36 m)	V/S_T = 44.5 ft (13.6 m)	V/S_A = 50.4 ft (15.4 m)
V/N = 369 ft³ (10.5 m³)	S_A/N = 7.32 ft² (0.68 m²)	S_a/N = 5.59 ft² (0.52 m²)
H/W = 0.53	L/W = 0.98	ITDG = 30 msec

Note: S_T = S_A + 1,940 ft² (180 m²); see Appendix 1 for definition of S_T.

NOTE: The terminology is explained in Appendix 1.

83

Tokyo

Tokyo Opera City, Concert Hall

*U*nquestionably a bold architectural form for the performance of music, the concert hall in Tokyo Opera City embodies a pyramidal ceiling with its apex reaching 92 ft (28 m) above the main floor. Seating 1636, this hall opened September 10, 1997, with a performance of Bach's *St. Matthew's Passion* by Japan's international Saito Kinen Orchestra under the direction of Seiji Ozawa. Everyone in attendance, including the Emperor and Empress of Japan, was awed by the soaring ceiling, the natural wood interior, the coordinated canopy and pipe organ, and the unusual lighting.

The TOC Hall is the first to embody a new set of acoustical parameters developed in the previous 40 years by psycho-acoustic laboratories and consulting firms around the world. This set of six primary acoustical factors promises to free architects from the age-old stricture that excellent acoustics can be obtained only by duplicating previously successful halls of the "shoebox" shape. In the process of achieving the desired acoustical results in this hall, a computer-assisted design (CAD) model was programmed, which enabled the tracing of sound rays in the "room" and determining their time delays and directions as they reached the "ears" of "listeners." The interior surfaces were adjusted to optimize these "sounds." Next followed a 10:1 scale wooden model in which the desired reverberation times, bass response, uniformity of sound throughout the hall, placement of the canopy over the orchestra, and the absence of echoes were resolved.

When measurements in the constructed TOC Concert Hall are compared with those in the three "superior" halls of Chapter 5, they confirm that it has optimum reverberation time (1.96 sec, full audience at mid-frequencies) and excellent clarity, intimacy, spaciousness, warmth, and strength of sound.

Performers and music critics uniformly say, "The sound is warm, enveloping, and reverberant, aids achievement of excellent ensemble, and is equal in acoustical quality to that of the best halls in the world."

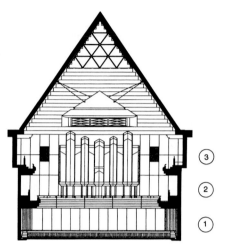

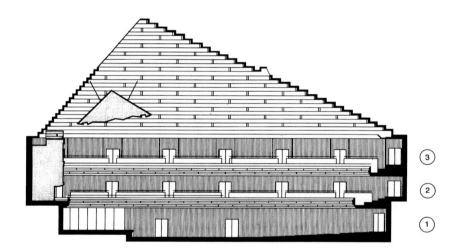

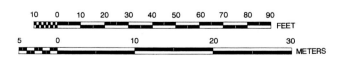

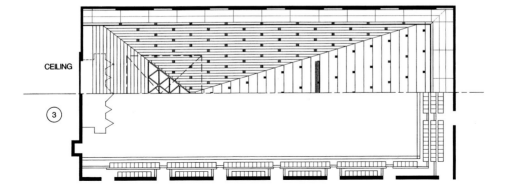

CEILING

③

SEATING CAPACITY 1632

① 974

② 356

③ 302

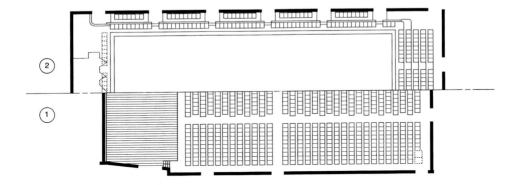

②

①

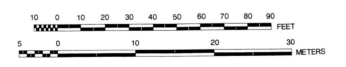

❧ ARCHITECTURAL AND STRUCTURAL DETAILS

Uses: Symphonic concerts, chamber music, and soloists. **Ceiling:** Three layers—two of glass fiber reinforced gypsum board, each 0.4 in. (1 cm) and one of silicated calcium board 0.24 in. (6 mm); quadratic residue diffusors (QRD) cover the longest surface of the pyramid. **Walls** (main and balcony fronts): Plywood 0.35 in. (0.9 cm) over wooden core (2 cm). **Balcony soffits:** Three layers—two of gypsum plasterboard, each 0.5 in. (1.25 cm), and one of glass-fiber-reinforced phenolic foam board, 0.31 in. (8 mm). **Floors** (audience): Two layers—one of tongue-and-groove wood flooring, 0.6 in. (1.5 cm), and one of plywood, 0.5 in. (1.2 cm), over concrete. **Stage floor:** Two layers—one of tongue-and-groove wood flooring, 0.6 in. (1.5 cm), over one of plywood, 0.5 in. (1.2 cm), over airspace. **Carpet:** None. **Stage height:** 31.5 in. (80 cm). **Seating:** 65% of backrest covered with cushion 0.8 in. (2.0 cm) thick; seat bottom covered with cushion 3-in. (8-cm)-thick.

ARCHITECT: Takahiko Yanagisawa + TAK Associated Architects, Inc. ACOUSTICAL CONSULTANTS: Leo Beranek and the Takenaka Research & Development Institute PHOTOGRAPHS: Asamu Murai C. REFERENCES: T. Hidaka, L. L. Beranek, S. Masuda, N. Nishihara, and T. Okano, "Acoustical design of the Tokyo Opera City (TOC) Concert Hall, Japan, *J. Acoust. Soc. Am.* **107**, 340–354 (2000).

TECHNICAL DETAILS

V = 540,000 ft³ (15,300 m³)	S_a = 8,533 ft² (793 m²)	S_A = 11,320 ft² (1052 m²)
S_o = 1808 ft² (168 m²)	S_T = 13,127 ft² (1220 m²)	N = 1,636
V/S_T = 41.1 ft (12.5 m)	V/S_A = 47.7 ft (14.5 m)	V/N = 330 ft³ (9.4 m³)
S_A/N = 6.92 ft² (0.64 m²)	S_a/N = 5.22 ft² (0.485 m²)	H = 70.5 ft (21.5 m)
W = 65.6 ft (20 m)	L = 111 ft (33.8 m)	D = 113.5 ft (34.6 m)
H/W = 1.08	L/W = 1.69	ITDG = 15 msec

NOTE: The terminology is explained in Appendix 1.

84

Kuala Lumpur

Dewan Filharmonik Petronas

*M*usic listeners in North America are so unacquainted with far distant Malaysia that they may be surprised to learn that Kuala Lumpur has a first-class orchestra, the Malaysian Philharmonic, created by the Petronas Oil Company. Well-known everywhere are the Twin Towers in Kuala Lumpur because they are billed as the tallest in the world.

As construction of the twin towers was just beginning, Petronas decided that the planned 400-seat meeting room nestled between the towers should be abandoned in favor of a concert hall, the first such edifice in Malaysia. Because of the nearby towers, the buildings' grand lobby, and a six-story mall, the only growth for the hall was upwards. The result is a 74 × 132-ft (22.6 × 40-m) rectangular footprint that includes a full concert stage and 850 seats on three levels. The hall is also used for recording. Because of the hall's busy and noisy surroundings, the auditorium had to be mounted on resilient pads and the building overall was structurally isolated from the rest of the complex.

On entering the hall one sees the enormous circular sound reflector above the stage and the pipes of the 55-rank Klais organ. The interior is a combination of light-colored wood and gold, with green upholstered seats. Owing to the small distance between attendees and the stage the performers seem only a short distance away.

The Petronas Oil Company created and continues lavishly to subsidize this excellent orchestra, and has booked a number of touring orchestras into the facility. Although the country has no tradition of Western symphonic music, audiences are enthusiastically attending the concerts, testifying to the hall's architectural and acoustical beauty.

To achieve the reverberation necessary for music of the Romantic period, the ceiling had to be high. However, the hall also had to accommodate chamber music and corporate meetings that, successively, demand lower reverberation times, and thus lower cubic volumes. The solution, urged by the acoustical consultant, was to

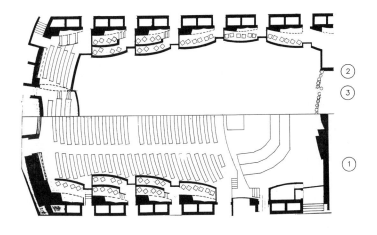

SEATING CAPACITY 850

① 588

② 138

③ 124

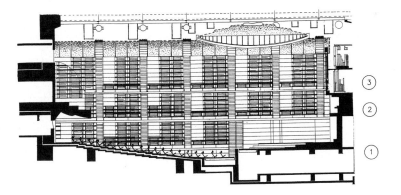

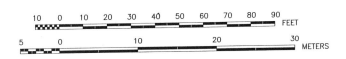

mount a heavy, movable acoustical ceiling on motor-operated screw jacks. The lowest setting for this ceiling is at 55 ft (17 m) above the stage, the highest at 75 ft (23 m). An acoustically transparent, perforated metal ceiling located below the lowest setting of the acoustical ceiling, finishes the room visually and accommodates the theatrical and stage lighting equipment. Deployable sound-absorbing panels are built into the sidewalls below the visual ceiling that further lower the reverberation times to make the hall optimum for meetings.

ARCHITECTURAL AND STRUCTURAL DETAILS

Uses: Symphonic and chamber music, recordings conferences.

ARCHITECT: Cesar Pelli & Associates, New Haven, Connecticut and Adamson Associates, Toronto. ACOUSTICAL CONSULTANT: Kirkegaard & Associates. PHOTOGRAPHS: Courtesy Kirkegaard & Associates.

TECHNICAL DETAILS

V = 630,600 ft³ (17,860 m³)	S_a = 4,143 ft² (385 m²)	S_A = 6,499 ft² (604 m²)
S_o = 2,399 ft² (223 m²)	S_T = 8,436 ft² (784 m²)	N = 850
H = 54 ft (16.5 m)	W = 74 ft (22.6 m)	L = 76 ft (23.2 m)
D = 77 ft (23.5 m)	V/S_T = 74.8 ft (22.8 m)	V/S_A = 97 ft (29.6 m)
V/N = 741 ft³ (21 m³)	S_A/N = 7.64 ft² (0.71 m²)	S_a/N = 4.87 ft² (0.453 m²)
H/W = 0.73	L/W = 1.03	ITDG = 23 msec

Note: $S_T = S_A + 180$ m²; see definition of S_T in Appendix 1.

NOTE: The terminology is explained in Appendix 1.

Sala Nezahualcoyotl

$\mathcal{S}$ala Nezahualcoyotl concert hall bears the name of one of Mexico's legendary poets and musicians. Built on the campus of the Universidad Nacional Atonoma de Mexico in 1976, seating 2,376, it is the home of one of Mexico City's professional orchestras. It boasts of being the first concert hall in North America to surround the stage with the audience. The arrangement of the audience into tiers creates side-wall reflecting surfaces that emulate seating in a narrower hall. An innovation is a large reverberant room located beneath the stage which is coupled to the hall by horizontal grilles located along the front and rear edges of the stage and by grillwork along the entire front face of the stage. This reverberant room also communicates with the hall through an open grille under the first rows of seats. The purpose is to enhance the lowest register tones, but it frequently becomes a convenient storage space.

An impressive part of the acoustical planning is the dominating arrangement of sound reflectors over the stage and the forward part of the main floor seating. In the ceiling is a reverse-pyramid structure arranged over the orchestra and the main-floor audience. Beneath it hang brown-tinted acrylic reflectors of two shapes—"watch crystals" and "prisms"—each designed to reflect and diffuse the high and upper-middle tones to orchestra and audience. The watch-crystals reflect over 360°, and the prisms are directed to the audience areas to shorten the initial-time-delay gap. The low frequencies pass through the arrays and are reflected, diffusely, by the ceiling structure. Of importance acoustically are the relatively high walls around the stage, 9 ft (2.7 m), with diffusing facias and an overhanging "shelf" at the top. Finally, on the wall behind the chorus seating, a large number of organ-pipe-like diffusing elements are placed to improve the quality of the reverberant sound field.

Is the Sala Nezahualcotyl successful? With the Cleveland Symphony Orchestra, under Lorin Maazel, performing Beethoven's *Ninth Symphony*, the hall received good reviews from music critics imported from New York, Chicago, and

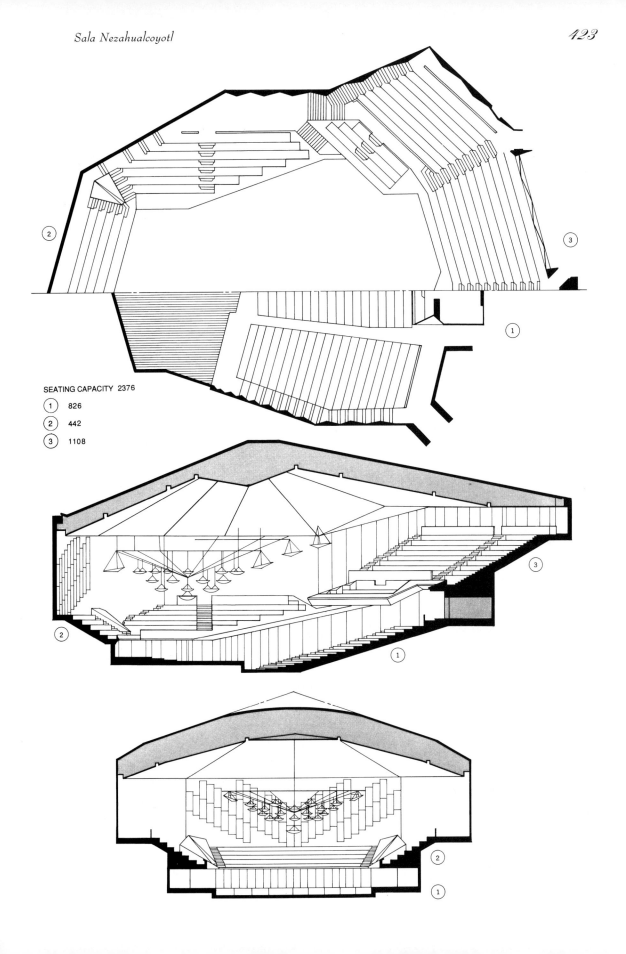

SEATING CAPACITY 2376

1 826
2 442
3 1108

Boston. When asked about newer halls by Chancellor Peter Czirnich of Munich in 1979, Maazel wrote, "The best new hall that I know of is the one built for the University of Mexico . . ."

ARCHITECTURAL AND STRUCTURAL DETAILS

Uses: Orchestral music and recitals. **Ceiling:** Two layers of splayed 1-in. (2.5-cm) gypsum board, with batt insulation above. **Walls:** Rear walls, 1-in. splayed wood boards with glass fiber batts behind; other walls, poured concrete with decorative splayed 1-in. wood boards. **Floors:** In main floor, 2-in. (5-cm) wood over 2-in. airspace; in balconies, concrete. **Carpet:** In aisles only. **Stage enclosure:** Walls, 1-in. splayed wood boards with glass fiber behind; floor, 2-in. wood over acoustic moat with 20% grille openings under audience seating. **Stage height:** 37 in. (94 cm). **Acoustic reflectors:** All reflectors are acrylic, 0.75–1 in. thick; the watchglasses are 6–10 ft (1.8–3 m) in diameter, convex in shape, and located over most of stage; the prisms are splayed and located over the near parts of the audience. **Sound-absorbing materials:** None. **Seating:** Front of seat back and top of seat bottom upholstered.

ARCHITECTS: Orso Nunez, Arcadio Artis, Manuel M. Ortiz, Arturo Trevino, and Roberto Ruiz. ACOUSTICAL CONSULTANT: Jaffe Holden Scarbrough. PHOTOGRAPHS: Courtesy of the architects. REFERENCES: "Acoustic devices enhance sound of music in a surround hall," *Architectural Record,* 125–128 (January 1978); "El Placer de Una buena Acustica," *Obras* (June 1979).

TECHNICAL DETAILS

$V = 1,082,000$ ft³ (30,640 m³)	$S_a = 15,877$ ft² (1,476 m²)	$S_A = 18,126$ ft² (1,684 m²)
$S_o = 2,906$ ft² (270 m²)	$S_T = 20,066$ ft² (1,864 m²)	$N = 2,376$
$H = 52$ ft (15.8 m)	$W = 134$ ft (40.8 m)	$L = 113$ ft (34.4 m)
$D = 138$ ft (42 m)	$V/S_T = 53.9$ ft (16.44 m)	$V/S_A = 59.7$ ft (18.2 m)
$V/N = 455$ ft³ (12.9 m³)	$S_A/N = 7.63$ ft² (0.71 m²)	$S_a/N = 6.68$ ft² (0.621 m²)
$H/W = 0.39$	$L/W = 0.84$	ITDG $= 16$ msec

Note: $S_T = S_A + 1,940$ ft² (180 m²); see Appendix 1 for definition of S_T.

NOTE: The terminology is explained in Appendix 1.

86

Amsterdam

Concertgebouw

The Amsterdam Concertgebouw, which opened in 1888 and seats 2,037 persons, is rated by almost anyone familiar with the concert halls of the Western world as one of the three best. These include Vienna's Grosser Musikvereinssaal and Boston's Symphony Hall, all of which are rectangular.

Several of the physical features of the Concertgebouw are different. It is wider than those two halls, 91 ft (27.7 m), compared, respectively, to 65 and 75 ft (19.8 and 22.9 m). Twenty percent of the audience is seated on steep stadium steps behind the orchestra. The stage is higher than that of any hall studied, 60 in. (153 cm).

The floor of the Concertgebouw is flat and the seats are removable. The irregular walls and deeply coffered ceiling produce excellent sound diffusion. The reverberation time at mid-frequencies, fully occupied, is about 2 sec.

Opinions of conductors of major orchestras and soloists obtained through the years include, "The Concertgebouw has marvelous acoustics. It is probably one of the best halls in the world." "I went into the audience and heard the Boston Symphony. It sounded excellent acoustically." "The reverberation in this hall gives great help to a violinist. As one slides from one note to another, the previous sound perseveres and one has the feeling that each note is surrounded by strength." "The sound is fabulous." "The Concertgebouw is arguably one of the best in the world."

I have attended many concerts in the Concertgebouw and from my notes, I glean, "The sound is well balanced, strong in bass. The reverberation sounds greater than that in other rectangular halls, a quality that generally pleases visiting conductors. The cello (in a concerto) sounds loud and luxurious. The full orchestra plays with rich tone."

For my taste, the sound in the balcony is better than on the main floor, probably because in those seats the articulation is somewhat better. But on the main floor, for those who love a full sumptuous sound with rich bass, and the feeling of being completely surrounded in an ocean of music, this hall has no superior.

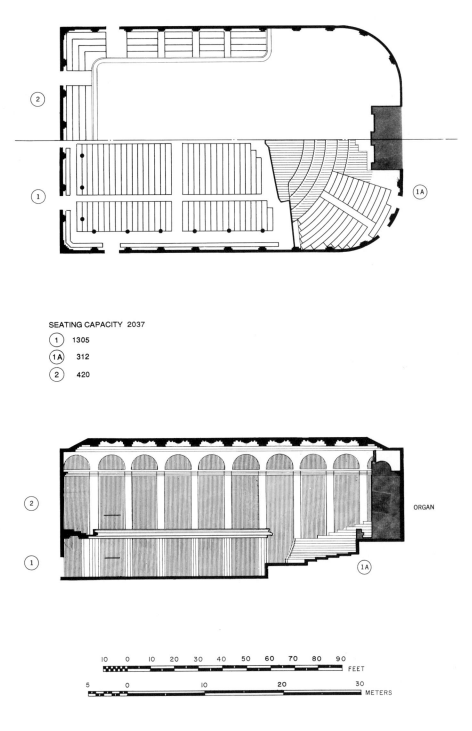

SEATING CAPACITY 2037

(1) 1305

(1A) 312

(2) 420

ORGAN

10 0 10 20 30 40 50 60 70 80 90
FEET

5 0 10 20 30
METERS

ARCHITECTURAL AND STRUCTURAL DETAILS

Uses: All types of music, with emphasis on orchestra. **Ceiling:** 1.5-in. (3.8-cm) plaster on reeds, coffered and with ornamentation; there are deep "window" recesses around the top edges. **Sidewalls and rear walls:** Below the balcony, plaster on brick; above the balcony, plaster on reed, which sounds dull or damped when tapped with the fingers. **Floors:** 5-in. (13 cm) concrete, on top of which hardwood boards are nailed to 2 × 3-in. (5 × 7.5-cm) wooden battens; cavity filled with 4-cm layer of sand. **Carpet:** On main floor aisles only. **Stage enclosure:** None. **Stage floor:** Heavy wood over deep airspace. **Stage height:** 59 in. (150 cm). **Added absorptive material:** 700 ft² (65 m²) of draperies over the front of the little room at the rear of the balcony and around the doorways. **Seating:** Upholstered in thick, hard-weave material.

ARCHITECT: A. L. van Gendt. DETAILS: Courtesy of the Council, Concertgebouw, Amsterdam. PHOTOGRAPHS: Stage view, Hans Samson; seating view, Robert Schlingemann.

TECHNICAL DETAILS

$V = 663{,}000$ ft³ (18,780 m³)	$S_a = 9{,}074$ ft² (843 m²)	$S_A = 12{,}110$ ft² (1,125 m²)
$S_o = 1{,}720$ ft² (160 m²)	$S_T = 13{,}830$ ft² (1,285 m²)	$N = 2{,}037$
$V/S_T = 48$ ft (14.6 m)	$V/S_A = 54.7$ ft (16.7 m)	$V/N = 325$ ft³ (9.2 m³)
$S_A/N = 5.94$ ft² (0.552 m²)	$H = 56$ ft (17.1 m)	$W = 91$ ft (27.7 m)
$L = 86$ ft (26.2 m)	$D = 84$ ft (25.6 m)	$S_a/N = 4.45$ ft² (0.414 m²)
$H/W = 0.62$	$L/W = 0.94$	ITDG $= 21$ msec

NOTE: The terminology is explained in Appendix 1.

87

Rotterdam

De Doelen Concert Hall

$\mathcal{D}$e Doelen Concert Hall was dedicated in 1966 at a time when acoustics was still based on visual observation of good and bad halls and on general rules-of-thumb. The reverberation time, the length of time it takes for a sound to die to inaudibility, was understood to be important and the quality of the reverberation was known to depend on large and small irregularities on the walls and ceiling, so-called sound-diffusing elements. Comparison of halls of different reputations had shown that a hall had to be narrow in order to keep the gap in time between the direct sound and the first reflected sound wave as small as possible. Yet to be discovered was that a large part of the early reflections should arrive at listeners' ears laterally—from the sides—for best sound quality. With the design adopted, these acoustical attributes were to a considerable extent achieved. A seating capacity of 2,242 is significantly more than in Vienna's Grosser Musikvereinssaal (1,680 seats) and Amsterdam's Concertgebouw (2,037 seats).

The central part of the main floor is a walled enclosure whose top is 10 ft (3 m) above stage level and whose average width in the first two-thirds of the area is 69 ft (21 m), making the main floor equivalent to a small concert hall (594 seats). A large number of sound-diffusing "boxes" and indentations line all sides of this enclosure, including the walls around the stage. These diffusers scatter the incident sound and send early lateral reflections to all seats. In the rear seats (facing the stage) outside the "inner hall" the earliest sound reflections arrive primarily by way of the sound diffusing ceiling. Later lateral reflections come from the sound-diffusing side walls. As usual, seats to the sides and rear of the stage are not favored with as pleasant a sound, but seeing the orchestra and conductor up close seems to compensate. There are no echoes and the noise level from machinery is very low. The musicians report that maintaining good ensemble playing is easy.

The acoustics of De Doelen have been praised by visiting conductors and musicians, because the sound on stage and in the "inner hall" is excellent.

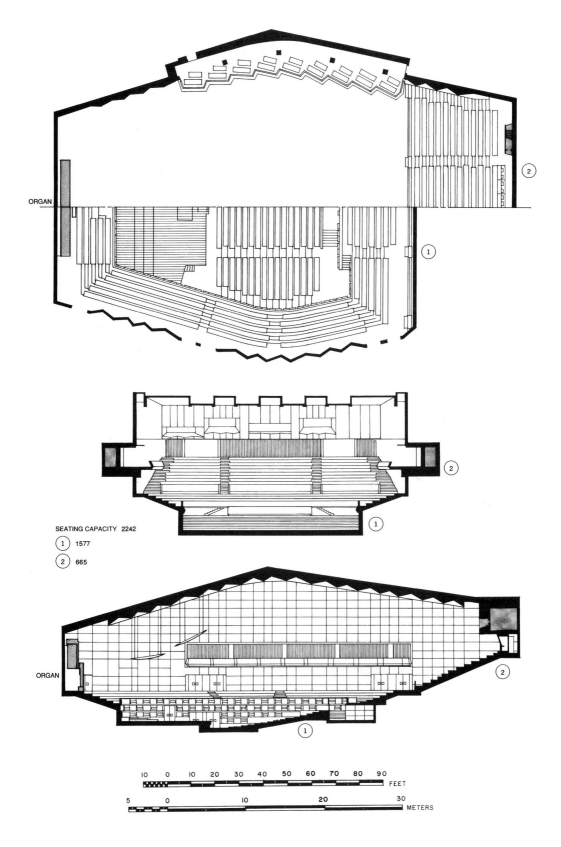

ORGAN

②

①

SEATING CAPACITY 2242

① 1577

② 665

ORGAN

②

①

10 0 10 20 30 40 50 60 70 80 90 FEET

5 0 10 20 30 METERS

ℐRCHITECTURAL AND STRUCTURAL DETAILS

Uses: Classical music, including large and small groups. **Ceiling:** Heavy concrete, 1.5 in. (4.5 cm) thick, of which visual layer is 0.4-in. (1-cm) plywood. **Upper walls:** Dense concrete to which is attached finished walls, including diffusing panels, constructed like ceiling. **Walls around "inner hall":** Marble, diffuse in two dimensions. **Balcony fronts:** Fibrous plaster, 1/2 in. (1.5 cm) thick. **Floors:** Concrete with linoleum surface. **Carpet:** None. **Stage:** Composite structure—top surface, 0.8-in. (2.4-cm) tongue-and-groove wood; three layers asphalt paper; 0.4-in. (1-cm) particle board; 0.65-in. (2-cm) elastic mounts, spaced 16 in. (40 cm) in one direction and 10 ft (3 m) other direction; airspace of about 2 ft (61 cm). **Stage height:** 30 in. (0.75 m). **Seating:** Front of seat back and top of seat bottom upholstered.

ARCHITECT: Kraaijvanger Architekten bv. ACOUSTICAL CONSULTANTS: C. W. Kosten and P. A. de Lange. PHOTOGRAPHS: Courtesy of the architects. REFERENCES: E. H. Kraaijvanger, H. M. Kraaijvanger, and R. H. Fledderus, "De Doelen, concert- en congresgebouw," Kraaijvanger Architekten bv, Watertorenweg 336, 3063HA Rotterdam (1966); C. C. J. M. Hak and H. J. Martin, "The acoustics of the Doelen Concert Hall after 30 years," Faculty of Architecture, Eindhoven University of Technology (April 1993).

TECHNICAL DETAILS

V = 849,900 ft³ (24,070 m³)	S_a = 13,000 ft² (1,208 m²)	S_A = 16,240 ft² (1,509 m²)
S_o = 2,100 ft² (195 m²)	S_T = 18,180 ft² (1,689 m²)	N = 2,242
H = 47 ft (14.3 m)	W (lower) = 80 ft (24.4 m)	W (upper) = 106 ft (32.3 m)
L (lower) = 49 ft (14.9 m)	L (upper) = 104 ft (31.7 m)	D = 126 ft (38.4 m)
V/S_T = 46.75 ft (14.25 m)	V/S_A = 52.3 ft (15.9 m)	V/N = 379 ft³ (10.7 m³)
S_a/N = 5.78 ft² (0.538 m²)	H/W (lower) = 0.59	H/W (upper) = 0.44
L/W (lower) = 0.61	L/W (upper) = 0.98	ITDG = 35 msec
Note: $S_T = S_A$ + 1,940 ft² (180 m²); see Appendix 1 for definition of S_T.		

NOTE: The terminology is explained in Appendix 1.

88

Christchurch

Christchurch Town Hall

*A*mong all concert halls worldwide, Christchurch Town Hall is the most startling visually. Here acoustical requirements had full swing in the design. The Town Hall, opened in 1972, was the first hall built following introduction of the concept by Harold Marshall that laterally reflected early sound energy was necessary to create spatial impression. Spatial impression was defined as the listener feeling immersed in the sound field rather than hearing the sound as though it were coming from directly ahead through a window. Subsequent studies have shown that for good spatial impression there must be a number of reflections in about the first 80 msec after arrival of the direct sound at a listener's ears and that a fair percentage of them should be lateral, but not necessarily all.

Advantages of the Town Hall are that no part of the audience of 2,662 is more than 92 ft (28 m) from the stage, there are good sight lines, and the potentially dangerous elliptical shape is prevented from creating focal spots in the hall by the overhead array of panels. A large sound-diffusing canopy is installed above the stage to provide good communication among the players.

The eighteen, very large, titled, free-standing reflecting panels that surround the audience are designed to spread the early sound uniformly throughout the seating areas. The balcony fronts and their soffits are shaped to provide additional early reflections. These areas accomplish their purpose—the sound is clear, the listener's impression is of running liveness, and the hall seems intimate, both visually and acoustically. The ensemble and orchestral balance are good.

As opposed to being in Boston Symphony Hall where one feels enveloped in the later reverberant sound field, in Christchurch the listener feels enveloped in a sound that is comprised of many closely spaced early reflections that arrive early. When the music comes to a sudden stop, the later reverberation sounds weak, although the reverberation time RT is 1.8 to 2.1 sec, measured fully occupied.

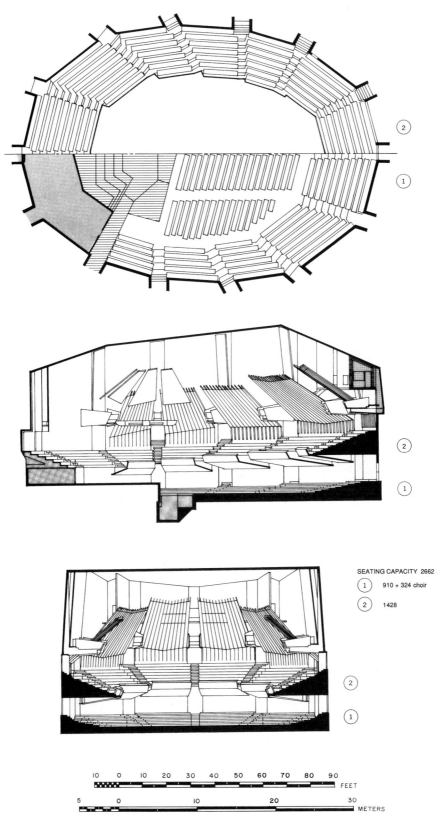

SEATING CAPACITY 2662

(1)   910 + 324 choir

(2)   1428

ARCHITECTURAL AND STRUCTURAL DETAILS

Uses: Orchestral music, recitals, choral, and speech events. **Ceiling:** Concrete. **Walls:** Concrete; free-standing reflectors are of wood, 1 in. (2.5 cm) thick, heavily braced. **Floors:** Concrete. **Carpet:** None. **Stage enclosure:** Walls, 1 in. wood; floor, 2 in. wood over airspace over concrete; 26 ft (7.9 m) irregular canopy hung 24 ft (7.3 m) above stage. **Stage height:** 52 in. (132 cm). **Sound-absorbing materials:** none. **Seating:** Front of seat back and top of seat bottom upholstered.

ARCHITECTS: Warren and Mahoney. ACOUSTICAL CONSULTANT: A. Harold Marshall. PHOTOGRAPHS: Mannering & Associates, Ltd. REFERENCE: A. H. Marshall, "Acoustical design and evaluation of Christchurch Town Hall, New Zealand," *J. Acoust. Soc. Am.* **65**, 951–957 (1979).

TECHNICAL DETAILS

V = 723,855 ft³ (20,500 m³)	S_a = 12,130 ft² (1,127 m²)	S_A = 15,240 ft² (1,416 m²)
S_o = 2090 ft² (194 m²)	S_T = 17,180 ft² (1,596 m²)	N = 2,662
H = 61 ft (18.6 m)	W = 96 ft (29.3 m)	L = 92 ft (28 m)
D = 93 ft (28.4 m)	V/S_T = 42.1 ft (12.9 m)	V/S_A = 47.5 ft (14.5 m)
V/N = 272 ft³ (7.70 m³)	S_A/N = 5.72 ft² (0.53 m²)	S_a/N = 4.56 ft² (0.423 m²)
H/W = 0.64	L/W = 0.96	ITDG = 11 msec

Note: $S_T = S_A$ + 1,940 ft² (180 m²); see Appendix 1 for definition of S_T.

NOTE: The terminology is explained in Appendix 1.

89

Trondheim

Olavshallen

Trondheim is one of the small, beautiful cities of Europe. It was founded by the Viking King Olav Tryggvason in 997 A.D., and it holds a special place in the country's history and culture. In the Middle Ages the city was one of the most visited sites of pilgrimage in the whole of Europe. Located on the Trondheim Fjord and also at the Nidelven river, it offers great attractions for summer visitors looking for boating and salmon fishing. Its population and more than 25,000 university students take great interest in the performing arts. To satisfy this need, the city built the Olavshallen as the venue for the Trondheim Symphony Orchestra. It opened in 1989 and seats 1,200. Because the orchestra and the city wanted a hall that possessed the acoustics of the famous old halls of Europe, the architects were urged to adopt a rectangular shape. The mid-frequency reverberation time, fully occupied, is 1.8 sec for concerts, but with provisions for reducing it down to as low as 1.3 sec for drama and congresses.

The hall has a pleasant, intimate feeling, partly because the distance between the farthest seat and the front of the stage is only 102 ft (31 m). Visually, two architectural features impress one when attending a concert. The first are the nine freely-hanging, 7.2-ft (2.2-m)-wide, long panels that serve both to augment the acoustics and to house the lighting. The other are the three "towers" on either side of the stage. The upper part of each tower contains a curved sound-reflecting surface, designed to reflect sound back to the musicians and thus enabling them to maintain good ensemble. To close off the fly tower, the orchestra enclosure also has a sound-reflecting fire curtain behind the towers and plywood elements above the ceiling. Plans are underway to replace the towers and the walls of the stage enclosure with a nicer-looking plywood shell.

For opera, the hall's acoustics are judged superb. The lower reverberation times for opera, chamber concerts, and congresses are achieved by lowering suitable amounts of heavy curtains into the space above the hanging panels and at the upper rear wall of the hall.

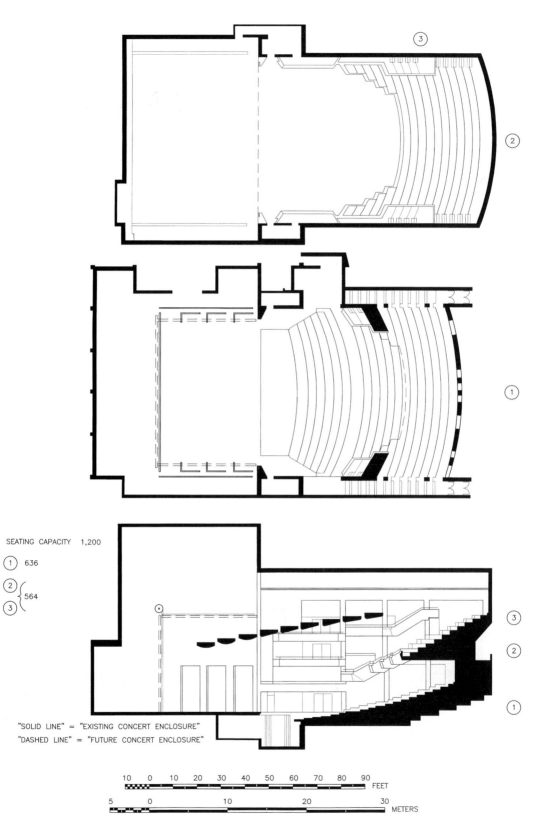

SEATING CAPACITY 1,200

① 636

② ⎱
⎰ 564
③

"SOLID LINE" = "EXISTING CONCERT ENCLOSURE"
"DASHED LINE" = "FUTURE CONCERT ENCLOSURE"

ARCHITECTURAL AND STRUCTURAL DETAILS

Uses: Primarily concerts, also opera, drama, meetings. **Upper ceiling:** 4-in. (6 cm) concrete. **Visual ceiling:** Freely hanging elements, 0.24-in. (0.6-cm)-thick glass-fiber-reinforced polyester. **Sidewalls:** 12–24-in. (30–60-cm)-thick sound-insulating gypsum plate construction, with three plates facing the hall. **Floor:** Oak parquet over precast slabs. **Seating:** Upholstered top of seat bottom and front of backrest. **Orchestra enclosure walls:** "Towers" made of 0.4-in. (1-cm) plywood, with upper curved part made from vibration-damped steel plate. **Variable sound-absorptive material:** Thick curtains stored in plywood boxes above ceiling and at rear wall.

ARCHITECT: Per Knudsen of Mellagerkvartalet. ACOUSTICAL CONSULTANT: Svein Strøm and Asbjørn Krokstad. PHOTOGRAPHS: Svein Strøm and Roar Øhlander. REFERENCE: ICA-95 International Congress on Acoustics, Trondheim, Norway; 661–664 (1993).

TECHNICAL DETAILS

Concerts

V = 458,900 ft³ (13,000 m³)	S_a = 7,736 ft² (719 m²)	S_A = 8,780 ft² (816 m²)
S_o = 2,517 ft² (234 m²)	S_T = 10,717 ft² (996 m²)	N = 1200
H = 59 ft (18 m)	W = 85.9 ft (26.2 m)	L = 98.4 ft (30 m)
D = 102 ft (31 m)	V/S_T = 42.8 ft (13 m)	V/S_A = 52.3 ft (15.9 m)
V/N = 382 ft³ (10.83 m³)	S_A/N = 7.32 ft² (0.68 m²)	S_a/N = 6.45 ft² (0.60 m²)
H/W = 0.69	L/W = 1.145	ITDG = 26 msec

Note: $S_T = S_A$ 180 m; see definition of S_T in Appendix 1.

Opera

V = 458,900 ft³ (13,000 m³)	S_a = 7,736 ft² (719 m²)	S_A = 8,780 ft² (816 m²)
S_{pit} = 717 ft² (67 m²)	Sp = 3185 ft² (296 m²)	S_T = 12,684 ft² (1,179 m²)
N = 1,200 H = 59 ft (18 m) W = 85.9 ft (26.2 m) L = 98.4 ft (30 m)		
D = 102 ft (31 m)	V/S_T = 36.2 ft (11 m)	V/S_A = 52.3 ft (15.9 m)
V/N = 382 ft³ (10.83 m³)	S_A/N = 7.32 ft² (0.68 m²)	S_a/N = 6.45 ft² (0.60 m²)
H/W = 0.69	L/W = 1.145	ITDG = 26 msec

NOTE: The terminology is explained in Appendix 1.

90

Edinburgh

Usher Hall

he beauty of Usher Hall is awe-inspiring since its 2000 refurbishing. New seats with deep red upholstering, a new stage floor, and gold leaf on vital decorative surfaces, combined with subtle changes in light-colored shades throughout the hall, leave one convinced that an exciting performance is to follow. Comfort is evident in the new seats on the main floor and the first balcony (called the Grand Circle). At the main floor level the seats close to a compact upright position when unoccupied and with a seat back that tilts as the seat base is lowered. The orchestra has a deepened stage, enlarged from 1,290 ft² (120 m²) to 1,775 sq² (165 m²) with a stage lift to the piano store below. With 29 seats removed, the stage can further be enlarged whenever a large orchestra is performing.

The main floor has been replaced, the center aisle eliminated, and sight lines slightly improved. Ventilation air now enters the hall from air grilles under each seat. The extensive choir seating (no upholstering), peculiar to the U.K., is also used for audience. The hall is partly multipurpose, all the main floor seats are removable for such events or concerts like promenade-type pop concerts and can be stored via the elevator under the stage.

Dedicated in 1914, and updated in 2000, the hall now seats 2,502, less, sometimes, the 29 seats mentioned above. The reverberation time has been substantially increased from a discouraging low, for orchestral music, of 1.5 sec up to 1.7 sec at mid-frequencies, fully occupied. This increase is attributable to removal of areas of carpet and the new seats that are slightly less absorbent than the previous ones. Some of the same problems attributable to its circular architecture persist as before, including focusing along the centerline, a deficiency in early lateral reflections, and some variability in seat location. Overall the sound is "brighter," owing to the increased reverberation time, and the orchestral tone and sectional balance are good in most parts of the hall. Critical reviews in the last concert season have been favorable.

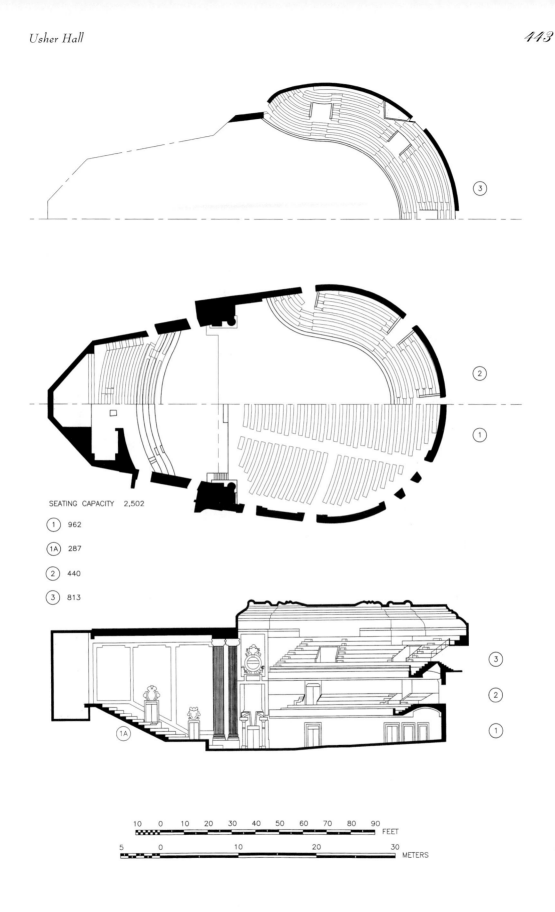

SEATING CAPACITY 2,502

① 962

①A 287

② 440

③ 813

10 0 10 20 30 40 50 60 70 80 90 FEET

5 0 10 20 30 METERS

Further refurbishment is planned, including secondary glazing to reduce external noise and introducing sound-diffusing panels, as well as improving stage and lighting equipment.

ARCHITECTURAL AND STRUCTURAL DETAILS

Uses: Mostly orchestral and popular music. **Ceiling:** Plaster on wood lath with thick plaster ornamentation. The center "saucer" is perforated by large holes that act as exhaust ports. **Walls:** Masonry finished with plaster on wood lath over batons with airspace behind. **Floors:** Timber. **Stage enclosure:** Plaster on wood lath. **Stage floor:** Planks with large airspace with wood boxes for risers. **Carpets:** None on main floor; runners on the walking areas in first balcony and in upper balcony, linoleum on boarding in the aisles. **Seats:** (see text).

ORIGINAL ARCHITECT: Stockdale Harrison and Sons, and H. H. Thomson. RENOVATION ARCHITECT: Architect Services Division, Corporate Services, City of Edinburgh. ACOUSTICAL CONSULTANT FOR RENOVATION: Sandy Brown Associates.

TECHNICAL DETAILS

V = 554,210 ft³ (15,700 m³)	S_a = 10,588 ft² (984 m²)	S_A = 12,955 ft² (1,204 m²)
S_o = 1775 ft² (165 m²)	S_T = 14,730 ft² (1,369 m²)	N = 2,502
H = 58 ft (17.7 m)	W = 78 ft (23.8 m)	L = 100 ft (30.5 m)
D = 114 ft (34.8 m)	V/N = 221.5 ft³ (6.27 m³)	V/S_T = 37.6 ft (11.5 m)
V/S_A = 42.8 ft (13.0 m)	S_A/N = 5.18 ft² (0.48 m²)	S_a/N = 4.23 ft² (0.393 m²)
H/W = 0.74	L/W = 1.28	ITDG = 33 msec

NOTE: The terminology is explained in Appendix 1.

91

Glasgow

Royal Concert Hall

Opening in 1990, the Glasgow Royal Concert Hall replaced the well-liked St. Andrew's Hall of 1877 that burned tragically in 1962. St. Andrew's Hall was a classical shoebox-type concert hall, which was generally acclaimed for its good acoustics. The plan of the new Royal Concert Hall is an elongated hexagon, seating 2,457, with two steeply raked balconies that cover a large percentage of the total sidewall area. The vertical wall space above the balconies was partly covered with 16 large, tilted, quadratic residue diffusers (QRDs). This QRD form of treatment was successfully embodied in the Christchurch Town Hall in New Zealand, to overcome the potentially fatal acoustics of an oval-shaped hall. Photographs of that Glasgow design are on the next page. Note also, that no suspended canopy was provided over the orchestra.

That design was not completely satisfactory. The complaints were as follows: The rich reverberation, that is to say, the fullness of tone that existed in the previous hall, was not present and the musicians complained that they could not hear each other well. Many listeners also felt that they were not getting sufficient support for the music from sidewall reflections. At this moment, an experiment, still in the trial stage, is in progress. The sixteen QRD reflectors have been removed and a canopy has been installed above the stage. Also the reflecting surfaces on the beams above the stage have been removed. The removal of the QRD's and the reflectors has made "around-the-hall" reverberation possible, thus adding "fullness of tone." In addition, sound-absorbing materials have been applied to parts of the rear wall and sidewalls to eliminate echoes and acoustic "flutter." These changes can be seen in the accompanying drawings.

The hall is partially multipurpose; the first 11 rows can be removed. Because all stage sections are on elevators that can be lowered, an extensive flat floor area can be made available for staging arena format productions. The extensive seating (264 chairs) behind the platform add to the arena format. The reverberation time,

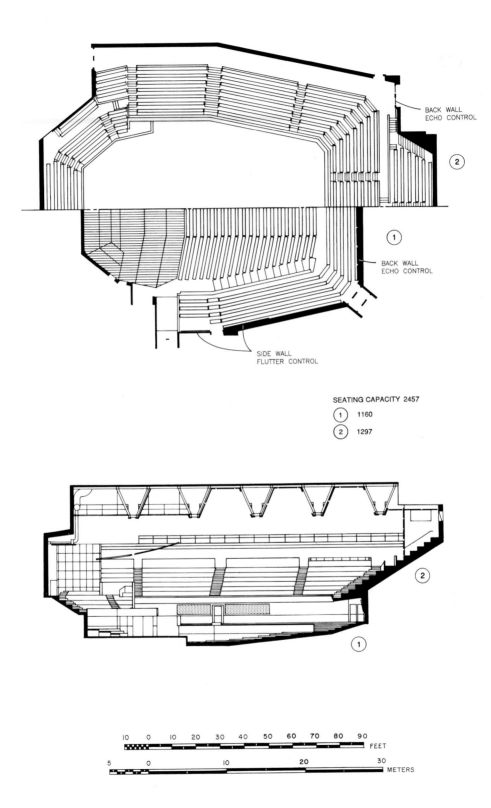

BACK WALL
ECHO CONTROL

2

1

BACK WALL
ECHO CONTROL

SIDE WALL
FLUTTER CONTROL

SEATING CAPACITY 2457

1   1160

2   1297

2

1

10 0 10 20 30 40 50 60 70 80 90
FEET

5   0   10   20   30
METERS

measured with full audience at mid-frequencies, is about 1.8 sec, and this can be lowered with extendable curtains on the upper sidewalls.

ARCHITECTURAL AND STRUCTURAL DETAILS

Uses: Events of all kinds, excluding opera and proscenium theater. **Main ceiling:** 4-in. (10-cm) concrete, clad underside with ash veneered chipboard, 0.7 in. (1.7 cm) thick, with 18-in. (45-cm) airspace above. The dropped ceiling over the promenade at the top of the upper balcony is 0.5-in. (1.25-cm) plasterboard. **Upper walls:** Painted concrete. **Walls around the main floor:** 0.8-in. (2-cm)-thick wood, over 1-ft (30-cm) airspace. **Balcony fronts:** Fibrous plaster, 0.5 in. (1.25 cm) thick. **Floors and seating steps:** Concrete. **Carpet:** In aisles, thin, cemented to concrete slab; in promenade, carpet over underpad. **Seating:** Front of seat back and seat upholstered.

ARCHITECT: Sir Leslie Martin and RMJM Scotland Ltd. ORIGINAL ACOUSTICAL CONSULTANT: Fleming & Barron and Sandy Brown Associates. RENOVATION ACOUSTICAL CONSULTANT: Kirkegaard & Associates. PHOTOGRAPHS: Courtesy of the Department of Architecture, City of Glasgow. REFERENCE: M. Barron and A. N. Burd, "The acoustics of Glasgow Royal Concert Hall," *Proceedings of Institute of Acoustics* **14**, 21–29 (1992).

TECHNICAL DETAILS

V = 810,135 ft³ (22,950 m³)	S_a = 12,342 ft² (1,147 m²)	S_A = 14,700 ft² (1,365 m²)
S_o = 2,346 ft² (218 m²)	S_T = 16,640 ft² (1,545 m²)	N = 2,457
H = 63 ft (19.2 m)	W = 108 ft (32.9 m)	L = 90 ft (27.8 m)
D = 111 ft (33.8 m)	V/S_T = 48.7 ft (14.86 m)	V/S_A = 55.2 ft (16.8 m)
V/N = 330 ft³ (9.34 m³)	S_A/N = 5.98 ft² (0.555 m²)	S_a/N = 5.02 ft² (0.466 m²)
H/W = 0.58	L/W = 0.83	ITDG = 20 msec

Note: $S_T = S_A + 1,940$ ft² (180 m²); see Appendix 1 for definition of S_T.

NOTE: The terminology is explained in Appendix 1.

92

Madrid

Auditorio Nacional de Música

adrid's old opera house, the Teatro Real, built in 1850 with very dry acoustics (low reverberation time), served as the venue for concert music until 1925, when it was closed because of poor maintenance. In the absence of a proper concert hall, classical music was then performed in the cinemas and theaters of Madrid. In 1962, the Teatro Real was renovated and fitted with a reasonably satisfactory orchestra enclosure. For the next 26 years it was Madrid's principal location for symphonic music. It was not until the early 1980s, with the new democracy, that the construction of a large modern concert hall in Madrid, along with a number of others in Spain, was contemplated. The site chosen is parallel to a new expressway in the northern part of the city. The Auditorio Nacional de Música was inaugurated in October 1988. The Spanish National Orchestra, forged out of two orchestras in March 1942, is its principal tenant, along with the Spanish National Choir.

The Auditorio Nacional is visually attractive. Four grandiose chandeliers and a spectacular pipe organ dominate over the primarily brown wood and white plaster interior. The seats number 2,293, of which about 20 percent, including those for the choir, are to the rear and the immediate sides of the stage. The hall is very wide, with an enormous, 3000 ft² (280 m²), stage which boasts of a system of adjustable tiered flooring and a number of hydraulically operated platforms to accommodate the needs of the orchestra and equipment changes. Four large reflectors over the stage provide communication between sections of the orchestra. The ceiling with large convex reflecting surfaces distributes reflected sound to the different audience areas, including those behind the orchestra.

The local music critics write favorably of the hall; acoustics: ". . . modern, nice and with splendid acoustics," ". . . majestic interior and with exceptional acoustics," ". . . the musical-acoustic result is totally positive in complete agreement with the first conductors who occupied the stage (Kurt Masur, Lorin Maazel, and Lopez Cobos)."

SEATING CAPACITY	2,293
(1) MAIN FLOOR	585
(2) FIRST BALCONY	476
(3) SIDE BALCONIES	201
(4) SECOND BALCONY (CENTER)	430
(5) SECOND BALCONY (LATERALS)	215
(6) BEHIND STAGE	194
(7) CHOIR	140
(8) UPPER GALARIES	52

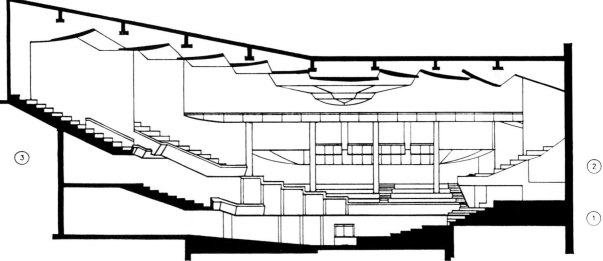

ARCHITECTURAL AND STRUCTURAL DETAILS

Uses: Concerts, broadcasting and recitals. **Ceiling:** 1-in. (2.5-cm) wood fiberboard finished in veneer of American walnut suspended from concrete roof with large air space. **Suspended stage reflectors:** Two layers of 0.63-in. (1.6-cm) laminated wood. **Side and rear walls:** 9.5-in. (24-cm) brick wall with 0.15-cm plaster finish. **Balcony fronts:** 1-in. (2.5-cm) wood over 2.0-cm battens. **Rear stage wall:** 0.75-in. (1.9-cm) wood over 0.8-in. (2-cm) battens. **Floors:** Sheet vinyl flooring on concrete slab. **Stage floor:** 0.9-in. (2.2-cm) wood on joists. **Stage height:** 39 in. (1 m). **Seating:** Laminated plywood for underseat and seatback; upholstered upperseat and backrest; wood arms.

ARCHITECT: Jose M. García de Paredes. ACOUSTICAL CONSULTANTS: Lothar Cremer/Thomas Futterer and GARCIA-BBM. PHOTOGRAPHS: Garcia-BBM Ltd. REFERENCES: A. García Senchermés, "Spain goes classical—10 years of concert halls and opera houses," Acoustics International Conference, Birmingham (UK), 1992.

TECHNICAL DETAILS

V = 706,000 ft³ (20,000 m³)	S_a = 15,086 ft² (1,402 m²)	S_A = 18,346 ft² (1,705 m²)
S_o = 2,991 ft² (278 m²)	S_T = 20,286 ft² (1,885 m²)	N = 2,293
H = 61.7 ft (18.8 m)	W = 92 ft (28 m)	L = 128 ft (39 m)
D = 131 ft (40 m)	V/S_T = 34.8 ft (10.6 m)	V/S_A = 38.4 ft (11.7 m)
V/N = 308 ft³ (8.72 m³)	S_A/N = 8.0 ft² (0.74 m²)	S_a/N = 6.58 ft² (0.61 m²)
H/W = 0.67	L/W = 1.39	ITDG = 45 msec
Note: $S_T = S_A + 180$ m²; see definition of S_T in Appendix 1.		

NOTE: The terminology is explained in Appendix 1.

93

Valencia

Palau de la Música

*V*alencia, the third-largest city in Spain, is proud of its historic past, dating from 137 B.C., and of its modern cultural achievements. Today it is a major tourist attraction, with eight principal museums, important public buildings, a cathedral, and the ocean front with magnificent beaches.

The Palau de la Música, opened in 1987, is a spectacular building, with an all-glass exterior, topped by a sun-screening canopy of interesting architectural design. It is located in the midst of the city's Gardens of Turia, which are filled with exotic fauna, tall palm trees, and large reflecting pools.

With 1,790 seats, the Palau was the first of a series of large concert halls built in Spain in the last 15 years. Valencia boasts of its high density of musicians per square kilometer, who were understandably excited by hearing classical music in the nation's first "2-sec" concert hall after a history of hearing music only in the dry acoustics of theaters and opera houses.

Because the hall is also used for recitals, opera in concert, conferences, and films, means have been provided for reducing the reverberation time. The upper parts of the walls contain a mechanized series of doors, each exposing an area of absorbent materials 31 in. (80 cm) thick that, when fully opened, reduce the reverberation time at mid-frequencies to 1.4 sec with full audience. During rehearsals, partially opening these doors reduces the reverberation time in the unoccupied hall to the expected 2 sec at concerts.

Since its seating capacity is similar to that of the Musikvereinssaal in Vienna, the audience relates to the orchestra intimately, and thus the hall has been a popular venue for visiting orchestras. Because of the wraparound seating, the distance from the front of the stage to the most distant listener is only 98 ft (30 m). Zarin Mehta, manager of the New York Philharmonic Orchestra, which performed there in 2000, and earlier, has spoken of the pleasure of not playing in a 3,000-seat hall and of the Palau's excellent acoustics.

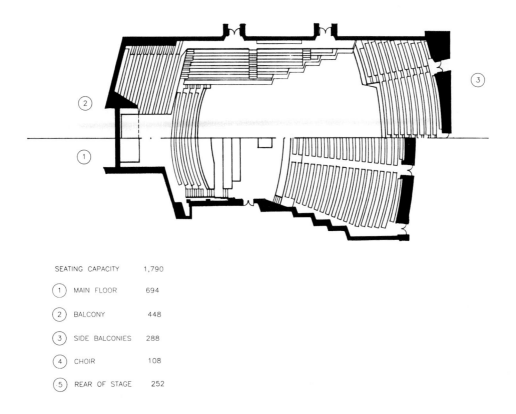

SEATING CAPACITY   1,790

(1) MAIN FLOOR   694

(2) BALCONY   448

(3) SIDE BALCONIES   288

(4) CHOIR   108

(5) REAR OF STAGE   252

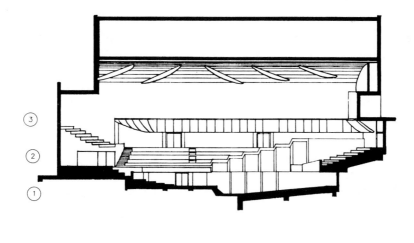

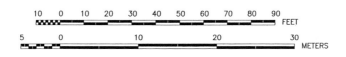

ARCHITECTURAL AND STRUCTURAL DETAILS

Uses: Symphonic, organ and choral concerts, jazz and flamenco amplified music, conferences and cinema projections. **Ceiling and audience reflectors:** 1-in. (2.5-cm) wood fiber with wood veneer. **Stage reflectors:** 0.5-in. (1.2-cm) glass. **Walls:** 0.6-in (1.5-cm) plaster over brick wall. **Floors:** Sheet vinyl over concrete. **Stage floor:** 0.9-in. (2.2-cm) wood on joists. **Balcony fronts:** 1-in. (2.5-cm) wood over 0.8-in. (2-cm) battens. **Rear stage wall:** 0.75-in (1.9-cm) wood over 0.8-in. (2-cm) battens. **Stage floor:** 0.9-in. (2.2-cm) wood over joists. **Stage height above front row of seats:** 49 in. (1.25 m). **Seating:** Plywood on metal frame, thin upholstering on front of seat back and top of seat bottom. Arms unupholstered.

ARCHITECT: J. M. Garcia de Paredes. ACOUSTICAL CONSULTANT: Garcia-BBM Ltd. PHOTOGRAPHS: Acoustical consultant and Jarque.

REFERENCE: A. Garcia Senchermés, "Spain goes classical—10 years of concert halls and opera houses," Acoustics International Conference, Birmingham (UK) 1992.

TECHNICAL DETAILS

V = 543,620 ft³ (15,400 m³)	S_a = 7,715 ft² (717 m²)	S_A = 8,737 ft² (812 m²)
S_c = 538 ft² (50 m²)	S_o = 1,915 ft² (178 m²)	S_T = 11,190 ft² (1,040 m²)
N = 1,790	H = 54 ft (16.5 m) W = 67 ft (20.5 m)	L = 61 ft (18.7 m)
D = 67 ft (20.5 m)	V/S_T = 48.6 ft (14.8 m)	V/S_A = 62.2 ft (19 m)
V/S_a = 70.5 ft (21.5 m)	V/N = 304.6 ft³ (8.6)	S_A/N = 4.88 ft² (0.454 m²)
H/W = 0.80	L/W = 0.91	ITDG = 29 msec

NOTE: The terminology is explained in Appendix 1.

94

Gothenburg

Konserthus

*G*othenburg relishes the Konserthus because of its intimacy, acoustical quality, and beauty. The hall opened in 1935, was renovated once in 1985, and again in 2000. It is finished entirely in wood. Having only 1,286 seats, it is small compared to halls of that period. There is no balcony, and elevated sections of seats at the sides and rear of the hall interestingly subdivide the floor. The warm color of the wood; the lighting, some introduced in 2000; the close relation of the audience to the stage; and the adequate reverberation time combine to make attendance at a concert a rewarding experience.

The 2000 renovation involved changes in the sending end of the hall. The canopy has been made 80 percent open, releasing more sound energy to the volume above the stage and improving the orchestra ensemble and balance. The decorative—nonfunctional—organ pipes have been removed from above the canopy. The result is an increase in the reverberation time, especially EDT. In the 1985 renovation the stage was enlarged to handle larger performing groups.

It must be emphasized that the wooden interior of the Konserthus does not damage the beauty of the sound. All good halls have interiors that are as heavy and strong as possible so that the surfaces do not resonate. Resonance would cause absorption of the bass tones. The interior of the Konserthus is made of panels about 1 in. (2.5 cm) thick, 30 in. (76 cm) wide, and 80 in. (203 cm) long. Each panel is heavily braced behind and securely held in place on heavy frames—all bolted to a 7-in. (18-cm)-thick concrete wall.

The Konserthus illustrates several basic tenets of acoustics: in a hall for music, small size means adequate loudness, a short initial-time-delay gap, and a high value for the Binaural Quality Index (BQI), which depends on lateral reflections from sidewalls. There is no balcony overhang and the reverberant sound has an "around-the-room" path to follow, thus providing "fullness of tone." The reverberation time, since the 2000 renovation, is about 1.65 sec, particularly suited to Classical and Contemporary music, and well in balance with the volume. This hall has always

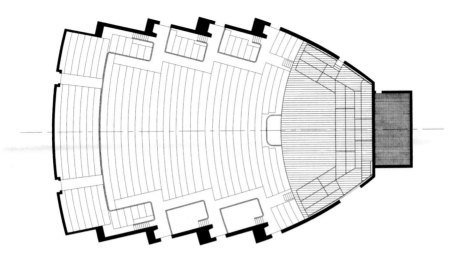

SEATING CAPACITY 1286

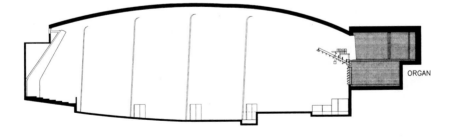

ORGAN

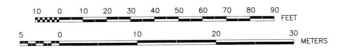

10   0   10   20   30   40   50   60   70   80   90
FEET

5   0   10   20   30
METERS

been judged a success, and the 2000 renovation has improved the conditions for the performers, both audibly as well as measurably.

ARCHITECTURAL AND STRUCTURAL DETAILS

Uses: Symphonic, 80%; soloists and chamber music, 5%; popular concerts, 10%; miscellaneous, 5%. **Ceiling:** Plywood affixed directly to a concrete sheet that is suspended from a steel overhead. **Side walls and rear walls:** (described in the text). **Floors:** Linoleum on concrete. **Stage enclosure:** Walls same as above, canopy 0.75-in. (2-cm) vertical plywood slats with 80% of the area open. **Stage floor:** Wood over small airspace with on-stage movable risers of wood. **Stage height:** 39 in. (99 cm). **Seating:** Wooden with fully upholstered seat back and upper seat.

ARCHITECT: Nils Einar Eriksson. ACOUSTICAL CONSULTANTS: Original, H. Kreuger; 1985 renovation, Ingemansson Acoustics; 2000 renovation, Niels V. Jordan. PHOTOGRAPHS: Front of hall, Jens Holger Rindel; rear, Max Plunger.

TECHNICAL DETAILS

V = 420,000 ft³ (11,900 m³)	S_a = 6,300 ft² (585 m²)	S_A = 7,170 ft² (666 m²)
S_o = 1,830 ft² (170 m²)	S_T = 9,000 ft² (836 m²)	N = 1,286
H = 45 ft (13.7 m)	W = 83 ft (25.3 m)	L = 100 ft (30.5 m)
D = 97 ft (29.6 m)	V/S_T = 46.7 ft (14.2 m)	V/S_A = 58.5 ft (17.8 m)
V/N = 326 ft³ (9.25 m³)	S_A/N = 5.58 ft² (0.52 m²)	S_a/N = 4.9 ft² (0.455 m²)
H/W = 0.54	L/W = 1.2	ITDG = 20 msec

NOTE: The terminology is explained in Appendix 1.

Stadt-Casino

uilt in 1776, the Stadt-Casino preceded the famous Leipzig "Neues" Gewandhaus (destroyed in World War II) by ten years. Seating 1,448, smaller by 112 seats, it bears all the distinguishing hallmarks of its former northern sister. It is rectangular with a flat coffered ceiling that connects to the side walls with a sloping cornice. Both halls resemble the ballrooms of their heritage, with flat main floors and shallow balconies. The widths of both are nearly the same, 69 ft (21 m), making the Stadt-Casino very intimate, acoustically.

Visually, one is struck by the large windows above the balcony, necessary then because of lack of artificial lighting and air conditioning. Four ornate chandeliers grace the ceiling. The beautiful organ stands above the highest stage riser on which the percussion or the organ console customarily sit.

The hall has been reseated since the photographs were taken, with upholstered backrests and seat bottoms. The reverberation time, about 1.8 sec at mid-frequencies, fully occupied, is nearly equal to that of Boston's Symphony Hall.

The interior is plaster, so that the bass is warm and the higher registers are brilliant. The hall responds well to the music and musicians love to perform in it.

Herbert von Karajan said of the Stadt-Casino (1960): "This is a typical rectangular hall—small, with a wonderfully clear and crisp resonance. It is almost perfect for Mozart. Although one can play nearly every kind of music in it, the volume of a very large orchestra is smashing." Dimitri Mitropoulos (1960) also thought the Stadt-Casino a very good hall but too small for full orchestra.

In my judgment, the pianissimo playing of the cellos and double basses is extraordinarily beautiful. Violin and bass parts are well balanced. The percussion and brass can be too loud. But, in brief, all music that is properly scaled to the size of the Stadt-Casino sounds wonderful there. The enthusiasm of the musicians who know the hall well is fully justified.

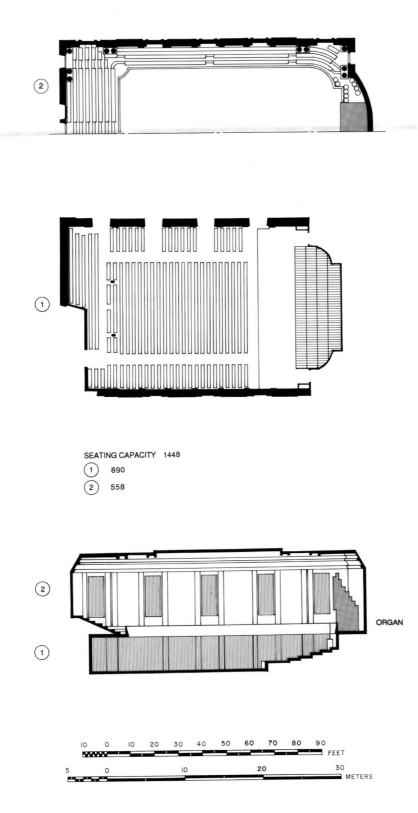

SEATING CAPACITY 1448

① 890

② 558

ORGAN

10 0 10 20 30 40 50 60 70 80 90 FEET

5 0 10 20 30 METERS

ARCHITECTURAL AND STRUCTURAL DETAILS

Uses: Mostly orchestra and recitals. **Ceiling:** Plaster, coffered. **Side walls and rear walls:** Plaster with a small amount of wood; baroque ornamentation, and columns. **Floors:** Parquet over solid base. **Carpet:** None. **Stage floor:** Heavy wood over airspace. **Stage height:** 36 in. (91 cm). **Seating:** Tops of seat bottoms and fronts of backrests upholstered; underseats solid metal.

ARCHITECT: Johann Jacob Stehlin-Burckhardt. PHOTOGRAPHS: Courtesy of Swiss National Tourist Office. REFERENCE: W. Furrer, *Raum- und Bauakustik, Larmabwehr,* 2nd Edition, Birkhauser Verlag, Basel and Stuttgart (1961).

TECHNICAL DETAILS

$V = 370{,}750 \text{ ft}^3 \ (10{,}500 \text{ m}^3)$	$S_a = 6{,}290 \text{ ft}^2 \ (584 \text{ m}^2)$	$S_A = 7{,}870 \text{ ft}^2 \ (731 \text{ m}^2)$
$S_o = 1{,}720 \text{ ft}^2 \ (160 \text{ m}^2)$	$S_T = 9{,}600 \text{ ft}^2 \ (891 \text{ m}^2)$	$N = 1{,}448$
$V/S_T = 38.6 \text{ ft} \ (11.8 \text{ m})$	$V/S_A = 47.1 \text{ ft} \ (14.4 \text{ m})$	$V/N = 256 \text{ ft}^3 \ (7.25 \text{ m}^3)$
$S_A/N = 5.44 \text{ ft}^2 \ (0.50 \text{ m}^2)$	$H = 50 \text{ ft} \ (15.2 \text{ m})$	$W = 69 \text{ ft} \ (21 \text{ m})$
$L = 77 \text{ ft} \ (23.5 \text{ m})$	$D = 80 \text{ ft} \ (24.4 \text{ m})$	$S_a/N = 4.34 \text{ ft}^2 \ (0.404 \text{ m}^2)$
$H/W = 0.72$	$L/W = 1.12$	ITDG $= 16 \text{ msec}$

NOTE: The terminology is explained in Appendix 1.

96

Lucerne

Culture and Congress Center Concert Hall

*T*he *pièce de résistance* of the Lucerne Culture and Congress Center situated on the Lake of Lucerne is its 1,892-seat concert hall. Opened in 1999, the center is composed of three main buildings, the concert hall, a multipurpose middle hall, and the Museum of Fine Arts. Lucerne has long been a center for tourists, who are attracted by the lake, the view of the Lower Alps, and the impressive accommodations. For 68 years, the International Festival of Music of Lucerne has attracted music aficionados from all over the world.

The concert hall is basically a rectangular room—based on the shoebox design of Vienna's Grosser Musikvereinssaal and similar halls around the world. Its interior space is augmented by 6,000 m³ of reverberation chamber, located on the two sides and the front of the hall. It is made available by opening all or part of 50 curved white doors spread out on three levels around the balconies. These doors are 8 ft (2.43 m) wide and vary in height from 10 ft (3 m) to 20 ft. Their surfaces are covered by hundreds of square "coffers," each 8 in. (21 cm) square and 2.4 in. (6 cm) deep, with small rectangular pits in the center. Five different patterns were chosen to avoid any predominance of a band of frequencies during the reflection of sound waves from their surfaces. Aside from the sheer beauty of the largely plaster and wood interior, the hanging acoustic canopy over the stage is impressive. This canopy is in two sections and its height can be adjusted over a wide range. The balconies extend over the sides of the stage. The distance from the farthest seat to the stage front is 124 ft (37.5 m).

Maestro Franz Welser-Möst, the Music Director of the Cleveland Orchestra, said, "We could do everything we wanted to do. In the beginning of the third movement, the pianissimos worked very well. In the loudest passages, at the podium it never went into overdrive. The clarity was always there. I have no wishes for any changes. Absolutely fantastic hall."

In my opinion, it is one of the best of the modern halls.

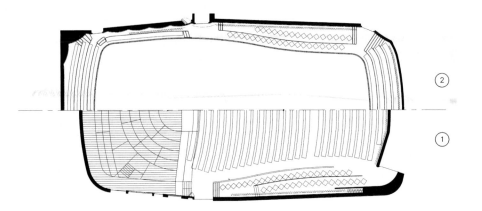

SEATING CAPACITY 1892

1   778
2   352
3   214
4   234
5   314

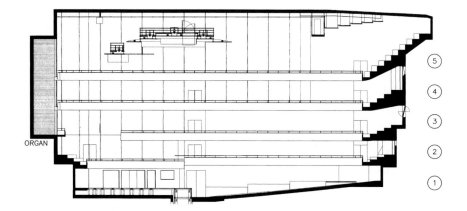

ORGAN

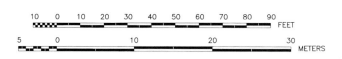

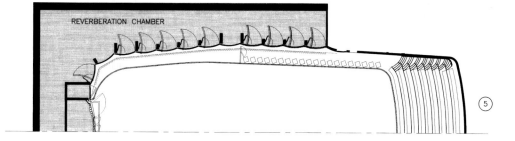

REVERBERATION CHAMBER

⑤

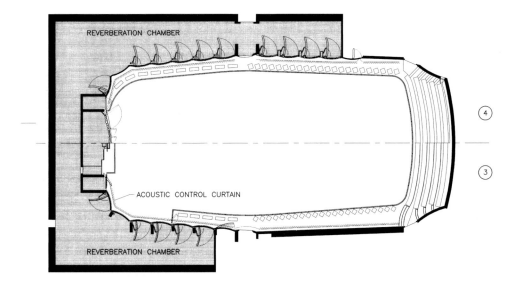

REVERBERATION CHAMBER

④

③

ACOUSTIC CONTROL CURTAIN

REVERBERATION CHAMBER

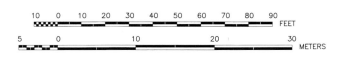

ARCHITECTURAL AND STRUCTURAL DETAILS

Uses: Orchestra, recitals, recordings, conferences, and pipe organ. **Ceiling:** Concrete, 6 in. (15 cm) thick. **Canopy:** Sound-reflecting surface, 0.75-in. (1.9-cm) finish wood attached to 1-in. (2.5-cm) honeycomb board attached to 1-in. plywood substrate: Normal range of travel is between 33 and 50 ft (10 and 15 m). **Sidewalls:** 12–16-in. (30–40-cm) poured concrete with applied-plaster sound-diffusing tiles. **Reverberation chamber doors:** Concrete, 3.6 in. (12 cm) thick, with 2.4-in. (6-cm)-thick sound-diffusing plaster tiles bonded to their convex surfaces. **Reverberation chamber:** Fifty operable concrete doors control access to 218,500-ft³ (6,190-m³) of volume. **Balcony faces:** Perforated plaster, 1.2 in. (3 cm) thick, backed by glass-fiber filled cavities. **Variable sound diffusion, retractable:** Maximum of 4,900 ft² (450 m²) acoustic materials in reverberation chambers and maximum of 12,900 ft² (1,200 m²) acoustic materials covering walls of the concert hall. **Main floor:** Tongue-and-groove wood planks, 1 in. (2.5 cm) thick, glued directly to concrete. **Stage floor:** Tongue-and-groove planks of solid Oregon pine, 1.4 in. (3.5 cm) thick. **Height of the stage:** 43 in. (1.1 m). **Seats:** Seat back and seat upholstered with 1.5 in. (3.7 cm) padding.

ARCHITECT: AJN Architectures Jean Nouvel. ACOUSTICAL CONSULTANT: ARTEC Consultants, Inc. PHOTOGRAPHS: Priska Ketterer, Lucerne.

TECHNICAL DETAILS

V = 629,150 ft³ (17,823 m³)	S_a = 9,286 ft² (863 m²)	S_A = 13,515 ft² (1,256 m²)
S_o = 2604 ft² (242 m²)	S_T = 15,450 ft² (1436 m²)	N = 1892
H = 77 ft (23.5 m)	W = 72 ft (22 m)	L = 90 ft (27.4 m)
D = 118 ft (36 m)	V/S_T = 40.7 ft (12.4 m)	V/S_A = 46.6 ft (14.2 m)
V/N = 332.5 ft³ (9.42 m³)	S_A/N = 7.14 ft² (0.664 m²)	S_a/N = 4.91 ft² (0.66 m²)
H/W = 1.07	L/W = 1.25	ITDG = 23 msec
Note: $S_T = S_A$ + 180 m²; see Appendix 1 for definition of S_T.		

NOTE: The terminology is explained in Appendix 1.

97

Zurich

Grosser Tonhallesaal

The Tonhalle, completed in 1895 and renovated in 1930, accommodates an audience of 1,546 persons. Two glittering chandeliers hang beneath a series of oil paintings on the ceiling that portray Gluck, Haydn, Schubert, Brahms, Mozart, and Handel. The walls are a grayish ivory abundantly decorated with gold. Beneath the balcony, natural birch paneling gives the hall an appearance of being all wood, but this is not the case. Plaster forms more than 80% of the walls and ceiling. The floors are wood parquet. The seats are not upholstered; during rehearsals the orchestra finds it necessary to lay thin layers of cloth over the seats on the main floor to control the reverberation.

The mid-frequency reverberation time, fully occupied, is about 2.0 sec, a value that is achieved by the majority of the best-liked halls in this study.

Three of the four conductors interviewed rated the Grosser Tonhallesaal as acoustically excellent. The fourth, who favors contemporary music, said, "This hall is not so good and not so bad. It is sort of in between." Music critics have generally given high praise to the acoustics.

My one musical experience with this hall was very pleasant. The program consisted of music of the Baroque and early classical periods, and the reverberation time of the hall seemed ideal. The bass was not as satisfactory in the main floor as in the galleries, possibly owing to some thin wood on the lower side walls. The violins sang forth clearly—their sound was intimate, live, and brilliant. Ensemble playing was good as was the balance among orchestral sections, with only a moderate tendency for the brass to dominate. The blend was excellent. The sound was not as good in the rear corners of the main floor and at the rear of the side balconies as on the main floor and in the rear balcony. All in all, it is an excellent hall.

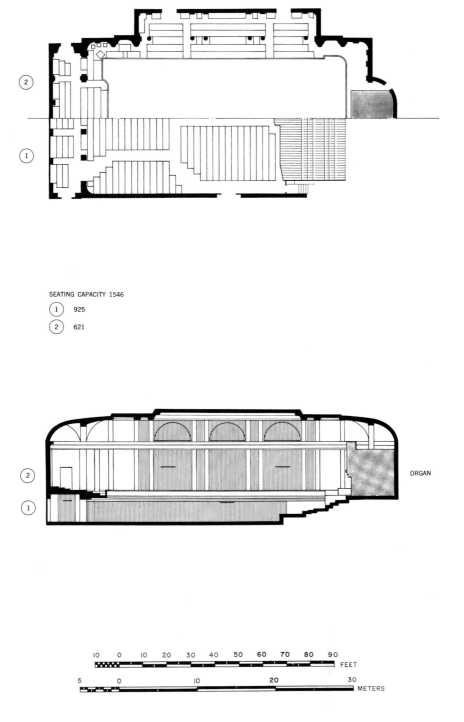

SEATING CAPACITY 1546

(1) 925

(2) 621

ORGAN

10 0 10 20 30 40 50 60 70 80 90
FEET

5 0 10 20 30
METERS

ARCHITECTURAL AND STRUCTURAL DETAILS

Uses: Symphonic music, 200 days a year; piano and violin recitals; oratorio, 12 days; and occasionally organ. **Ceiling:** Plaster on wire lath with deep irregularities in the ceiling and rounded cornices, interrupted by clerestory windows. **Sidewalls:** Wood paneling beneath the balcony and on the soffit under the balcony; walls and decorations are of plaster above the balcony; balcony fronts are of plaster. **Floors:** Wood parquet over solid base on the main floor and linoleum over solid base in balcony. **Carpet:** None. **Stage floor:** Heavy wood over airspace. **Stage height:** 48 in. (122 cm). **Added absorptive material:** Cloth curtains hang over the clerestory windows; during rehearsals large areas of cloth are placed over half the seats on the main floor. **Seating:** All wood. (NOTE: the doors at the rear of the main floor may be opened, exposing 200 additional seats in the Kleiner Tonhallesaal.)

ARCHITECT: Fellner and Helmer of Vienna. PHOTOGRAPHS: Courtesy of Swiss National Tourist Office. DETAILS AND SEATING PLAN: From the General Manager. REFERENCE: Drawings courtesy of W. Furrer.

TECHNICAL DETAILS

$V = 402{,}500$ ft^3 (11,400 m^3)	$S_a = 7{,}550$ ft^2 (702 m^2)	$S_A = 9{,}440$ ft^2 (877 m^2)
$S_o = 1{,}560$ ft^2 (145 m^2)	$S_T = 11{,}000$ ft^2 (1,022 m^2)	$N = 1{,}546$
$V/S_T = 36.6$ ft (11.2 m)	$V/S_A = 42.6$ ft (13 m)	$V/N = 260$ ft^3 (7.37 m^3)
$S_A/N = 6.11$ ft^2 (0.57 m^2)	$H = 46$ ft (14 m)	$W = 64$ ft (19.5 m)
$L = 97$ ft (29.6 m)	$D = 98$ ft (29.9 m)	$S_a/N = 4.88$ ft^2 (0.455 m^2)
$H/W = 0.72$	$L/W = 1.5$	ITDG $= 14$ msec

NOTE: The terminology is explained in Appendix 1.

Cultural Centre Concert Hall

The Taipei Concert Hall may be the most stunning building in the world for music performance. Strongly reminding one of the palace buildings in the Forbidden City of Beijing, its majestic roof lines and column-enclosed veranda make for a breath-taking appearance, especially when illuminated at night. Inside, the concept is more subdued, more contemporary in design.

The concert hall is rectangular in shape with two balconies. The horizontal ceiling is deeply coffered and the sidewalls are irregular to promote diffusion of the sound field, essential to a good singing tone. Basically the hall resembles Boston Symphony Hall, but wider. The stage is very large, 270 m^2 (2,900 ft^2), which should make ensemble playing for a small orchestra somewhat difficult. The stage ceiling is a canopy designed partially to send sound to the audience and partially to return sound to the players. The sidewalls are planned to do likewise. A large organ covers the entire back wall of the stage. The seating capacity is 2,074, with 862 seats in the balconies. The acoustical design goal, which called for a reverberation time of 2 sec, makes the hall ideally suited to the Romantic repertoire.

To achieve the least loss of sound as it travels over the audience rows, the floor contour was sloped in the form of a logarithmic spiral. The heights of the steps increase from front to rear, ascending gradually to 10 ft (3 m). There are no sound-absorbing materials in the hall. The floors are parquet without any carpet. The walls provide some bass absorption, meaning that the reverberation time does not rise at lower frequencies above that at middle frequencies. The finish is dark mahogany.

The Taipei Concert Hall opened in 1987 with the Cleveland Orchestra and Taipei's own two symphony orchestras. Isaac Stern wrote afterwards, "Yo Yo Ma, Emanuel Ax and I had the pleasure of performing in [this new hall]. We . . . took the trouble, with great care, to go into the auditorium and to listen to the sound of the house both empty and full . . . We all found the acoustics of the auditorium absolutely wonderful, clearly world class."

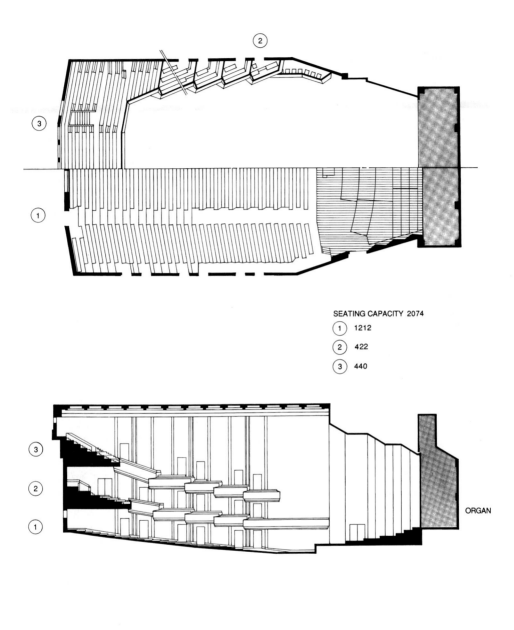

SEATING CAPACITY 2074

① 1212

② 422

③ 440

ORGAN

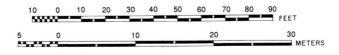

ARCHITECTURAL AND STRUCTURAL DETAILS

Uses: Primarily orchestra. **Ceiling:** 117 quadratic coffered spaces, each 18 in. (45 cm) deep, with upper surfaces chipboard, 1 in. (2.5 cm) thick. **Sidewalls and rear walls:** 0.9-in. (2-cm)-thick chipboard backed by an airspace 3–7 in. (8–18 cm) deep. **Floors:** Parquet on concrete. **Carpet:** None. **Stage enclosure:** Sidewalls and ceiling, 1.7-in. (4-cm) chipboard; back wall: pipe organ, which can be closed by a special organ door; the back wall beneath the organ reflects sound back to the musicians. **Stage floor:** Heavy wood over airspace. **Stage height:** 37.5 in. (95 cm). **Seating:** Tops of seat bottoms and fronts of backrests upholstered, with upholstering ending at a height of 33.5 in. (85 cm) above the floor to avoid sound absorption; underseats solid.

ARCHITECT: C. C. Yang. ACOUSTICAL CONSULTANT: K. Heinrich Kuttruff. PHOTOGRAPHS: Courtesy of the architect. REFERENCE: H. Kuttruff, "Acoustical design of the Chiang Kai Shek Cultural Centre in Taipei," *Applied Acoustics* **27**, 27–46 (1989).

TECHNICAL DETAILS

V = 590,000 ft³ (16,700 m³)	S_a = 11,000 ft² (1,022 m²)	S_A = 13,570 ft² (1,261 m²)
S_o = 2,900 ft² (269 m²)	S_T = 15,510 ft² (1,441 m²)	N = 2,074
H = 59 ft (18 m)	W = 88 ft (26.8 m)	L = 106 ft (32.3 m)
D = 116 ft (36.4 m)	V/S_T = 38.0 ft (11.59 m)	V/S_A = 43.5 ft (13.2 m)
V/N = 284 ft³ (8.05 m³)	S_A/N = 6.54 ft² (0.61 m²)	S_a/N = 5.30 ft² (0.493 m²)
H/W = 0.67	L/W = 1.2	ITDG = 29 msec

Note: $S_T = S_A$ + 1,940 ft² (180 m²); see Appendix 1 for definition of S_T.

NOTE: The terminology is explained in Appendix 1.

99

Aula Magna

Fantastico is the word that aptly describes the Great Hall of the University of Caracas. It is the only hall anywhere in which the work of a sculptor is so much in evidence. How that came about is an interesting story.

The architect won a competition with his design for the hall. His plans called for a very broad, fan-shaped room with a domed ceiling and a rear wall that formed a sector of a circle with its center of curvature at the rear of the stage. When the preliminary drawings were complete, the acoustical firm of Bolt Beranek and Newman was consulted. The curved ceiling and the circular rear wall presented serious problems of focused echoes, dead spots, and a general lack of uniformity in sound distribution—in all, an acoustical nightmare. The acoustical consultant urged a major modification of the shape of the hall, but by that time the plans had already been fixed.

An alternative solution was then explored, and this solution must be credited to the late Robert B. Newman. He recommended that sound-reflecting panels, equal in area to about 70 percent of the ceiling, be hung below the ceiling and on the side walls. These panels were originally recommended to be rectangular in shape. But the designers of the Aula Magna, hoping for a more satisfying solution, contacted Alexander Calder in Paris, a sculptor whose mobiles had by that time made him famous. He was invited to participate in an unusual and rewarding collaboration of sculptor with architect, engineers, and acoustical consultant. The result is beautiful—in both form and color—an exciting array of "stabiles" suspended from the ceiling and standing away from the side walls. No photograph can do it justice. One must be inside the hall—inside the sculpture—to feel its rhythm and color. I was with the architect when he first entered the hall after the color was added to the stabiles, he spread his arms in an overhead "V" and shouted the word, "Fantastico."

The hall is a university auditorium and seats 2,660 people. It was dedicated in 1954 and its first use was by the International Pan-American Congress. I was present during the summer of that year when the initial acoustical tests were being

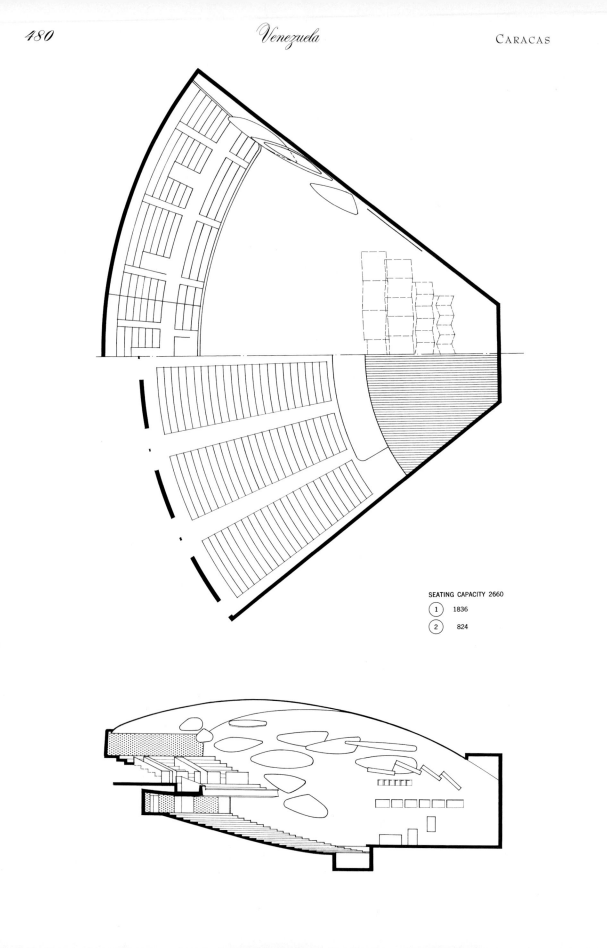

SEATING CAPACITY 2660

(1) 1836

(2) 824

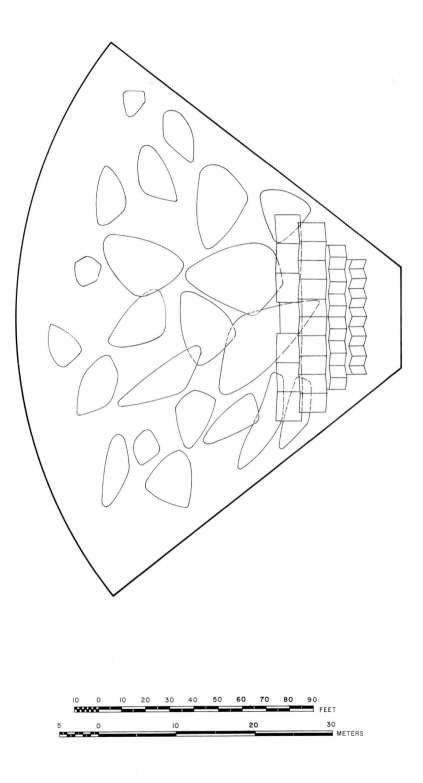

conducted. There was a just noticeable echo, in the front of the hall from the focusing rear wall above the balcony. It was necessary to add the sound-absorbing material shown on the drawing to eliminate this problem. All other surfaces are plaster on a concrete backing. No other adjustment was necessary. Because of the solid walls and ceiling construction and the thick stabiles, two 0.5-in. thick layers (1.25 cm each) of plywood glued together, the reverberation time lengthens at low frequencies and gives the sound a warm character.

The Aula Magna serves many purposes for the university, and though its basic shape is potentially unfavorable for music, it has received favorable comments from conductors and music critics. Music played there is clear and distinct. The string tone is brilliant. Both on stage and in the audience the bass is rich and warm. Short initial-time-delayed early reflections are provided by the hanging panels. Because the edges of the hanging panels are frames 4- to 8-in. wide, sound is reflected laterally, giving a distinct feeling that the orchestra is wider and fills the room, even with 2,660 seats. The distance from the front of the stage to the farthest listener is only 116 ft (35.4 m) compared to Boston's 133 ft (40.5 m), which also helps the feeling of intimacy.

The reverberation time of the hall was optimized for the Pan-American Congress at 1.35 sec with full occupancy, by adding 2,770 ft² (257 m²) of 1 in. (2.5 cm) of glass fiber blankets above the hanging panels. When this is removed, the reverberation is about 1.7 sec, which is more nearly optimum for symphonic concerts. Thus, as measured, the hall is excellent for piano, modern music, and chamber music—that is, for all music where clarity of detail is desired. In any case, the music lacks the singing tone of the classical rectangular halls because the very wide fan shape prevents a general mixing of the sound from cross reflections between the side walls.

On stage the sound is excellent, as a result of an effective canopy over the performers. Just after a tour around South America with the New York Philharmonic Orchestra in 1957, the late Leonard Bernstein said to me, "This hall was the best in which I conducted in South America. After the concert I told reporters that on the stage the sound is excellent. I wished that I could take that part of the hall back to New York for the Philharmonic to use [in Carnegie Hall]."

Architectural and Structural Details

Uses: University auditorium, used for lectures, convocations, music, and drama. **Ceiling:** Plaster on metal lath. **Sidewalls:** Plaster on concrete. **Rear walls:** Highly absorbent to prevent echo. **Floors:** Quarry tile. **Stage enclosure:** Canopy of plywood, about 1 in. (2.5 cm) thick; rear wall of the stage and about 20 ft (6.1 m) of the two side walls contiguous to that wall are of wood on solid concrete. **Stage floor:** Wood over airspace. **Stage height:** 40 in. (102 cm). **Carpets:** Main aisles only. **Added absorbing material:** At the time the reverberation data were taken, 2,770 ft² (257 m²) of 1-in. (2.5-cm) glass fiber blanket had been installed on top of the stabiles to make the hall optimum for the Pan-American congress; without this material, the mid-frequency reverberation time, fully occupied, increases to about 1.7 sec, more nearly optimum for music. **Seating:** Front of backrest and top of seat bottom are upholstered; underseat is perforated with rock wool inside. **Hanging reflectors:** 30, each of which is made of two layers of 0.5-in. (1.25-cm) laminated wood, glued together, on heavy framing, 4–8 in. (10–20 cm) thick.

ARCHITECT: Carlos R. Villanueva. ASSOCIATED DESIGNERS: Santiago Briceno-Ecker and Daniel Ellenberg. ACOUSTICAL CONSULTANT: Bolt Beranek and Newman, now ACENTECH, Cambridge, MA. PHOTOGRAPHS: Courtesy of S. Briceno.

TECHNICAL DETAILS

V = 880,000 ft³ (24,920 m³)	S_a = 17,000 ft² (1,580 m²)	S_A = 20,300 ft² (1,886 m²)
S_o = 2,200 ft² (204 m²)	S_T = 22,240 ft² (2,066 m²)	N = 2,660
H = 58 ft (17.7 m)	W = 189 ft (57.5 m)	L = 102 ft (31.1 m)
D = 116 ft (35.4 m)	V/S_T = 39.6 ft (12.06 m)	V/S_A = 43.3 ft (13.2 m)
V/N = 331 ft³ (9.37 m³)	S_A/N = 7.63 ft² (0.71 m²)	S_a/N = 6.38 ft² (0.594 m²)
H/W = 0.31	L/W 0.54	ITDG = 30 msec

Note: $S_T = S_A$ + 1,940 ft² (180 m²); see Appendix 1 for definition of S_T.

NOTE: The terminology is explained in Appendix 1.

100

Cardiff

St. David's Hall

*C*ardiff, the Welsh capital, has an exciting world-class concert hall. Opened
in 1982, it is located in St. David's Centre, a major shopping area in the
heart of the city. From the start, it was decided not to duplicate the classical shoebox
hall, but rather to make it more "theatrical," involving the public more, with no
one seated in excess of 125 ft (38 m) from the stage. There were other challenges.
The building had to be shoe-horned into the available ground space, and it was
desired not to make it too high. These criteria were achieved by wrapping the audience around the stage, and by choosing its shape as an elongated hexagon, tapered
in at the rear. Again, the vineyard style of the Berlin Philharmonie hall is suggested,
with 16 tiers in all, each comparable in size to the orchestra. St. David's Hall seats
1,682 plus 270 choir, totaling 1,952.

Walls between tiers do not provide early reflections to listeners, as are found
in Berlin and the Costa Mesa, California, halls, because they are parallel to the
sight lines. A mid-frequency reverberation time of 2.0 sec, fully occupied, was
achieved by incorporating the trusses for the roof and the ventilating system ducts
into the room by an acoustically transparent ceiling. In St. David's Hall, a season
of orchestral concerts is staged using the BBC National Orchestra of Wales and
international/regional touring orchestras. For a Cardiff Festival of Music it contributes a broad-based program and, in July, the Hall offers the Welsh Proms.

A subjective survey of 11 British concert halls (not including more recent
halls, Birmingham and Glasgow) at public concerts by a group of mainly acoustical
consultants seated throughout the halls concluded that St. David's Hall was at or
near the top in the characteristics of clarity, reverberance, intimacy, orchestral balance, and envelopment of the listener by the sound. Loudness was also optimum.
Why? Even though the reverberation is high, there seems to be ample early sound
energy reflections from the myriad of trusses, suspended ceilings, lighting grid, back-
stage wall, and balcony fronts, as well as the ceiling in the upper levels, to achieve
the desired qualities of clarity and intimacy. The judgment of good envelopment is

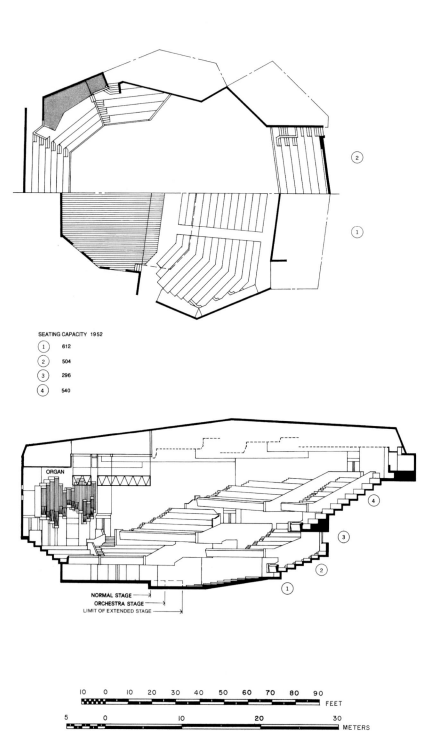

SEATING CAPACITY 1952

1.   612
2.   504
3.   296
4.   540

ORGAN

NORMAL STAGE
ORCHESTRA STAGE
LIMIT OF EXTENDED STAGE

10   0   10   20   30   40   50   60   70   80   90
FEET

5   0   10   20   30
METERS

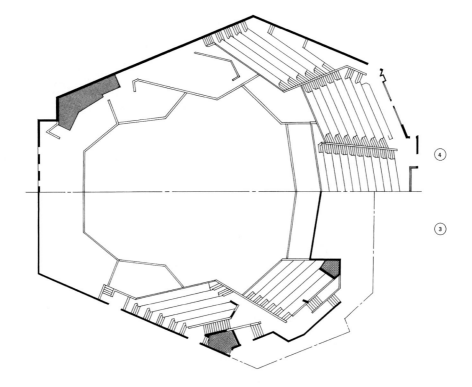

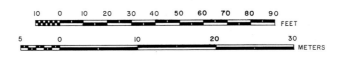

surprising because of an apparent lack of surfaces to create early lateral reflections. Some believe that this may only be possible with a reverberation of 2 sec or more.

ARCHITECTURAL AND STRUCTURAL DETAILS

Uses: Symphonic music. **Ceiling:** The roof is 8-in. (20-cm) pre-cast foamed concrete slabs, sealed with paint. Open steel walkways with plywood flooring. Ventilation ducts are lined with sound-absorbing material. **Suspended ceilings:** Visual only, 4 × 4-in. (10-cm) open cells; a large suspended metal space frame incorporates stage lighting, loudspeakers, etc., and profiled plywood sound reflecting panels within the depth of the frame aimed at selected positions (see the photo). **Walls:** 8-in. (20-cm) brick with 0.5-in. (1.25-cm) plaster; behind, choir seating, profiled panels with varying depths of painted plasterboard (see the photo); above, suspended ceiling level, dense concrete block, surface sealed; rear wall around stage, plywood with ash veneer; balcony fronts, the same. **Main floors:** 0.75-in. (1.9-cm) wood over concrete ramp with 10–20-cm airspace beneath. Mushroom ventilation openings in floors. **Tier floors:** Oak strips solidly imbedded in screed. **Carpets:** On perimeter aisles only. **Stage floor:** 0.75-in. (1.9-cm) oak screwed to 0.75-in. plywood, supported by joists on 0.2-in. resilient pads. **Stage height:** 32 in. (80 cm). **Seating:** seat backs and underseats, molded plywood. Upholstering on two surfaces.

ARCHITECT: Seymour Harris, Partnership. ACOUSTICAL CONSULTANT: Sandy Brown and Associates. PHOTOGRAPHS: Gwyn Williams, courtesy of St. David's Hall. REFERENCES: A. N. Burd, "St. David's Hall, Cardiff," *Proc. Inst. of Acoustics and Electro-Acoustics meeting,* Edinburgh, Sept. 1982; M. Barron, *Auditorium Acoustics and Architectural Design,* E & FN Spon, Chapman & Hall, pp. 173–181 (1993).

TECHNICAL DETAILS

V = 777,000 ft³ (22,000 m³)	S_a = 10,760 ft² (1,000 m²)	S_A = 13,300 ft² (1,235 m²)
S_o = 2,000 ft² (186 m²)	S_T = 15,300 ft² (1,420 m²)	N = 1,952
H = 59 ft (18 m)	W = 90 ft (27.4 m)	L = 90 ft (27.4 m)
D = 110 ft (33.5 m)	V/S_T = 50.7 ft (15.5 m)	V/S_A = 58.4 ft (17.8 m)
V/N = 397 ft³ (11.2 m³)	S_A/N = 6.8 ft² (0.63 m²)	S_a/N = 5.51 ft² (0.512 m²)
H/W = 0.66	L/W = 1.0	ITDG = 25 msec

NOTE: The terminology is explained in Appendix 1.

FOUR

ACOUSTICS

OF

CONCERT HALLS

$\mathcal{C}$harles Garnier, designer of the Opera Garnier in Paris, said in his book, *The Grand Opera in Paris,* that he had pursued diligently the elusive factors of good acoustics, but he confessed that he finally trusted to luck, ". . . like the acrobat who closes his eyes and clings to the ropes of an ascending balloon.

Eh bien!" he concludes, "Je suis arrive!" He went on, The credit is not mine. I merely wear the marks of honor. It is not my fault that acoustics and I can never come to an understanding. I

gave myself great pains to master this bizare science, but after fifteen years of labor, I
found myself hardly in advance of where I stood on the first day. . . . I read diligently in
my books, and conferred industriously with philosophers—nowhere did I find a positive
rule of action to guide me; in the contrary, nothing but contradictory statements. For
long months, I studied, questioned everything, but after the travail, finally I made this
discovery. A room to have good acoustics must be either long or broad, high or low, of
wood or stone, round or square, and so forth. . . . Chance seems as dominant in the
theatrical [opera house] world as it is in the dream world in which a child enters
Wonderland!

Wallace Clement Sabine is acclaimed as the first person to apply scientific
principles to the design of a concert hall. A young assistant professor at Harvard
University, he had studied the sound in 11 halls and rooms at Harvard University
between 1893 and 1898 as a basis for correcting the acoustics of a university lecture
hall. With accumulated piles of data on reverberation time versus number of seat
cushions (from a nearby theater) introduced into each hall, he sought a formula to
encapsulate his findings. One day, with the equivalent of Aristotle's utterance
"Eureka!" he suddenly perceived a mathematical equation for calculating the reverberation time. This formula yields the length of time it takes in seconds for a loud
sound in a room, suddenly cut off, to die down to inaudibility. Valid today, it simply
says that the reverberation time of the sound in a room is directly related to the
room's cubic volume, i.e., twice the volume yields twice the reverberation time, and
inversely related to the amount of sound absorption in it—double the absorption,
half the reverberation time.

  Sabine had learned from tests in a Harvard lecture room that an audience
absorbs sound in proportion to its size. Similarly, his laboratory tests showed that
each wall and ceiling surface, carpet, and drapery-like material absorbs sound in an
amount proportional to its area. Of paramount importance, his room and laboratory
measurements had enabled him to determine the amount of sound absorption attributable to 100 m² of each sound-absorbing entity, i.e., plaster, wood, concrete,
rugs, curtains, audience, etc.

  When the president of Harvard University, Charles Eliot, communicated Sabine's discovery to Major Henry Higginson, chairman of the building committee
for Boston's newly contemplated Symphony Hall, Sabine was invited to apply his
formula to its architectural design. But more than a formula was needed. Here he
was lucky. The planners had already chosen the most celebrated concert hall of that
period as a model—the Gewandhaus in Leipzig, Germany, unfortunately the victim
of aerial bombing in World War II. From drawings and photographs of the Gewandhaus, Sabine determined its volume, audience area and other areas (walls, ceiling,

balcony surfaces, etc.). He then inserted his Harvard measurements of the amount of sound absorbed by each of those entities into his formula and calculated the reverberation time of the Gewandhaus. He said that if the new hall were to be as successful as the Leipzig hall it would have to have the same reverberation time. If achieved, he stated that he could "guarantee" the success of the new Boston hall. The result? Boston Symphony Hall is judged to have "superior" acoustics by musicians and music critics throughout the world. But more than his formula was necessary.

The success of a concert hall does not depend on its reverberation time alone, but also on its size, shape, the materials and irregularities of its walls and ceiling, and on the upholstering of the audience chairs. Here again, he was fortunate. The Gewandhaus is rectangular in shape, which we know today is the most likely shape for good acoustics. Higginson had previously decided on the width, because he had told the architect that he didn't want the hall any wider than Boston's existing Music Hall. Today we know that top acoustics are not achievable in too wide a hall (today's Boston Symphony Hall has a width of 75 ft). Also, Higginson and the committee had insisted on masonry construction to make it fireproof, which meant preservation of the bass sounds. Equally fortunate for Sabine, the architect, Charles McKim, for aesthetic reasons, wanted the interior of the hall to have many irregularities, e.g., coffers on the ceiling, niches and statues on the sidewalls and shallow balconies, also necessary features for top acoustics.

McKim's next drawings then followed these requirements exactly. But Sabine and Higginson did not accept them. McKim's design called for a single balcony, as in the Gewandhaus, but because the building committee wanted a hall that that seated 2,600 instead of 1,560, the hall became very long, which Sabine and Higginson feared would cause a "tunnel effect." Sabine then strongly recommended that a second balcony be added and Higginson reduced the row-to-row spacing of the seats, both shortening the hall. How the seats were chosen is not known, but their thin upholstering and the hall's masonry construction give the room good bass response. Finally, Sabine, with the cooperation of the organ builder, designed a successful stage enclosure without formula or previous experience. Finally, to get the reverberation time right, he used his new formula to determine the precise height of the ceiling, which yielded the needed cubic volume. Sabine was soon hailed for his success. Obviously, he was aided by common sense, good judgments by the building committee, and a little luck, in addition to his formula, and, without doubt, his persuasive manner.

The purpose of this chapter and the next is to examine a set of acoustical parameters which are now known to affect architectural design, and to establish optimum values for each based on the data accumulated on many halls by a bevy of acoustical scientists throughout the world, and reported in this book.

The oldest of concert halls used today by principal orchestras date from the last half of the nineteenth century. When asked to rate the acoustics of halls in the Western World, most conductors, musicians, and well-informed listeners name as "superior" Grosser Musikvereinssaal in Vienna, Concertgebouw in Amsterdam, and Symphony Hall in Boston. The first opened in 1870, the second in 1888, and Symphony Hall in 1900. Each has a reverberation time of 1.9 to 2.1 sec and is rectangular, that is to say, a "shoebox" shape. We must enquire, is shape a paramount ingredient of excellent concert hall acoustics? Is a reverberation time of about 2.0 sec (occupied, at mid-frequencies) of these halls crucial to their long-time success? "Does aging make a hall superior? Or was the legendary conductor Wilhelm Furtwaengler right when he remarked to the music critic of the *Chicago Tribune*, "The hall with the best acoustics is the hall with the best performances"?

RANK-ORDERINGS OF ACOUSTICAL QUALITY OF 58 CONCERT HALLS DEVELOPED FROM INTERVIEWS AND QUESTIONNAIRES

An essential ingredient in choosing which objective acoustical measurements are best suited to estimate the subjective acoustical quality of a concert hall is a list of halls, rank-ordered according to their acoustical quality. To develop such a list, the author has conducted interviews and made questionnaire surveys of conductors, music critics, and aficionados of concert music over a period of years. The process was easier when there were fewer halls, most of which were well known by many of the persons contacted. A plethora of halls has been built since 1980, and the assessment of their acoustical quality has become ever more difficult.

One problem is that no one of those interviewed is well acquainted with more than one-third of the 65 or so concert halls that I have recently asked about; most know only about ten halls, so each person's rating is for a different set. I am well aware that the combination of their remarks, each on a limited number of halls, and my interpretation of how they should be combined, or overlapped, does not constitute a scientific survey of expert opinions. Written questionnaires sent to everybody would seem to be a better alternate. Questionnaires are successful for the rating of a limited number of halls, say, 25 or less, that have been in existence for many years. Such a questionnaire study was reported recently for opera houses

(Hidaka and Beranek, 2000). But a questionnaire with 100 halls on it, of which the recipient is familiar with only a small fraction, would probably not even be acknowledged.

From the overlapping responses to my solicitations, I feel that I have made an overall ranking that is adequate to serve the purpose of evaluation of objectively measured parameters, but it is not meant to be an accurate depiction of what an in-depth scientific survey might produce.[1]

The results for 58 halls are shown in Table 4.1. Caveats applying to the data in this table are given in its caption. None of the halls has bad acoustics, that is to say, none is unsuited for musical performance. It also must be understood that the ratings apply to the halls in their architectural condition before about 1995. For the purposes of this chapter, the ratings and reverberation data go together, but *my ratings must be applied with great caution to those halls where changes have been made recently.* Such halls are marked "br" in the list.

It is safe to say that 15–20 halls in Table 4.1 are indisputably in the highest-rated group. Similarly, there are 15 or so that seem properly to lie in the lowest part of the list. The evidence for the latter statement is that all but one of them are marked "br" (before renovation) and are undergoing or have undergone extensive renovations to improve their acoustics. The list of Table 4.1 is divided into three parts: the 20 halls that the author believes were satisfactorily rank-ordered by the interviewees; 19 halls that were not clearly separated from each other, and that did not fall into the upper group; and 20 halls that are believed to be satisfactorily rank-ordered by the interviewees and that were judged to lie below the intermediate group.

REVERBERATION TIME: MUSICIANS' PREFERENCES

In the past three decades, numerous attempts have been made to find measures of acoustical quality in concert halls. From the Boston Symphony Hall history given above and from Chapter 2 we have observed that there are a number of acoustical attributes that contribute to the overall quality of the acoustics in a concert hall. We know from the Boston Hall that if the reverberation time in a symphonic concert hall is in the range of 1.8 to 2.1 sec, if it is rectangular in shape and has plenty of surface irregularities to diffuse the sound; if it has heavy walls,

---

[1] The "non-parametric" (Webster definition: Estimates of quantities determined from observations) procedure was used in developing the relative rankings in Table 4.1. Ref: L. L. Beranek, *ACTA Acustica–Acustica*, Vol. 89, pp. 494–508 (2003).

The rank-ordering presented here is based on interviews and questionnaires involving conductors, music critics and concert aficionados. No one interviewed expressed opinions on more than 15 halls. The list is compiled by overlapping these subjective judgments. The list is only made to assist in judging the efficacy of the different objective measures of the sound fields in the halls. All of the halls are regularly used for concerts and the audiences are generally satisfied with their acoustics.

Perception of acoustical quality differs from one person to another and different parts of a hall may have different acoustics. The author does not recommend the use of this list by any party for purposes of comparing halls other than for research, or listing any hall as superior or inferior to any other. Further, the author does not claim that the results below are the same as those that would be obtained by a scientifically rigid procedure.

The halls 21 to 39 were judged to lie below the first 20 halls in acoustical quality, but were not clearly separated from each other by those questioned. They were judged superior to those after No. 39. An alphabetical list of the halls in that group is given in the table. The unrevealed listings are the author's best judgment of how these halls should be ranked based on the scrambled evidence.

Note that (br) indicates that both the interviews and measurements were made before recently planned or completed renovations.

VM	Vienna, Grosser Musikvereinssaal
BO	Boston, Symphony Hall
BA	Buenos Aires, Teatro Colón (Concert Shell)
BZ	Berlin, Konzerthaus (Schauspielhaus)
AM	Amsterdam, Concertgebouw
TN	Tokyo, Tokyo Opera City TOC Concert Hall
ZT	Zurich, Grosser Tonhallesaal
NY	New York, Carnegie Hall
BC	Basel, Stadt-Casino
CW	Cardiff, St. David's Hall
DA	Dallas, McDermott/Meyerson Hall
BN	Bristol, Colston Hall
SO	Lenox, Seiji Ozawa Hall
CM	Costa Mesa, Segerstrom
SL	Salt Lake City, Abravanel Symphony Hall
BP	Berlin Philharmonie
TS	Tokyo, Suntory Hall
TB	Tokyo, Bunka Kaikan (Orchestra Shell)
BR	Brussels, Palais des Beaux-Arts (Renovated)
BM	Baltimore, Meyerhoff Symphony Hall

21

to

39

Bonn, Beethovenhalle
Chicago, Civic Center
Chicago, Orchestra Hall (br)
Christchurch, Town Hall
Cleveland, Severance Hall (br)
Gothenburg, Konserthus
Jerusalem, Binyanei Ha'Oomah
Kyoto, Concert Hall
Leipzig, Gewandhaus
Lenox, Tanglewood Music Shed

Munich, Philharmonie Am Gasteig
Osaka, Symphony Hall
Rotterdam, De Doelen Concertgebouw
Tokyo, Metropolitan Art Space
Tokyo, Orchard Hall, Bunkamura
Toronto, Roy Thompson Hall (br)
Vienna, Konzerthaus (br)
Washington, JFK Concert Hall (br)
Washington, JFK Opera House (set)

SA	Salzburg, Festspielhaus
ST	Stuttgart, Liederhalle, Grosser Saal
AF	New York, Avery Fisher Hall
CR	Copenhagen, Radiohuset, Studio 1
EB	Edinburgh, Usher Hall (br)
GL	Glasgow, Royal Concert Hall (br)
LF	London, Royal Festival Hall (br)
LV	Liverpool, Philharmonic Hall (br)
MA	Manchester, Free Trade Hall (Replaced)
PP	Paris, Salle Pleyel (br)
ED	Edmonton, No. Alberta Jubilee Auditorium (br)
MP	Montreal, Salle Wilfrid-Pelletier (br)
TK	Tokyo, NHK Hall (3,677 seats)
SH	Sydney, Opera House Concert Hall (br)
SF	San Francisco, Davies Symphony Hall (br)
TE	Tel Aviv, Fredric R. Mann Auditorium (br)
LB	London, Barbican, Large Concert Hall (br)
BU	Buffalo, Kleinhans Music Hall (br)
LA	London, Royal Albert Hall (5,080 seats) (br)

Note: (br) means before recent renovations. These halls may have greatly changed acoustics since renovations.

balcony faces and soffits to preserve the bass; and if it is not too large, a hall is on the way to being a contender for excellent acoustical quality. Architecturally, other shapes are possible. Special sound-reflecting and diffusing panels suspended overhead or attached as an integral part of the sidewalls can be used to enhance the acoustics on stage or in the audience. Variable acoustics can extend the hall's fitness to a greater range of musical compositions.

Reverberation Time for Occupied Halls

The portrayal of the design process for Boston Symphony Hall, given above, has illustrated the importance of reverberation time in a concert hall. In most halls used by orchestras performing today's symphonic music repertoire the reverberation time at mid-frequencies with full occupancy is classified as "live." There are some halls, usually designed originally for opera, in which the sound is "dry" or "dead."

In my early interviews, before 1962, I asked for permission to use the names of those interviewed. Later, I learned that this led to some biased answers, because of ties the persons had to particular halls or managements. Since then, I have not requested the use of names.

LIVE HALLS. The following are comments from the great conductors and musicians of the pre-1962 era in regard to the three most highly rated "live" halls.

George Szell observed, "The Musikvereinssaal is just right acoustically. Boston has one of the world's best halls. Carnegie is not as reverberant as suits my taste."

Bruno Walter said, "The Musikvereinssaal has beauty and power, and I consider it better than Carnegie. Boston is very fine, more live than Carnegie. I like the Concertgebouw, it is really a live hall."

Herbert von Karajan said of the Grosser Musikvereinssaal, "The sound in this hall is so full that the technical attack of instruments—bows and lips—gets lost. Also successive notes merge into each other. I consider Symphony Hall a little better than the Musikvereinssaal."

Dimitri Mitropoulos commented, "The Musikvereinssaal rings too much. Carnegie Hall is good; it is not overly reverberant. Symphony Hall has good acoustics."

Eugene Ormandy said, "The Musikvereinssaal is just right acoustically. Boston has one of the world's best halls. Carnegie is not as reverberant as it might be."

Pierre Monteux remarked, "The Concertgebouw has marvelous acoustics. The Musikvereinssaal too is very good. I find Carnegie a little dead with audience. Symphony Hall I like very, very much.

DRY HALLS. Liveness (reverberance) was consistently mentioned in the interviews as an important aspect of the acoustics of a concert hall. A room that is not live enough is usually described as "dry" or "dead." In six dry halls investigated, the *occupied* RTs lay between 1.0 and 1.4 sec. Most of those interviewed considered London's Royal Festival Hall too dry (1.45 sec). For example, Sir John Barbirolli commented, "Everything is sharp and clear and there is no impact, no fullness on climaxes." Suffice it to say that, for the repertoires of today's symphony orchestras, conductors and trained listeners consider a reverberation time of 1.5 sec or less to be too short, and they call its acoustics "dry."

Architectural Basics

Age

The three halls in the topmost rating category were built in 1900 or before. Others in the top 20 were built before 1900. Most likely, these halls have survived because they long have housed world-class orchestras whose musicians probably would not have tolerated them if they had inferior acoustics. There are numerous examples where concert halls with poor acoustics have been destroyed and halls with good acoustics saved. As proof are Carnegie Hall, which was saved from the wrecking ball by those who loved its acoustics, and several halls in New York City that were destroyed earlier in the twentieth century because of poor acoustics. Even recently, concert halls in three cities in Canada have been abandoned by the resident orchestras in favor of new halls built with the expectation of superior acoustics. The same is true in Great Britain. In addition, numerous renovations of halls with unsatisfactory acoustics have been and are being undertaken in recent years.

But do the acoustics of concert halls mellow with age? Logic alone says that there should be no change in the acoustics of a hall that has not been modified. A hall's acoustical characteristics depend on cubic volume; interior shape; density of materials that form the surfaces; and kind, size, spacing, and number of the chairs and carpets or other sound-absorbing materials. There may be some difference in the clothing worn or brought into the hall, but the customs for checking outer clothing have generally not changed in any given hall in the twentieth century.

There are only nine halls in this book that have not been modified during the past 30 years or so and for which comparable acoustical data exist. The only acoustical quantity that could be measured 30 years ago was the reverberation time. In these nine halls, measurements of reverberation times in five octave-frequency bands, 125, 250, 500, 1,000 and 2,000 Hz, reveal that the 30-year differences among the 45 measured reverberation times are on average separated by only 0.08 sec. This difference is within the 0.1-sec accuracy of the measurements themselves.

One must conclude that halls do not change with age *per se*—only with modifications. In the majority of halls over 25 years old, the chairs have been changed, usually for ones that are more sound absorbent. Other common changes include enlargement of the stage and removal or addition of carpets or pipe organs. In a number of halls, like those in San Francisco, Chicago, Cleveland, Washington, D.C., Liverpool, London-Barbican, and Glasgow, significant architectural changes have been made with the specific purpose of improving the acoustics.

Shape

Of the top 15 halls, determined by my interviews and Haan/Fricke's questionnaire survey to be "excellent", two-thirds are "shoebox" shaped. Certainly, the shoebox shape, provided the hall is not too wide, is a safe acoustical design. Parallel sidewalls assure early lateral reflections to the audience on the main floor, essential to the desired acoustical attribute "spaciousness." But, as demonstrated in several recent halls, spaciousness can also be achieved by one of three means: (1) some combination of suspended or sidewall-splayed panels and by taking steps to preserve bass energy, (2) shaping of the sidewalls near the proscenium and the sides of the performing space so as to direct the sound more uniformly to the audience areas, or (3) interspersing seating areas with "walls" that are located to provide lateral reflections as are found in several "surround" halls.

Fan-shaped halls have not been as successful acoustically, although the overall design of the Lenox, Massachusetts, Tanglewood Music Shed has pleased audiences, musicians, and music critics as a place for summer concerts (See Chapter 3, Hall 12). The ceiling is flat and high enough to provide an acceptable reverberation time (2.1 sec at 500 Hz). The strength G_{mid} in the front half of the hall is equal to that in the best halls, but the levels fall off toward the rear because the sidewalls are open to the outdoors. The Saturday night attendance is often 5,000 indoors and another 10,000 outdoors on the lawn. Loudspeakers, with time delays, augment the sound outdoors. The G_{low} (average of 125 and 250 Hz) level (3.8 dB) in the front half is slightly stronger than that in Boston Symphony Hall (3.2 dB). The reverberation time at low frequencies is about 2.8 sec (fully occupied), compared to Boston's 1.9 sec, Amsterdam's 2.6 sec, and Vienna's 3.0 sec. The bass is perceived as being as loud as that in any of those halls. This is partly due to the fact that the chairs are not upholstered, which means that the audience absorbs less of the bass sound.

The Tanglewood Music Shed has a short initial-time-delay gap owing to the suspended panels, triangular in shape, of different sizes, and 50% open area, that hang over the stage and the front part of the audience. The panels are transparent to low-frequency sounds, so the early bass reflections arrive from the ceiling. Those reflections are delayed in comparison with the violin tones that reflect from the panels, by only 10 m of travel distance. The time difference introduced by 10 m of travel is about the same as that between the bass and violin players on stage. It has also been shown that lateral reflections are generated by diffraction of the sound from the edges of the panels, although the reflections are not as strong as in a rectangular hall. That lateral reflections exist, although not to the extent found in smaller rectangular halls, is proven by the value of the binaural quality index (BQI), that when averaged over audience area, equals 0.35, only a little lower than that in the Berlin Philharmonie Hall. The sloping panels around the upper periphery of the curved rear wall reflect sound back onto the audience in the back half of the hall, augmenting both the direct and reverberant sound. No fully enclosed fan-shaped hall seating a substantial number of people has been built incorporating the design of the Tanglewood Shed, which seats 5,000. Thus its success has not been validated for enclosed spaces.

The most successful non-rectangular hall, seating 2,325, is the Berlin Phil-harmonie. The orchestra is seated near the center of the hall, and the audience is situated on 14 "trays," each at a different level and different in configuration. The acoustical consultant believed it to be important that early reflections come from overhead, so that the ceiling is tent-shaped. There are some exposed walls between the trays that reflect early lateral sound to some parts of the audience. An array of panels hangs high above the stage. The musical quality varies from one "tray" to another, as would be expected, because of the directivity of instruments, piano and voice. There are some excellent seats, acoustically, in the hall, all located in the front quadrant of the orchestra.

The architectural effect is breathtaking on first entrance to the hall, so that part of its success is due to its architectural design. Also, the Berlin Philharmonic Orchestra is among the best in the world, so that Maestro Furtwängler's quote may have something to do with this hall's success. The architect said that his goal was to bring the audience into closer relation to the performers than is possible in a shoebox hall. And this fact, too, may add to the hall's appeal. Those in the audience behind the orchestra enjoy viewing the conductor's face and gestures. A number of terraced, surround halls have been built, though none have been as acclaimed as the Berlin Philharmonie.

Music Power

The total music power created by the performing group that reaches a listener has three parts—the direct sound, the sound from the early lateral reflections, and the reverberant sound. It is important in design that the early lateral reflections be preserved, but not emphasized to the point that there is insufficient power in the reverberant sound. Three examples of overemphasis on the early reflections to the extent that the reverberant sound does not carry "punch" are the Christchurch Town Hall in New Zealand, the Costa Mesa Segerstrom Hall in the United States, and the Glasgow Royal Concert Hall in Scotland (before recent renovations). In Segerstrom Hall, for example, the early decay time EDT is equal to 2.2 sec, which is equal to the average of the EDT's in five most highly rated halls. However, the reverberation time RT (occupied) for Segerstrom is 1.6 sec and the average for the other five halls is 2.1 sec. The music in Segerstrom Hall is beautiful, but it lacks listener envelopment (LEV) that is so prized in the famous halls like Boston and Vienna. The principal goal in the revisions of the Glasgow Hall is to increase the LEV. During one visit to Christchurch, I personally enjoyed the music, but following stop chords, the reverberation was almost inaudible. Other visitors have made this same comment.

Audience Absorption and Type of Chairs

AUDIENCE DENSITY. We have already emphasized the importance of keeping the area covered by the audience as small as possible, consistent with reasonable comfort. In Symphony Hall in Boston (1900), 445 people, and in the Leipzig (1981) Gewandhaus, 369 people, sit over the same floor area (2,500 ft², 232 m²). An audience area that is divided into a number of small seating blocks absorbs more sound than if it comprised only a few blocks. This occurs because of the edge absorption—that is to say, the sound that reaches the "sides" of each audience block is absorbed about as well as that which reaches an equivalent area on the "top." In every description of Chapter 3, the true area over which the audience sits is listed as "S_a." The "acoustic area" in each hall, which includes absorption at the edges of the blocks, is called "S_A." The ratio of S_A to S_a ranges from 1.1 to 1.4 in the different halls, depending on the size of the blocks, the number and length of the aisles, and the number of access places in the hall. When the area over which the musicians sit is added, the total "acoustical area" is called "S_T." In Chapter 3 and Appendix 3 the acoustical area of the orchestra is limited to 1,940 ft² (180 m²), and the area beyond that is added to the area of sidewalls. To preserve loudness, the total area S_T must not become too large, because to a first approximation, the power available to each person (i.e., per unit area) is equal to the total power radiated by the performing group divided by the total audience area S_T.

CHAIRS. Obviously, comfort is important—the row-to-row spacing of the seats in Boston Symphony Hall, which was adequate in 1900, is uncomfortable for the over-six-foot-tall people of this century. Widely spaced seats are more luxurious, but they come at the expense of acoustical quality and high building costs in halls with a large seating capacity. If the seats in a large hall are too generously spaced, the architect is likely to design a wide hall in order to obtain the necessary floor area. If the architect selects the alternative of adding depth to the balconies, the balcony overhangs may shield a significant percentage of the audience from the desired reverberant sound field; or, if not, the back-row listeners in the balcony will be very far from the stage, thus diminishing the strength of the direct and early sound. Finally, to keep the seating area from becoming too large, the owner should not request a hall that will be fully occupied only a few times a year.

AUDIENCE ABSORPTION OWING TO TYPE OF CHAIRS. As is shown by the table of sound absorption coefficients in Appendix 3, people seated in heavily upholstered chairs absorb more sound than those seated in a medium, lightly or non-upholstered chair. The difference is particularly noticeable at bass frequencies. A common cause of bass deficiency in concert halls is overly sound-absorbent chairs. It is strongly recommended that a chair be made of molded material, such as plywood, and that the upholstering on the top of the seat bottom be no thicker than 2 in. (5 cm) and, on the seat back, no thicker than 1 in., and, if comfortable, cover only two-thirds of the seat back. Also, the armrest and the rear of the seat back of a chair should not be upholstered. These requirements rule out thick seat bottoms containing springs.

Materials for Walls, Ceiling and Stage

Except around the stage, the three famous halls in the top category are constructed of plaster on wire or wood lath for the ceiling, and stucco plaster on brick or on concrete block for the walls. In the Boston hall, the floors are wooden, because the seats are mounted on a removable floor that converts a raked floor for regular concerts to a flat, hard, floor for pops concerts. When the audience sits over the raised floor, their weight suppresses some of the vibration and the loss of bass is not excessive. In most modern halls where the bass response is good the floors are concrete, covered with either wooden parquet or some synthetic material that is cemented to the concrete, and the walls and ceiling are constructed with materials that have a large weight per square foot.

Thin wood paneling strongly absorbs bass energy, where "thin" means 1 in. (2.5 cm) or less. For a hall that is lined with wood, it should be as near 2 in. in thickness as is possible. For the sidewalls of many halls, wood veneer ("wallpaper") on solid (plaster) backing is employed to give the hall a warm, traditional appearance. An example of proper usage of wood is the Sibelius/talo in Lahti, Finland, where a large percentage of the interior surfaces are built from 2.7-in. (6.9-cm)-thick laminated wood.

Wood is used for sides and ceiling of a stage enclosure in a number of halls, in a thickness of 0.75 in. (1.9 cm), because musicians in general believe that "wood is good" and feel more comfortable when surrounded by it. This is permissible, because the surface area around a stage is small compared to that of the other surfaces in a hall. The stage floor is different. It is usually constructed of flexible wood over a large airspace, which will augment the sound of the double basses and cellos that stand on pins.

In summary, to preserve bass, all surfaces, except the stage floor, should be of heavy dense material. The seats should be lightly upholstered, compatible with comfort. Sound-absorbing materials and carpet must be used sparingly.

PHYSICAL MEASURES OF ACOUSTICAL QUALITY

It is impossible to name all the entrepreneurs of acoustics who have contributed to our present state of knowledge. Their works pertinent to this text are referenced at the end and include such names as Schroeder, Marshall, Barron, Blauert, Bradley, Cremer, Ando, Gade, Potter, Hidaka, Okano, Johnson, Fricke, Kirkegaard, Pompoli, Jaffe, Ianniello, Morimoto, Mueller, Tachibana, and Beranek.

Reverberation Time (RT)

Reverberation times in halls with a full audience are hard to measure. Most halls do not permit recordings during concerts, and audiences will not sit through a series of extensive acoustical measurements. When possible, measurements of RT's are made in pauses after stop-chords in the music by a recorder with a single microphone. Some compositions have only one or two stop chords and thus, during a concert, only six or so stop-chords are found. If the timpani do not ring during the pause, good occupied reverberation times in six frequency bands may be obtained. However, measurements are usually possible at only one or two seats instead of 8 to 20 in unoccupied halls. RT's are not the same in all parts of the hall, and that is the reason why it is customary when possible to measure at a number of seats and average the results. Because most halls are symmetrical, such measurements are only necessary in one-half of the hall, which is why only eight positions are acceptable.

RT's IN 40 HALLS. The occupied reverberation times at mid-frequencies (average of values at 500 and 1,000 Hz) for 40 concert halls are shown versus subjective ratings in Fig. 4.1. The best halls have reverberation times between 1.8 and 2.0 sec, while the lowest-rated halls range from 1.5 to 1.8 sec. In-between, the range is from 1.5 to 2.1 sec. Some of the halls that have lower RT's cater to chamber music and recitals; hence, a reverberation of 1.5 to 1.7 sec is entirely appropriate. Those with optimum reverberation times that have intermediate rankings owe their reduced ratings to other acoustical problems.

The principal conclusions to be drawn from this chart are that for symphonic music the reverberation times for the best halls are near 2.0 sec, while the lowest-rated ones have reverberation times in the vicinity of 1.5 sec.

VARIABLE REVERBERATION TIMES. Halls are being built with variable reverberation times. The best-known examples are the Meyerson Center Hall in Dallas, Texas; Symphony Hall in Birmingham, England; and the Culture and Congress Center Concert Hall in Lucerne, Switzerland. In all these cases, the main hall is essentially shoebox shaped. Surrounding the main hall, at different levels in

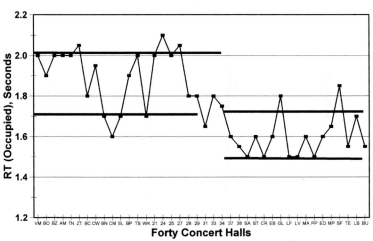

FIGURE 4.1. Mid-frequency reverberation times for 40 concert halls, measured with full occupancy, plotted versus the subjective rank orderings of acoustical quality listed in Table 4.1. Average standard deviation, measured with loudspeakers on stage 0.05 sec; from stop-chords 0.11 sec.

each, are a number of concrete chambers with large cubic volumes that can be opened or closed by heavy motorized doors. Sabine's reverberation formula says that by increasing the volume the reverberation time is increased in direct proportion. But, except for Lucerne, this appears not to have happened to the extent predicted. In Lahti and Philadelphia, the RT has changed only a few percent as the doors are opened. In Dallas, the increase is especially noticeable in the lower three frequency bands, 125, 250 and 500 Hz, and the change is less than expected in the upper bands. The reason is that at the higher frequencies it is more difficult for the shorter wavelengths to enter and exit the openings. The principal negative comment I have heard about Dallas is that an increase in the reverberation time at low frequencies may affect a musical composition in ways that the composer did not have in mind. In most halls today, relatively large areas of sound absorbing drapes or other materials can be brought from enclosed spaces into the audience chamber to reduce the reverberation times. Certainly, these features enable the conductor to adjust a long reverberation that is suitable for music of the Romantic period, to a shorter one for music of, say, the Baroque period. It also gives the conductor opportunities to make original interpretations of contemporary music. In addition, shorter reverberation times make a hall suitable for conferences.

One has to assume that because such added reverberation chambers are being built into successive halls, the latest in the Philadelphia Verizon Hall in the Kimmel Center, that musicians judge this innovation desirable. But the addition is very costly, and a building committee should investigate how often the variable reverberation is varied in practice—if not often, a less costly hall built like one of the "superior" halls may be a better choice. It is possible to build a hall with acoustics that equal those in the classical shoebox-shaped halls and that has a wide range of variability. This has been accomplished in the Lucerne Concert Hall, which combines a classical shape with a reasonable seating capacity (1,900 seats) and a width that is not large (22 meters; 70 ft). See discussion on p. 552.

Early Decay Time (EDT)

The early decay times EDT for 36 *unoccupied* concert halls are shown in Fig. 4.2. The heavy lines depict the expected range in the data, because the points that lie above the lines are for halls that have seats with very light upholstering. In the best halls, assuming upholstered seats, the EDT's lie between 2.25 and 2.75 sec, and in the lowest-rated halls, between 1.4 and 2.0 sec.

The closer spacing between the heavy lines on this graph shows that the EDT's for unoccupied halls are a better guide to their acoustical qualities than are the RT's for occupied halls, provided the seats in the different halls have reasonably equivalent upholstering.

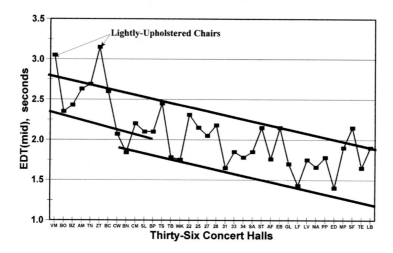

FIGURE 4.2. Early decay times EDT for 36 concert halls, measured without audience, plotted versus the subjective rank orderings of acoustical quality listed in Table 4.1. Average standard deviation 0.18 sec.

In unoccupied halls, the early decay time EDT is closely related to the reverberation time RT, for all frequencies, as shown in Fig. 4.3. The correlation coefficient (r) for the six octave bands with center frequencies at 125, 250, 500, 1,000, 2,000, and 4,000 Hz are, respectively, 0.97, 0.98, 0.99, 0.98, 0.97, and 0.97.

The relations between early decay times (EDT's) in occupied and unoccupied halls are linearly related for the eight halls for which accurate data are available. The graphs for low (125/250 Hz), mid (500/1,000 Hz), and high (2,000/4,000 Hz) are shown in Fig. 4.4.

Binaural Quality Index (BQI)

The Binaural Quality Index (BQI) was described in Chapter 2 and the formulas for its determination are shown in Appendix 2. As we shall see shortly, it is one of the most effective indicators of the acoustical quality of concert halls. Keet first proposed it (1968), with subsequent studies by Schroeder *et al.*, Ando, Potter, Hidaka, Okano, and Beranek (see References). It is a measure of the *differences* in the musical sounds that reach the two ears within 80 msec after arrival of the direct

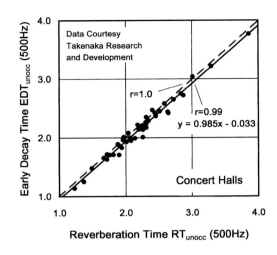

𝒯IGURE 4.3. For unoccupied halls, at all frequencies, the early decay times EDT's are almost exactly the same as the reverberation times RT's.

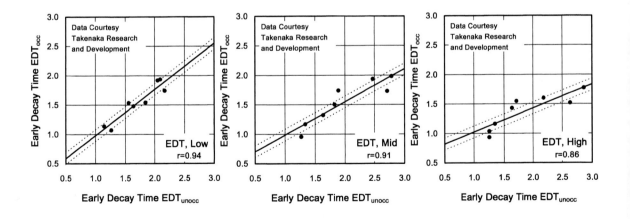

𝒯IGURE 4.4. Graphs of the relations between early decay times in occupied halls versus those in unoccupied halls. The three graphs are for low, mid and high frequencies.

sound. This difference can only be created by sound waves arriving at the listener's head from lateral directions. Sounds that arrive from straight ahead, are exactly alike at the two ears; hence, they do not contribute to BQI.

When a sound wave approaches a listener, say, from the right side of the head, the wave at the right ear is exactly like that reflected from a wall surface on that side. As the wave crosses to the other side of the head, it reaches the left ear a thousandth of a second later and it is less strong because it is "shaded" by the head. The BQI assembles all of the differences at the ears from all arriving lateral waves in the first 80 msec after the arrival of the direct sound. In addition, it measures these differences in the frequency region of the hearing mechanism that is most sensitive, from 350 Hz to 2,850 Hz.

The physical measurement is called "the interaural cross-correlation coefficient IACC" (see Appendix 2). The binaural quality index BQI is defined as [1 − $IACC_{E3}$] where "E" designates early sound, i.e., less than 80 msec and "3" indicates the average of the $IACC_E$ values in the 500-, 1,000-, and 2,000-Hz bands. The "1 minus" is introduced so that the index is zero for the case of no lateral reflections—direct sound only. BQI has been measured in many concert halls and opera houses for which subjective rank-orderings exist. In a typical satisfactory hall, the value of BQI was found to be above 0.5, and in the highest-rated halls above 0.6, indicating that in very good halls the similarity of the sounds at the two ears is surprisingly low.

In Fig. 4.5, the measured BQI's (averaged over 8–20 seats) are compared with the rank orderings of twenty-five halls taken from Table 4.1. It is seen that BQI correctly rank-orders all but three of the halls within a narrow range of 0.1, which is 12% of the highest value and 25 % of the lowest value. The two surround halls, Berlin Philharmonie Hall (BP) and the Tokyo Suntory Hall (TS) do not have as high BQI's as the results of the subjective interviews would indicate, possibly because the conductors and music critics interviewed customarily occupy seats in front of the orchestra. Many seats in both halls are to the sides and rear of the performance stage, where the BQI values are lower. Also, the photographs in Chapter 3 of these halls show that there are fewer surfaces in them available for producing lateral reflections than in more conventionally shaped halls.

The BQI is impressively effective in rating the acoustical quality of a concert hall, provided the other important acoustical parameters are within reasonable limits, that is to say, the reverberation time at mid-frequencies, fully occupied, lies between 1.7 and 2.1 sec, the initial-time-delay gap is less than 35 msec, there is satisfactory bass response, the hall is not too large so that the loudness is satisfactory, and the hall has irregularities on the walls and ceiling so that the reverberant sound field is pleasant.

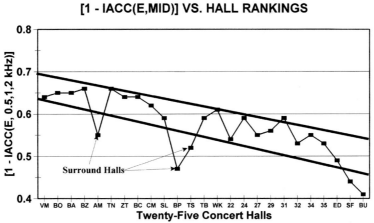

BINAURAL QUALITY INDEX (BQI)
[1 - IACC(E,MID)] VS. HALL RANKINGS

𝒯IGURE 4.5. Binaural Quality Index (BQI) for 25 concert halls, measured when unoccupied, plotted versus the subjective rank orderings of acoustical quality listed in Table 4.1. Average standard deviation 0.11 sec.

Another outstanding feature of BQI is that it permits one to estimate the acoustical quality of a fully occupied hall when the BQI has only been measured in the unoccupied hall. Every acoustician has learned that a hall's acoustical quality can usually only be determined at an opening (or pre-opening) concert when the first full audience is present. This advantage of BQI is clearly demonstrated in Fig. 4.6, where its change from unoccupied to occupied state is only about 10%. This finding means that the values of BQI in Fig. 4.5 are no more than 10 percent higher than those that would be measured in the same halls occupied.

Loudness, Strength of the Sound (G)

The strength of the sound in a concert hall is very important. The thrill of hearing Bach's *B-Minor Mass*, Beethoven's *Ninth Symphony,* or Mahler's *Eighth Symphony* is not only associated with the quality of the orchestra and the interpretation of the conductor, but is enhanced immeasurably by the dynamic response of the concert hall. Dynamic response means both quiet support for the pianissimo parts and majestic levels at the fortissimos. The greatest ovations in a concert hall usually follow a rousing fortissimo at the close of a piece. Such a fortissimo is only possible if the hall is not too large and if there is a minimum of carpets, draperies

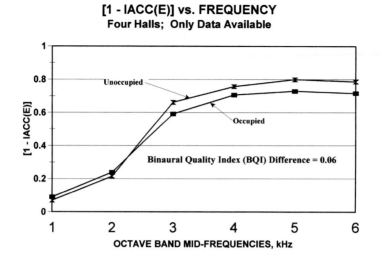

FIGURE 4.6. Plots of the Binaural Quality Index (BQI) versus frequency for occupied and unoccupied concert halls. Data are available for only four halls.

and a lack of overly upholstered seats. For some compositions, a small hall can be too loud. Herbert von Karajan said that he liked the sound in Symphony Hall a little better than that in the smaller Musikvereinssaal. Of course, he could be referring mainly to the reverberation time, but the reverberation time (occupied) of Symphony Hall is not that much different, 1.9 sec for it and 2.0 sec for the Musikvereinssaal. But their sound levels are significantly different—Vienna is about 2.5 dB louder.

Loudness is affected by four architectural features: (1) The greater the *distance* of the listener from the stage the less the loudness; (2) those surfaces that reflect *early sound* energy to the audience, preferably from lateral directions, increase the loudness; (3) the larger the acoustical area S_T (audience plus orchestra), and the larger the cubic volume, the less the loudness; and (4) added materials that absorb sound, such as carpets, draperies, heavily upholstered seats, and pipe organs also reduce the loudness.

The loudness of the direct sound decreases as it travels from front to rear of the hall. Substantial early reflections (within 80 msec after arrival of the direct sound) will reduce this decrease, which is automatically provided by a narrow rectangular hall, like the Musikvereinssaal in Vienna or Symphony Hall in Boston. Otherwise, to some extent, special panels installed on the sidewalls or hung from the ceiling overhead, which will reflect early sound energy toward the listeners, can counteract this drop in level. An example of successful augmentation of the sound levels in a large wide concert hall is in the Costa Mesa (California) Segerstrom Hall, where large, slanted, sound-reflecting panels are located on the upper sidewalls. The mid-frequency sound levels there are nearly the same as in Boston Symphony Hall, in spite of its 2,900 seats versus Boston's 2,600 (actually Boston Symphony Hall would only seat about 2,200 if reseated with modern seats, wider row-to-row spacing, and wider exit aisles).

The third component of loudness is the size of the audience. An orchestra produces a certain amount of sound energy at any playing level. If this energy is distributed equally over the audience, the greater the number of listeners, the less energy per listener.

Figure 4.7 is a plot of the sound level in decibels as a function of the distance from the stage in Meyerhoff Symphony Hall in Baltimore (hall unoccupied). It shows the combined effects of the decrease in level of the direct sound and the augmentation in the level by early reflections and reverberant sound. A typical decrease in sound level from center to back is 3–5 dB in a well designed hall, say, with 2,400 seats and $RT_{occ} = 2$ sec.

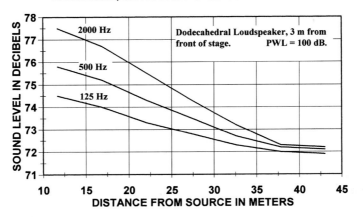

SOUND LEVEL DECREASE WITH DISTANCE
BALTIMORE, MEYERHOFF SYMPHONY HALL

*F*IGURE 4.7. Sound level in the Meyerhoff Symphony Hall in Baltimore at different distances from a non-directional sound source that is located 3 m from the front of the stage, hall unoccupied. The decrease is greater at the higher frequencies because the absorption by the seats is greater than at lower frequencies.

A sound level meter that has a modified reference scale is used to measure the *strength G of the sound in decibels.* (See Appendix 2 for the formula.) The measured values of G_{mid} for a number of halls are shown in Figs. 4.8 and 4.9. G_{mid} is the average of the measurements of G in the 500- and 1,000-Hz octave frequency bands. The former figure is for measurements made by the Takenaka Research and Development Institute in Chiba, Japan, and the latter is for measurements made in other countries. Figure 4.8 shows that the 5,000-seat Royal Albert Hall has a G that is about 8.5 dB lower than that in Vienna's Grosser Musikvereinssaal (labeled "Vienna"). If one considers only the halls in the highest-rated category, the G levels are within about 3.0 dB of each other, with Zurich's hall, seating 1,546, the loudest, and Boston's hall, seating 2,625, at the lower level.

The readers should be aware that, because of a difference in the calibration techniques for the omnidirectional source used in a hall to produce the impulsive sound during a test, the values of the strength G are greater for Takenaka (Japan) measurements by about 1.2 decibels than for measurements made in Europe and elsewhere, as shown by comparison of the two graphs, Figs. 4.8 and 4.9.

For almost all listings in this book, the values of G are for unoccupied halls. As can be ascertained from Appendix 4, there are only seven halls where G was measured both unoccupied and occupied (Nos. 9, 31, 53, 76, 77, 79, 83). The average differences between unoccupied and occupied values of G as a function of frequency are shown in Fig. 4.10. For low frequencies the difference is about 1.2 dB. For middle frequencies, the difference is about 1.5 dB.

Warmth, Bass Ratio (BR), and Bass Strength (G_{low})

When heard in a hall, music that is rich in bass is said to have "warmth." There are at least three ways of measuring bass sounds as possible indicators of warmth. The measure that has been around longest is the bass ratio (BR). It is the ratio of the reverberation times (RT's) in *occupied* halls at low frequencies to those at mid-frequencies. By definition it is the sum of the RT's at 125 and 250 Hz divided by the sum at 500 and 1,000 Hz. In my previous observations of BR's in concert halls, I found unexpectedly that, "It is immediately apparent that BR does not correlate strongly with the rating categories. . . ." Hence, other measures have been sought.

A second method is to investigate the strength of the bass sounds by subtracting the measured strengths (in decibels) at mid-frequencies (hall unoccupied) from the strengths at low frequencies. In other words, $(G_{125} + G_{250})$ minus $(G_{500} + G_{1,000})$. When this was tried for 38 concert halls, the conclusion was that it also was not a useful measure.

The third method is to measure the absolute strength of the sound in the lower frequency bands, say, G_{low} or G_{125}, in decibels. Bradley and Soulodre (see References) concluded from subjective laboratory experiments that BR was not a meaningful measure of warmth, but that G_{low} and especially G_{125} were meaningful. First let us look at BR.

In Fig. 4.11 a graph is shown of the bass ratio BR plotted as a function of the ranking of 45 halls *measured with full occupancy*. It is seen that there is no correlation between BR and the rank-orderings. This finding agrees with the Bradley/Soulodre laboratory subjective tests. Next, we investigate their alternate finding.

The strength G_{125} in decibels, in 31 *unoccupied* halls, measured at 125 Hz is shown in Fig. 4.12. Note that the difference in calibration techniques among countries has been taken care of by subtracting 1.2 dB from the Takenaka values and using the values as measured for other data. Except in halls where the upholstering is very light, this figure indicates a helpful correlation between G_{125}, in decibels, and the rank-orderings. The high values for G_{125} measured in those halls with light upholstering would not measure that way in occupied halls, because people cover the lightly upholstered seats. Thus, one might expect the G_{125}'s for the *occupied* halls to lie about 1.2 dB below the range indicated by the heavy lines.

The values of G, in decibels, shown in Figs. 4.8, 4.9 and 4.12, indicate that G_{mid} (occupied) and G_{125} (unoccupied) increase with hall quality by about the same amount for the same halls. A separate list made of G_{low} minus G_{mid} for unoccupied halls showed that, for 37 halls, the average difference was -0.9 dB, and that there was no correlation of the individual values with acoustical quality. It seems pure chance that the difference between G_{mid} and G_{125} should be nearly the same for all halls, at least on average, because, logically, they are determined by differences in the type of seat and to some extent by the residual absorption (i.e., the sound absorption by the walls, ceiling, carpets, chandeliers, etc.).

Intimacy, Initial-Time-Delay Gap (ITDG)

The subjective impression of listening to music in a large room and its sounding as though the room were small is one definition of intimacy. A deaf person can sense the size of a room by listening to the sound when standing in the center of the main floor. I originated the use of initial-time-delay gap as a measure of concert hall intimacy. This decision was based, in part, on experience in three halls. The first two were the Tanglewood Music Shed and the Caracas, Aula Magna. Both halls are fan shaped, so that there is a minimum of lateral reflections in the first 80 msec after the arrival of the direct sound. The music in the Tanglewood Shed was grossly unsatisfactory. After the introduction of horizontal suspended panels,

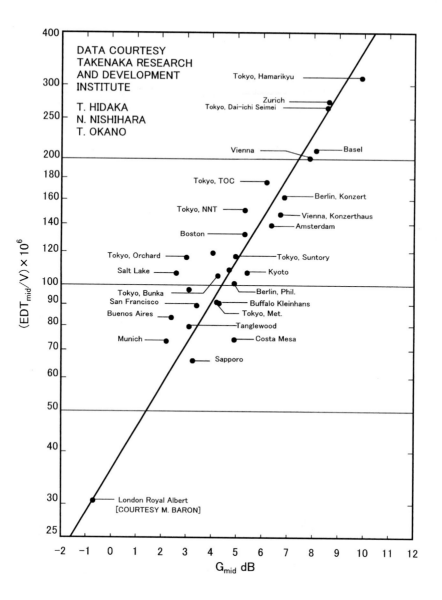

𝒯IGURE 4.8. Measurements of the strength of the sound at mid-frequencies G_{mid} versus the ratio EDT/V, where both G_{mid} and EDT were determined in un-occupied halls. All data taken by Takenaka Research and Development Institute. The definition of the reference for G (i.e., G = 0) is the sound pressure measured at a point 10 m from the center of a non-directional sound source located in an anechoic chamber. Throughout this chapter, the sound source used was a regular dodecahedral "box" with a cone loudspeaker in each of its 12 faces. The halls that deviate most from the 3 dB per doubling of the ordinate values have architectural features that either heavily concentrate the early sound energy on the main floor (i.e., there is less energy in the reverberant sound, as in Costa Mesa) or the opposite (as in Tokyo, Orchard Hall). Two methods of calibration, including that used by Takenaka, are discussed in Appendix 3.

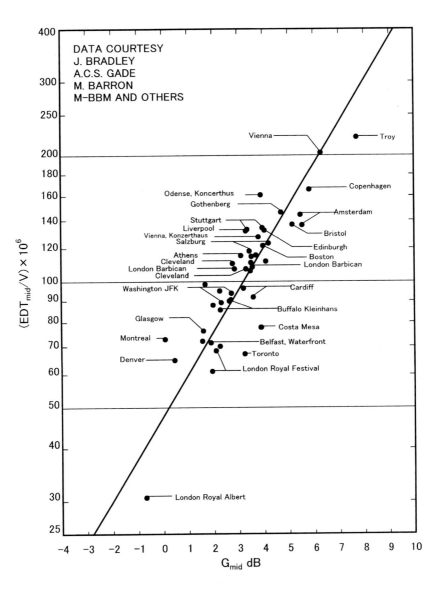

$\mathcal{F}$IGURE 4.9. Same as Fig. 4.8 except the data were obtained by researchers in other countries. The possible method of calibration used by these specialists is discussed in Appendix 3. The stated difference of 1.2 dB between the two graphs is based on the difference between measurements made by the two different parties in the same halls.

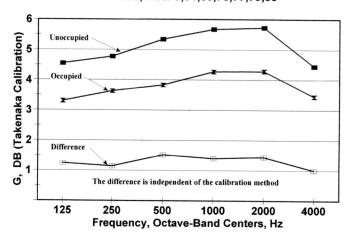

$\mathcal{F}$IGURE 4.10. The difference in the strength of the sound G as a function of frequency between unoccupied and occupied halls. Data are available for only seven halls. This difference is independent of the calibration method.

which reduced the ITDG from 45 msec to 19 msec, the music became unbelievably better. "Stabiles" were built into the Aula Magna from the beginning, yielding an ITDG equal to 30 msec. Musical performances there have been well accepted. The third experience was with a very wide, rectangular hall in Chicago (destroyed by fire) that seated 5,000. There were no reflecting surfaces near the proscenium. The initial-time-delay gap in the center of the main floor was 40 msec. Although the reverberation time was 1.7 sec (occupied), sound in the hall was unsatisfactory, like that in the Tanglewood Shed before the panels were introduced.

In this text the initial-time-delay gap is given only for one position near the center of the main floor, about halfway between the stage and the first balcony front and about one yard (one meter) off the centerline. This corresponds to the blind-person experience and reflections from nearby walls or surfaces are avoided.

A long ITDG is encountered in a hall that is very wide and has a high ceiling. The difference between the sound in a hall with an initial-time-delay gap of 20 msec and one in a much larger hall with a ITDG equal to 40 msec is illustrated by the reflectograms in Fig. 4.13. To produce such time patterns, a non-directional sound source that can emit a short, intense "beep" is located on the stage. For the two halls shown in this figure, the microphone was in a seat about 60 percent of

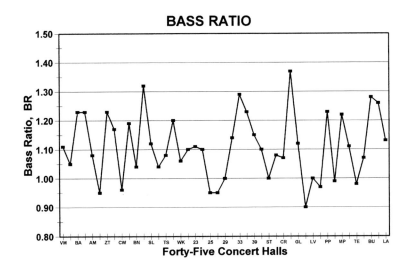

𝒯IGURE 4.11. Bass ratio for 45 concert halls, measured in occupied halls.

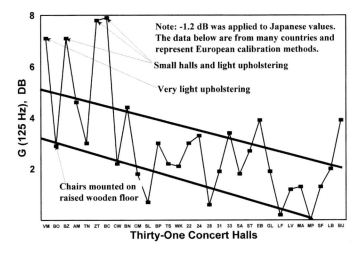

𝒯IGURE 4.12. Strength G at 125 Hz for 31 concert halls. Measured with halls unoccupied. G_{125} is larger in small halls, and in halls with light upholstering.

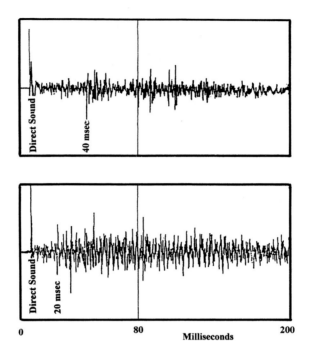

𝒯IGURE 4.13. Reflectograms showing two halls, one, narrower, with an initial-time-delay gap of 20 msec. and the other, wider, with an ITDG of 40 msec. Note the number and strength of the early reflections between the first reflection and the 80-msec mark. The jumble preceding the first reflection is reflections off of the seats between the stage and the point of measurement.

the way between the stage and the rear wall. The long vertical line near the left of each pattern indicates the sound that reaches the listener directly from the source. The reflectograms of Fig. 4.13 differ in two ways. First, in the hall with a 40-msec ITDG, there are fewer early reflections (those occurring before 80 msec) than in the hall where the ITDG is 20 msec. Second, in the case of a 40-msec ITDG, the amplitudes of the early reflections are weaker because the early reflections have traveled farther.

The initial-time-delay gaps for 39 halls are shown in Fig. 4.14. In most of the highest-rated halls, the ITDG is 21 msec or less. Amsterdam (AM) is a wide hall. It is famous for its high and enveloping reverberation time, which tends to mask the longer initial-time-delay gap. In most of the other halls the average ITDG is about 30 msec. No hall among the 39 in Fig. 4.14 is a poor hall. In a very large

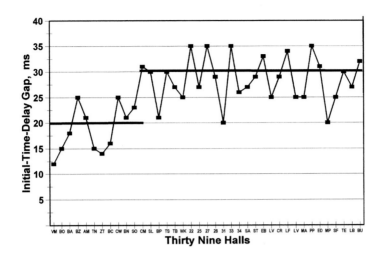

Initial-Time-Delay Gap, Various Halls

IGURE 4.14. Plot of ITDG for 39 concert halls. Of the top 14 halls, only two have ITDG's greater than 25 msec.

hall, particularly one that is fan shaped, the initial-time-delay gap is as high as 55 msec. One conclusion to draw from this graph is that it is not too difficult to obtain a reasonable ITDG in a hall that is not too large, provided it is not fan shaped.

Lateral Fraction (LF_{E4})

Marshall is credited with recognizing that lateral reflections in a concert hall are an important factor in determining its acoustical quality (1968). Barron and Marshall in 1981 devised a measure of the strength of lateral reflections by a quantity LF, which equals the ratio of the energy in lateral reflections to the total energy arriving at a seat in a hall. This measure correlates highly (in laboratory tests) with the broadening of a source on stage beyond its visual width owing to lateral reflections. That is to say, LF correlates well with the apparent-source-width (ASW). The mathematical definition for LF is given in Appendix 2.

The version of LF that is commonly published is LF_{E4}, where "E" indicates the early sound and "4" indicates the average of the LF values in the four octave-frequency bands, 125, 250, 500, and 1,000 Hz. Reliable measurements of LF have been made in 22 concert halls for which there are also subjective acoustical quality ratings. The comparison between LF_{E4} and the rank-ordering of these 22 halls is shown in Fig. 4.15. Marshall and Barron have shown from their laboratory tests that LF, or ASW, is strongly related to measurements in low-frequency bands. Potter has shown that measurement in the 500-Hz octave band contributes most to ASW. To check whether the low frequency bands have more influence on LF values than the middle frequency bands, LF was determined from data in Appendix 4 for the same halls but averaging the measured values for the three octave bands 500, 1000 and 2000 Hz. It is seen in Fig. 4.15 that there is no appreciable difference between low-band and middle-band measurements of LF, thus making it obvious that LF is not attuned particularly to low frequencies.

From Fig. 4.15, we see that LF separates the six halls with top ratings from the six halls with lowest ratings, but does not separate those in between well. The general trend would indicate that one could rank-order the halls within the two heavy solid lines. But the spread between the two lines is 0.06, which is 25% of the highest value and six times the lowest value. As an example, it is inconceivable that the acoustical rating of London's Barbican Hall (LB), before its recent reno-

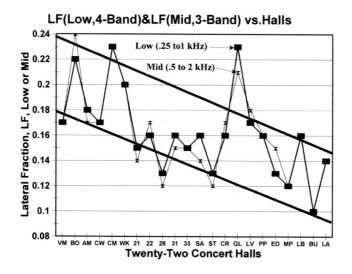

$\mathscr{F}$IGURE 4.15. Plot of the early lateral fraction LF_E vs. the quality ratings of 22 concert halls. The L_E was measured in two ways: "low," the average of the values in the four lowest bands; and "mid" the average in the 500-, 1,000-, and 2,000-Hz bands. There is no significant difference. Average standard deviation 0.08 sec.

vations, should be higher than that of Hall 21, or that the Glasgow Hall (GL) should be equal to the very best. The Glasgow hall is currently undergoing major changes to improve its acoustics.

The major conclusion from comparison of Figs. 4.5 and 4.15 is that BQI is a superior measure for estimating the acoustical quality of a concert hall.

Acoustical "Glare" and Surface Diffusivity Index (SDI)

Concert music sounds better to a listener when the early sound is not "glary" and the late reverberant sound seems to arrive from many directions, i.e., from the sides, from the overhead, as well as from the front.

Many of the finest concert halls, e.g., those in Amsterdam, Boston, Vienna, Berlin Konzerthaus, Leipzig, Basel and Zurich (most built in the nineteenth century) have coffers, beams or curved surfaces on the ceiling, and columns, niches, irregular boxes, and statues on the upper side walls. In addition, the surfaces of the lower sidewalls often have fine-scale ornamentation. These irregularities and ornamentation "diffuse" the sound when reflected and give the music a mellow (non-glary) tone.

Diffusion must be thought of in relation to two different venues—those portions of the hall associated with the early reflections and those portions associated with the reverberant sound.

DIFFUSION OF EARLY SOUND. If one listens to music in a rectangular hall with flat, smooth sidewalls, the sound takes on a brittle, hard, or harsh sound, analogous to optical glare, sometimes called frequency coloration. Such a harsh sound was heard in the original New York Philharmonic Hall, because the principal early lateral reflections came from smooth plaster sidewalls. The sensation there was that the upper tones from the violins contributed most to the "glare."

To reduce acoustic glare related to the string tones that are caused by flat sidewalls or flat suspended panels, irregularities of the order of 1–2 in. (2.5–5 cm) deep should be embossed into the reflecting surfaces. *Detailed designs are discussed below.*

A further advantage of the diffusion provided by the irregularities on the sidewalls is shown in Fig. 4.16. It is seen that diffusion added to a wall distributes the early reflections from all instruments across a wide area of the sidewall, rather than from a different point for each instrument.

DIFFUSION OF LATE (REVERBERANT) SOUND. Since the early days of music in rooms, it has been recognized that the reverberant tone quality in a concert hall is improved by irregularities on the upper wall surfaces and the ceiling.

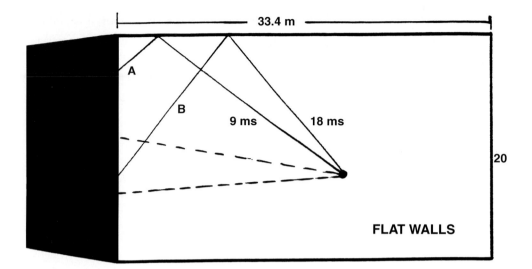

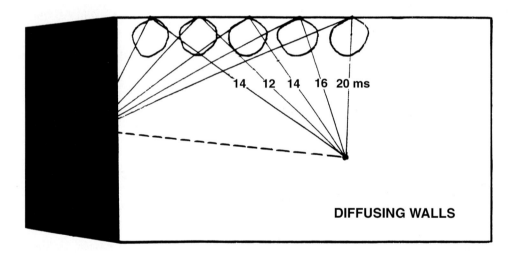

𝒯IGURE 4.16. One effect of adding sound diffusion to the sidewalls is illustrated roughly by the sketches above. The upper sketch shows sound A coming
from a flute on the stage and arriving at the listener's position from a single
point on the sidewall, well ahead of the sound B from a violin which is perceived
as coming from a point farther back on the sidewall. In the lower sketch, the
sound from any source on the stage is reflected to the listener from points all
along the wall because of its diffusivity. These distributed reflections from all
the instruments in the orchestra will be perceived by the listener as reflecting
from the entire sidewall, not just from a different place for each instrument.

Such irregularities scatter a sound wave during each reflection, so that, after many reflections, they add a homogenizing effect to the reverberant sound. This homogenizing effect is called "diffusivity of the sound field," and for the surfaces themselves we say they have "surface diffusivity." No physical measurement exits for determining the optimum amount or type of irregularities; instead we rely on visual means.

SOUND DIFFUSIVITY INDEX (SDI). Haan and Fricke (1993) have devised a visual procedure for classifying the "acoustical quality" of the surface irregularities in a hall based on inspection, or on photographs and drawings. The visual criteria they have promulgated for categorization of the "degree of diffusivity" of surfaces are as follows:

HIGH DIFFUSIVITY. For high diffusivity, the ceiling is coffered or checker-designed with deep recesses or beams [greater than 4 in. (10 cm) in depth, but not so great as to block ongoing sound waves, that is to say, not greater than about 10 in. (25 cm)]. The upper sidewalls should have random diffusing elements of sizable depth, or frequent niches and columns oriented vertically, over them. To avoid acoustical "glare," from early lateral reflections, the lower side walls should have fine scale diffusion, say, irregularities of about 2 in. (10 cm) depth. No area should absorb sound appreciably except for duct openings for air conditioning. Examples are Boston Symphony Hall (coffered ceiling, statues, niches, and ornamentation on sidewalls), Bonn Beethoven Hall (specially designed ceiling for random diffusion, and sidewalls with diffusing panels), Salzburg Festspeilhaus (special geometrical construction on ceiling and sidewalls meeting the above description), Vienna Grosser Musikvereinssaal (coffered ceiling, irregular sidewalls created by windows, deep plaster ornamentation, ornamentation on balcony fronts, and statues on lower sidewalls.)

MEDIUM DIFFUSIVITY. Broken surfaces of varying depth on the ceiling and sidewalls. Otherwise, ornamentally decorative treatment applied with shallow recesses [on the upper sidewalls, not greater than 2 in. (5 cm) in depth]. No absorbing surfaces, as above. Examples are Christchurch Town Hall (relatively smooth sound-reflecting panels, smooth balcony fronts, regularly spaced slats on side walls); London, Royal Festival Hall (relatively smooth sidewalls); Stuttgart, Liederhalle (relatively smooth side walls); Toronto (relatively smooth sidewalls before renovations).

LOW DIFFUSIVITY. Large separate paneling, or smoothly curved surfaces, or large flat and smooth surfaces, or heavy absorptive treatment applied. Examples: Buffalo, Kleinhans Hall (relatively smooth sidewalls and ceiling); Liverpool, Philharmonic Hall (relatively smooth sidewalls and ceiling) (both before renovations completed or under consideration).

Haan and Fricke then assigned a numerical rating to each degree of surface diffusivity: "high" equals 1.0; "medium" equals 0.5; and "low" equals 0. They then weighted diffusivity of each area of surface; for example, if a ceiling had a total area of 3,200 m² and 1,000 m² of it was smooth and the remainder deeply coffered, it would be rated "$0 \times 1,000 + 1 \times 2,200$" = 2,200 area points. This number was added to the area points for the sidewalls (they neglected the end walls) to get the total area points for the room. The "surface diffusivity index (SDI)" is obtained by dividing the total diffusing area points (just determined) by the actual total area of the ceiling and sidewalls. For the Vienna Musikvereinssaal they determined an overall SDI rating of 0.96. At the other extreme, the London Barbican Hall (before recent renovations) received a SDI rating of only 0.23.

In retrospect, a principal defect in the ill-fated New York Philharmonic Hall, as mentioned above, was the lack of surface irregularities. The SDI for that hall would have equaled "0." Parenthetically, the reason the hall was built without surface irregularities, even though the need for them was already appreciated as early as the nineteenth century, was that the building committee, to reduce costs, eliminated all sound diffusion from the plans in the closing year and hired an interior decorator to hide their lack. The hall opened with sidewalls covered with blue paint and illuminated by blue lights.

A plot of the *visually determined* SDI against the rank-orderings of 31 concert halls is shown in Fig. 4.17. The best halls have SDI's of 1.0, the medium quality halls, between 0.5 and 0.9, and the lower quality halls between 0.3 and 0.7. Obviously, the other acoustical parameters, BQI, RT, ITDG, G_{125}, good stage design, etc., must be taken into account simultaneously with SDI to obtain a meaningful relation between overall estimated acoustical quality and the rank-orderings of Table 4.1.

Listener Envelopment

A diffuse reverberant sound field is said to create "listener envelopment"; that is to say, the reverberant sound seems to arrive at the listener's ears from all directions, thus "enveloping" him/her. There have been suggestions in the literature that the *late* interaural cross-correlation coefficient might be used as a measure of listener envelopment. The quantity $IACC_{L3}$ measures the correlation between the sound waves that arrive at the two ears in the time period starting at 80 msec after arrival of the direct sound and extending up to 1 sec where "L" indicates this time period and "3" is the average of the sound in the 500-, 1,000- and 2,000-Hz frequency bands.

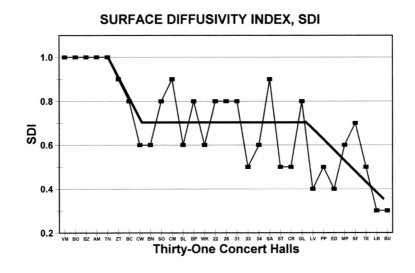

SURFACE DIFFUSIVITY INDEX, SDI

SDI

Thirty-One Concert Halls

$\mathcal{F}$IGURE 4.17. The surface diffusivity index (SDI) as determined from visual inspections of photographs or visits to 31 concert halls. The highest rated halls have SDI's of 1.0, while those of the lowest rated halls fall in the range of 0.3 to 0.7. The intermediate halls have SDI's of the order of 0.7 $\pm$ 0.1.

From Fig. 4.18 it is seen that all of the concert halls have nearly the same average values for $[1 - \text{IACC}_{L3}]$. In only one hall, which has fairly smooth sidewalls and ceiling, does its value deviate noticeably from those of the other halls. Unfortunately, this type of measurement has never been made in a hall with no sound-diffusing surfaces. However, as a measure of sound diffusion in normal halls it appears to have little value.

Clarity

Except in a few halls, clarity decreases with increased reverberation and *vice versa,* as was discussed in Chapter 2. Exceptions include Birmingham, Costa Mesa, Dallas, and Lucerne, which have reverberation chambers located outside the audience space. Here, we will treat its measurement and how it varies from one class of hall to another.

The usual physical measurement of clarity is the ratio of the energy in the early sound to that in the reverberant sound, a ratio that is expressed in decibels (dB) by C_{80}.

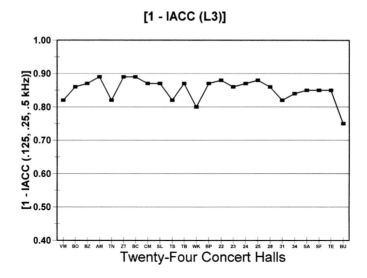

*F*IGURE 4.18. Plot of the late (after 80 msec) quantity [1 − IACC$_{L3}$] against the 24 rated concert halls.

The measurement of C_{80} requires two pieces of equipment—one to produce an impulse sound, such as the bursting of a balloon, a pistol shot, or a short intense beep from a loudspeaker; and the other, a tape recorder that is moved about the hall to record successive impulses at a number of seats. With a computer, two quantities are taken from the tape: (i) the energy of the sound that arrives directly from the loudspeaker plus all reflections from surfaces in the hall that occur within 80 msec thereafter, and (ii) the energy of the sound that arrives after 80 msec, usually up to 1 or 2 sec. The ratio of (i) to (ii), expressed in decibels, is C_{80}. If there is no reverberation—the room is very dead—the music will be very clear and C_{80} will have a large *positive* value (in decibels). If the reverberation is very large—such as exists in a huge cathedral—the music will be unclear and C_{80} will take on a large *negative* dB value. C_{80} equals 0 dB when the early energy is equal to the reverberant energy.

Published C_{80}'s are usually the average of the C_{80} values in the 500-, 1,000- and 2,000-Hz octave bands and at a number of seats in a hall. The symbol $C_{80}(3)$ is used to designate the time considered and the three bands. In Appendix 3, numerical values for C_{80} in unoccupied halls are listed for each of the eight frequency bands.

A interesting observation about clarity is that different amounts of it are desirable in different situations. During rehearsals, a conductor will often express satisfaction with a rehearsal hall that has a $C_{80}(3)$ of $+1$ to $+5$ dB (hall unoccupied), so that all the details of the music can be heard. But at a concert, whether conducting or listening in the audience, the same person will usually prefer a more reverberant space, i.e., with $C_{80}(3)$ equal to -1 to -4 dB.

In the interviews that were conducted for this study, the persons were asked to judge the halls as an audience would. As expected, the halls with clarities between -1 and -5 dB were judged the best. Of course, in no situation would a hall for symphonic music be acceptable if it had cathedral-like reverberation, which would yield a very high negative value of C_{80}.

A good feature of $C_{80}(3)$ is that it can be judged qualitatively by an experienced listener. It is easy to hear whether the music is too clear, or sounds "muddy" (the reverberant energy is too strong), or the balance between the reverberant and the early sound is not satisfactory for a particular kind of music. However, in normal concert halls, C_{80} is so highly correlated (inversely) with the reverberation time that it cannot be used as an additional way to estimate the acoustical quality of a concert hall.

The Dallas, Birmingham (England), and Lucerne halls are special. They are basically rectangular halls, not great in width. But they have extra reverberation chambers at the sides (and in one case the top) that create higher reverberation times at low frequencies with full occupancy. Because clarity is needed for instruments that play rapidly, particularly the violins, the C_{80} measure in those halls is similar to that in Boston Symphony Hall or Carnegie Hall in New York. However, at lower frequencies, 500 Hz and below, the longer reverberation there adds a new character to the music, not related to clarity.

Texture

A sound quality parameter that is difficult to measure is "texture." It involves the number and nature of the early sound reflections, those that arrive at a listener's ears soon after the direct sound. Even when all other parameters are acceptably near optimum, unsatisfactory texture may create remarks such as "the music is missing something," or "the music has an edge," or it is "a bit off."

Texture is related to a time pattern, a "reflectogram," at a listener's position like those shown in Fig. 4.13. As was already stated, the initial-time-delay gap (ITDG) for the upper hall is about 40 msec, and for the lower hall about 20 msec. For good "texture," not only should ITDG be short, but the reflections before the 80-msec marker should be large in number and relatively uniformly spaced like those in the lower graph.

Visual inspection of a reflectogram is a very important index of sound quality. In principle, rectifying and smoothing the reflectogram should facilitate this procedure, but the physical constant needed, the number of reflections in the early sound, is inaccurate. One way to eliminate the uncertainty in the count is to introduce a mathematically well-defined procedure to the reflectogram, namely to form its "Envelop Function, EF" (Kuttruff, 2000).

T. Hidaka (2002) advocates using a band width extending from 353 Hz to 2.8 kHz for determining EF and the usual monaural impulse response. Also, the range of 0 to −25 dB should be used. Hidaka applied this method to the opera houses where the rank ordering was the results of a questionnaire study (See Chapter 5). He found that the most highly rated houses had more than 17 reflection peaks before the 80 msec marker, a middle group, 10 to 16 peaks, and the lowest group, less than 10. This method has not been tried on concert halls, but there is no reason why it would be less effective.

Orthogonality of the Objective Acoustical Measures

Any list of acoustical parameters based on physical measurements that purport to be of assistance in judgments of concert hall quality must be independent of each other. In Table 4.2 the correlations among the physical measurements that are made in actual concert halls are portrayed. In this table, if the correlation between two quantities is greater than about 0.6, the quantities are not independent of each other. It is seen that the reverberation time (RT), early decay time (EDT), and the clarity factor (C_{80}) are highly correlated and therefore are not independently useful in judging the acoustical quality in concert halls. There is a reasonable correlation between LF_{E4} and $[1 - IACC_{E3}]$ (i.e., BQI) because both are responsive only to sound reflections from lateral directions and both measure the sound in the first 80 msec after arrival of the direct sound. However, as we showed above, BQI is more accurate in judging acoustical quality than LF. As one would expect, there is reasonably high correlation between volume and seating capacity.

Special Structures for Reducing Acoustical "Glare" and for Diffusing Sound

REDUCING ACOUSTICAL GLARE. Two 1-in.-deep designs that were used in the Tokyo NNT Opera House and the Tokyo TOC Concert Hall for the surfaces that controlled the early reflections are shown in Figs. 4.19 and 4.20. In those examples, the diffusion starts at about 2,000 Hz and reaches a maximum of about 8 decibels at 5,000 Hz. If the depth were 2 in., the diffusion would start at 1,000 Hz. The 1.0-in. version satisfied listeners in the Tokyo halls.

TABLE 4.2. Correlations among physical quantities measured in 42 concert halls. Correlations greater than 0.6 are listed in bold type. A low correlation means the two parameters are independent of each other.

	RT_M	EDT_M	$C_{80,3}$	G_M	$1\text{-}IACC_{E3}$	LF_{E4}	BR	ITDG	V	N
RT_M	—									
EDT_M	**0.99**	—								
$C_{80,3}$	**−0.84**	−0.88	—							
G_M	0.29	0.27	−0.30	—						
$1\text{-}IACC_{E3}$	0.15	0.17	−0.33	0.49	—					
LF_{E4}	0.23	0.25	−0.27	0.33	**0.71**	—				
BR	0.08	0.04	0.03	0.05	−0.13	−0.38	—			
ITDG	−0.48	−0.50	0.57	−0.43	−0.12	−0.20	−0.04	—		
V	0.31	0.27	−0.06	−0.57	−0.53	−0.09	0.20	0.25	—	
N	0.12	0.11	0.02	−0.55	−0.57	−0.28	0.27	0.18	**0.83**	—

DIFFUSING SOUND FIELDS (QRD DIFFUSERS). QRD diffusers are a design of irregular flat surfaces for the purpose of creating diffusion of the sound on reflection, usually over a wide frequency range (see References). A QRD diffuser when used to reduce a strong "echoey" sound has the advantage over using thin layers of sound absorbing material in that only a portion of the sound energy is lost. Rather, with a QRD reflector, the incident sound energy is "scattered" in many directions and takes part in the general reverberation.

One basic shape of a QRD is shown in Fig. 4.21. It consists of a series of parallel "wells" of varying depths. These wells are strips in a panel that could, for the example shown, be 9.4 in. (0.24 m) deep and 4.4 ft wide (1.34 m) (two repeated sections in width are shown).

The diffusion pattern for this panel is also shown in Fig. 4.21. The radii of the circles on the graphs are separated 10 dB from each other. At 3,200 Hz, the right-hand graph, it is shown that when a sound wave incident at 135 degrees (arriving from the upper left) strikes the horizontal QRD surface (at the centerpoint of the circles) the QRD reflects sound energy off, as shown by the solid curve, in almost equal intensity at all angles between 30 degrees and 150 degrees. In other words, it diffuses the incident sound wave. The dotted curve shows that if the wave

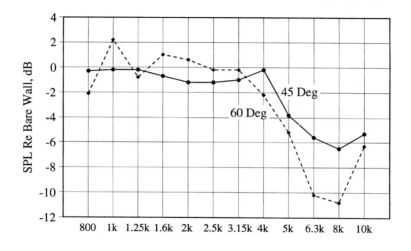

1/3 Octave Band Mid-Frequency, Hz

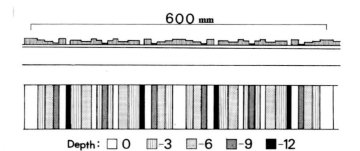

Depth: □ 0 ▥ -3 ▦ -6 ▨ -9 ■ -12

𝒯IGURE 4.19. Fine-scale diffusing surfaces on the side and rear walls, below the first balcony, of the Tokyo Opera City Concert Hall. Bottom: Depth in mm of the grooves. Next up: Section view of the pattern of the grooving. Middle: Photograph of the diffusing panel as tested in the laboratory. When installed on the sidewalls the grooves are vertical. Top: The sound pressure level relative to that from a hard flat surface, with the sound wave incident at 45 and 60 degrees. The microphone was located at the optical reflection angle.

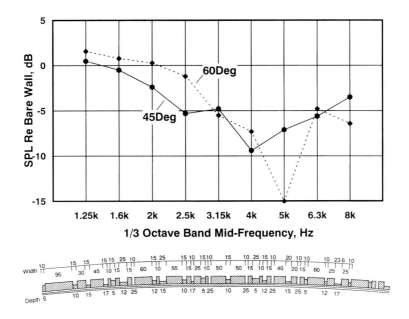

$\mathcal{F}$IGURE 4.20. Fine scale diffusing surfaces used in the "first-reflection" areas of the Tokyo New National Theater Opera House. The description is similar to that of Fig. 4.19.

were instead incident on a smooth flat panel, also at 135 degrees, almost all the energy would be reflected at 45 degrees (angle of reflection equal to the angle of incidence). Comparison of the dotted line with the solid line reveals that the energy reflected from the flat panel at 45 degrees would be 10 to 14 dB greater than that from the QRD panel. Thus, a QRD can be used both as a diffusing and an echo eliminating structure.

The highest frequency at which diffusion can take place is determined by the width of the wells in the QRD diffuser and the lowest frequency by the depth of the cells, but 16 in. (41 cm) is about the greatest practical depth. The negative side of the QRD diffusers is that some designs absorb a significant amount of sound and therefore cannot be used in large areas. The manufacturers supply both the diffusion patterns and the sound absorption coefficients for each type sold.

One important test of a QRD-type diffuser was in the Tokyo's Opera City TOC Concert Hall (No. 83 in Chap. 3). The architect created a hall with a pyramidal "ceiling"—he did not want a flat ceiling as in the conventional shoebox hall.

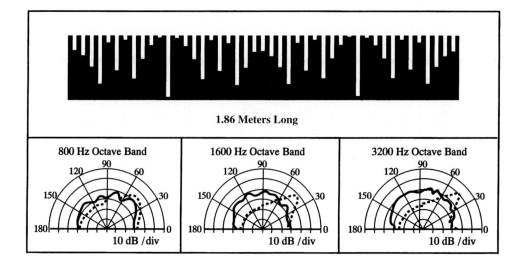

*F*IGURE 4.21. Structure and performance of a QRD diffuser. The QRD unit is about 4.85 ft (1.48 m) wide, 5 ft long, and 9 in. deep. The lower graphs show the diffusion patterns at 800, 1,600, and 3,200 Hz, as measured in octave bands. The reason for the unusual frequencies for the bands is that the measurements were made at one-fifth scale, which requires test bands with mid-frequencies of 4,000, 8,000, and 16,000 Hz. (Courtesy RPG Diffusor Systems, Inc.)

The peak of the pyramid is moved forward toward the stage end of the hall so that a long, sloping, triangular part of the ceiling hangs above the heads of the audience on the main floor. If this surface were to have remained flat, the ceiling would have returned a strong echo to the stage and to the front part of the main floor.

The design of the TOC QRD reflector and its performance characteristic for sounds striking perpendicular to its surface are shown in Fig. 4.22. The QRD reflector scatters most of the sound laterally. The sound that would have been reflected directly back to the orchestra and audience is reduced by more than 12 dB at frequencies above 500 Hz, eliminating any chance of a disturbing echo.

QRD's were tried around the sides of the stages in three American concert halls, (Baltimore, Chicago, and San Francisco) with the hope that their diffusing properties would enable the players to hear each other better. But the musicians did not like what they heard—the sound seemed muddled—and the QRD's were removed in all three cases.

Brilliance

Reverberation is the result of sound waves traveling around inside a hall following the early sound. As stated several times, early sound is the sequence of reflected sound waves that reaches a listener's ears in the first 80 msec after the direct sound has arrived. These early sound waves continue to reflect back and forth in the hall until they die out to inaudibility. That is to say, after 80 msec they are perceived as reverberation.

Let us consider a hall whose dimensions on average are about 80 ft in any direction. Because sound travels about 1,128 ft (344 m) per second, a wave will

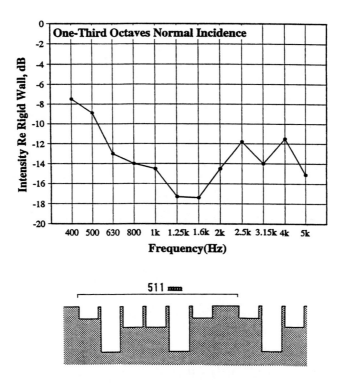

 ℱ IGURE 4.22. Bottom: Design of Schroeder QRD sound diffuser installed horizontally on the rear wall of the pyramid facing the stage of the Tokyo Opera City Concert Hall. Top: Reflected intensity measured in one-third-octave bands at perpendicular incidence. The sound reflected back to the stage is reduced by more than 12 dB in the frequency bands from 630 to 5,000 Hz. The QRD directs a considerable part of the diffused sound energy to the sidewalls of the pyramid that, on reflection, strengthens early lateral reflections.

strike a wall, ceiling, or the audience about 30 times in 2 sec. Each time a wave is reflected from a wall or a ceiling surface, a small amount is absorbed. Of course, the waves strike the audience areas some of the time, and those areas absorb a high percentage of the incident sound. But there is another way that a sound wave loses energy. This energy loss occurs in the air itself as a wave travels through it, just as a person loses energy when he wades through deep sand adjacent to a beach. In air, the energy loss is caused by friction among the air particles and resonant interactions among the molecules as they are shaken by the sound wave.

The type of absorption that occurs when a sound wave travels around a room is appreciable only at frequencies above 1,500 Hz. This means that the diminution in the strength of the sound due to natural causes will just become noticeable in the 2,000-Hz band and will be much more noticeable in the 4,000-Hz band. The loss is greater in dry air than in humid air. Because in a concert hall strict control of humidity is not usually maintained, sound absorption must be kept to a minimum at every surface the sound strikes. Failure to do so means the loss of high-frequency energy, that is to say, *loss of brilliance*. Numerical calculation of sound attenuation in air and its effect on reverberation time is available in Appendix 3.

Draperies and carpets should be strictly avoided or limited in area. If carpets are necessary, they should only be employed in aisles. Even there they should be as thin as possible and be cemented directly to a solid backing. Acoustical sound absorbing materials should only be used when necessary to eliminate echoes or acoustical glare and then in areas as small as possible.

Noise, Vibration and Echo

NOISE AND VIBRATION. A motion picture artfully conceived and movingly acted would never be submitted to the Cannes Festival if the film were overexposed. Nor would the noblest gastronomic effort of the chef at the Tour d'Argent surmount the addition of an extra spoonful of salt. The whitened scene and the briny taste of acoustics are echoes, noise, tonal distortion, and non-uniformity of hearing conditions. These attributes can only detract from the beauty of music; they add nothing. It is beyond the scope of this book to discuss noise and vibration control. Large texts exist on this subject. Suffice it to say that the noise levels in the hall should be less than the NCB-20 criterion and preferably less than NCB-15. If the hall is located over a railway or subway line or adjacent to heavy street traffic, the hall proper may have to be vibration isolated using springs or elastic layers in the foundations, depending on the magnitude of the vibrations.

ECHO. An echo in a hall for music is generally caused by a reflection that is returned to the front part of the hall from a rear wall or a rear balcony face. Tilting

spacing of the early reflections (see Fig. 4.13 for examples). For greatest accuracy, a wooden model, at 1/10th or 1/20th scale, is best. A miniature loudspeaker can be moved around on the stage to represent the different sections of an orchestra and a small sphere with miniature microphones on two sides representing ears can be used to represent a listener. The "listener" can be moved about the seating areas in the model and BQI can be measured accurately. An important advantage of a "real" model is that the balcony fronts, wall diffusion, and irregularities throughout the room can be adjusted both to satisfy the architect's visual demands and to obtain uniformity of BQI over the seating areas. If the proper equipment is provided, a person using earphones can actually hear the sound as it would be heard in the completed hall. A good feature of the Binaural Quality Index is that its value is nearly the same whether the hall is occupied or unoccupied (see Fig. 4.6).

The formula for calculating BQI from impulse sounds generated by a non-directional loudspeaker on stage is given in Appendix 3.

G_{MID}. We have already learned in Figs. 4.8 and 4.9 that the loudness (which relates largely to G_{mid}) can be too little, assuming a full orchestra is performing, in a very large hall like London's Royal Albert Hall, or too great, as in halls seating less than 1,000. These graphs show that G_{mid} is related to the early decay time $EDT_{unoccup}$ divided by the cubic volume V of the hall. The EDT (unoccupied) approximately equals RT (occupied) times 1.15 to 1.2 in modern halls with medium upholstering. The values of (EDT/V) $\times$ 10^6 plotted in those two graphs are the values that were actually measured in the halls, regardless of the upholstering. Therefore, for most of the halls, a portion of the scattering of the data points may be attributed to variations in sound absorption of the chairs. In a hall like the Segerstrom Hall in Costa Mesa, California, or the Town Hall in Christchurch, New Zealand, large reflecting panels around the upper part of the hall send a large part of the early sound directly to the audience, making the levels of G_{mid} higher than would be expected from the ratio of EDT/V.

G_{125}. As detailed earlier, the bass response is tied to the strength of the sound at low frequencies, especially the strength of the sound in the 125-Hz octave frequency band.

ITDG. The initial-time-delay gap (ITDG) is usually quoted for an audience position near the center of the main floor, halfway between the stage front and the first balcony front, or the rear wall if there is no balcony. The first reflection in shoebox halls usually comes from a balcony front—otherwise from a lower side wall. In a fan-shaped hall the first reflection may come from suspended panels or the ceiling. In the best concert halls and opera houses the ITDG is less than 25 msec.

S D I. Details on the determination of surface diffusion index (SDI) were given earlier. The more and varied the surface irregularities in a hall, the better. Experience in five concert halls that the author has been associated with indicates that the depth of the irregularities on the lower sidewalls can be of lesser magnitude than those on the upper walls. The irregularities on the lower walls are designed to remove acoustical "glare." Those on the upper walls primarily affect the reverberation time and need be large to cover a wide frequency range and to thoroughly homogenize the sound.

$C_{80,3}$. This term relates to clarity. If it is positive, as in opera houses, the sound is much clearer than if it is negative. Actually, $C_{80,3}$ correlates highly (inversely) with the reverberation time and is hard to control separately.

S T 1. There is no question but that the degree of stage support is important. SD1 measures how well a player hears himself and other players near him. When a canopy is used to create a favorable ST1 on stage, its height should be between 7 and 13 m, adjusted according to the orchestra's preference. Depending upon what energy is reflected from other surfaces this height will make ST1 equal approximately to -12 to -15 dB (see Fig. 4.23). The famous Concertgebouw in Amsterdam (AM) is an unusual example. It has a high ceiling over the stage, and part of the audience sits in steep formation to either side of the orchestra. Thus, there are no reflections from stage sidewalls. The conductor and musicians have trouble hearing each other and thus must closely watch the conductor's baton. From the standpoint of the quality of the sound in the audience this makes no difference.

S P E C I A L   N O T E: The values given above the different acoustical attributes were measured in halls with acoustical ratings ranging from "fair" to "excellent." When using these numbers in design, observe that a large percentage of the "excellent" halls are "shoebox-shaped." Halls of different shapes could have the same measured numbers but might not sound exactly the same. In other words, there may be other less prominent attributes that contribute to the "excellence" of those halls. Further, some believe, and with some justification, that non-acoustical factors, such as the beauty of the architecture and the quality of the performances, affect the "acoustical quality" judgments of listeners.

Preliminary Design Procedures

D E T E R M I N A T I O N   O F   A U D I E N C E   S I Z E   A N D   C U B I C   V O L U M E   O F   A   H A L L   G I V E N   S E L E C T E D   V A L U E S   F O R G_{MID} a n d RT_{OCC}. Before the detailed design of a concert hall is undertaken, the owner and acoustician should do some preliminary determinations. Because three factors are immediately involved—

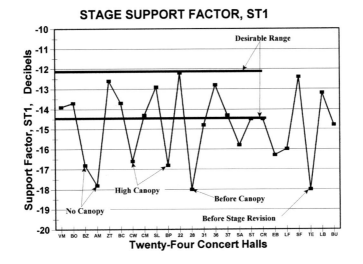

STAGE SUPPORT FACTOR, ST1

$\mathcal{F}$IGURE 4.23. The stage support factor (ST1) plotted against the subjective ranking for 24 concert halls. The method of determining it is described in Appendix 2. Generally speaking, halls with high ceilings and no canopy make it more difficult for an orchestra to play in good ensemble. It is seen that even in the highest-ranked halls, there may be no canopy, the usual reason for not installing one is that it is unsightly. A desirable range of -12 to -14.4 dB is indicated by the heavy lines.

number of seats, reverberation time, and the approximate cubic volume—a simple procedure for guidance is presented here. The assumptions underlying this exercise are given in Appendix 3, under the title "Derivatives of Sabine Equation." There are two ways to start. One is to begin with the number of seats desired. The other is to begin with the strength of sound that is desired in the hall. As we can see from Fig. 4.9, the strength factor G_{mid} at mid-frequencies for the best halls lies between 2 and 5 dB (European calibration). The design chart in Fig. 4.24 shows that if we pick the number of seats we can estimate the value of G_{mid}, or if we start with the desired G_{mid}, we can obtain S_T, and with it determine the number of seats.

Let us start with G_{mid}, the strength of sound in the hall. Assume, from Fig. 4.9, that 3 dB is satisfactory. From Fig. 4.24, for 3 dB, the approximate value for S_T is 1750 m². Because S_T includes the area of the orchestra it must be factored out, i.e., $S_A = S(T) - 180$ equals 1,570 m². Dividing 1,570 by 0.645 (approx-

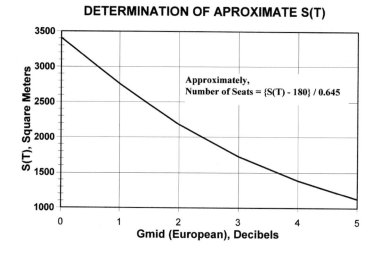

$\mathcal{F}$IGURE 4.24. Approximate chart for determining the acoustical audience area or the number of seats once the strength of sound (see Fig. 4.9) is agreed on. Conversely, if the number of seats N is chosen first, the acoustical area $S_T = (N \times 0.645) + 180$ m². Values of G_{mid} in different halls is taken from the chart in Fig. 4.9.

imate acoustical area per seat) gives us 2,434 seats. Next, assume we want a reverberation time RT = 1.9 sec at mid-frequencies. The cubic volume that will be needed is found from Fig. 4.25 after we determine the product of S_T and RT_{occup}, which is $1,750 \times 1.9 = 3,325$. From the chart we see that this will require a volume of about 21,000 m³. A hall that approximately has these physical characteristics is the Meyerhoff Concert Hall in Baltimore, and another that comes close is the Berlin Philharmonie. With these approximate numbers—1.9 sec, 2,400 seats, and 21,000 m³—one can decide, for example, what happens if fewer seats are desired, in which case the value of G_{mid} will increase. Or, what about a higher reverberation time? This would require an increase in the cubic volume.

Having these estimates at hand, more precise planning can begin, with the shape of the hall being the first consideration, and, next, all the other considerations that have been discussed in this chapter. Obviously, the correct reverberation time can only be calculated after the cubic volume and the acoustical characteristics of the walls, ceiling and chairs are known. With these known, the formula in Appendix 3 is used to calculate RT_{occup}.

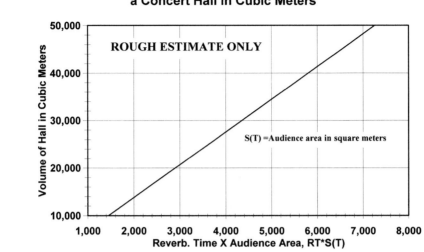

Determination of the Volume of
a Concert Hall in Cubic Meters

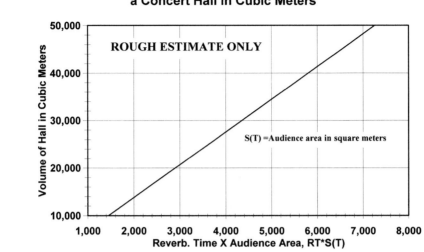

𝒯IGURE 4.25. This chart gives the reverberation time in seconds (hall fully occupied) when the reverberation time RT_{occup} is chosen and the audience acoustical area S_T has been determined from Fig. 4.24.

If the preferred values produce an audience size that is too small, the owner and architect will have to decide how far they wish to deviate from the preferred loudness G_{mid}. Once that decision is made, the volume can be determined based on the preferred RT_{occup}.

Stage Design

Balance, blend, and ensemble are important to the players' performance. Good "balance" of a symphonic performance means that no instrumental group dominates or is overpowered by another. Satisfactory "blend" means that the different sections of an orchestra sound as one tightly coupled body, not as several sections striving to play together. "Ensemble" means that the players can play in unison without watching every movement of the conductor.

These attributes are brought together by the acoustics of the "sending end of the hall," which include the sound-reflecting surfaces at the sides, the rear and above the stage, and those portions of the ceiling and sidewalls near the front part of the audience. Let us discuss these aspects of player performance by studying the surfaces surrounding the stages of the three top-rated halls in Boston, Vienna, and Amsterdam.

BOSTON, SYMPHONY HALL (NO. 3 IN CHAPTER 3). Musicians are fond of the stage enclosure in this hall because it offers the best in balance, blend, and ensemble. The stage is small by modern standards, 1,600 ft² (149 m²), which means that with a full-sized orchestra, risers are nearly impossible. For large orchestra and chorus, the stage must be extended into the hall by 5 ft (1.52 m). Occasionally, with orchestra, chorus, soloists, and risers, another 5 ft is added. In Boston, the average distance between the splayed sidewalls is about 50 ft (15.2 m) and the angle of splay is about 20 degrees. The construction comprises wood paneling in sections about 3 ft (0.91 m) square: the frames in each section are about 6 in. wide and 1 in. thick (15 and 2.5 cm) and the centers are about 0.5 in. thick. From the stage floor up to a height of about 14 ft (4.3 m), the paneling is about 1 in. thick.

There is a soffit (a shelf) beneath the pipe organ, about 1 m deep in the middle third of the stage width and 0.5 m on either side that acts to reflect sound from one side of the stage to the other and from front to back. The stage ceiling has an average height above the stage of 43 ft (13 m) and slopes at an angle of 15 degrees. Since the photograph in Chapter 3 was taken, the two panels, perpendicular to the ceiling, which acted as light shields, have been removed and various openings have been closed so that it more actively exchanges sounds between the back rows and the front rows of the orchestra and into the front rows of the main-floor audience. Although no measurement of the sound absorption by the organ has been made, it is presumed that it absorbs sound, but not excessively.

I had the opportunity to conduct Sousa's *Stars and Stripes Forever* with the Boston Pop's orchestra (regular Boston Symphony Orchestra minus the principal player from each orchestral section) and a full audience. During the piece, I purposely changed the tempo, loudness, and sectional balance. The composition contains parts that range from piccolo solos to full orchestra and is a useful selection for exploring stage acoustics, which was my intention. At the conductor's position, the acoustical support to the music seemed to come from the stage enclosure itself—the hall's acoustics were almost unnoticeable, remaining in the background. Every instrumental section sounded in good balance with the others. Also, the bass/treble balance was ideal to my ear. Most noticeably, the orchestra seemed small, partly owing to the small stage area and its shape—no instrument was more than 30 ft (9.2 m) distant from the podium. During one phrase of the Sousa piece, I turned and faced the audience, while they were rhythmically clapping. The reverberant sound in the hall, enhanced by their clapping, sounded absolutely natural.

Although the hall has a 0.1 sec lower reverberation time and a larger cubic volume than the Vienna and Amsterdam halls, the sound level on stage is higher by 1 or 2 dB, owing to the reverberation in the orchestra enclosure.

VIENNA, MUSIKVEREINSSAAL (No. 30 IN CHAPTER 3). This hall lacks a stage enclosure, but the orchestra is surrounded on three sides by an over-hanging balcony that sends strong reflections across the orchestra. The stage, with an area of 1,754 ft^2 (163 m^2), is steeply raked and 41 listeners (not counted in the area just quoted) are seated in each of the rear corners. The average height of the ceiling above the players is about 50 ft (15.2 m). The most distant player is seated about 30 ft from the podium. The sidewalls of the stage are only about 3 ft (1 m) more distant from the center line of the stage than the average stage width in Boston. As in Symphony Hall, an organ at the upper part of the back wall of the stage absorbs some of the sound. Beneath the balcony, the walls around the stage are wooden. By comparison, because there is no explicit stage enclosure, one would expect the players to be more aware of the acoustics of the hall as a whole than in Boston.

Acoustical measurements (Gade, 1989b; Bradley, 1994) show that the sound returned from the walls to a player in the center of the orchestra by a player seated 1 m to his/her side is about the same as that returned in Boston. These comparisons would indicate that both the conductor and the players experience acoustical con-ditions on stage that are much the same as in the Boston hall.

AMSTERDAM, CONCERTGEBOUW (No. 96 IN CHAPTER 3). The sending end of Amsterdam's Concertgebouw is significantly different from that of the previous two halls. The platform area is about the same (160 m^2) and the distance from the conductor to the farthest player is also about 30 ft (9 m). The average ceiling height above the players is about 55 ft (16.8 m), 12 ft more than in Boston. The distance to either of the two sidewalls from the centerline of the stage is 48 ft (14.6 m), whereas in Symphony Hall, the distance to the sidewalls is about half as great. The sound returned by the hall to a player at the center of the orchestra originating from a player one meter to his/her side is about 4 dB less in Amsterdam than that in the other two halls, a large difference.

In the Concertgebouw, the sound heard by the conductor is less clear than that in either the Boston or the Vienna hall because of the lack of strong early reflections, although it has the same small size of the stage. Maestros Bernard Haitink and the late Eugene Ormandy both have expressed to me the difficulty that players have in maintaining good ensemble, Ormandy remarking, "There is a jumble of sound and poor orchestral balance," and Haitink saying, "A visiting orchestra must have a special rehearsal to accommodate itself to the stage acoustics." Good ensemble depends on close attention to the conductor's beat.

AREA OF CONCERT STAGE. Barron (1993) states that it is desirable to provide the following areas per player for different groups of instruments:

13.5 ft² (1.25 m²) for upper string and wind instruments

21.6 ft² (2 m²) for cello and larger wind instruments

19.5 ft² (1.8 m²) for double bass

10.8 ft² (1.0 m²) for each tympani and double that for other percussion instruments

For a 100-piece symphony orchestra, these requirements set a stage area of about 1,615 ft² (150 m²). The stage at Boston Symphony Hall has an area of 1,600 ft² that is just adequate for a 100-piece orchestra if they are seated on a flat stage. If the platform has an area of 180 m² (1,940 ft²) there is ample space for risers and more space for soloists. Barron adds, "For seated choirs, 0.5 m² (5.4 ft²) per person is needed, so that a 100-person choir requires 50 m² (540 ft²) of space . . . [in the Royal Festival Hall] choir seating is in a separate elevated choir balcony . . . when not required as choir seating, these seats are sold to concertgoers . . . "

STAGE SHAPE. Very wide or very deep stages have serious disadvantages. When the stage is too wide, a listener seated on either side and near the front of the hall hears the instruments near him before he hears the sound from the other side of the stage. The difference in time may be great enough to affect the blend adversely. On the other hand, when the stage is very deep, the sound from the instruments at the back of the stage will arrive at a listener's ears a detectable instant after sound from the front of the stage, with similar adverse effects, and this fault can be heard by the entire audience. In addition, a very wide stage makes it difficult for the conductor to hold the sections of the orchestra in good ensemble. Recently, the stage in Chicago's Orchestra Hall (see Hall No. 5 in Chapter 3) was narrowed from 74 ft to 67 ft, partly to make it easier for the players on the two sides to hear each other. Also, the stage was made deeper bringing the total area up to 2,502 ft² (232.5 m²). Choral space, seating two rows of audience, is provided around the periphery of the stage 10 ft (3 m) above the stage level. A hanging canopy and reflecting surfaces at the edges of the reverberant space above the stage improve communication among the musicians and enable each player to hear his own instrument.

From the above evidence and other experiences, it is concluded from an acoustical standpoint that when the average width of the stage is greater than about 60 ft (18.3 m) and the area is greater than 2,000 ft² (186 m²), special means must be provided to create early reflections to the players so that they can hear each other during performances. Some combination of reflecting soffits, stage ceiling, suspended panels, canopies, shelves, and diffusing irregularities must be incorporated into the architectural design of the sending end of the hall. It goes almost without saying that whatever stage design is selected, it must be integrated acoustically into the overall architecture of the hall so that it projects the sound with proper balance and blend to the audience.

Balconies

Balconies are used in large halls to reduce the distance between the stage and the farthest row of seats. The architect must choose between one large balcony or a number of smaller ones. Whichever is chosen, the people seated beneath a deep overhang do not receive sound reflected from the upper part of the hall, thus reducing the perceived reverberation. If the balcony is deep and the mouth of the opening is not very high, even the sound that travels directly from the stage will be muffled.

The balconies shown in Fig. 4.26 represent satisfactory designs. The ideal arrangement is an overhang that involves only one to three rows of seats, like those in Tel Aviv and Canada (upper sketches). Boston's balconies have overhangs that are deep enough to cause some dissatisfaction, but they illustrate the limit to which one should go. There, the middle balcony overhangs five rows of seats and the lower one overhangs eight rows. Those people seated after the third row receive less and less of the reverberant sound the farther back they are located.

Lesser quality balconies are shown in Fig. 4.27. In all three cases, large numbers of seats are overhung, and those near the rear participate little in the reverberant sound. In the Royal Festival Hall, even the direct sound is attenuated in the rear seats. Because the balcony shown in the lower sketch is from the former Metropolitan Opera House (no longer existent), the loss of reverberation was of less concern, both because the reverberation time was low and because the listeners' attention during an opera was focused on the early sound (direct plus early reflections).

As a general principle, the D dimension shown in Fig. 4.28 should not exceed the H dimension. An even better design is to restrict the opening to the height given by the angle $\theta = 45°$. With this restriction the early decay time EDT_{mid} is decreased by only about 10%, as measured in a number of halls. In addition, the soffit of the balcony should be shaped to reflect the early sound to the heads of the listeners.

Models

An exact copy of a successful hall uses the original as the model. Any new design, and especially any radical design, should be modeled early in the project's

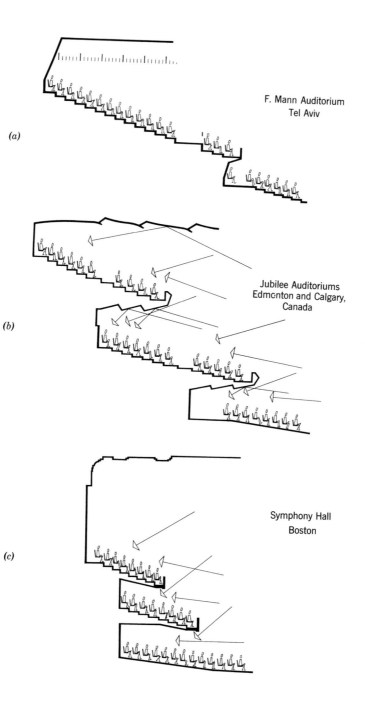

F. Mann Auditorium
Tel Aviv

(a)

Jubilee Auditoriums
Edmonton and Calgary,
Canada

(b)

Symphony Hall
Boston

(c)

𝒯IGURE 4.26. Three relatively satisfactory balcony designs: (a) excellent design, no overhang; (b) very good, little overhang and wide openings; (c) center
balcony is good acoustically except for the last three rows of seats that do not
receive sound from the upper part of the hall. The last five rows of seats in the
lower balcony are also less good.

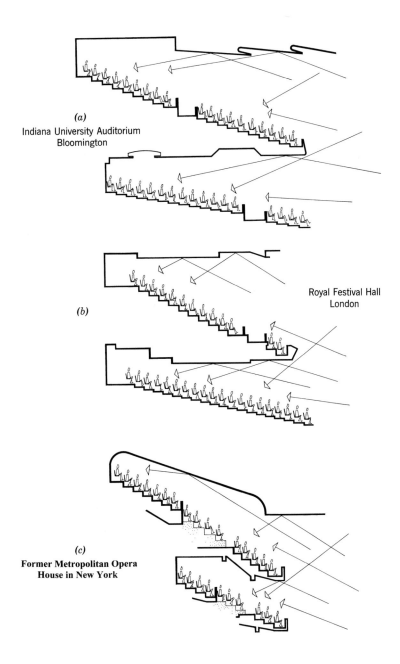

(a)
Indiana University Auditorium
Bloomington

Royal Festival Hall
London

(b)

(c)
**Former Metropolitan Opera
House in New York**

 FIGURE 4.27. Three balcony designs of lesser quality: (a) large overhang but fair acoustics because of large openings; (b) large overhang, small opening; (c) lower balcony is poor, receives little sound from upper hall and the opening is small. [Note: The (c) design was taken from the demolished Metropolitan Opera House in New York.]

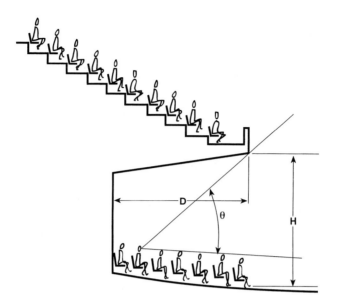

$\mathcal{F}$IGURE 4.28. Recommended designs for excellent balcony overhangs in concert halls. As a general principle, D should not exceed H. Angle θ is a better measure and should not be less than 45 degrees. The underbalcony soffit and rear wall should be shaped to reflect the direct sound toward the heads of the listeners. Shaping is specific to each hall and depends on the location of the stage and the rake of the floor beneath the seats, and whether the overhang is for upper balconies.

life. A computer model is most desirable at the beginning stage; it can be changed or discarded and an entirely new approach considered, without a great loss in time or cost. Using the latest computer modeling techniques for room acoustics, which improve yearly, the first 10–20 reflections arriving at a listener's position in the "hall" from an on-stage source can be determined. A model is able to show whether the design meets the basic requirements for a high value of the Binaural Quality Index (BQI), and a satisfactorily low initial-time-delay gap (ITDG).

The value of a wooden model cannot be overstressed. Obviously, it should be constructed only after a near-final architectural design has been selected from the computer modeling. With the wooden model, individual reflecting surfaces can be adjusted to their optimum shapes for distributing early reflections uniformly over the audience areas, echoes can be observed and corrected, and so on. It size should be 10:1 ideally, but not smaller than 20:1. The reason is that in a 10:1 model, a tiny loudspeaker can be used to radiate sound from all parts of the stage, and a listener can be simulated by a sphere about 1 in. (2.5 cm) in diameter, with two tiny microphones for ears. With this set-up, the actual sound in the "hall" can be heard, and the variation in acoustical quality from one part of the hall to another can be compared. Further, the BQI and ITDG in various parts of the hall can be determined accurately. Thus, the final architectural plans can be produced with the assurance that only small adjustments might be necessary in the completed hall.

For complete advanced assurance, a trial (tuning) concert with audience (usually invited) is made in the finished hall about 3 months ahead of its opening. The author's experience in five recent concert halls has shown the 10:1 model to be enormously helpful in the shaping of all surfaces and then assuring the owner and architect that the design contains no features that might be troublesome. In three of the halls some minor adjustments were made after the tuning concert. In each hall, some accessible places were left available for adding some acoustical materials to control echoes or to decrease the reverberation time slightly and in one case, some acoustical material which we thought was necessary from the model tests was removed.

Multipurpose Halls

Many halls among the 100 detailed in Chapter 3 are multipurpose. The owner or architect who is faced with such a hall should study the different approaches presented. It is generally agreed that, if economically possible, a city should have a separate hall for each major purpose: symphonic concerts, chamber music, opera and ballet, and conferences. Otherwise, a multipurpose hall is necessary. There are at least three approaches to the design of a multipurpose hall:

DESIGN 1. For this alternative the hall is built with a fly (scenery) tower. A demountable orchestra enclosure (shell) is provided for concerts, which, when in place, will result in a reverberation time that is on the lower edge of the recommended values of Table 4.3, say, 1.7 sec. Each of the other acoustical factors should be as near those recommended in Table 4.3 as possible. These include the strength of the sound, which depends on keeping the audience size as small as feasible; the Binaural Quality Index (BQI), which depends on an abundance of early lateral reflections and a sufficiently low ITDG; and, finally, adequate irregularities on the ceiling and side walls. For opera use, the fly tower should have a low reverberation time, which means that the house, with the shell stowed, would have a reverberation time of 1.4–1.6 sec. The highest-rated opera house in the world, the Buenos Aires Teatro Colón, has a RT_{occup} equal to 1.6 sec. For Italian opera alone 1.4 sec is best. For speech events, large areas of sound-absorbent draperies or banners that are otherwise concealed in pockets in the sidewalls or the ceiling, would be exposed (lowered). If the total area is large enough, the RT can be reduced by 0.2 to 0.3 sec. For amplified speech, using column-type (directional) loudspeakers, an RT of less than 1.5 sec is a satisfactory environment for lectures. Without loudspeakers, in smaller halls, the reverberation should be 1.2 sec or less.

DESIGN 2. A second approach is to design a hall like that in Lucerne, Switzerland. This hall is not meant for opera, as no fly tower is provided. It is rectangular and has an extra 350,000 ft³ (10,000 m³) of reverberation chamber located on the two sides of the hall, and accessed by heavy operable doors. The reverberation time can be adjusted by opening part or all of the doors. To make the hall satisfactory for conventions, retractable curtains, as described in the previous paragraph, are provided both in the main hall and in the chambers. When the doors are closed and some curtains are exposed, the hall will be satisfactory for intimate (chamber) music.

DESIGN 3. A third method, suitable for opera, ballet, and intimate music, has a fly tower, but is made into a concert hall with a "concert hall shaper" and hanging panels like those shown for the Bass Performance Hall in Fort Worth, Texas. For concerts in Bass Hall the RT_{occup} is about 1.9 sec. Without the shaper, with a low reverberation time in the upper part of the fly tower, and with 6,600 ft² (613 m²) draperies deployed, the reverberation time is reduced to 1.6 sec. If the volume of the hall were chosen smaller, say, with an RT for concerts equal to about 1.7 sec, then with the shaper stowed and the curtains deployed, the hall would also be suitable for conferences, with an amplifying system.

Other variations on the above three are possible. The success depends, in part, on not making the basic hall too large.

ARCHITECTURAL DESIGN OF CHAMBER MUSIC HALLS

As this is written, the first modern acoustical data on a significant number (18) of chamber music halls has been published by Hidaka and Nishihara (see References). The study includes halls with seating capacities of less than approximately 800, in comparison with the study in the earlier sections of this chapter that treat symphonic concert halls with seating capacities in excess of that number. The published data are in Table 4.4.

The reverberation times RT_{occup} at mid-frequencies are between 1.0 and 2.0 sec, with the median at 1.65 sec. Half of the halls have reverberation times between 1.65 and 1.76 sec, which is consistent with the preferred times shown in Table 4.3.

The strength factors are higher than those for large halls, which is to be

$\mathscr{T}$ABLE 4.4. Accoustical parameters in chamber music halls for which extensive objective measurements are available (courtesy Takayuki Hidaka and Horiko Nishihara).

	V m^3	N	V/S$_A$ m	RT$_{occ,M}$ sec	EDT$_{unocc,M}$ sec	BR$_{occ}$ —	C80$_{3B}$ dB	G$_L$ dB	G$_M$ dB	1-IACC$_{E3}$ BQI	ITDG msec
Amsterdam, Kleinersaal in Concertgebow	2,190	478	9.4	1.25	1.49	1.21	1.5	13.7	12.9	0.69	17
Berlin, Kleinersaal in Schauspielhaus	2,150	440	9.0	1.08	1.33	1.24	2.0	12.2	10.9	0.67	11
Kanagawa, Higashitotsuka Hall	3,576	482	8.6	1.18	1.11	0.87	3.1	5.4	8.7	0.72	10
Kirishima, Miyama Conceru	8,475	770	15.8	1.84	1.80	1.12	−0.1	8.2	8.3	0.75	26
Prague, Martinc Hall	2,410	201	18.4	1.76	2.19	1.12	−1.9	12.6	12.6	0.68	11
Salzburg, Grossersaal in Mozarteum	4,940	844	11.5	1.66	2.06	1.07	−1.6	9.9	9.6	0.69	27
Salzburg, Wiennersaal in Mozarteum	1,070	209	8.4	1.11	1.33	1.09	1.7	14.9	14.3	0.77	15
Tokyo, Casals Hall	6,060	511	17.8	1.67	1.79	1.00	−1.3	7.6	9.4	0.71	15
Tokyo, Dai-Ichi Seimei Hall	6,800	767	13.3	1.66	1.83	1.09	−0.1	9.8	10.8	0.71	24
Tokyo, Hamarikyu Asahi Hall	5,800	552	14.7	1.67	1.82	0.93	0.0	7.1	8.8	0.71	15
Tokyo, Ishibashi Memorial Hall	5,450	662	14.9	1.70	1.84	1.10	−0.8	9.2	10.8	0.75	19
Tokyo, Mitaka Arts Center	5,500	625	13.3	1.73	2.28	1.02	−2.2	9.1	11.1	0.75	17
Tokyo, Sumida Small-Sized Hall	1,460	252	9.7	0.93	1.08	1.03	2.8	8.1	10.6	0.73	8
Tokyo, Tsuda Hall	4,500	490	12.5	1.33	1.42	0.90	0.8	7.6	10.7	0.71	20
Vienna, Brahmssaal	3,390	604	10.0	1.63	2.37	1.16	−2.8	12.8	13.6	0.77	7
Vienna, Mozartsaal in Konzerthaus	3,920	716	9.1	1.49	1.79	1.14	−0.2	11.6	10.8	0.70	11
Vienna, Schubertsaal in Konzerthaus	2,800	336	15.6	1.98	2.54	1.14	−3.3	14.7	13.6	0.77	12
Zurich, Kleinersaal in Tonhalle	3,234	610	9.3	1.58	2.11	1.18	−1.8	14.1	13.2	0.70	18

expected, because the halls have about half to one-third the number of seats. The measured differences are 3 to 4.5 dB larger, consistent with this fact. However, the performing groups are smaller, so that this difference is generally not significant.

As would be expected, with the smaller volumes, the initial-time-delay gaps (ITDG's) are short, ranging from 8 to 27 msec, with the median at 15 msec.

There is one important difference. The value of BQI, as defined earlier in this book, is nearly the same for all halls and is higher than that for symphonic halls. Here the range of BQI = $[1 - \mathrm{IACC_{E3}}]$ is 0.7 to 0.75. Because the ITDG's are about 0.6 of those in symphonic halls, it seems sensible to limit the time for measuring BQI's to 0.6 of 80 msec, i.e., to a span of 50 msec. With this change, the range of values for $\mathrm{BQI_{50}}$ expands to 0.58 to 0.7.

CONCLUDING REMARKS

As the complexity of material in this chapter illustrates, an experienced acoustical consultant should be engaged with the joint approval of the owner and the architect. Careful planning before an architectural design is fixed is almost a necessity. A hall that is shaped so as to create echoes and that has a minimum of early lateral reflections is almost impossible to repair after opening. The owner and architect should not put originality in the architectural design as the highest priority. A round violin can never be made to sound like a Stradivarius. Use of a computer to model a hall is almost essential today, because various architectural designs can be studied and absolute failures prevented. For complete assurance before construction, a 10:1 or 20:1 wooden model will enable final shaping of balcony fronts, diffusion, rear wall designs, proper cubic volume, and sight lines. These steps are costly, but in no way as costly as postconstruction changes.

Additional comments are in order about variable acoustics. The most complete data available are for the Concert Hall in Lucerne (No. 96). There are three independently variable features in this hall: (1) 6,000 m³ of added reverberation chamber accessed in whole or parts by heavy, moveable doors; (2) 1,200 m² of retractable sound-absorbent curtains that can cover part of the inside walls; and (3) a large canopy variable in height from 10 to 15 meters. This amount of variability provides the performers with a wide range of acoustic conditions: (a) mid-frequency, occupied reverberation times from 1.5 to 2.15 s; (b) low-frequency reverberation times from 2 to 3 seconds; and (c) a canopy height suitable for large orchestra (15 m), or for chamber groups (10 m). The conductor has these choices available, which can be made at his/her will during a musical event. The hall is shoebox shaped and has an initial-time-delay gap of 23 ms, which combine to give it a high rating.

FIVE

ACOUSTICS

OF

OPERA HOUSES

Opera is said to have originated in Italy in the royal courts at the end of the fifteenth century, but opera as we know it came to life a hundred years later in a Florence palace, evolving from a collection of short vocal pieces with melodies that were accompanied by instrumental chords (recitativo). The first "opera" was staged in Florence in 1598—*Dafne* by the poet Rinuccini, set to music by Jacopo Peri. The first opera house, built for the purpose, was the Teatro di San Cassiano that opened in Venice in 1637. For the first time the orchestra was placed between the stage and the audience. The oldest opera house in use today is the Teatro di San Carlo, which opened in 1737, burned in 1816, and was restored exactly as before in 1817. Verdi suggested the addition of the orchestra pit in 1872. The second-oldest standing opera house in Italy is the Teatro alla Scala that opened in August 1778. Every great singer and conductor of opera since that time has appeared in the beautiful La Scala.

In the mid-eighteenth century, opera was going strong in Paris under Louis XV. The royalty and members of the cabinet sat in boxes at the sides of the stage. There were three levels of boxes, usually in horsehoe plan, with partitions between so that patrons had to stand to see the stage. Those in the parterre had the misery of standing up for hours, compounded by routine overcrowding. In the nineteenth century Paris fostered "grand opera" with elaborate scenery, costumes, scores of supernumeries, and large orchestras. In 1875, France's famous opera house, Opéra Garnier, was opened.

Today's opera houses: Hidaka and Beranek (2000) decided to make objective and subjective evaluations of 23 opera houses in Europe, Japan, and the Americas. The houses that were investigated were selected according to the following criteria: (1) they should be widely known as venues for classical opera; (2) their architectural characteristics should be available; and (3) they should represent a reasonable range in size and shape.

QUESTIONNAIRE RANK-ORDERINGS OF ACOUSTICAL QUALITY OF 21 OPERA HOUSES

Questionnaires were mailed to 67 important opera conductors: 22 responded; one response not usable. They were asked to rate the acoustics of the opera houses that they knew well on scales that had five steps: *Poor, Passable, Good, Very Good,* and *One of the Best,* which were assigned the numbers 1 to 5 for analysis.

The conductors' responses are given in Fig. 5.1. The median of the standard deviations was 0.8. Some conductors (3) used the full range of 1–5 for their ratings, while some others (3) used as little as 2–4. The main group (15) used 2–5. A nonparametric rank-order comparison of the houses, disregarding the numerical ratings given them by the conductors, was made. By that method, the halls fell into the following order BA, DS, MS, TT, MB, NS, PG, PS, VS, SG, NM, AS, LO, SW, HS, RO, BD, CC, PB, TK, BK. There are some reversals, but they are for houses that had almost the same numerical ratings, so that the rank ordering

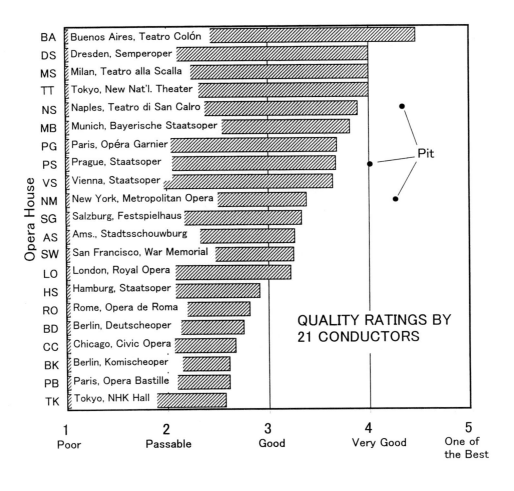

The chart shows "Opera House" on the vertical axis with the following entries:

BA — Buenos Aires, Teatro Colón
DS — Dresden, Semperoper
MS — Milan, Teatro alla Scalla
TT — Tokyo, New Nat'l. Theater
NS — Naples, Teatro di San Calro
MB — Munich, Bayerische Staatsoper
PG — Paris, Opéra Garnier
PS — Prague, Staatsoper
VS — Vienna, Staatsoper
NM — New York, Metropolitan Opera
SG — Salzburg, Festspielhaus
AS — Ams., Stadtsschouwburg
SW — San Francisco, War Memorial
LO — London, Royal Opera
HS — Hamburg, Staatsoper
RO — Rome, Opera de Roma
BD — Berlin, Deutscheoper
CC — Chicago, Civic Opera
BK — Berlin, Komischeoper
PB — Paris, Opera Bastille
TK — Tokyo, NHK Hall

Pit

QUALITY RATINGS BY 21 CONDUCTORS

Horizontal axis:
1 Poor — 2 Passable — 3 Good — 4 Very Good — 5 One of the Best

FIGURE 5.1. Acoustical quality ratings in the audience areas of 21 opera houses by 21 opera conductors. The rating sheets contained scales for 24 houses. Only houses that received six or more ratings are included in this figure. All conductors rated the acoustics of the pits (conductor's position) as well, but for only three houses were their ratings significantly different (higher) than the ratings for the audience areas, as shown by the large dots. A non-parametric rank ordering, without reference to the rating numbers, yields nearly the same sequence—the differences are only among those with almost the same numerical ratings. The rating for TT—the New National Theater Opera House in Tokyo, which opened in October 1997—was obtained from two opera conductors, two opera singers, two opera directors, two visiting listeners, and four music critics, all with world opera experience.

of Fig. 5.1 is as valid for the purposes of this chapter as the non-parametric comparison.

The conductors were also asked, "What mostly makes you judge the sound in some opera houses better than in others?" Their answers boil down to (1) hall support for singers; (2) uniformity of singer projection from a wide area on the stage; (3) good balance between orchestra and singer; and (4) clarity and richness of the orchestral and singing tones.

The authors analyzed the statistics of opera performances in the 1997–98 seasons at 32 major opera houses around the world. Italian and Mozart operas were performed about three-fourths the time. German operas were performed only 13 percent of the time. This indicates that there is little demand for opera houses like the Festspielhaus at Bayreuth, Germany. That house is especially suited to the performance of Wagner's compositions—particularly his *Parsifal*. It has a covered pit with a slit open at the top that allows the music to be heard by the audience in an eerie form. Intimacy and clarity of both the singers and the orchestra are vital for Italian and Mozart operas. Hence, those needs are probably why horse–shoe-shaped halls of compact size with an open pit and relatively low reverberation times have predominated throughout much of operatic history.

Objective Measurements of the Acoustical Properties of 23 Opera Houses

Measurement Procedure

MEASUREMENT SYSTEM AND SOURCE AND RECEIVER POSITIONS. The measurement system was the same as that used for concert halls. An impulse sound (like a controlled handclap) was emitted by a non-directional loudspeaker five to ten times. People with small microphones taped to the entrances to their ear canals sat in a house at each of various positions for each set of impulse repetitions. The electrical outputs of the microphones were recorded on digital (DAT) recorders, and the tapes were brought back to the laboratory for analysis. Typical source and receiver positions are shown in Fig. 5.2. Four positions were on stage and three in the pit. In a number of the halls there was insufficient time available to make more than one or two positions on the stage and in the pit and half of the positions in the audience. For each parameter, the hall-wide data were averaged to yield a single number.

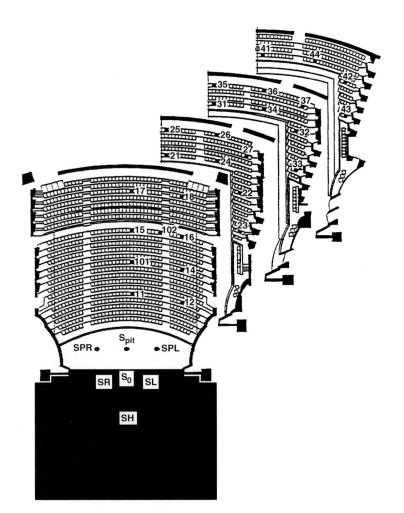

FIGURE 5.2. Typical source and receiver positions in opera houses. The floor plans are for the Tokyo New National Theater Opera House. When time did not permit, the number of audience locations for measurements may have been reduced to as few as 10 to 15. Positions S_o and S_{pit} represent the source positions most often referred to in this book, and Position 102 represents the main-floor position most quoted because it is off-center and is usually a choice seat.

MEASUREMENT RESULTS. The results of the physical measurements are listed in Table 5.1. It is seen that they include the same physical parameters as for concert halls as discussed in Chapter 4. They were executed under the following conditions: (1) without audiences; (2) with fire and performing curtains open;

TABLE 5.1. Opera houses of the world for which objective measurements are available. The source was at S_o and the microphone was at 10 to 24 positions. Data averaged to get a single number. All data are directly comparable because they were taken in recent years by the Takenaka Research and Development Laboratory in Chiba, Japan, except for one hall, LO.

	Hall Name	V (m³)	N	V/S$_T$ (m)	RT$_{occ, M}$ (sec)	EDT$_{unocc, M}$ (sec)	C$_{80, 3}$ (dB)	G$_{125}$ (dB)	G$_M$ (dB)	1-IACC$_{E3}$ (BQI)	ITDG (msec)	Stage set
AM	Amsterdam, Music Theater	10,000	1,689	—	1.30	1.30	1.9	1.8	1.7	0.55	32	n
BD	Berlin, Deutscheoper	10,800	1,900	7.5	1.36	1.60	0.7	2.1	1.2	0.39	33	n
BK	Berlin, Komischeoper	7,000	1,222	7.1	1.25	1.23	3.1	1.7	6.0	0.62	20	y
BE	Budapest, Erkel Theater	17,000	2,340	—	1.30	1.40	3.8	1.6	3.3	0.45	17	y
BS	Budapest, Staatsoper	8,900	1,227	—	1.34	1.37	1.9	1.5	4.4	0.65	15	y
BA	Buenos Aires, Teatro Colón	20,570	2,487	9.6	1.56	1.72	1.1	2.0	2.4	0.65	18	y
CC	Chicago, Civic Opera House	23,000	3,563	9.1	1.51	1.49	2.1	1.9	0.3	0.53	41	n
DS	Dresden, Semperoper	12,480	1,284	10.3	1.60	1.83	0.8	2.0	2.7	0.72	20	n
EO	Essen, Opera House	8,800	1,125	—	1.61	1.90	1.3	2.2	−0.4	0.54	16	n
HS	Hamburg, Staatsoper	11,000	1,679	7.4	1.23	1.35	2.2	1.5	1.3	0.46	34	y
LO	London, Royal Opera House*	12,250	2,157	7.7	1.20	—	4.5	1.3	0.7	0.53	18	n
MS	Milan, Teatro alla Scala***	11,252	2,289	6.9	1.24	1.14	3.6	1.4	−0.3	0.48	16	y
NM	N.Y., Metropolitan Opera**	24,724	3,816	9.1	1.47	1.62	1.7	1.6	0.5	0.62	18	n
PG	Paris, Opéra Garnier	10,000	2,131	6.9	1.18	1.16	4.6	1.4	0.7	0.50	15	y
PS	Prague, Staatsoper	8,000	1,554	—	1.23	1.17	3.1	1.7	2.2	0.64	16	y
RE	Rochester, Eastman Theater	23,970	3,347	10.2	1.63	1.90	0.8	2.3	3.6	0.54	22	y
SG	Salzburg, Festspielhaus	14,020	2,158	8.9	1.50	1.80	1.5	1.6	1.2	0.40	27	n
SO	Seattle, Opera House	22,000	3,099	11.2	2.02	2.50	−0.4	2.9	2.7	0.48	25	n
TB	Tokyo, Bunka Kaikan	16,250	2,303	9.8	1.51	1.75	1.1	2.0	0.3	0.56	14	n
TT	Tokyo, New National Theater	14,500	1,810	9.9	1.49	1.70	1.6	1.6	1.7	0.65	20	n
NT	Tokyo, Nissei Theater	7,500	1,340	7.4	1.11	1.06	4.4	1.5	5.3	0.58	17	y
VS	Vienna, Staatsoper	10,665	1,709	7.3	1.36	1.43	2.7	1.6	2.8	0.60	17	y
WJ	Washington, JFK Center, Opera House	13,027	2,142	8.2	1.28	1.27	4.3	1.4	3.1	0.53	15	y

*Renovated in 2000, data preceded.
**Steel shutter behind main stage closed.
***In renovation.

(3) with major musical instruments and chairs in the orchestra pit (except for the Essen, the Tokyo New National Theater, and the Seattle Opera House); and (4) with an orchestral enclosure at Rochester.

SIZE AND SHAPE. The sizes of the houses ranged from 1,125 to 3,816 seats; volumes from 247,000 to 873,000 ft³ (7000 to 24,724 m³); and S_T from 10,500 to 29,250 ft² (980 to 2,718 m²). S_T is the area over which the audience sits plus edge corrections, and plus the area of the proscenium when it is not closed by a fire curtain. With a large number of seats, assuming evenly distributed sound, there is less soloist energy per person. For the architect, a large house with modern seats occupying a larger area per person necessitates special construction around the proscenium to project voices efficiently to the audience areas.

Of the top nine houses only one, TT, is not horseshoe shaped (see Fig. 5.2). It is more like a conventional theater with three shallow balconies—and with special "trumpet" shaping at the front of the main floor and each balcony to enhance the strength of the singers' voices. The acoustical measurements in this hall indicate that the horseshoe shape is not necessary to obtain a good opera house; indeed that shape is outperformed by the TT shape. Visual factors certainly have entered into the decisions to build multi-tiered, horseshoe-shaped houses, because this construction reduces the distances between the stage and most of the listeners compared to those distances in conventional auditoriums. The negative feature of the horseshoe-shaped houses is that the vision from the boxes nearest the stage is usually bad.

Sound Quality Parameters

BQI, BINAURAL QUALITY INDEX. The binaural quality index, measured by [1 − IACC$_{E3}$], has the highest correlation of all the physical measures with the subjective judgments of acoustical quality of opera houses by the conductors. The results for those houses for which BQI data are available are shown in Fig. 5.3. For the top-rated houses, except for the Paris Garnier PG, the BQI is above 0.6, similar to that found in concert halls. An interesting case is the famous La Scala in Milan. The measurements in the boxes (at positions for those seated nearest the box opening) was 0.63, but on the main floor the BQI was only 0.38. The reason for this difference is that there are no sound-reflecting surfaces near the proscenium that could direct early reflections to the main floor. The conductor's ratings for this house would seem to be for the box seats, probably because they would normally be seated in the management's box when they are not conducting. Only one conductor mentioned the main floor and he said it was "quite disappointing." The main floor seats are generally those sold to visitors, the boxes being occupied by season-ticket holders. However, if one sits in the middle of the main

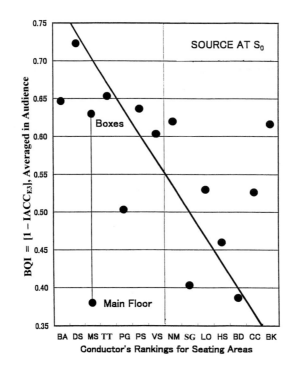

𝒯IGURE 5.3. The Binaural Quality Index (BQI = [1 − IACC$_{E3}$]), averaged for all house positions with the source at S$_o$, is plotted against the names of the halls (see Table 5.1) rank-ordered according to the conductors' ratings of the acoustical qualities for the audience areas. The higher ratings are toward the left end of the abscissa. The Milan, La Scala, opera house was the only one of those shown where the measurements on the main floor were noticeably different from the measurements in the boxes. La Scala is being renovated, to be opened in the Fall of 2004. A spokesperson said (*N.Y. Times*, Dec. 12, 2001) "... the acoustics are not that great now."

floor, the distance to the stage is only 40 ft (12 m) and the direct sound is loud enough to make attendance pleasurable. For this reason, there are few complaints. This intervening distance is the same in the Opéra Garnier.

The major discrepancy between BQI and the conductors' rating is for the Berlin Komischeoper. This hall is circular rather than horseshoe shaped. The result is that the sounds from different parts of the stage are reflected to different parts of the house, meaning that for each listener a singer has a different "hot spot" on stage.

Intimacy, the Initial-Time-Delay Gap (ITDG). The ITDG is generally measured in the center of the main floor, because it is there that the "size" of the house can most easily be estimated by the ear. It is shorter in small houses than in large ones unless, in the latter, special means have been taken to provide early first reflections. The results in Table 5.2 establish that in the best houses, the ITDG is 20 msec or less. In the lowest-rated houses, the ITDG exceeds 30 msec. This measure of intimacy, following BQI, most closely correlates with the ratings of these 21 opera houses.

TABLE 5.2. Initial-time-delay gaps in msec for 19 opera houses for which data are available. The gap is the average of measurements at Positions 101 and 102, near the center of the main floor.

Opera House	ITDG (msec)
WJ Washington, JFK Center, Opera House	15
BS Budapest, Staatsoper	15
PG Paris, Opéra Garnier	15
MS Milan, Teatro alla Scala	16
PS Prague, Staatsoper	16
VS Vienna, Staatsoper	17
BE Budapest, Erkel Theater	17
NM New York, Metropolitan Opera	18
BA Buenos Aires, Teatro Colón	18
BK Berlin, Komischeoper	20
TT Tokyo, New National Theater	20
DS Dresden, Semperoper	20
SO Seattle, Opera House	25
RE Rochester, Eastman Theater	26
TB Tokyo, Bunka Kaikan	26
AM Amsterdam, Music Theater	32
BD Berlin, Deutscheoper	33
HS Hamburg, Staatsoper	34
CC Chicago, Civic Opera House	41

REVERBERATION. From the measured data shown in Table 5.1, it was found that for both the top ten houses and the bottom eleven, the median reverberation time (at mid-frequencies, occupied) is ca. 1.5 sec. It is generally agreed that a reverberation time between 1.4 and 1.6 sec is a good design goal.

STRENGTH FACTORS, G_{MID} AND G_{125}. The strength factors at mid-frequencies (500- and 1,000-Hz frequency bands) are plotted in Figs 5.4 and 5.5. It is observed in Fig. 5.4 that there are appreciable differences in the strengths in the top seven houses. But these can partly be explained. The BA house had a closed set on stage (like a room in a home) during the measurements, hence there was no loss in the fly tower, and the reverberation time was high. DS is a small house. NM and CC are very large houses. MS had a set that communicated easily with the fly tower, so there were losses, and the reverberation time was low. A general conclusion is that in the top half of the houses, except for MS, the levels tend to be higher than in the bottom five.

It was observed in concert halls that the strength of the sound at 125 Hz correlated well with the subjective ratings; that is, the greater the strength at 125 Hz, the higher the rating. For opera houses, this finding is not supported by the data in Fig. 5.5. The two highest-rated and two of the three lowest-rated halls have

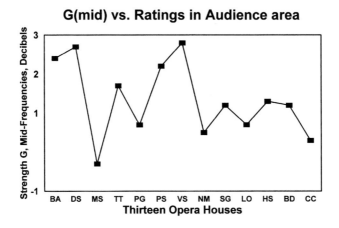

FIGURE 5.4. Plot of the strength factor at mid-frequencies, G_{mid}, in dB versus the ratings in the audience area for 13 houses for which reliable data were available. It would appear that for comfortable singing, the strength factor should be greater than 1 dB. The spread in the data are about 3.5 dB.

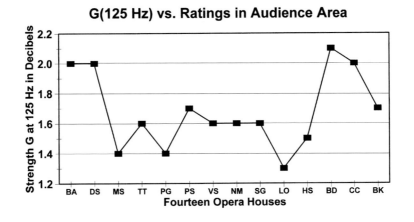

$\mathscr{F}$IGURE 5.5. Plot of the strength factor at 125 Hz, G_{125}, in dB versus the ratings in the audience area for 14 houses for which reliable data were available. There seems little correlation between G_{125} and acoustical quality. However, the spread in the data is only about 0.7 dB, and the accuracy of the data is about 0.5 dB.

G_{125} values of about 2.0 dB. The average of the values in the nine intermediate halls is about 1.5 dB. For the 14 opera houses the G_{125} and the G_{mid} averages (from Table 5.1) are the same, about 1.7 dB. The difference in the average G_{mid} minus G_{125} levels, equal to 0 dB, is smaller than the difference in concert halls, which was found to be about 0.9 dB. The reason, probably, is that materials in the fly tower and the carpets and upholstering of chairs in the audience chamber of opera houses absorb low-frequency sound better than the thinner carpets and more lightly upholstered chairs in concert halls.

The Metropolitan Opera House in New York is the only large house that is ranked in the top half (the tenth) of the 21 houses for which both data and ratings exist. Fredric Dannen, writing in *The New Yorker,* October 3, 1994, speaks of the Met:

> It can also be argued that the Met's most profound artistic problem is beyond [the Artistic Director, James] Levine's control: the sheer size of the theatre. The house at Lincoln Center has thirty-eight hundred seats and one of the largest stages in the United States. It is virtually too big to accommodate operas such as Mozart's *Cosi Fan Tutte* and *La Clemenza di Tito,* and even Rossini's *Il Barbiere di Siviglia.* Ironically, these are the operas that can be better cast today than in [Rudolf] Bing's time, while the opposite

is true of the larger-scale Wagner, Verdi and Puccini works, around which the new house was designed. During Bing's last season at the old Met, in 1965–66, he presented four different Toscas—Renata Tebaldi, Birgit Nilsson, Dorothy Kirsten, and Régine Crespin. Now Tosca is a difficult role to cast at all."

The Met's answer to size is to engage singers with the most powerful voices. It is for this reason that the Met is only as low in ratings as the tenth position.

TEXTURE. Texture is the subjective impression that listeners derive from the patterns in which the sequence of early sound reflections arrives at their ears. In an excellent hall, those reflections that arrive soon after the direct sound follow in a more-or-less uniform sequence. In other halls there may be a considerable interval between the first and the following reflections or one reflection may overly dominate. Good texture requires a large number of early reflections, reasonably strong in amplitude, uniformly but not precisely spaced apart, and with no single reflection dominating the others.

Texture can only be determined visually from "reflectograms" like those shown in Fig. 5.6 and 5.7. The reflectograms for the 21 halls were examined individually and the three best and the three of lower quality are those in those two figures. The others are in between. All were measured with the source of sound at position S_o on stage, and with the microphone at position 102 on the main floor (see Fig. 5.2). The Tokyo house was specially designed to reflect early reflections to all seats and it ranks "very good" in Fig. 5.1. The Buenos Aires, and Vienna houses also rank "very good." Those two are horseshoe in shape and have similar appearances. The early reflections in the houses of Fig. 5.7 are low in amplitude.

A new method for measuring texture is presented in Chapter 4.

ORCHESTRA PITS

For non-Wagnerian opera, the acoustical requirements of an orchestra pit are for clear, undistorted projection of the music into the hall, in good balance and blend, without tonal distortion. In order for the vocalists to sing in good ensemble with the orchestra, they must be able to hear a clear and balanced orchestral sound, so that they can adjust their voice levels properly. The musicians in the pit should be able to hear other sections of the orchestra without the undesirably long time delays that result from too long a pit, and the musicians also need to hear the singers in order to maintain good ensemble. As for visual requirements, it is necessary for the singers and players to be able to see the conductor easily. It is more important that the orchestra players and the singers hear each other than see each other, and this favors the open pit, neither too deep nor overhung. The data in Fig. 5.1 show that in three of the halls, Naples, Prague, and New York, the conductors rated the sound as heard in the pit better than they rated sound of the opera in the audience. Only in Buenos Aires was the sound in the pit rated as high as that in Naples and New York.

For Wagnerian opera, on the other hand, the creation of a "mystical" sound by an "invisible" orchestra was an important element in the composer's dramatic conception, at least for his later works. This requirement led Wagner to develop the sunken, covered pit whose acoustical behavior is quite different from the open pit (Chapter 3, No. 54). Some pits combine features of both of these plans, but generally with only moderate success.

Types of Orchestra Pits

For purposes of discussion, orchestra pits are classified in three ways:

1. Open pit (e.g., Vienna Staatsoper, No. 32, in Chap. 3)
2. Sunken pit, covered (e.g., Bayreuth Festspielhaus, No. 54)
3. Sunken pit, open (e.g., Eastman Theatre in Rochester, No. 19)

OPEN PIT. The first objection to an entirely open pit is the disturbing visual effect of the lights from the conductor's and orchestra's music stands on the audience in the upper rings. A second objection is that, in order to accommodate a large orchestra of 80 to 110 pieces, a fully open pit must be 22 to 30 ft (6.7 to 9.1 m) wide, measured from the stage to the railing on the audience side at the centerline. A sizable gulf is thereby created between the singers and the audience. The pit depth is usually between 8 and 11.5 ft (2.5 to 3.5 m). Some conductors prefer a shallow pit so that the audience can see them easily. Some singers say that they can hear the orchestra best without the feeling that it is interfering with their singing when the curves of the double basses are one foot (30 cm) below the level of the stage.

A slight overhang of the pit by the forestage is not objectionable acoustically and has the advantage of increasing the reflecting area of the stage between the singers and the audience, a very desirable feature in a large hall. An overhang of about 3 ft (1 m) is found in many of the German opera theaters built since World War II and in the present Metropolitan Opera House. La Scala in Milan has a sliding forestage with a minimum overhang of about 3.5 ft (1.1 m) and a maximum overhang of about 12 ft (3.7 m). Several of the conductors who were interviewed expressed themselves strongly against an overhang at La Scala of more than 3.5 ft.

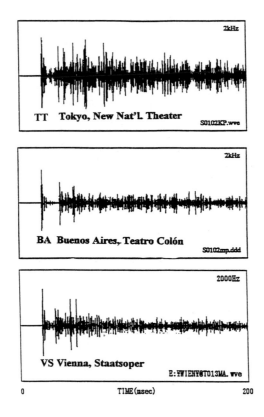

$\mathcal{F}$IGURE 5.6. Reflectograms. Each reflectogram is the result of an impulse sound produced by a loudspeaker on stage, recorded at a microphone position in the auditorium. The loudspeaker was at S_o, the microphone at 102 (See Fig. 5.2) and the frequency was 2,000 Hz. A reflectogram is a plot against time showing the direct sound at the left and the succession of reflections that follow. These three reflectograms were chosen by visual judgments of the "texture," and represent those of the 22 houses that were judged best. The initial-time-delay gaps are short and there are a sizable number of reflections in the first 80 msec after arrival of the direct sound.

The pit in the Metropolitan Opera House has received high praise from all those interviewed. Its width (fore-aft) at the center is 25 ft (6.6 m) and its length is about 70 ft (21 m). Its area is about 1,420 ft² (132 m²), which for 90 musicians is 15.8 m² per musician, an acceptable figure. There are surfaces overhead that reflect the sound of the musicians back to themselves. Some opera houses have placed reflecting surfaces at the top and inner (stage) side of the proscenium, that reflect singers' voices to the orchestra and vice versa.

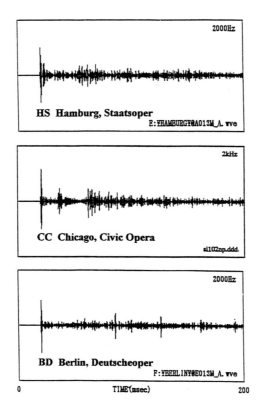

HS Hamburg, Staatsoper

E:¥HAMBURG¥@A013M_A. ¥ve

2kHz

CC Chicago, Civic Opera

sl102np.ddd.

2000Hz

BD Berlin, Deutscheoper

F:¥BERLIN¥@E013M_A. ¥ve

0   TIME(msec)   200

𝓕IGURE 5.7. Same as Fig. 5.6, except for the three houses for which the visual judgments indicated the "texture" to be the least good among the 22 houses.

Recent measurements in 12 opera houses (fully occupied) show that the sound from the orchestra in the pit that reaches the audience has a strength factor G_{mid} equal to about 0.4 dB with the curtain open, which is about 1.5 dB lower than the sound from position S_o on the stage. Only in the Vienna Staatsoper is the sound from the pit greater than that from the stage, by about 1.5 dB. The author has sat in the Vienna pit (second violin section, second stand, next to the pit railing) during an opera and saw and heard nothing unusual. The probable explanation is that the ceiling above the pit is shaped to reflect pit sounds to the main floor, while this ceiling surface reflects singers' voices into the lower boxes.

SUNKEN PIT, COVERED. The antithesis of the open pit is the covered pit designed by Wagner for the Festspielhaus in Bayreuth, Germany. It is partly buried beneath the stage and the remaining portion is nearly completely covered by an overhang. The Bayreuth pit is used with a very large orchestra in a hall that has a relatively long reverberation time, 1.55 sec. Except for Wagner's music, the Bayreuth pit is not considered satisfactory because the string tones are muffled and the orchestra takes on an eerie sound.

SUNKEN PIT, OPEN. In a sunken pit, the floor area is usually at least twice as great as the open area. To the best of my knowledge, no existing pit that is in large part buried under a platform yields the natural kind of sound heard in the opera houses of Vienna, Buenos Aires, Naples, and Prague.

BOXES AND BALCONIES

BOXES. The earliest opera houses already featured boxes, for example, the Teatro di San Cassiano that opened in Venice in 1637. It was said that they were designed to show off their inhabitants, particularly the jewelry and dress of the female attendees, while the men sat in the seats behind the opening or retreated to the bars. A photograph that I have of La Scala, in Milan, during a gala performance shows more men than that statement indicates, but there are more women framed in the boxes at the lower levels.

The construction of the 155 boxes in La Scala is shown in Fig. 5.8, each with six chairs, or upholstered stools. The box openings constitute only 43% of the surface of the horseshoe, thus reflecting considerable sound to the center space, producing more lively acoustics for those sitting on the main floor and for those at the box openings. Visually, some opera house boxes have a mirror on the front part of the wall farthest from the stage. It is obvious that those sitting back in the boxes on the sides of the horseshoe, particularly near the proscenium, would probably enjoy the opera more at home on TV.

The boxes in the Royal Opera House in London are more hospitable to the music (see Fig. 5.9). Not only are they more spacious, but the openings are larger and the partitions do not extend to the front of the box, making the view better.

In the Philadelphia Academy of Music (No. 17 in Chap. 3), the partitions between boxes have been removed, so that everybody receives nearly the same music, and there is no obstruction to the view. In the Philadelphia Academy the box openings occupy about 75% of the sidewall area.

BALCONIES. Reverberation is less important in an opera house than in a concert hall. Thus, balcony overhangs are less of a problem, because listeners' attention is focused more on the intelligibility of the voice. Two general recommendations for balcony overhangs in opera houses are illustrated in Fig. 5.10.

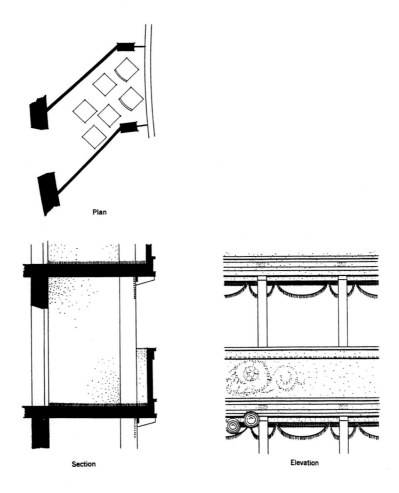

Plan

Section

Elevation

*F*IGURE 5.8. Boxes in La Scala opera house in Milan. La Scala is undergoing reconstruction and the boxes may be altered.

1. As a general rule, the distance D should not exceed twice the height H. In addition, the balcony soffit and the rear and sidewalls should be shaped to reflect the direct and early lateral reflections to the heads of the listeners.
2. As a more sensitive measure than the D over H ratio, the angle θ should preferably be 25 degrees or greater.

Good and bad designs of balconies are shown in Figs. 4.26 and 4.27 of Chapter 4.

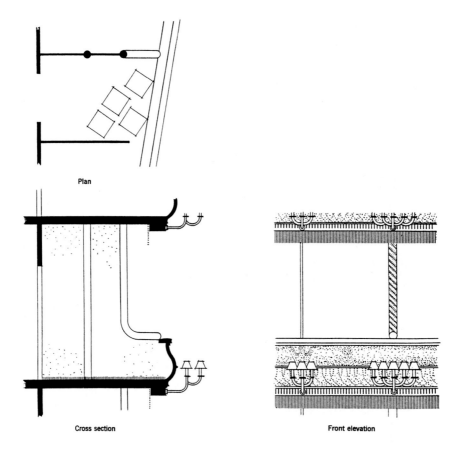

Plan

Cross section

Front elevation

FIGURE 5.9. Boxes in the Royal Opera House in London.

ECHO AND DISTORTION

ECHO. In opera houses, a small "echo" heard on the stage by a singer may be desirable, while that same "echo" heard on the stage of a concert hall might be mildly troublesome to the players. Soloists in an opera house particularly want to hear that they are "filling the hall." They desire acoustic feedback, if it is at the right strength, because their voices are directed outwards more than would be true for string instruments on the stage. A soloist stands near the front of the stage where he or she gets little audible support from surfaces in front of or behind the proscenium.

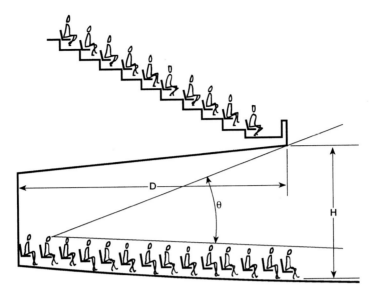

FIGURE 5.10. Suggested rules governing acceptable balcony overhangs. (1) The distance D should not exceed twice H, or, more stringently, the angle θ should not be less than 25 degrees.

In opera houses some acoustical experts agree that some degree of adjustability of the sound reflected from surfaces at or near the rear of the hall should be provided. For example, the rear wall on the main floor, which otherwise may return too strong a reflection, can be "toned down," by adding sound-absorbing materials to it or by tilting it backward. In thickness, the sound-absorbing material can be between 0.75 and 1.0 in. (1.9 to 2.5 cm) and it should be placed behind a removable, acoustically transparent screen so that varying amounts of it can easily be added or subtracted. That thickness of material will absorb sound at frequencies of 500 Hz and higher. If the decision is to tilt a hard rear wall backward, the sound absorbing material must be placed on the underside of the balcony above (soffit) so that the wave reflected from the wall is reduced in magnitude on reflection before it travels back to the stage.

How much sound energy should be returned to the stage to accomplish desirable feedback to singers without creating an annoying echo? Since no means has been devised to predict this amount accurately, the sound-absorbing material should be added incrementally to the wall or soffit until the correct amount of "toning down" has been accomplished. The procedure of tilting the rear wall backward and adding sound-absorbing material to the soffit of the balcony above was followed in the design of the New National Theatre Opera House in Tokyo (Beranek *et al.*, 2000). Male and female singers moved about the stage and decided on the amount of sound-absorbing material that suited their desire for feedback from the house.

Nakamura *et al.* (1991) determined that for a delay time of 140 msec (reflection from a surface about 80 ft. (24 m.) distant, the level of the frontal reflection should lay 25–35 dB below the level of the direct sound (both measured with a microphone 1 m in front of the singer). Okano (1994) performed somewhat the same experiment with a group of soprano singers and asked them for their judgment of the tolerance level of the wall reflection. For delay times of 120, 170, and 220 msec (reflecting surface 68, 95, and 124 ft [21, 29, and 38 m] distant, respectively) the most sensitive of the vocalists reported that the reflected levels (measured 1 m. in front of the singer) should be, respectively, 36, 38, and 42 dB below the direct sound.

DISTORTION. There are three types of distortion to be guarded against: (1) selective sound absorption by some surface or by the seats; (2) focusing of parts of the performance to certain parts of the house; and (3) rattles or sympathetic resonances from some objects or grilles.

The actual seats should always be tested in a reverberation chamber before installation. Protective sprays applied to the underside of the fabric upholstery covering should not be permitted, because with it, the seats would reflect much more sound when unoccupied than with audience present and make rehearsals difficult. The tests should clearly show that there is no "bubble" resonance of the sitting part, which can occur at frequencies below 300 Hz.

Focusing can occur owing to the shape of the room. Some focusing effects are hard to find from drawings of the hall, and a model is recommended for this purpose. In the Tokyo Dai-ichi Seimei Hall, even the model test failed to show a narrow line of focusing from the rear sidewalls of this oval-shaped hall. This focusing effect was only heard along a strip about 1 ft (30 cm) wide down the centerline of the hall for a sound source located anywhere on the centerline, either on stage or in the audience. The application of some sound-absorbing material on two surfaces near the rear of the hall removed the echo (Hall No. 76 in Chapter 3).

Rattles and resonances occur occasionally, usually from grillework in front of ventilation ducts. When this occurs, the grille components have to be vibration-dampened to remove the resonance.

CONCLUDING REMARKS

Opera houses are no simpler to design than concert halls. Actually they are more complicated, because they involve both an orchestra in one location and singers in another. The problems are different. It is important in an opera house to obtain a high value of binaural quality index (BQI), a short initial-time-delay gap (ITDG), and good reflection patterns for sources on the stage like those in Fig. 5.6.

There should be reflecting surfaces near the proscenium (audience side), as are found in the Tokyo, New National Theater Opera House (No. 79 in Chapter 3), or as a minimum, those found in the Teatro Colón (No. 27 in Chapter 3). The balcony fronts and the ceiling surfaces must be shaped to send early reflections uniformly to all parts of the audience area. To achieve the optimum results, a 10:1 wooden acoustic model of the audience chamber is invaluable. A qualified acoustical expert should be engaged to plan the basic shaping of the room and the reverberation time, to make the model tests, and to tend to all of the details that lead to good texture and a high quality of sound.

1

Terminology, Definitions, and Conversion Factors

$\mathcal{D}$IMENSIONAL QUANTITIES IN CHAPTER 3 AND APPENDIX 2

In Chapter 3 and the tables of Appendix 2 various terms are used for the architectural and acoustical parameters of concert halls and opera houses. The meanings of those terms as used in that chapter and elsewhere in the text are as follows:

N = number of seats in the hall (usually wheelchair space is not counted).

V = Volume of the hall in cubic feet (cubic meters).

In *concert halls,* V includes the volume of air in the main hall and in the orchestra enclosure. If there is a stagehouse, the volume V does not include that volume of the stagehouse that lies outside the orchestra enclosure unless the construction is like that of Hall No. 10 in Chapter 3. V is measured as though there were no seats in the hall. The volume occupied by the solid balcony structures is not included.

In *opera houses,* V includes the volume of air contained in the house forward of the main curtain. It does not include the volume of air in the stagehouse or the volume occupied by the solid balcony structures.

S_a = area of floor space over which the audience chairs are located. The seating areas given in this book are projected areas. That is to say, in sloped (raked) floors, such as in balconies, the slant area is not measured. Instead, the area is that shown on the drawing.

S_A = acoustical audience area. It includes the sum of (a) the area S_a (see above) and (b) the areas of strips 20 in. (0.5 m) wide around the separated blocks of the seating area, except that such strips are neither included at the front edge of a balcony where the audience is seated against a balcony rail nor where the seats abut a wall. In this book, the range of S_A/S_a for the 100 halls is 1.1 to 1.5. The average for European and American halls built since 1975 is 1.25. The largest ratios are often found in Japanese halls, where the building codes require more aisles.

In boxes, the same rule applies, unless the seats are far apart or the box is much larger than the seating area S_a. In that case, use S_A (per chair) = 7.5 ft^2

(0.7 m²). However, the number of seats in the box times 7.5 (or 0.7) must not exceed the area of the floor of the box.

S_c = area occupied by the chorus, or by the audience if seated in the chorus area. It is not considered as part of audience area if closed off or unused during non-choral performances.

S_o = area of stage. When the stage area exceeds 1,940 ft² (180 m²), S_o is limited to that value. This numerical area is deemed the acoustical area of a 100-piece orchestra. No side strips are added.

S_{pit} = area of the open surface of the pit.

S_p = area of the proscenium curtain. It is assumed that the fire curtain is pulled up during the acoustical measurements and that the proscenium curtain is present and sound absorbent. The reason for needing this area is for calculating the reverberation of the audience space without including the volume and absorption of the fly (scenery) tower.

$S_T = S_A + S_o \, (+S_c)$ (for concert halls). The acoustical absorption area S_T is used in formulas for calculating reverberation times. It includes the surfaces that are highly sound absorbent. The other areas—namely, ceilings, sidewalls, and those floor areas that are not included in S_T—usually contribute between 15 and 25 percent of the total sound absorption. S_c is included when it is normally used for audience seating.

$S_T = S_A + S_{pit} + S_p$ (for opera houses). The proscenium area is included when the main performance curtain is lowered. If the measurements are made without lowering the curtain, the fly (scenery) tower will affect the measured reverberation times in unknown ways.

H = average room height, measured from main floor to ceiling in that part of the main-floor audience area not covered by balconies. This height is needed to determine the time delay of the first ceiling reflection, which may be long relative to the arrival of the direct sound when the reflection goes to a main floor seat and much shorter when it goes to a balcony seat.

W = average width, measured between sidewalls in the audience area on the main floor, disregarding any balcony overhang. This width is an indication of the "intimacy" of the hall. The first reflection after the direct sound that reaches a listener on the main floor may come from a balcony face. However, most side balconies are only a few rows deep, so that the average width between sidewalls is a more general indication of the "intimacy" factor.

L = average room length, measured from the stage front to the average of the back wall positions at all levels. This length is a general indication of the magnitude of the fall-off in loudness with distance from the stage. The maximum fall-off is determined by D.

It is not intended that $H \times W \times L$ should be the exact cubic volume of the auditorium.

D = distance from the front of the stage to the most remote listener, measured on the centerline, unless the rear wall is not flat or curved outward, in which case a more suitable location on the rear wall must be chosen.

SD = average stage depth.

SW = average stage width

SH = mean ceiling height above the stage area, measured relative to the front of the stage.

$\mathcal{A}$COUSTICAL PARAMETERS IN APPENDIX 2

Appendix 2 contains a detailed listing of the data obtained from acoustical measurements in concert halls and opera houses as outlined in Chapter 2. The terminology and definitions of the parameters are covered here.

RT = reverberation time in sec. It is defined as the time, multiplied by a factor of 2, that it takes for the sound in a hall to decay from -5 to -35 dB below its steady-state value. The factor of 2 is necessary because RT must conform to the original definition of sound decay that was from 0 to -60 dB. RT is usually measured in octave or one-third octave bands and the sound source is usually a pink (random) noise or a sound impulse. Originally, RT was determined from a plot of sound pressure level vs. time as recorded on the moving paper of a graphic level recorder. Today it is determined by the Schroeder (1965) method which involves computer integration of the equivalent of a backward-played tape-recording of the decaying signal. The mid-frequency RT is the average of the RT's at 500 and 1,000 Hz. The measurement is generally made in both occupied and unoccupied halls, the former being more important because it is the sound field in which a listener actually participates. When time permits, measurements are made for several sound source positions on the stage (and the pit in an opera house) and at 16 to 24 positions in a hall. When a hall is symmetrical the number of positions may be halved and the measurements made in only half of the hall. The data at each frequency for the various positions are generally averaged to obtain an average value of the RT for the hall.

EDT = early-decay-time in sec. EDT is measured in the same fashion as RT except that it is the time it takes for a signal to decay from 0 to -10 dB relative to its steady state value. A multiplying factor of 6 is necessary to make EDT time comparable to RT. There are a limited number of halls in Appendix 2 for which EDT was measured with full occupancy.

$IACC_A$ = measure of the difference in the sounds arriving at the two ears of a listener facing the performing entity in a hall. It is called the "interaural cross-correlation coefficient." For the data reported in this book the source of sound is an omnidirectional (dodecahedron) loudspeaker fed by an impulse sound. To measure IACC a digital recorder is connected to the outputs of two tiny microphones located at the entrances to the ear canals of a person or a dummy head, and quantifying the two ear differences with a computer program that performs the operations of Eqs. (A3.1) and (A3.2) of Appendix 3. $IACC_A$ is determined with a frequency bandwidth of about 100 to 8,000 Hz and for a time period of 0 to about 1 sec. No frequency weighting is used. There are a limited number of halls in Appendix 2 for which the IACC's were measured with full occupancy.

IACC (bands) = the same as above except determined in each of the six frequency bands with mid-frequencies of 125, 250, 500, 1,000, 2,000, and 4,000 Hz. The time periods and band-combinations are selected as follows:

$IACC_{E3}$ = the interaural cross-correlation coefficient determined for a time period of 0 to 80 msec, where 0 msec is the time at which the direct impulse sound from the omnidirectional source reaches the tiny microphones. It is the average of the values measured in the three octave bands with mid-frequencies of 500, 1,000, and 2,000 Hz. It is demonstrated in Chapter 4 to be fairly highly correlated with the subjective ratings of acoustical quality as expressed by qualified judges seated in the audience. A particularly favorable feature of $IACC_{E3}$ is that its measured values are only 8 to 10 percent lower in fully occupied halls than its values measured in unoccupied halls.

BQI = binaural quality index. It equals $[1 - IACC_{E3}]$.

$IACC_{L3}$ = the interaural cross-correlation coefficient determined for a time period of 80 to 750 msec; the average of the values measured in the three frequency bands with mid-frequencies of 500, 1,000, and 2,000 Hz.

$C_{80}(3)$ = the clarity factor. It is the ratio, expressed in decibels, of the energy in the first 80 msec of an impulse sound arriving at a listener's position divided by the energy in the sound after 80 msec. The divisor is approximately the total energy of the reverberant sound. The symbol (3) indicates the average of the C_{80} values in the 500-, 1,000-, and 2,000-Hz bands. C_{80} is given by Eq. (A3.3).

D = the distinctness (Deutlichkeit) ratio. It is the ratio of the sound energy in the first 50 msec after arrival of the direct sound at a listener's position to the total sound energy arriving. It is usually expressed as a percentage and is determined from the impulse response of the hall. This factor is not used in this text. D is given by Eq. (A3.9) in Appendix 3.

LF_{E4} = the lateral energy fraction. It is determined by the ratio of the output of a figure-8 microphone with its null axis pointed to the source of the sound, divided by the output of a non-directional microphone at the same position. LF_{E4} covers the time period of 0 to 80 msec and is the average of the LF's in the four frequency bands—125, 250, 500, and 1,000 Hz. It is equal to the ratio of the energy in the sound at a listener's position that does not come from the direction of the source (nor from 180° behind the listener) to that which comes from all directions including that of the source. LF_{E4} correlates with the "apparent source width ASW", as determined in laboratory measurements. It is given by Eq. (A3.4).

LG = the late lateral energy. It is the ratio, expressed in decibels, of (a) the output of a figure-8 microphone(with its null direction aimed at the source) in the time range of 80 msec to 2 sec after the arrival of the direct sound at a listener's position in a hall, to (b) the total output of a non-directional microphone, where its output is measured in an anechoic chamber at a distance of 10 m from the acoustical center of an omnidirectional source operating at the same acoustical power output as was used for determining (a). There is no published body of data to demonstrate its usefulness as a measure of sound diffusivity or as an additional parameter for use in estimating the acoustical quality of a hall for music. LG is given by Eq. (A3.8) in Appendix 3.

G = the "strength factor." It is the ratio, expressed in decibels, of the sound energy at a seat in a hall that comes from a non-directional source (usually located successively at one to three different positions on the stage) to the sound energy from the same source when measured in an anechoic room at a distance of 10 m. G is measured in the usual six-octave-frequency bands. In Appendix 2, it is the average of the G-values at all seats in a hall at which data were taken and all source positions. It is given by Eq. (A3.5).

G_{mid} = the mid-frequency strength factor, in decibels. This means that the decibel levels are the average of the G's measured in the 500- and 1,000-Hz frequency bands.

G_{low} = the low-frequency strength factor, in decibels. Here, the decibel levels are the average of the G's measured in the 125- and 250-Hz frequency bands.

G_{125} = the value of G measured in the 125-Hz frequency band.

BR = the "bass ratio" equal to the ratio of the average of the RT's at 125 and 250 Hz to the average of the RT's at 500 and 1,000 Hz. It is determined only for a hall when fully occupied. It is given by Eq. (A3.6).

ITDG = the "initial-time-delay gap," the time interval in msec between the arrival, at a seat in the hall, of the direct sound from a source on stage to the arrival of the first reflection (see Fig. 2.2 in Chapter 2). In this text, its value is generally given only for a position near the center of the main floor. It correlates with the subjective impression of "intimacy."

ST1 = the degree of support that the hall gives to the players on stage, owing to reflections of the sound from the walls and ceiling of the hall and of the enclosure immediately surrounding the players. It is the difference, in decibels, between (a) the impulse sound energy from an omnidirectional sound source that arrives at a player's position within the first 10 msec, measured at a distance of 1 m from the sound source, and (b) that which arrives in the time interval between 20 and 100 msec at the same position. The sound arriving in the later interval has been reflected from one or more surfaces back to the player's position on the stage in the time interval indicated. That is to say, the difference between (a) and (b) is a measure of the hall's acoustical support to a musician's performance. The measurements are generally made with no musicians on the stage, but with the chairs, music stands, and the percussion instruments in place. Those items that are near the source and receiver are set aside. The measurements are made at several positions on the stage and the data are averaged. ST1 is given by Eq. (A.3.7) in Appendix 3.

TABLE A1. Conversion Factors

To Convert	Into	Multiply by	Conversely Multiply by
inches (in.)	centimeters (cm)	2.54	0.394
	feet (ft)	0.0833	12
	meters (m)	0.0254	39.4
feet (ft)	centimeters (cm)	30.5	$3.28 \text{ times } 10^{-2}$
	meters (m)	0.305	3.28
square feet (ft²)	square centimeters (cm²)	929	$1.076 \text{ times } 10^{-3}$
	square meters (m²)	0.0929	10.76
cubic feet (ft³)	cubic centimeters (cm³)	$2.83 \text{ times } 10^4$	$3.53 \text{ times } 10^{-5}$
	cubic meters (m³)	0.0283	35.3
pounds (lb)	ounces (oz)	16	0.0625
	kilograms (kg)	0.454	2.205
yards (yd)	inches (in.)	36	0.0278
	meters (m)	0.914	1.094
pounds per square foot (lb/ft²)	kilograms per square meter (kg/m²)	4.88	0.2048
pounds per cubic foot (lb/ft³)	kilograms per cubic meter (kg/m³)	16.0	$6.24 \text{ times } 10^{-2}$

NOTE: $10^{-2} = 0.01$; $10^{-3} = 0.001$; $10^{-4} = 0.0001$; $10^{-5} = 0.00001$; $10^2 = 100$; $10^3 = 1,000$; $10^4 = 10,000$

2

Acoustical Data for Concert Halls & Opera Houses

Sequential Index

Sequential Index, *continued*

$\mathscr{T}$**ABLE A2.** Measured acoustical attributes of concert halls and opera houses (s.c. = stop chords; b.r. = before renovations)

Attribute	Measured by	Year of Data	Center Frequencies of Filter Bands					
			125	250	500	1,000	2,000	4,000
			Hertz					
1. ASPEN, COLORADO, BENEDICT MUSIC TENT (Opened 2000, 2,050 seats)								
RT, unoccupied	Kirkegaard							
	Louvers Open	2000	3.20	3.20	3.40	2.90	2.10	1.55
	Louvers Closed	2000	3.30	3.30	3.60	3.15	2.50	1.80
2. BALTIMORE, JOSEPH MEYERHOFF SYMPHONY HALL (Opened 1982, 2,467 seats)								
RT, unoccupied	Takenaka	1991	2.67	2.36	2.31	2.31	2.12	1.82
	Bradley	1992	2.85	2.44	2.40	2.37	1.95	1.54
	Gade	1992	2.83	2.44	2.36	2.32	1.98	1.58
	Average		2.78	2.41	2.36	2.33	2.02	1.65
RT, occupied	Calc.: JASA **109**: 1028	2001	2.30	2.10	2.00	2.00	1.65	1.35
EDT, unoccupied	Takenaka	1991	2.61	2.41	2.38	2.31	2.07	1.71
	Bradley	1992	2.55	2.35	2.28	2.27	1.84	1.39
	Gade	1992	2.63	2.38	2.30	2.31	1.95	1.46
	Average		2.60	2.38	2.32	2.30	1.95	1.52
$IACC_A$, unoccupied	Takenaka	1991	0.93	0.76	0.27	0.19	0.20	0.16
	Bradley	1992	0.94	0.73	0.26	0.21	0.22	0.16
	Average		0.94	0.75	0.27	0.20	0.21	0.16
$IACC_E$, unoccupied	Takenaka	1991	0.95	0.84	0.51	0.45	0.44	0.32
	Bradley	1992	0.96	0.83	0.51	0.45	0.38	0.28
	Average		0.96	0.84	0.51	0.45	0.41	0.30
$IACC_L$, unoccupied	Takenaka	1991	0.91	0.70	0.18	0.10	0.07	0.06
	Bradley	1992	0.93	0.70	0.18	0.09	0.11	0.09
	Average		0.92	0.70	0.18	0.10	0.09	0.08
LF, unoccupied	Bradley	1992	0.18	0.15	0.13	0.15	0.18	0.13
	Gade	1992	0.16	0.20	0.20	0.18	0.18	0.18
	Average		0.17	0.18	0.17	0.17	0.18	0.16
C_{80}, dB, unoccupied	Takenaka	1991	−4.10	−3.90	−2.80	−1.90	−1.30	−0.10
	Bradley	1992	−3.90	−3.20	−2.00	−1.30	−0.20	0.80
	Gade	1992	−2.05	−2.83	−2.13	−1.88	−1.17	0.36
	Average		−3.35	−3.31	−2.31	−1.69	−0.89	0.35
G, dB, unoccupied	Takenaka	1991	4.30	3.60	4.60	4.80	4.90	4.80
	Bradley	1992	2.71	2.77	3.73	3.22	1.43	−0.95

Attribute	Measured by	Year of Data	Center Frequencies of Filter Bands					
			125	250	500	1,000	2,000	4,000
			Hertz					

3. BOSTON, SYMPHONY HALL (Opened 1900, 2,625 seats)

Attribute	Measured by	Year of Data	125	250	500	1,000	2,000	4,000
RT, unoccupied	M.A.A.	1958	2.50	2.80	2.70	2.85	2.85	2.10
	Schultz	1963	2.70	2.70	2.60	2.50	2.30	1.95
	Mastracco	1981	2.60	2.40	2.70	3.00	2.80	
	Average before 1982		2.60	2.63	2.67	2.78	2.65	2.03
	Takenaka	1991	2.11	2.19	2.33	2.69	2.80	2.47
	Bradley	1992	2.17	2.38	2.56	2.70	2.67	2.38
	Gade	1992	2.12	2.30	2.39	2.56	2.59	2.26
	Selected average after 1988		2.13	2.29	2.40	2.63	2.66	2.38
RT, occupied	Hidaka/Beranek	1992	1.80	1.90	1.80	1.80	1.70	1.40
	Griesinger/Kirkegaard	1993	2.10	1.80	1.90	1.90	1.60	1.20
	Beranek (s.c.)	1997	1.95	1.90	1.90	1.95	1.59	1.43
EDT, unoccupied	Takenaka	1991	2.13	2.23	2.28	2.68	2.85	2.41
	Bradley	1992	2.00	2.08	2.19	2.38	2.43	2.14
	Gade	1992	2.00	2.12	2.26	2.45	2.57	2.14
	Average		2.04	2.14	2.24	2.50	2.62	2.23
$IACC_A$, unoccupied	Bradley	1992	0.94	0.76	0.27	0.24	0.21	0.22
	Takenaka	1991	0.91	0.73	0.29	0.14	0.11	0.11
	Average		0.91	0.74	0.28	0.19	0.16	0.17
$IACC_E$, unoccupied	Bradley	1992	0.93	0.82	0.40	0.41	0.43	0.43
	Takenaka	1991	0.95	0.82	0.49	0.30	0.27	0.23
	Average		0.94	0.82	0.45	0.36	0.35	0.33
$IACC_L$, unoccupied	Takenaka	1991	0.89	0.70	0.23	0.11	0.07	0.06
	Bradley	1992	0.93	0.74	0.25	0.14	0.07	0.07
LF, unoccupied	Bradley	1992	0.23	0.19	0.17	0.22	0.25	0.18
	Gade	1992	0.15	0.25	0.28	0.25	0.24	0.25
	Average		0.19	0.22	0.23	0.24	0.25	0.22
C_{80}, dB, unoccupied	Takenaka	1991	−2.90	−3.50	−3.10	−3.40	−3.80	−2.90
	Bradley	1992	−2.87	−2.46	−2.25	−1.98	−1.91	−2.00
	Gade	1992	−1.50	−1.92	−2.94	−2.18	−3.21	−2.04
	Average		−2.42	−2.63	−2.76	−2.52	−2.97	−2.31
G, dB, unoccupied	Takenaka	1991	2.90	3.60	5.00	5.80	6.50	6.50
	Bradley	1992	1.43	2.61	3.70	4.27	3.70	2.34

4. BUFFALO, KLEINHANS MUSIC HALL (Opened 1940, 2,839 seats)

Attribute	Measured by	Year of Data	125	250	500	1,000	2,000	4,000
RT, unoccupied	Takenaka	1998	2.50	2.06	1.94	1.71	1.69	1.65
	Bradley	1992	2.60	2.01	1.95	1.70	1.60	1.49
	Gade	1992	2.46	2.02	1.91	1.68	1.63	1.50
	Average		2.52	2.03	1.93	1.70	1.64	1.55
RT, occupied	Calc.: JASA **109**: 1028	2001	2.15	1.65	1.60	1.40	1.27	1.23
EDT, unoccupied	Bradley	1992	1.92	1.78	1.72	1.39	1.26	1.00
	Gade	1992	1.97	1.78	1.85	1.42	1.37	1.06
	Takenaka	1998	1.94	1.88	1.83	1.46	1.44	1.15

Attribute	Measured by	Year of Data	Center Frequencies of Filter Bands					
			125	250	500	1,000	2,000	4,000
			Hertz					

4. BUFFALO, KLEINHANS MUSIC HALL, *continued*

Attribute	Measured by	Year	125	250	500	1,000	2,000	4,000
$IACC_A$, unoccupied	Bradley	1992	0.96	0.85	0.60	0.51	0.27	0.30
$IACC_E$, unoccupied	Bradley	1992	0.97	0.90	0.79	0.61	0.37	0.38
$IACC_L$, unoccupied	Bradley	1992	0.95	0.80	0.40	0.30	0.12	0.12
LF, unoccupied	Bradley	1992	0.12	0.09	0.08	0.10	0.12	0.07
	Gade	1992	0.10	0.11	0.10	0.09	0.09	0.08
C_{80}, dB, unoccupied	Bradley	1992	−0.91	1.10	1.99	3.59	3.73	4.73
	Gade	1992	0.49	1.60	1.62	3.85	3.57	4.67
	Takenaka	1998	−0.60	−0.40	0.80	3.50	3.70	4.30
G, dB, unoccupied	Bradley	1992	3.34	2.44	2.72	1.83	0.45	−1.08
	Gade	1992	4.83	3.39	2.49	2.78	2.58	−0.23
	Takenaka	1998	4.70	4.60	4.50	4.10	4.00	3.40

5. CHICAGO, ORCHESTRA HALL (Opened 1904; renovated 1997; 2,530 seats)

Attribute	Measured by	Year	125	250	500	1,000	2,000	4,000
RT, unoccupied	Before renovation: Kirkegaard	1994	1.73	1.63	1.46	1.45	1.32	1.01
	After renovation: Kirkegaard	1998	2.70	2.20	2.00	1.80	1.70	1.50
RT, occupied	Before renovation: Kirkegaard	1994	1.44	1.29	1.21	1.16	1.10	0.98
	After renovation: Beranek (s.c.)	1999	2.15	2.00	1.80	1.65	1.55	1.30

6. CLEVELAND, SEVERANCE HALL (Opened 1931; stage house renovated 2000; 2,101 seats)

Attribute	Measured by	Year	125	250	500	1,000	2,000	4,000
RT, unoccupied	Before renovation: Various	1987	2.00	1.80	1.80	1.75	1.60	1.50
	After renovation: Jaffe, Holden	2000	2.20	2.00	2.10	2.20	2.10	1.40
RT, occupied	Before renovation: Griesinger	1995	1.70	1.60	1.50	1.45	1.35	1.30
	After renovation: Beranek (s.c.)	2000	1.75	—	1.65	1.55	1.40	1.30
EDT, unoccupied	Before renovation: Bradley/Gade	1992	1.70	1.70	1.70	1.70	1.65	1.45
	After renovation: Jaffe, Holden	2000	2.50	2.10	2.40	2.10	1.90	1.50

7. COSTA MESA, SEGERSTROM HALL, ORANGE COUNTY PERFORMING ARTS CENTER
   (Opened 1986, 2,903 seats)

Attribute	Measured by	Year	125	250	500	1,000	2,000	4,000
RT, unoccupied	Takenaka	1991	2.27	2.14	2.14	2.27	2.10	1.81
	Hyde	1993	2.52	2.25	2.27	2.46	2.26	1.97
	Barron	1986	2.43	2.37	2.41	2.47	2.27	—
	Average, Barron and Hyde		2.48	2.31	2.34	2.47	2.27	1.97

Attribute	Measured by	Year of Data	\multicolumn{6}{c}{Center Frequencies of Filter Bands}
			125	250	500	1,000	2,000	4,000
			\multicolumn{6}{c}{Hertz}

7. COSTA MESA, SEGERSTROM HALL, ORANGE COUNTY PERFORMING ARTS CENTER, *continued*

Attribute	Measured by	Year	125	250	500	1,000	2,000	4,000
RT, occupied	Takenaka & Hyde	1995	2.33	1.89	1.62	1.57	1.44	1.16
EDT, unoccupied	Takenaka, all seats	1991	2.16	2.02	2.01	2.01	1.90	1.56
	Hyde, all seats	1993–94	2.28	2.10	2.10	2.22	1.97	1.58
	Takenaka, w/o under balc. seats	1991	2.19	2.09	2.07	2.06	1.95	1.59
	Hyde, w/o underbalcony seats	1993–94	2.32	2.14	2.14	2.28	2.05	1.66
	Average, all seats		2.22	2.06	2.06	2.14	1.96	1.58
EDT, occupied	Takenaka & Hyde	1995	2.67	2.43	2.25	2.12	1.86	1.84
$IACC_A$, unoccupied	Takenaka	1991	0.90	0.71	0.26	0.20	0.16	0.19
$IACC_E$, unoccupied	Takenaka	1991	0.92	0.80	0.47	0.37	0.31	0.30
$IACC_L$, unoccupied	Takenaka	1991	0.88	0.64	0.19	0.10	0.09	0.07
LF, unoccupied	Barron	1986	0.27	0.19	0.23	0.22	—	—
C_{80}, dB, unoccupied	Barron	1986	−3.70	−1.90	−0.70	−0.40	1.20	—
	Takenaka	1991	−3.10	−1.60	−0.90	−0.50	−0.20	0.60
	Hyde	1993	−3.20	−1.35	−1.05	−0.50	0.00	1.00
	Average		−3.15	−1.48	−0.98	−0.50	−0.10	0.80
G, dB, unoccupied	Barron	1986	0.80	3.30	3.70	4.10	3.70	—
	Takenaka	1991	2.90	3.50	4.70	5.10	4.70	4.70
	Hyde	1993	1.80	2.40	3.80	3.95	3.80	2.30

8. DALLAS, McDERMOTT CONCERT HALL IN MEYERSON SYMPHONY CENTER (Opened 1989, 2,065 seats)

Attribute	Measured by	125	250	500	1,000	2,000	4,000
RT, unoccupied	(1992) ARTEC: Canopy-mid; Rev. chambers open	3.44	3.16	2.93	2.87	2.58	2.08
	(1992) ARTEC: Canopy-low; Rev. chambers closed	3.22	2.94	2.70	2.50	2.37	1.97
RT, occupied	(1994) Anon.–I: Canopy-high; Rev. chambers open	3.53	3.21	2.85	2.92	2.46	2.22
	(1995) Anon.–2: Canopy-high; Rev. chambers open	—	3.15	2.73	2.68	2.27	1.71
	Average: Canopy-high, reverberation doors open	3.50	3.18	2.79	2.80	2.37	1.97
EDT, unoccupied	(1992) ARTEC: Canopy-mid; Rev. chambers open	2.00	2.00	1.80	2.00	1.80	1.60
	(1992) ARTEC: Canopy-low; Rev. chambers closed	2.20	2.10	1.80	2.00	1.90	1.60
EDT, occupied	(1995) Anon:–2: Canopy-high; Rev. chambers open	—	3.16	2.31	2.07	1.96	1.81

			Center Frequencies of Filter Bands					
Attribute	**Measured by**	**Year of Data**	**125**	**250**	**500**	**1,000**	**2,000**	**4,000**
			Hertz					

8. DALLAS, McDERMOTT CONCERT HALL IN MEYERSON SYMPHONY CENTER, *continued*

C_{80}, dB, unoccupied	(1992) ARTEC: Canopy-mid; Rev. chambers open		−0.90	−4.20	0.70	−0.70	0.20	0.40
	(1992) ARTEC: Canopy-low; Rev. chambers closed		−4.40	−5.30	−0.60	−1.40	−0.20	1.10
C_{80}, dB, occupied	(1995) Anon.−2: Canopy-high; Rev. chambers open		—	−3.90	−2.60	0.30	−2.10	−1.70

9. DENVER, BOETTCHER HALL (Opened 1978, 2,750 seats)

RT, unoccupied	Madaras, Bradley & Jaffe	1997	2.90	3.30	2.80	2.55	2.20	1.70
RT, occupied	(All data presented at	1997	2.70	2.70	2.50	2.30	2.05	1.85
EDT, unoccupied	Spring meeting Acous.	1997	2.90	3.00	2.55	2.35	2.05	1.70
EDT, occupied	Soc. Am. '97)	1997	2.75	2.50	2.30	1.90	1.65	1.40
$IACC_E$, unoccupied		1997	0.92	0.84	0.79	0.71	0.60	0.62
LF_E, unoccupied		1997	0.14	0.11	0.10	0.12	0.20	0.15
LF_E, occupied		1997	0.27	0.14	0.11	0.11	0.18	0.15
C_{80}, dB, unoccupied		1997	−2.70	−0.70	0.00	1.10	1.20	2.40
C_{80}, dB, occupied		1997	−2.70	0.00	1.00	2.20	2.40	4.00
G_E, dB, unoccupied		1997	2.00	2.10	1.00	2.00	−1.00	1.00
G_E, dB, occupied		1997	1.70	1.80	0.00	1.00	−2.00	0.40
G_{LL}, dB, unoccupied; "LL" = late lateral levels		1997	−8.00	−8.30	−10.00	−8.30	−9.00	−8.50

10. FORT WORTH, BASS PERFORMANCE HALL (Opened 1998, 2,072 seats)

RT, unoccupied	[C. Jaffe, director &	1998	2.30	2.37	2.45	2.75	2.55	2.30
RT, occupied, Est.	Rensselaer Polytechnic	1998	2.10	2.05	1.95	1.95	1.85	1.65
EDT, unoccupied	Institute student group,	1998	2.00	2.30	2.40	2.60	2.40	1.90
$IACC_A$, 0–1 sec	A. Case, D. Lennon,	1998	0.95	0.82	0.55	0.20	0.20	0.18
$IACC_E$, unoccupied	A. Stefaniw, J. Summers,	1998	0.95	0.88	0.63	0.45	0.35	0.38
$IACC_L$, unoccupied	and B. Viehland]	1998	0.95	0.80	0.47	0.12	0.15	0.20
C_{80}, dB, unoccupied		1998	−2.00	−2.00	−2.00	−2.00	−1.50	−1.00

11. LENOX, MASSACHUSETTS, SEIJI OZAWA HALL (Opened 1994, 1,180 seats)

RT, unoccupied (rear wall closed)	Kirkegaard	1995	2.34	2.36	2.21	2.28	2.16	1.70
RT, occupied (rear wall open)	Takenaka/Beranek	1994	2.24	2.16	1.82	1.50	1.35	1.28

			Center Frequencies of Filter Bands					
Attribute	Measured by	Year of Data	125	250	500	1,000	2,000	4,000
			Hertz					

12. LENOX, MASSACHUSETTS, TANGLEWOOD, KOUSSEVITZKY MUSIC SHED (Opened 1938, 5,121 seats)

Attribute	Measured by	Year of Data	125	250	500	1,000	2,000	4,000
RT, unoccupied	Takenaka/Beranek	1993	4.29	4.06	3.86	3.09	2.68	2.32
	Bradley	1993	4.13	4.00	3.65	3.09	2.65	2.29
	Average		4.21	4.03	3.76	3.09	2.67	2.31
RT, occupied	Takenaka/Beranek	1994	2.85	2.63	2.09	1.69	1.50	1.42
EDT, unoccupied	Takenaka/Beranek	1993	4.02	3.81	3.77	2.98	2.51	2.22
	Bradley	1993	3.42	3.87	3.60	2.98	2.49	2.16
	Average		3.72	3.84	3.69	2.98	2.50	2.19
$ICCA_A$, unoccupied	Takenaka/Beranek	1993	0.94	0.79	0.50	0.33	0.29	0.22
	Bradley	1993	0.97	0.80	0.39	0.30	0.22	0.19
	Average		0.96	0.80	0.45	0.32	0.26	0.21
$IACC_E$, unoccupied	Takenaka/Beranek	1993	0.96	0.89	0.76	0.65	0.57	0.44
	Bradley	1993	0.97	0.89	0.73	0.58	0.47	0.38
	Average		0.97	0.89	0.75	0.62	0.52	0.41
$IACC_L$, unoccupied	Takenaka/Beranek	1993	0.92	0.75	0.40	0.15	0.09	0.06
	Bradley	1993	0.95	0.76	0.27	0.13	0.07	0.06
	Average		0.94	0.76	0.34	0.14	0.08	0.06
LF, unoccupied	Bradley	1993	0.10	0.14	0.10	0.12	0.15	0.12
C_{80}, dB, unoccupied	Takenaka/Beranek	1993	−3.99	−4.83	−5.03	−3.44	−2.14	−1.32
	Bradley	1993	−0.62	−3.92	−3.85	−2.76	−1.29	−0.65
	Average		−2.31	−4.38	−4.44	−3.10	−1.72	−0.99
G, dB, unoccup.	Taka/Ber Hall Average	1993	0.00	1.10	3.10	3.10	−1.70	−2.50
	Bradley Hall Average	1993	0.18	−0.16	−0.90	−1.92	−3.21	−5.29
	Taka/Ber 10th row, off center line	1993	2.9	4.6	6.5	6.5	2.7	2.4
	Taka/Ber 20th row, off center line	1993	1.1	2.3	4.2	5.0	0.2	−0.3

13. MINNEAPOLIS, MINNESOTA ORCHESTRA HALL (Opened 1974, 2,450 seats)

Attribute	Measured by	Year of Data	125	250	500	1,000	2,000	4,000
RT, unoccupied	Schultz	1979	2.35	2.30	2.35	2.20	2.15	1.90
RT, occupied	Harris	1974	2.10	1.95	1.90	1.80	1.70	1.50

14. NEW YORK, AVERY FISHER HALL (Opened 1976, 2,742 seats)

Attribute	Measured by	Year of Data	125	250	500	1,000	2,000	4,000
RT, unoccupied	Takenaka	1986	—	2.24	2.26	2.20	2.03	1.48
	Beranek (s.c., w/orchestra)	1994	1.93	1.98	2.15	2.18	2.06	1.80
RT, occupied	Beranek (s.c.)	1994	1.60	1.76	1.78	1.74	1.55	1.46
EDT, unoccupied	Takenaka	1986	—	2.00	2.01	1.90	1.64	1.24
$IACC_A$	Takenaka	1986	—	0.70	0.22	0.17	0.32	0.21
LF, unoccupied	Takenaka	1986	—	—	0.13	0.11	0.12	—
C_{80}, dB, unoccupied	Takenaka	1986	—	−2.80	−2.30	−2.00	−0.20	0.80

Attribute	Measured by	Year of Data	Center Frequencies of Filter Bands					
			125	250	500	1,000	2,000	4,000
			Hertz					

15. NEW YORK, CARNEGIE HALL (Opened 1891; renovated 1986 and 1989; 2,804 seats)

Attribute	Measured by	Year	125	250	500	1,000	2,000	4,000
RT, unoccupied	Anon.	1968	2.20	2.15	2.10	2.00	1.80	1.40
RT, occupied	Beranek (s.c.)	2001	2.12	2.09	1.83	1.75	1.57	1.28

16. NEW YORK, METROPOLITAN OPERA HOUSE (Opened 1966, 3,816 seats)

Attribute	Measured by	Year	125	250	500	1,000	2,000	4,000
RT, unoccupied	Takenaka Source on stage	1998	1.74	1.64	1.73	1.76	1.69	1.51
RT, occupied	Calc.: JASA **109**, p. 1028	2001	1.65	1.55	1.55	1.55	1.40	1.25
EDT, unoccupied	Takenaka	1998	1.61	1.43	1.60	1.63	1.46	1.30
IACC$_A$, unoccupied	Takenaka	1998	0.92	0.75	0.30	0.22	0.22	0.24
IACC$_E$, unoccupied	Takenaka	1998	0.95	0.81	0.46	0.35	0.33	0.34
IACC$_L$, unoccupied	Takenaka	1998	0.90	0.69	0.21	0.14	0.12	0.13
C$_{80}$, dB, unoccupied	Takenaka	1998	0.0	0.8	1.4	1.6	2.2	2.0
G, dB, unoccupied	Takenaka	1998	−2.0	−0.1	0.3	0.7	1.1	−0.6

17. PHILADELPHIA, ACADEMY OF MUSIC (Opened 1857, 2,827 seats for opera)

Attribute	Measured by	Year	125	250	500	1,000	2,000	4,000
RT, unoccupied	M.A.A. (BBN & Potwin)	1959	2.00	1.80	1.60	1.50	1.40	1.30
(There were changes in the hall after 1959, including new seats and added carpets.)								
	Kirkegaard	1992	1.50	1.30	1.30	1.20	1.20	1.10
	Bradley	1992	1.50	1.32	1.35	1.30	1.21	1.02
	Marshall	1992	1.40	1.40	1.40	1.30	1.30	1.10
RT, occupied	(M.A.A.)	1958	1.40	1.70	1.45	1.35	1.25	1.15
(Note the changes after 1959 as above.)								
	Kirkegaard	1992	1.40	1.30	1.20	1.20	1.10	1.00
EDT, unoccupied	Marshall	1992	1.30	1.20	1.20	1.20	1.20	1.10
	Bradley	1992	1.23	1.11	1.20	1.21	1.13	0.89
IACC$_A$, unoccupied	Bradley	1992	0.94	0.79	0.44	0.41	0.35	0.30
IACC$_E$, unoccupied	Bradley	1992	0.94	0.81	0.58	0.54	0.48	0.41
IACC$_L$, unoccupied	Bradley	1992	0.90	0.73	0.25	0.18	0.12	0.12
LF, unoccupied	Bradley	1992	0.19	0.16	0.14	0.17	0.19	0.12
C$_{80}$, dB, unoccupied	Marshall	1992	−1.20	−0.20	2.00	0.30	2.30	2.10
	Bradley	1992	1.70	2.34	1.97	2.31	2.45	4.09
G, dB, unoccupied	Bradley	1992	−1.87	−0.12	1.77	1.13	−0.53	−2.35

18. PHILADELPHIA, VERIZON HALL IN KIMMEL CENTER (Opened 2001, 2,519 seats)

Attribute	Measured by	Year	125	250	500	1,000	2,000	4,000
RT, unoccupied	doors 40% open, ARTEC	2002	2.63	2.15	2.12	2.10	2.00	1.68
RT, occupied	same, ARTEC, s.c.	2002	2.05	2.01	1.70	1.60	1.45	1.23
RT, occupied	same, Beranek, s.c.	2002	1.99	1.93	1.75	1.66	1.59	1.46
	doors closed, Beranek, s.c.	2002	1.84	1.70	1.62	1.53	1.38	1.14
EDT, unoccupied	20% open, ARTEC	2002	2.74	1.98	1.95	2.26	2.19	1.89
EDT, occupied	doors closed, Beranek, s.c.	2002	1.55	1.69	1.50	1.61	1.34	1.09

Attribute	Measured by	Year of Data	Center Frequencies of Filter Bands					
			125	250	500	1,000	2,000	4,000
			Hertz					

19. ROCHESTER, NY, EASTMAN THEATRE (Opened 1923, 3,347 seats)

Attribute	Measured by	Year	125	250	500	1,000	2,000	4,000
RT, unoccupied	M.A.A. (BBN)	1959	3.10	2.50	1.90	1.75	1.70	1.55
RT, occupied	M.A.A. (BBN)	1959	2.30	1.85	1.75	1.55	1.45	1.30

20. SALT LAKE CITY, ABRAVANEL SYMPHONY HALL (Opened 1979, 2,812 seats)

Attribute	Measured by	Year	125	250	500	1,000	2,000	4,000
RT, unoccupied	Takenaka	1991	2.08	1.92	2.03	2.03	1.92	1.72
RT, occupied	Calc.: JASA **109**, p. 1028	2001	1.90	1.80	1.80	1.70	1.60	1.40
EDT, unoccupied	Takenaka	1991	2.05	1.99	2.09	2.07	1.94	1.70
$IACC_A$, unoccupied	Takenaka	1991	0.91	0.73	0.28	0.19	0.17	0.14
$IACC_E$, unoccupied	Takenaka	1991	0.93	0.81	0.48	0.40	0.35	0.26
$IACC_L$, unoccupied	Takenaka	1991	0.90	0.69	0.22	0.10	0.08	0.07
C_{80}, dB, unoccupied	Takenaka	1991	−4.60	−3.90	−2.10	−1.80	−1.40	−0.90
G, dB, unoccupied	Takenaka	1991	1.90	0.30	2.40	2.70	2.50	1.60

21. SAN FRANCISCO, DAVIES SYMPHONY HALL (Opened 1980, 2,743 seats) (Data before renovations)

Attribute	Measured by	Year	125	250	500	1,000	2,000	4,000
RT, unoccupied	Takenaka	1991	2.58	2.41	2.29	2.24	1.96	1.63
(Since 1993, the volume is 5% less, Est.)		2000	2.45	2.29	2.18	2.13	1.86	1.55
RT, occupied	Kirkegaard	1992	—	1.90	1.90	1.80	1.70	1.40
EDT, unoccupied	Takenaka	1991	2.69	2.51	2.26	2.04	1.72	1.38
$IACC_A$, unoccupied	Takenaka	1991	0.93	0.77	0.39	0.30	0.29	0.27
$IACC_E$, unoccupied	Takenaka	1991	0.95	0.87	0.65	0.54	0.50	0.40
$IACC_L$, unoccupied	Takenaka	1991	0.92	0.72	0.23	0.09	0.09	0.08
C_{80}, dB, unoccupied	Takenaka	1991	−5.10	−3.90	−2.30	−0.60	0.10	1.30
G, dB, unoccupied	Takenaka	1991	2.50	3.00	3.30	3.40	2.90	2.90

22. SAN FRANCISCO, WAR MEMORIAL OPERA HOUSE (Opened 1932, 3,252 seats)

Attribute	Measured by	Year	125	250	500	1,000	2,000	4,000
RT, occupied	Beranek, stop cords	2001	1.59	1.58	1.53	1.46	1.39	1.26

23. SEATTLE, BENAROYA HALL (Opened 1998; 2,500 seats)

Attribute	Measured by	Year	125	250	500	1,000	2,000	4,000
RT, unoccupied	Harris	1998	2.29	2.24	2.23	2.17	2.12	1.93
RT, occupied	Harris (s.c., 1 concert)	1998	2.20	2.10	2.04	1.90	1.87	1.67
(Above, before organ installed; below, after installation)								
RT, occupied	Anon (s.c., 3 concerts)	2001	2.15	1.90	1.80	1.75	1.65	1.55

24. WASHINGTON, D.C., JFK CENTER, CONCERT HALL (Opened 1971; renovation 1997; 2,448 seats)

Attribute	Measured by	Year	125	250	500	1,000	2,000	4,000
RT, unoccupied	Jaffe, Holden Acoustics	1997	2.20	2.00	2.00	1.85	1.80	1.35
RT, occupied	Calc.: JASA **109**, p. 1028	2001	2.00	1.80	1.75	1.50	1.50	1.20

25. WASHINGTON, D.C., JFK CENTER, OPERA HOUSE (Opened 1971, 2,142 seats)

Attribute	Measured by	Year	125	250	500	1,000	2,000	4,000
RT, unoccupied	Harris Fire curtain down	1971	2.20	2.00	1.80	1.60	1.50	1.40
RT, occupied	Harris (s.c.)	1971	2.00	1.90	1.60	1.40	1.20	1.20

			Center Frequencies of Filter Bands					
Attribute	**Measured by**	**Year of Data**	**125**	**250**	**500**	**1,000**	**2,000**	**4,000**
			Hertz					

26. WORCESTER, MASSACHUSETTS, THE GRAND HALL, MECHANICS HALL (Opened 1857, 1,343 seats)

RT, unoccupied	Bradley	1993	2.10	2.40	2.30	2.00	1.60	1.40
RT, occupied	Bradley	1993	1.70	1.90	1.60	1.50	1.40	1.30
EDT, unoccupied	Bradley	1993	2.10	2.30	2.30	2.00	1.70	1.40
EDT, occupied	Bradley	1993	1.80	1.90	1.70	1.50	1.40	1.30
$IACC_A$, unoccupied	Bradley	1993	0.90	0.70	0.20	0.20	0.20	0.20
$IACC_E$, unoccupied	Bradley	1993	0.95	0.81	0.47	0.45	0.44	0.42
$IACC_L$, unoccupied	Bradley	1993	0.92	0.70	0.22	0.21	0.22	0.24
LF, occupied	Bradley	1993	0.20	0.20	0.20	0.20	0.20	0.30
LF, unoccupied	Bradley	1993	0.30	0.20	0.20	0.20	0.20	0.30
C_{80}, dB, unoccupied	Bradley	1993	−1.00	−2.00	−2.00	−1.00	0.00	1.00
G, dB, unoccupied	Bradley	1993	4.00	6.00	5.50	4.50	2.00	−1.50

27. BUENOS AIRES, TEATRO COLÓN (Opened 1908; 2,487 seats)

RT, unoccupied	Takenaka Source on stage	1998	2.20	2.09	2.04	1.76	1.61	1.37
RT, occupied	Calc.: JASA **109**, p. 1028	2001	2.10	1.90	1.75	1.50	1.40	1.20
EDT, unoccupied	Takenaka Source on stage	1998	2.00	1.97	1.87	1.57	1.34	1.12
$IACC_A$, unoccupied	Takenaka Source on stage	1998	0.90	0.72	0.28	0.20	0.20	0.24
$IACC_E$, unoccupied	Takenaka Source on stage	1998	0.92	0.81	0.46	0.30	0.30	0.31
$IACC_L$, unoccupied	Takenaka Source on stage	1998	0.90	0.69	0.27	0.13	0.10	0.09
C_{80}, dB, unoccupied	Takenaka Source on stage	1998	−1.3	−0.8	0.2	1.3	1.9	2.6
G, dB, unoccupied	Takenaka Source on stage	1998	0.4	1.9	2.5	2.3	2.0	0.3
ST1, unoccupied	Takenaka Source on stage	1998	−11.6	−13.7	−16.2	−16.3	−15.1	−15.6
RT, unoccupied	Takenaka Source in pit	1998	2.02	2.03	1.96	1.73	1.60	1.38
RT, occupied	Calc.: JASA **109**, p. 1028	2001	1.90	1.85	1.70	1.50	1.35	1.20
EDT, unoccupied	Takenaka Source in pit	1998	1.92	1.80	1.85	1.62	1.41	1.23
$IACC_A$, unoccupied	Takenaka Source in pit	1998	0.90	0.73	0.27	0.16	0.13	0.12
$IACC_E$, unoccupied	Takenaka Source in pit	1998	0.89	0.77	0.48	0.31	0.24	0.19
$IACC_L$, unoccupied	Takenaka Source in pit	1998	0.91	0.72	0.28	0.13	0.13	0.10
C_{80}, dB, unoccupied	Takenaka Source in pit	1998	−4.3	−2.3	−3.0	−3.1	−1.7	−0.8
G, dB, unoccupied	Takenaka Source in pit	1998	1.0	2.4	2.1	1.6	1.1	−0.6
ST1, unoccupied	Takenaka Source in pit	1998	−7.7	−11.9	−12.5	−13.1	−11.8	−11.0

28. SYDNEY OPERA HOUSE, CONCERT HALL (Opened 1973, 2,696 seats)

RT, unoccupied	Jordan	1973	2.45	2.46	2.45	2.55	2.60	2.56
	Yamasaki	1992	2.25	2.40	2.30	2.65	2.60	2.20
RT, occupied	Jordan	1973	2.10	2.20	2.10	2.30	2.20	2.00
EDT, unoccupied	Jordan	1973	2.18	2.28	2.16	2.22	2.18	2.10
EDT, occupied	Jordan	1973	2.05	2.13	1.98	2.16	2.12	2.02

Attribute	Measured by	Year of Data	Center Frequencies of Filter Bands					
			125	250	500	1,000	2,000	4,000
			Hertz					

29. SALZBURG, FESTSPIELHAUS (Opened 1960, 2,158 seats)

Attribute	Measured by	Year of Data	125	250	500	1,000	2,000	4,000
RT, unoccupied	M.A.A. (Schwaiger)	1960	1.90	2.30	2.25	2.10	2.10	1.80
	Gade	1987	1.70	2.00	2.00	2.00	1.80	1.60
	Bradley	1987	1.69	1.93	1.91	1.82	1.74	1.50
	Gade/Bradley Average		1.70	1.97	1.96	1.91	1.77	1.55
RT, Occupied	M.A.A. (Schwaiger, BBN)	1960	1.70	1.60	1.50	1.50	1.40	1.30
EDT, unoccupied	Bradley	1987	1.65	1.89	1.88	1.78	1.71	1.40
	Gade	1987	1.61	1.98	2.04	1.78	1.85	1.59
LF, unoccupied	Bradley	1987	0.17	0.16	0.14	0.12	0.11	0.19
	Gade	1987	0.18	0.16	0.15	0.16	—	—
C_{80}, dB, unoccupied	Bradley	1987	−1.61	−1.32	−0.50	0.65	−0.04	1.45
	Gade	1987	−2.66	−2.10	−1.29	0.28	−0.79	0.16
	Average		−2.14	−1.71	−0.90	0.47	−0.42	0.81
G, dB, unoccupied	Bradley	1987	1.57	2.89	3.24	3.63	3.15	2.16
	Gade	1987	2.02	4.18	4.53	3.88	1.86	1.36

30. VIENNA, GROSSER MUSIKVEREINSSAAL (Opened 1870, 1,680 seats)

Attribute	Measured by	Year of Data	125	250	500	1,000	2,000	4,000
RT, unoccupied	Various agencies	1958	3.10	3.30	3.60	3.50	3.10	2.20
(There were changes in the seats between 1960 and 1980.)								
RT, unoccupied	Gade	1987	3.00	3.20	3.20	3.20	2.60	2.10
	Tachibana	1987	2.85	2.85	2.90	2.95	2.70	2.00
	Takenaka	1993	2.97	2.95	3.04	2.99	2.67	2.21
	Bradley	1987	3.05	3.11	3.11	3.04	2.72	2.09
	Average		2.97	3.03	3.06	3.05	2.67	2.10
	Matsuzawa (rehearsal)	1992	2.61	2.62	2.56	2.42	2.22	1.91
RT, occupied	M.A.A. (various)	1959	2.40	2.20	2.10	2.00	1.90	1.60
	Tachibana	1987	2.20	2.20	2.00	1.95	1.80	1.70
	Matsuzawa	1992	2.14	2.15	2.01	1.94	1.71	1.55
	Average		2.25	2.18	2.04	1.96	1.80	1.62
EDT, unoccupied	Gade	1987	2.93	3.37	3.31	3.18	2.77	2.08
	Bradley	1987	2.93	3.06	3.06	3.03	2.70	2.00
	Takenaka	1993	2.98	3.01	3.03	2.99	2.71	2.17
	Average of Taka & Bradley		2.96	3.04	3.05	3.01	2.71	2.09
$IACC_A$, unoccupied	Takenaka	1993	0.89	0.68	0.20	0.11	0.17	0.26
$IACC_E$, unoccupied	Takenaka	1993	0.92	0.76	0.42	0.32	0.34	0.40
$IACC_L$, unoccupied	Takenaka	1993	0.88	0.66	0.17	0.09	0.07	0.07

Attribute	Measured by	Year of Data	Center Frequencies of Filter Bands					
			125	250	500	1,000	2,000	4,000
			Hertz					
30. VIENNA, GROSSER MUSIKVEREINSSAAL, *continued*								
LF, unoccupied	Bradley	1987	0.14	0.17	0.19	0.16	0.12	0.21
	Gade	1988	0.12	0.16	0.19	0.17	—	—
	Average		0.13	0.17	0.19	0.17	—	—
C_{80}, dB, unoccupied	Bradley	1987	−5.71	−4.74	−4.00	−3.18	−2.98	−1.18
	Takenaka	1993	−5.10	−5.10	−4.70	−4.00	−3.10	−1.60
	Gade	1987	−5.02	−6.56	−5.45	−4.68	−3.88	−1.93
	Average		−5.28	−5.47	−4.72	−3.95	−3.32	−1.57
G, dB, unoccupied	Takenaka	1993	8.10	7.40	7.80	7.90	6.80	6.10
	Bradley	1987	6.10	6.04	5.97	6.57	6.04	4.51
31. VIENNA KONZERTHAUS (Opened 1913; renovated 2000; 1,865 seats)								
RT, unoccupied	Müller-BBM, Takenaka (b.r.)	1991–97	3.64	3.24	2.55	2.45	2.32	1.90
	Müller-BBM	2001	2.92	2.89	2.30	2.09	1.96	1.69
RT, occupied	Müller-BBM	1997	2.73	2.41	2.03	1.80	1.69	1.42
	Müller-BBM	2001	2.31	2.14	1.96	1.79	1.62	1.36
EDT, unoccupied	Muller-BBM, Takenaka	1991–97	3.16	3.04	2.46	2.42	2.30	1.76
	Müller-BBM	2001	2.60	2.64	2.18	2.05	1.88	1.47
EDT, occupied	Müller-BBM	1997	2.32	2.49	2.01	1.95	1.82	1.62
	Müller-BBM	2001	2.20	1.98	1.94	1.97	1.72	1.44
C_{80}, dB, unoccupied	Muller-BBM, Takenaka	1991–97	4.0	4.8	0.5	−2.2	0.6	−0.1
	Müller-BBM	2001	−2.5	−2.7	−1.3	−1.0	−0.3	1.2
C_{80}, dB, occupied	Müller-BBM	1997	−3.5	−2.2	−2.0	−0.5	0.0	1.9
	Müller-BBM	2001	−2.0	−0.9	−0.6	−0.3	−0.7	1.1
G, dB, unoccupied	Müller-BBM	1997	3.8	6.0	5.3	4.7	5.8	4.3
	Müller-BBM	2001	6.6	4.6	4.6	3.0	5.6	1.0
G, dB, occupied	Müller-BBM	1997	2.3	3.4	2.4	2.0	3.3	3.5
	Müller-BBM	2001	4.8	3.7	3.9	2.2	5.4	1.9
$IACC_E$, unoccupied	Müller-BBM	1997				0.43		
	Müller-BBM	2001				0.34		

Notes: Hall renovated in 2000. RT (occupied) without orchestra is approx. 0.1–0.2 sec higher, especially at high frequencies. G (occupied) approx. 1–2 dB higher without orchestra. $IACC_E$ (above), 2 octaves stradling 1,000 Hz.

Attribute	Measured by	Year of Data	Center Frequencies of Filter Bands					
			125	250	500	1,000	2,000	4,000
			Hertz					

32. VIENNA, STAATSOPER (Opened 1869, 1,709 seats)

Attribute	Measured by	Year	125	250	500	1,000	2,000	4,000
RT, unoccupied	M.A.A. (Bruchmeyer)	1955	2.00	1.90	1.80	1.80	1.70	1.50
	TAK (Source So)	1988	1.84	1.61	1.55	1.54	1.44	1.28
	TAK (Source pit)		1.67	1.52	1.53	1.54	1.44	1.27
RT, occupied	M.A.A. (various agencies)	1960	1.40	1.45	1.40	1.20	1.20	1.15
EDT, unoccupied	TAK (Source So)	1988	1.56	1.43	1.38	1.47	1.34	1.07
	TAK (Source pit)		1.51	1.31	1.35	1.43	1.39	1.13
$IACC_A$, unoccupied	TAK (Source So)	1988	0.90	0.72	0.36	0.28	0.24	0.20
	TAK (Source pit)		0.93	0.77	0.34	0.18	0.15	0.13
$IACC_E$, unoccupied	TAK (Source So)	1988	0.92	0.80	0.49	0.38	0.32	0.24
	TAK (Source pit)		0.94	0.82	0.47	0.31	0.26	0.19
$IACC_L$, unoccupied	TAK (Source So)	1988	0.89	0.65	0.23	0.15	0.13	0.09
	TAK (Source pit)		0.91	0.74	0.27	0.13	0.11	0.09
C_{80}, dB, unoccupied	TAK (Source So)	1988	−0.20	1.50	2.50	2.80	2.70	4.30
	TAK (Source pit)		−2.10	−0.70	−0.70	−0.60	−0.20	1.20
G, dB, unoccupied	TAK (Source So)	1988	−1.40	0.70	3.10	2.50	1.30	−1.80
	TAK (Source pit)		0.20	2.70	4.30	3.60	2.50	−0.90

33. BRUSSELS, PALAIS DES BEAUX-ARTS (Opened 1929; modified between; renovated 2000; 2,150 seats)

Attribute	Measured by	Year	125	250	500	1,000	2,000	4,000
RT, unoccupied	Raes and RTB	1961	2.20	2.40	2.00	1.90	1.75	1.60
	Commins, before renovation	1999	1.87	1.69	1.63	1.53	1.43	1.22
	Commins, after renovation	2000	2.09	2.09	1.77	1.75	1.67	1.44
RT, occupied	Raes and BBN	1961	1.90	1.75	1.50	1.35	1.25	1.10
	Est.: JASA **109**, p. 1028, after renovation	2001	2.00	2.00	1.70	1.55	1.50	1.25
EDT, unoccupied	Commins, before renovation	1999	1.82	1.70	1.62	1.50	1.24	0.95
	Commins, after renovation	2000	2.46	2.09	2.01	1.72	1.59	1.25
C_{80}, dB, unoccupied	Commins, before renovation	1999	−1.72	−0.88	0.52	1.58	1.69	2.99
	Commins, after renovation	2000	−1.41	−0.99	−0.29	−0.75	−0.97	1.13

34. SÃO PAULO, SALA SÃO PAULO (Opened 1999, 1,610 seats)

Attribute	Measured by	Year	125	250	500	1,000	2,000	4,000
RT, occupied	ARTEC (s.c.)	1999	1.90	1.90	2.00	2.10	2.20	1.80
EDT, occupied	ARTEC (s.c.)	1999	0.80	0.80	0.80	1.40	1.50	1.90

35. MONTREAL, SALLE WILFRID-PELLETIER (Opened 1963, 2,982 seats)

Attribute	Measured by	Year	125	250	500	1,000	2,000	4,000
RT, unoccupied	Bradley	1989	2.49	2.10	1.98	1.87	1.69	1.29
	Acentech	1989	—	2.47	2.25	1.95	1.68	1.30
RT, occupied	Schultz	1963	2.20	1.85	1.73	1.60	1.50	1.20
EDT, unoccupied	Bradley	1989	2.47	2.18	1.98	1.88	1.61	1.19
LF, unoccupied	Bradley	1989	0.09	0.12	0.15	0.12	0.09	0.16
C_{80}, dB, unoccupied	Bradley	1989	−6.50	−2.44	−1.04	0.39	−0.37	1.60
G, dB, unoccupied	Bradley	1989	−0.41	−0.94	−0.37	0.53	−0.47	−3.22

Attribute	Measured by	Year of Data	Center Frequencies of Filter Bands					
			125	250	500	1,000	2,000	4,000
			Hertz					

36. TORONTO, ROY THOMPSON HALL (Opened 1982, 2,613 seats) (Renovated 2002)

RT, unoccupied	Bradley	1988	2.37	2.22	2.21	2.11	1.94	1.49
RT, occupied	Schultz, extrapolated	1982	1.97	1.86	1.83	1.78	1.60	1.38
EDT, unoccupied	Bradley	1988	2.23	2.01	1.97	1.84	1.56	1.18
LF, unoccupied	Bradley	1988	0.10	0.16	0.16	0.16	0.13	0.19
C_{80}, dB, unoccupied	Bradley	1988	−4.48	−1.58	0.01	0.95	0.74	2.92
G, dB, unoccupied	Bradley	1988	2.84	2.93	3.20	3.29	3.28	2.44

(No postrenovation data available.)

37. HONG KONG, CULTURAL CENTRE CONCERT HALL (Opened 1989, 2,019 seats)

RT, unoccupied	Marshall, no A.B.*	1997	2.12	1.95	1.93	2.07	1.95	1.73
	Marshall, with A.B.	1997	1.88	1.83	1.84	1.99	1.89	1.69
EDT, unoccupied	Nielson, no A.B.	1997	1.99	1.83	1.84	1.94	1.84	1.63
	Nielson, with A.B.	1997	1.86	1.72	1.76	1.89	1.80	1.60
LF, unoccupied	Halstead, no A.B.	1997	0.17	0.16	0.11	0.11	0.12	0.13
	Halstead, with A.B.	1997	0.18	0.16	0.11	0.11	0.11	0.12
C_{80}, dB, unoccupied	Halstead, no A.B.	1997	−3.29	−1.80	−0.26	−0.08	−0.47	0.02
	Halstead, with A.B.	1997	−2.91	−1.54	−0.01	−0.02	−0.50	0.16
G, dB, unoccupied	Halstead, no A.B.	1997	3.06	1.49	1.57	2.44	3.30	2.28
	Halstead, with A.B.	1997	2.74	1.20	1.59	2.32	3.24	2.29
LG, dB, unoccupied	Halstead, no A.B.	1997	−5.93	−6.76	−7.17	−6.67	−6.08	−6.97
	Halstead, with A.B.	1997	−6.33	−7.04	−7.45	−6.80	−6.22	−7.16

*Absorbent Blinds

38. SHANGHAI, GRAND THEATRE (Opened 1998, 1,676 seats, opera; 1,895 seats, concerts)

RT, unoccupied	Opera, drapery behind grille	1998	2.40	1.60	1.40	1.35	1.35	1.30
RT, unoccupied	Concert, no drapery	1998	2.30	1.90	1.80	1.85	1.80	1.70
RT, occupied	Est.: JASA **109**: 1028	2001	2.20	1.80	1.65	1.65	1.55	1.50

39. COPENHAGEN, RADIOHUSET, STUDIO 1 (Opened 1945, 1,081 seats)

RT, unoccupied	Jordan, V. L.	1945	1.60	1.70	2.00	2.00	1.90	1.20
	Gade	1993	1.75	1.74	1.86	1.99	2.15	1.95
	Average		1.68	1.72	1.93	2.00	2.03	1.58
RT, occupied	Jordan, V. L.	1945	1.60	1.60	1.50	1.50	1.50	1.20
EDT, unoccupied	Gade	1993	1.59	1.72	1.91	2.02	2.19	1.86
LF, unoccupied	Gade	1993	0.11	0.17	0.17	0.17	0.15	0.19
C_{80}, dB, unoccupied	Gade	1993	0.83	−0.36	−0.35	−0.26	−1.53	−0.52
G, dB, unoccupied	Gade	1993	8.26	4.76	5.40	6.24	6.17	5.38

Attribute	Measured by	Year of Data	Center Frequencies of Filter Bands					
			125	250	500	1,000	2,000	4,000
			Hertz					

40. ODENSE, KONCERTHUS, NIELSEN HALL (Opened 1982, 1,320 seats)

Attribute	Measured by	Year of Data	125	250	500	1,000	2,000	4,000
RT, unoccupied	Jordan	1982	2.4	2.3	2.3	2.2	2.1	1.8
EDT, unoccupied	Jordan	1982	2.4	2.4	2.3	2.2	2.2	1.7
C_{80}, dB, unoccupied	Jordan	1982	-2.7	-2.2	-2.4	-1.1	-2.9	-0.7
G, dB, unoccupied	Jordan	1982	1.8	3.2	3.6	4.2	3.3	1.5
LF, unoccupied	Jordan	1982	0.35	0.16	0.28	0.16	—	—
ST1, unoccupied	Gade	1982	—	-18	-14	-17	-17	—

41. BIRMINGHAM, SYMPHONY HALL (Opened 1991, 2,211 seats)

Attribute	Measured by	Year of Data	125	250	500	1,000	2,000	4,000
RT, unoccupied	ARTEC Rev. doors open	1992	2.42	2.81	2.56	2.45	2.63	1.85
	ARTEC Rev. doors closed		2.35	2.40	2.40	1.94	1.91	—
RT, occupied	Kimura Some doors open	1992	2.05	1.95	1.80	1.90	1.85	1.65
EDT, unoccupied	ARTEC Rev. doors open	1992	2.20	2.20	2.00	1.80	1.70	1.40
	ARTEC Rev. doors closed		2.40	2.20	2.20	1.80	2.10	—
C_{80}, dB, unoccupied	ARTEC Rev. doors open	1992	-4.80	-1.30	-0.60	1.10	0.90	1.90
	ARTEC Rev. doors closed	1992	-3.70	-1.30	-1.40	-1.00	-0.20	—

42. GLYNDEBOURNE, OPERA HOUSE (Opened 1994, 1,243 seats)

Attribute	Measured by	Year of Data	125	250	500	1,000	2,000	4,000
RT, unoccupied	Barron, lightly draped stage	1999	2.10	1.77	1.55	1.52	1.50	—
	Harris, typical box stage set	1994	1.95	(1.66)	1.46	1.43	(1.41)	—
RT, occupied	Harris, orchestra & typical set	1994	1.65	(1.42)	1.26	1.23	(1.21)	—
EDT, unoccupied	Barron, lightly draped stage	1999	1.96	1.41	1.27	1.19	1.26	—
C_{80}, dB, unoccupied	Barron, lightly draped stage	1999	0.1	2.3	4.3	4.7	4.0	—
G, dB, unoccupied	Barron, lightly draped stage	1999	0.0	-0.2	1.9	2.9	1.5	—
LF, unoccupied	Barron, lightly draped stage	1999	0.23	0.16	0.17	0.14	0.17	—

43. LIVERPOOL, PHILHARMONIC HALL (Opened 1939; renovated 1995; 1,803 seats)

Attribute	Measured by	Year of Data	125	250	500	1,000	2,000	4,000
RT, unoccupied	Kirkegaard, pre-renovation	1991	1.57	1.65	1.59	1.54	1.44	1.00
	Kirkegaard, after-renovation	1995	2.01	1.87	1.71	1.63	1.52	1.37
RT, occupied	Est. JASA **109**, p. 1028	2002	2.00	1.77	1.60	1.50	1.45	1.25

44. LONDON, BARBICAN, LARGE CONCERT HALL (Opened 1982; renovated 2002; 1,924 seats)

Attribute	Measured by	Year of Data	125	250	500	1,000	2,000	4,000
RT, unoccupied	Kirkegaard, after-renovation	2002	1.74	1.61	1.70	1.79	1.63	1.37
RT, occupied	Kirkegaard, after-renovation	2002	1.50	1.39	1.40	1.41	1.32	1.13
EDT, unoccupied	Kirkegaard, after-renovation	2002	1.64	1.60	1.80	1.81	1.64	1.39
EDT, occupied	Kirkegaard, after-renovation	2002	1.54	1.36	1.46	1.62	1.42	1.22
C80, dB, unoccupied	Kirkegaard, after-renovation	2002	-1.04	1.14	0.55	-0.02	1.73	3.31
C80, dB, occupied	Kirkegaard, after-renovation	2002	-0.22	1.43	1.40	1.38	2.15	3.94

Attribute	Measured by	Year of Data	Center Frequencies of Filter Bands					
			125	250	500	1,000	2,000	4,000
			Hertz					

45. LONDON, ROYAL ALBERT HALL (Opened 1871, 5,222 seats)

Attribute	Measured by	Year of Data	125	250	500	1,000	2,000	4,000
RT, unoccupied	Barron	1982	2.92	2.85	3.03	2.99	2.96	—
	Tachibana	1986	2.75	3.20	3.15	3.15	3.30	2.50
	Average		2.84	3.03	3.09	3.07	3.13	—
RT, occupied	Barron	1982	2.80	2.64	2.42	2.40	2.27	1.81
EDT, unoccupied	Barron	1982	2.53	2.54	2.67	2.63	2.62	—
LF, unoccupied	Barron	1982	0.13	0.16	0.14	0.13	0.14	—
C_{80}, dB, unoccupied	Barron	1982	−1.30	−1.10	0.00	1.00	0.60	—
G, dB, unoccupied	Barron	1982	−1.70	−1.90	−0.50	−0.90	−0.80	—

46. LONDON, ROYAL FESTIVAL HALL (Opened 1951, 2,901 seats)

Attribute	Measured by	Year of Data	125	250	500	1,000	2,000	4,000
RT, unoccupied	Barron, assisted resonance off	1982	1.43	1.44	1.62	1.76	1.80	1.70
	Gade assisted resonance off	1986	1.35	1.45	1.55	1.60	1.60	1.53
	BDP Acoustics, assisted resonance off	1994	1.60	1.50	1.60	1.70	1.70	1.50
	Average, assisted resonance off		1.46	1.46	1.59	1.69	1.70	1.58
	Barron, assisted resonance on	1982	2.55	2.00	1.96	1.79	1.87	1.70
	BDP Acoustics, assisted resonance on	1994	2.40	2.00	1.80	1.70	1.70	1.60
	Average, assisted resonance on		2.48	2.00	1.88	1.75	1.79	1.65
RT, occupied	M.A.A., assisted resonance off	1959	1.35	1.35	1.45	1.50	1.40	1.30
	BRS, assisted resonance off	1970	1.35	1.35	1.36	1.51	1.46	1.32
	Average, assisted resonance off		1.35	1.35	1.41	1.51	1.43	1.31
	Kimura, assisted resonance on	1992	1.95	1.60	1.50	1.50	1.40	1.25
	BDP Acoustics assisted resonance on	1994	1.75	1.70	1.50	1.45	1.45	1.30
	Average, assisted resonance on		1.85	1.65	1.50	1.48	1.43	1.28
EDT, unoccupied	Barron, assisted resonance off	1982	1.33	1.37	1.43	1.57	1.66	—
	Gade assisted resonance off	1986	1.15	1.41	1.37	1.32	1.42	1.37
	Barron assisted resonance on	1982	1.92	1.92	1.75	1.57	1.70	—
LF, unoccupied	Barron, assisted resonance on	1982	0.18	0.14	0.19	0.20	—	—
C_{80}, dB, unoccupied	Barron, assisted resonance off	1982	−0.70	−0.30	0.80	1.10	0.60	—
	Gade, assisted resonance off	1986	−0.33	−0.11	0.40	1.64	0.76	0.51
	Barron, assisted resonance on	1982	−3.90	−0.60	0.60	0.90	0.20	—
G, dB, unoccupied	Barron, assisted resonance on	1982	−0.50	0.00	1.90	2.20	3.00	—
	Gade, assisted resonance off	1986	0.18	0.96	2.10	1.79	1.59	—

Attribute	Measured by	Year of Data	125	250	500	1,000	2,000	4,000
					Hertz			

47. LONDON, ROYAL OPERA HOUSE (Opened 1858; renovated 2000; 2,157 seats)

Attribute	Measured by	Year	125	250	500	1,000	2,000	4,000
RT, unoccupied	Newton, set on-stage & hangings	2000	1.55	1.40	1.35	1.30	1.20	1.05
RT, occupied	Est. JASA **109**: 1028	2001	1.55	1.40	1.25	1.15	1.00	0.95

48. MANCHESTER, BRIDGEWATER HALL (Opened 1996, 2,357 seats)

Attribute	Measured by	Year	125	250	500	1,000	2,000	4,000
RT, unoccupied	Harris	1996	3.20	2.55	2.41	2.46	2.36	2.00
	Barron, drapes-balcony faces	1999	3.11	2.58	2.41	2.47	2.37	2.00
RT, occupied	Harris, 100% occupied	1996	2.60	2.12	2.00	2.03	1.96	1.70
	Est. JASA **109**: 1028	2001	2.60	2.20	2.00	2.00	1.90	1.65
EDT, unoccupied	Harris	1996	3.19	2.63	2.36	2.45	2.32	1.90
	Barron, drapes-balcony faces	1999	2.86	2.49	2.25	2.28	2.19	—
EDT, occupied	Harris	1996	2.52	2.13	2.04	2.02	1.96	1.69
C_{80}, dB, unoccupied	Barron, drapes-balcony faces	1999	−6.00	−3.20	−1.70	−0.80	−0.80	—
	Harris	1996	—	—	−1.87	−1.67	—	—
LF, unoccupied	Barron, drapes-balcony faces	1999	0.25	0.27	0.28	0.23	0.26	—
G, dB, unoccupied	Barron, drapes-balcony faces	1999	5.00	2.20	2.80	3.60	2.50	—

49. LAHTI, SIBELIUS/TALO, (Opened 2000, 1,250 seats)

Attribute	Measured by	Year	125	250	500	1,000	2,000	4,000
RT, occupied	ARTEC, doors open, added V	2000	2.50	2.50	2.50	2.30	1.90	1.40
RT, occupied	ARTEC, doors closed	2000	2.30	2.30	2.30	2.10	1.90	1.30

50. PARIS, OPÉRA BASTILLE (Opened 1989, 2,700 seats)

Attribute	Measured by	Year	125	250	500	1,000	2,000	4,000
RT, unoccupied	CSTB	1992	1.80	1.70	1.70	1.70	1.70	1.50
	Commins	1992	1.55	1.55	1.70	1.70	1.65	1.40
	Commins (stage set)	1992	1.59	1.60	1.69	1.74	1.66	1.33
	Average		1.65	1.62	1.70	1.71	1.67	1.41
RT, occupied	CSTB	1992	1.70	1.60	1.55	1.60	1.40	1.25
	Commins (calculated)	1992	1.45	1.45	1.50	1.50	1.45	1.30
EDT, unoccupied	CSTB	1992	1.75	1.65	1.55	1.65	1.55	1.20
	Commins	1992	1.56	1.59	1.52	1.52	1.37	1.20
EDT, occupied	CSTB	1992	1.60	1.40	1.30	1.35	1.20	1.00
C_{80}, dB, unoccupied	CSTB	1992	−0.50	2.00	3.50	3.30	4.40	7.00
	Commins	1992	−0.85	0.63	1.25	2.24	2.59	2.29
G, dB, unoccupied	CSTB	1992	−30.00	−30.00	−28.00	−26.50	−27.00	−28.00
G, dB, occupied	CSTB	1992	−32.00	−31.00	−29.00	−27.50	−27.50	−29.50

Attribute	Measured by	Year of Data	Center Frequencies of Filter Bands					
			125	250	500	1,000	2,000	4,000
			Hertz					

51. PARIS, OPÉRA GARNIER (Opened 1875, 2,131 seats)

Attribute	Measured by	Year	125	250	500	1,000	2,000	4,000
RT, unoccupied	TAK (Source So)	1988	1.84	1.40	1.26	1.18	1.14	1.02
	TAK (Source pit)	1988	1.48	1.32	1.20	1.15	1.11	1.01
RT, occupied	Calc.: JASA **109**, p. 1028, So	2001	1.80	1.35	1.10	1.10	0.09	0.09
EDT, unoccupied	TAK (Source So)	1988	1.43	1.30	1.20	1.12	1.12	1.04
	TAK (Source pit)	1988	1.21	1.39	1.33	1.12	1.04	0.93
$IACC_A$, unoccupied	TAK (Source So)	1988	0.94	0.80	0.46	0.37	0.32	0.30
	TAK (Source pit)	1988	0.93	0.79	0.34	0.23	0.19	0.15
$IACC_E$, unoccupied	TAK (Source So)	1988	0.96	0.86	0.58	0.48	0.43	0.40
	TAK (Source pit)	1988	0.93	0.86	0.45	0.32	0.28	0.24
$IACC_L$, unoccupied	TAK (Source So)	1988	0.91	0.74	0.30	0.13	0.13	0.12
	TAK (Source pit)	1988	0.91	0.74	0.30	0.17	0.15	0.10
C_{80}, dB, unoccupied	TAK (Source So)	1988	1.40	1.50	3.60	5.10	5.00	5.50
	TAK (Source pit)	1988	−0.80	−1.30	−1.10	0.20	1.80	2.50
G, dB, unoccupied	TAK (Source So)	1988	−2.40	0.20	1.10	0.30	−1.50	−3.40
	TAK (Source pit)	1988	−3.50	0.20	0.50	−0.40	−2.20	−4.00

52. PARIS, SALLE PLÉYEL (Opened 1927, 2,386 seats)

(All data taken before revisions of 1994.)

Attribute	Measured by	Year	125	250	500	1,000	2,000	4,000
RT, unoccupied	Bradley	1987	3.11	2.49	2.05	1.83	1.61	—
	Barron	1981	3.10	2.50	2.10	1.80	1.60	—
	CSTB, Paris		3.10	2.40	2.20	2.00	1.95	1.70
	Average		3.10	2.46	2.12	1.88	1.72	0.57
RT, occupied	CSTB		2.10	1.60	1.55	1.40	1.30	1.20
EDT, unoccupied	CSTB		2.40	2.00	1.90	1.80	1.75	1.50
	Bradley	1981	2.81	1.97	1.86	1.60	1.57	1.34
	Average		2.61	1.99	1.88	1.70	1.66	1.42
LF, unoccupied	Bradley	1987	0.10	0.15	0.19	0.15	0.11	0.18
	Barron	1981	0.23	0.11	0.16	0.17	—	—
	Average		0.17	0.13	0.18	0.16	—	—
C_{80}, dB, unoccupied	Barron, w/o overhung seats	1981	−5.90	−0.40	0.00	1.40	2.30	—
	Barron, all seats	1981	−5.90	0.10	0.50	2.00	2.90	—
	CSTB		−7.00	−2.00	2.00	4.50	2.50	2.00
	Bradley	1987	−2.84	−0.83	−0.72	0.95	0.43	1.11
	Average		−5.41	−0.78	0.45	2.21	2.03	0.78
G, dB, unoccupied	CSTB		5.00	2.50	1.00	0.05	0.00	−1.00
	Bradley	1987	5.96	3.88	3.94	4.25	3.99	3.34
	Barron	1981	5.00	2.50	3.60	3.80	3.30	—

Attribute	Measured by	Year of Data	Center Frequencies of Filter Bands					
			125	250	500	1,000	2,000	4,000
			Hertz					

53. BADEN-BADEN, FESTSPIELHAUS (Opened 1998, 2,300 seats)

Attribute	Measured by	Year of Data	125	250	500	1,000	2,000	4,000
RT, unoccupied	Müller-BBM, partly closed set	1998	2.52	2.23	2.12	2.05	1.94	1.74
RT, occupied	Müller-BBM, partly closed set	1998	2.30	2.05	1.97	1.86	1.75	1.52
EDT, unoccupied	Müller-BBM, partly closed set	1998	2.19	1.93	1.82	1.83	1.78	1.57
EDT, occupied	Müller-BBM, partly closed set	1998	2.14	1.83	1.80	1.77	1.71	1.54
C_{80}, dB, unoccupied	Müller-BBM, partly closed set	1998	-2.00	-1.60	-0.60	0.20	0.60	1.40
C_{80}, dB, occupied	Müller-BBM, partly closed set	1998	-1.90	-0.40	-0.10	0.30	1.10	2.40
G, dB, unoccupied	Müller-BBM, partly closed set	1998	-0.30	1.10	3.10	3.60	1.90	-1.00
G, dB, occupied	Müller-BBM, partly closed set	1998	-1.20	0.40	2.20	3.40	1.10	-1.60

54. BAYREUTH, FESTSPIELHAUS (Opened 1876, 1,800 seats)

Attribute	Measured by	Year of Data	125	250	500	1,000	2,000	4,000
RT, occupied	Various agencies	1960	1.75	1.70	1.60	1.50	1.40	1.30

55. BERLIN, KAMMERMUSIKSAAL DER PHILHARMONIE (Opened 1987, 1,138 seats)

Attribute	Measured by	Year of Data	125	250	500	1,000	2,000	4,000
RT, unoccupied	BeSB, Berlin	1987	2.18	2.00	2.07	2.20	2.11	2.19
RT, occupied	BeSB (2 positions)	1987	1.70	1.55	1.74	1.89	1.72	—
EDT, unoccupied	BeSB, Berlin	1987	2.04	2.01	1.99	2.21	2.11	1.64
EDT, occupied	BeSB (2 positions)	1987	1.40	1.55	1.48	1.88	1.65	—
C_{80}, dB, unoccupied	BeSB, Berlin	1987	—	—	-1.17	-2.37	—	—

56. BERLIN, KONZERTHAUS (SCHAUSPIELHAUS) (Opened 1986, 1,575 seats)

Attribute	Measured by	Year of Data	125	250	500	1,000	2,000	4,000
RT, unoccupied	Takenaka	1993	2.85	2.79	2.51	2.43	2.24	1.92
(Bass resonators added 1990)	Fasold, w/90-piece orchestra	1990	2.12	2.30	2.16	2.10	2.00	1.75
	Matsuzawa, rehearsal	1992	2.75	2.53	2.30	2.18	2.04	1.76
RT, occupied	Fasold, before resonators	1986	2.60	2.50	2.15	2.00	1.90	1.60
	Fasold after resonators	1990	2.20	2.10	2.00	2.00	1.90	1.60
	Kimura, after resonators	1992	2.65	2.35	2.10	2.00	1.85	1.60
	Matsuzawa	1992	2.53	2.34	2.05	1.87	1.74	1.59
	Selected average		2.20	2.10	2.00	2.00	1.80	1.60
EDT, unoccupied	Takenaka	1993	2.87	2.71	2.47	2.39	2.19	1.83
$IACC_A$, unoccupied	Takenaka	1993	0.88	0.69	0.20	0.13	0.20	0.20
$IACC_E$, unoccupied	Takenaka	1993	0.92	0.80	0.37	0.29	0.42	0.36
$IACC_L$, unoccupied	Takenaka	1993	0.87	0.65	0.20	0.10	0.08	0.06
C_{80}, dB, unoccupied	Takenaka	1993	-4.20	-4.30	-3.90	-2.30	-1.30	-0.50
G, dB, unoccupied	Takenaka	1993	8.30	7.10	6.70	7.00	5.90	5.40

Attribute	Measured by	Year of Data	Center Frequencies of Filter Bands					
			125	250	500	1,000	2,000	4,000
			Hertz					

57. BERLIN, PHILHARMONIE (Opened 1963, 2,218 seats)

Attribute	Measured by	Year of Data	125	250	500	1,000	2,000	4,000
RT, unoccupied	BeSB, Berlin	1990	2.20	1.90	2.10	2.20	2.10	1.70
	BeSB	1992	2.40	1.90	2.20	2.10	2.10	1.80
	Takenaka	1993	2.06	1.94	2.20	2.24	2.20	1.94
RT, occupied	Cremer,							
	300 musicians & chorus	1964	2.40	2.00	1.90	2.00	1.95	1.70
	Tachibana	1986	1.90	2.00	1.85	1.95	1.90	1.80
	Matsuzawa	1989	2.20	1.81	1.79	1.71	1.71	1.61
	BeSB, 80% occupancy	1992	2.20	1.90	1.90	2.00	1.80	1.50
	Selected average		2.10	1.85	1.85	1.95	1.80	1.60
EDT, unoccupied	BeSB	1990	2.00	1.70	1.90	2.00	1.90	1.30
	BeSB	1992	2.10	1.60	1.80	1.90	1.90	1.60
	Takenaka	1993	2.05	1.92	2.09	2.14	2.09	1.82
$IACC_A$, unoccupied	Takenaka	1993	0.93	0.76	0.31	0.22	0.27	0.27
$IACC_E$, unoccupied	Takenaka	1993	0.96	0.88	0.60	0.50	0.53	0.45
$IACC_L$, unoccupied	Takenaka	1993	0.90	0.68	0.18	0.11	0.10	0.08
C_{80}, dB, unoccupied	Takenaka	1993	−2.20	−0.70	−0.70	−0.60	−0.50	0.00
G, dB, unoccupied	Takenaka	1993	4.20	3.40	4.90	4.90	4.10	3.70

58. BONN, BEETHOVENHALLE (Opened 1959, 1,407 seats)

Attribute	Measured by	Year of Data	125	250	500	1,000	2,000	4,000
RT, unoccupied	M.A.A. (Meyer & Kuttruff)	1959	2.20	2.10	2.00	1.90	2.10	1.80
	(After a fire, new seats and some sound-absorbing material were installed.)							
	Kuttruff	1984	1.80	1.80	1.80	1.80	1.80	1.50
RT, occupied	M.A.A. (Meyer & Kuttruff)	1959	2.00	1.65	1.70	1.70	1.75	1.65
	Estimate after revisions	1995	1.80	1.70	1.65	1.65	1.5	1.3

59. DRESDEN, SEMPEROPER (Opened 1869; bombed 1945; reopened 1985; 1,284 seats)

Attribute	Measured by	Year of Data	125	250	500	1,000	2,000	4,000
RT, unoccupied	TU Dresden	1985	2.60	2.60	2.40	2.00	1.70	1.30
	Takenaka (30 dB decay)	1993	2.30	2.13	1.98	1.95	1.80	1.57
	Müller-BBM (20 dB decay)	1996	2.08	2.00	1.89	1.90	1.82	1.47
	Average	2001	2.19	2.06	1.94	1.92	1.81	1.52
RT, occupied	Est. JASA **109**: 1028	2001	2.00	1.90	1.75	1.60	1.50	1.30
EDT, unoccupied	Takenaka	1993	2.17	1.94	1.95	1.76	1.62	1.35
	Müller-BBM	1996	2.17	2.06	1.70	1.66	1.57	1.20
C_{80}, dB, unoccupied	Takenaka	1993	−2.00	−0.30	0.30	1.30	1.70	2.50
	Müller-BBM	1996	—	—	−2.10	−1.90	—	—
$IACC_E$, unoccupied	Takenaka	1993	—	—	0.35	0.31	0.33	0.35
	Müller-BBM	1996	2 octaves centered around 1,000 Hz, 0.26					
G, dB, unoccupied	Takenaka	1993	2.70	2.50	2.40	2.90	2.20	1.30
	Müller-BBM	1996	—	—	2.00	2.20	—	—

Attribute	Measured by	Year of Data	Center Frequencies of Filter Bands					
			125	250	500	1,000	2,000	4,000
			Hertz					

60. LEIPZIG, GEWANDHAUS (Opened 1981, 1,900 seats)

Attribute	Measured by	Year of Data	125	250	500	1,000	2,000	4,000
RT, unoccupied	Fasold	1982	1.95	2.00	2.20	2.20	2.00	1.70
RT, occupied	Fasold	1982	1.95	2.00	2.00	2.05	1.90	1.70

61. MUNICH, HERKULESSAAL (Opened 1953, 1,287 seats)

Attribute	Measured by	Year of Data	125	250	500	1,000	2,000	4,000
RT, unoccupied	M.A.A. (various agencies)	1956	2.60	2.00	2.20	2.40	2.30	1.90
RT, occupied	M.A.A. (Müller & BBN)	1960	2.00	1.75	1.85	1.85	1.80	1.65
	Matsuzawa	1985	2.04	2.01	1.88	1.63	1.76	1.37

62. MUNICH, PHILHARMONIE AM GASTEIG (Opened 1985, 2,487 seats)

Attribute	Measured by	Year of Data	125	250	500	1,000	2,000	4,000
RT, unoccupied	Müller	1986	2.35	2.10	2.30	2.35	2.40	2.10
	Bradley	1987	2.28	2.10	2.18	2.16	2.18	1.88
	Gade	1987	2.40	2.15	2.20	2.20	2.20	1.90
	Tachibana	1986	2.45	2.25	2.20	2.25	2.35	2.00
RT, occupied	Müller	1986	2.00	2.00	2.10	2.10	1.90	1.20
	Tachibana	1986	1.90	2.00	1.85	1.95	1.90	1.80
	Matsuzawa	1989	1.93	1.94	1.87	1.84	1.92	1.94
EDT, unoccupied	Bradley	1987	2.32	2.03	2.16	2.13	2.14	1.74
	Gade	1987	2.28	2.26	2.09	2.08	2.12	1.80
LF, unoccupied	Bradley	1987	0.13	0.13	0.14	0.12	0.10	0.18
	Gade	1987	0.12	0.13	0.11	0.08	—	—
C_{80}, dB, unoccupied	Bradley	1987	−4.51	−1.11	−0.44	0.38	−0.71	0.61
	Gade	1987	−5.02	−2.80	−0.51	−0.82	−1.95	−0.30
G, dB, unoccupied	Bradley	1987	0.04	0.42	1.28	1.85	1.65	0.51
	Gade	1987	1.16	1.48	2.72	1.81	3.33	1.72

63. STUTTGART, LIEDERHALLE, GROSSER SAAL (Opened 1956, 2,000 seats)

Attribute	Measured by	Year of Data	125	250	500	1,000	2,000	4,000
RT, unoccupied	M.A.A. (ITA)	1956	2.00	2.00	2.20	2.20	2.10	1.80
	Gade	1987	1.80	1.70	2.00	2.20	2.10	2.00
	Bradley	1987	1.73	1.80	2.05	2.13	2.13	1.88
	Average (Gade, Bradley)		1.77	1.75	2.03	2.17	2.12	1.94
RT, occupied	M.A.A. (large chorus)	1960	1.60	1.60	1.60	1.65	1.60	1.40
EDT, unoccupied	Gade	1987	1.65	1.65	2.09	2.21	2.18	1.78
	Bradley	1987	1.55	1.76	2.06	2.20	2.17	1.83
LF, unoccupied	Gade	1987	0.19	0.13	0.13	0.13	—	—
	Bradley	1987	0.08	0.12	0.13	0.12	0.09	0.16
C_{80}, dB, unoccupied	Gade	1987	−2.72	−0.36	2.09	2.01	−3.10	−0.54
	Bradley	1987	−1.85	−0.99	−1.04	−1.03	−1.15	−0.34
G, dB, unoccupied	Gade	1987	2.85	2.91	3.80	4.11	3.55	2.77
	Bradley	1987	2.63	3.34	3.63	3.08	1.10	−0.40

Attribute	Measured by	Year of Data	Center Frequencies of Filter Bands					
			125	250	500	1,000	2,000	4,000
			Hertz					

64. ATHENS, MEGARON THE ATHENS CONCERT HALL (Opened 1991, 1,962 seats)

Attribute	Measured by	Year	125	250	500	1,000	2,000	4,000
RT, unoccupied	Müller-BBM	1991	2.10	2.20	2.30	2.30	2.20	2.00
RT, occupied	Beranek (s.c.)	1998	2.40	2.00	1.90	1.80	1.70	1.50
C_{80}, dB, unoccupied	Müller-BBM	1991	—	—	−0.80	−0.20	—	—
G, dB, unoccupied	Müller-BBM	1991	—	—	4.20	2.00	—	—

65. BUDAPEST, MAGYAR ALLAMI OPERAHAZ (Opened 1884; rebuilt 1984; 1,277 seats)

Attribute	Measured by	Year	125	250	500	1,000	2,000	4,000
RT, unoccupied	Kotschy	1971	1.90	1.82	1.60	1.50	1.36	1.10
RT, occupied	Est.: JASA **109**: 1028	2001	1.90	1.75	1.40	1.30	1.20	1.00

66. BUDAPEST, PÁTRIA HALL IN CONVENTION CENTER (Opened 1985, 1,750 seats)

Attribute	Measured by	Year	125	250	500	1,000	2,000	4,000
RT, unoccupied	Fasold	1985	2.10	2.39	1.81	1.81	1.67	1.47
RT, occupied	Fasold	1985	2.10	1.90	1.80	1.60	1.45	1.30

67. BELFAST, WATERFRONT HALL (Opened 1997, 2,250 seats)

Attribute	Measured by	Year	125	250	500	1,000	2,000	4,000
RT, unoccupied	Sandy Brown Assoc.	1997	2.50	2.30	2.40	2.30	1.83	1.50
	S. Brown, adj. drapes exposed	1997	2.30	2.20	2.20	1.80	1.70	1.40
	Barron, adj. drapes exposed	1999	2.50	2.27	2.20	2.03	1.76	—
RT, occupied	Sandy Brown Assoc.	1997	2.10	2.40	2.20	2.12	1.83	1.50
	Barron, adj. drapes exposed	1999	2.12	2.26	2.00	1.75	1.65	—
EDT, unoccupied	Barron, adj. drapes exposed	1999	2.26	2.35	2.10	1.94	1.55	—
C_{80}, dB, unoccupied	Barron, adj. drapes exposed	1999	−4.20	−2.60	−0.40	0.40	1.50	—
LF_E unoccupied	Barron, adj. drapes exposed	1999	0.19	0.21	0.20	0.19	0.20	—
G, dB, unoccupied	Barron, adj. drapes exposed	1999	3.00	1.90	2.50	2.50	1.70	—

68. JERUSALEM, BINYANEI HA'OOMAH (Opened 1960, 3,142 seats)

Attribute	Measured by	Year	125	250	500	1,000	2,000	4,000
RT, unoccupied	M.A.A. (BBN)	1960	2.70	2.40	2.30	2.20	2.00	1.80
	Klepper/Beranek	1994	2.36	1.83	1.88	1.71	1.32	1.18
RT, occupied	M.A.A. (BBN)	1960	2.20	2.00	1.75	1.75	1.65	1.50
	Klepper/Beranek	1995	2.20	2.10	1.75	1.75	1.50	1.40
EDT, unoccupied	Klepper/Beranek	1994	2.30	1.93	1.89	1.80	1.40	1.12
$IACC_A$, unoccupied	Klepper/Beranek	1994	—	—	0.34	0.22	0.24	0.34
$IACC_E$, unoccupied	Klepper/Beranek	1994	—	—	0.55	0.40	0.40	0.48
$IACC_L$, unoccupied	Klepper/Beranek	1994	—	—	0.22	0.11	0.11	0.16
C_{80}, dB, unoccupied	Klepper/Beranek	1994	−0.80	−2.20	−1.10	0.30	1.50	3.00

Attribute	Measured by	Year of Data	Center Frequencies of Filter Bands					
			125	250	500	1,000	2,000	4,000
			Hertz					

69. TEL AVIV, FREDRIC R. MANN AUDITORIUM (Opened 1957, 2,715 seats)

Attribute	Measured by	Year of Data	125	250	500	1,000	2,000	4,000
RT, unoccupied	M.A.A. (BBN)	1957	1.80	1.65	1.95	2.00	1.85	1.60
	Klepper/Beranek	1994	1.62	1.56	1.67	1.67	1.56	1.35
RT, occupied	M.A.A. (BBN)	1957	1.55	1.50	1.55	1.55	1.50	1.30
	Klepper	1995	1.70	1.50	1.50	1.50	1.30	—
EDT, unoccupied	Klepper/Beranek	1994	1.54	1.60	1.66	1.73	1.57	1.36
$IACC_A$, unoccupied	Klepper/Beranek	1994	—	—	0.46	0.31	0.32	0.22
$IACC_E$, unoccupied	Klepper/Beranek	1994	—	—	0.71	0.55	0.50	0.36
$IACC_L$, unoccupied	Klepper/Beranek	1994	—	—	0.25	0.11	0.09	0.08
C_{80}, dB, unoccupied	Klepper/Beranek	1994	0.20	−0.50	−1.30	−0.50	0.50	0.60

70. MILAN, TEATRO ALLA SCALA (Opened 1778, 2,289 seats) (Closed 2002–2004)

Attribute	Measured by	Year of Data	125	250	500	1,000	2,000	4,000
RT, unoccupied	M.A.A. (Paolini)	1947	1.85	1.50	1.35	1.35	1.20	1.15
	Takenaka	1993	1.81	1.57	1.40	1.31	1.22	1.11
RT, occupied	M.A.A. (BBN, Furrer & Reichardt)	1959	1.50	1.40	1.25	1.15	1.10	1.00
EDT, unoccupied	Takenaka	1993	1.47	1.22	1.20	1.17	1.15	1.05
$IACC_A$, unoccupied	Takenaka	1993	0.94	0.79	0.42	0.35	0.41	0.41
$IACC_E$, unoccupied	Takenaka	1993	0.96	0.84	0.54	0.49	0.53	0.48
$IACC_L$, unoccupied	Takenaka	1993	0.92	0.73	0.36	0.16	0.13	0.10
C_{80}, dB, unoccupied	Takenaka	1993	1.00	1.40	2.00	3.80	4.70	4.40
G, dB, unoccupied	Takenaka	1993	−1.30	−1.80	−1.70	−1.10	−1.50	−3.00

71. NAPLES, TEATRO DI SAN CARLO (Opened 1737; burned 1816; reopened 1817; changes since; 1,414 seats)

Note: Sound source in pit. The first RT, EDT, C_{80}, and G below are measurements on main floor; second, in boxes.

Attribute	Measured by	Year of Data	125	250	500	1,000	2,000	4,000
RT, unoccupied	DETEC, U. of Naples Federico	2000	2.25	1.65	1.35	1.20	1.20	1.00
	Same, in boxes		2.20	1.65	1.10	1.05	1.05	1.00
RT, occupied	Est.: JASA **109**: 1028, main floor	2001	2.15	1.55	1.25	1.05	1.00	0.90
EDT, unoccupied	DETEC, U. of Naples Federico	2000	1.50	1.40	1.20	1.10	1.10	1.05
	Same, in boxes		1.40	1.20	1.00	0.90	0.85	0.75
C_{80}, dB, unoccupied	DETEC, U. of Naples Federico	2000	−2.30	−2.30	−0.25	−0.10	0.00	0.10
	Same, in boxes		0.70	1.10	2.60	3.70	3.80	5.00
G, dB, unoccupied	DETEC, U. of Naples Federico	2000	−1.30	−2.80	−0.90	−0.85	−0.75	−1.80
	Same, in boxes		0.80	−0.40	−0.40	−0.15	0.05	−0.50

Attribute	Measured by	Year of Data	Center Frequencies of Filter Bands					
			125	250	500	1,000	2,000	4,000
			Hertz					

72. KYOTO, CONCERT HALL (Opened 1995, 1,840 seats)

Attribute	Measured by	Year	125	250	500	1,000	2,000	4,000
RT, unoccupied	Nagata Acoustics	1995	2.30	2.20	2.20	2.30	2.20	1.90
RT, occupied	Est.: JASA **109**: 1028	2001	2.20	2.00	2.00	1.90	1.80	1.65
EDT, unoccupied	Nagata Acoustics	1995	2.10	2.20	2.20	2.00	2.00	1.70
C_{80}, dB, unoccupied	Nagata Acoustics	1995	−2.23	−1.66	−0.88	−0.47	−0.47	−0.47

73. OSAKA, SYMPHONY HALL (Opened 1982, 1,702 seats)

Attribute	Measured by	Year	125	250	500	1,000	2,000	4,000
RT, unoccupied	Tachibana	1986	2.05	1.95	2.20	2.20	2.15	1.85
RT, occupied	Anonymous	1990	500−1,000 Hz = 1.8 sec					
EDT, unoccupied	Tachibana	1986	300−1,400 Hz = 2.1 sec					
C_{80}, dB, unoccupied	Tachibana	1986	300−1,400 Hz = −1.18 dB					
$IACC_A$, unoccupied	Tachibana	1986	300−1,400 Hz = 0.22					

74. SAPPORO CONCERT HALL (Opened 1997, 2,008 seats)

Attribute	Measured by	Year	125	250	500	1,000	2,000	4,000
RT, unoccupied	Nagata Acoustics	1997	2.40	2.10	2.20	2.20	1.90	1.40
	Takenaka	1997	2.27	2.03	2.10	2.17	2.13	1.90
RT, occupied	Est.: JASA **109**: 1028	2001	2.00	1.95	1.80	1.80	1.75	1.40
EDT, unoccupied	Takenaka	1997	2.22	1.92	1.90	1.89	1.92	1.64
C_{80}, dB, unoccupied	Takenaka	1997	−4.70	−1.10	0.30	1.00	0.30	1.30
G, dB, unoccupied	Takenaka	1997	2.00	2.30	3.00	3.50	3.60	2.90
$IACC_A$, unoccupied	Takenaka	1997	0.93	0.77	0.37	0.31	0.29	0.29
$IACC_E$, unoccupied	Takenaka	1997	0.95	0.86	0.59	0.53	0.47	0.42
$IACC_L$, unoccupied	Takenaka	1997	0.91	0.71	0.23	0.12	0.10	0.07
LF, unoccupied	Takenaka	1997	0.11	0.13	0.12	0.12	0.14	0.15
ST1, unoccupied	Takenaka	1997	−10.9	−12.5	−14.5	−14.1	−13.1	−13.1

75. TOKYO, BUNKA KAIKAN (Opened 1961, 2,327 seats)

Attribute	Measured by	Year	125	250	500	1,000	2,000	4,000
RT, unoccupied	Nagata (concert)	1993	2.20	1.90	2.00	2.00	2.00	1.80
	TAK (no stage enclosure)	1989	2.15	1.85	1.75	1.87	1.75	1.35
	Takenaka (concert)	1995	1.92	1.79	1.89	1.99	1.95	1.78
RT, occupied	Nagata	1993	1.90	1.50	1.50	1.50	1.40	1.30
	Anonymous	1994	2.05	1.81	1.58	1.48	1.37	1.30
$IACC_A$, unoccupied	Takenaka	1995	—	—	0.30	0.20	0.18	0.18
$IACC_E$, unoccupied	Takenaka	1995	—	—	0.51	0.38	0.34	0.30
$IACC_L$, unoccupied	Takenaka	1995	—	—	0.20	0.10	0.09	0.08
LF, unoccupied	Takenaka	1995	0.17	0.17	0.19	0.19	0.22	0.25
C_{80}, dB, unoccupied	Nagata	1993	−2.90	−1.24	−0.08	−1.72	−1.23	−0.08
	Takenaka	1995	−1.30	−0.80	−0.30	−1.00	−0.80	0.00
G, dB, unoccupied	Takenaka	1995	3.30	3.40	4.20	4.30	4.40	5.50

Attribute	Measured by	Year of Data	\multicolumn{6}{c}{Center Frequencies of Filter Bands}
			125	250	500	1,000	2,000	4,000
			\multicolumn{6}{c}{Hertz}

76. TOKYO, DAI-ICHI SEIMEI HALL (Opened 2001, 767 seats)

Attribute	Measured by	Year	125	250	500	1,000	2,000	4,000
RT, unoccupied	Takenaka	2001	2.02	1.87	1.78	1.89	1.88	1.72
RT, occupied	Takenaka	2001	1.68	1.66	1.52	1.59	1.58	1.42
EDT, unoccupied	Takenaka	2001	1.90	1.79	1.75	1.85	1.84	1.70
EDT, occupied	Takenaka	2001	1.56	1.51	1.46	1.52	1.51	1.36
C_{80}, dB, unoccupied	Takenaka	2001	−2.20	−1.20	−0.70	−0.30	−0.30	−0.80
C_{80}, dB, occupied	Takenaka	2001	−1.30	−0.50	0.80	0.60	0.90	1.10
$IACC_A$, unoccupied	Takenaka	2001	—	—	0.26	0.14	0.10	0.17
$IACC_A$, occupied	Takenaka	2001	0.89	0.76	0.26	0.14	0.12	0.13
$IACC_E$, unoccupied	Takenaka	2001	—	—	0.17	0.12	0.06	0.09
$IACC_E$, occupied	Takenaka	2001	0.91	0.80	0.42	0.25	0.21	0.22
$IACC_L$, occupied	Takenaka	2001	0.88	0.74	0.20	0.13	0.08	0.06
G, dB, unoccupied	Takenaka	2001	8.80	8.80	8.10	9.00	9.90	9.40
G, dB, occupied	Takenaka	2001	7.20	7.70	6.90	7.60	8.50	8.30

77. TOKYO, HAMARIKYU ASAHI HALL (Opened 1992, 552 seats)

Attribute	Measured by	Year	125	250	500	1,000	2,000	4,000
RT, unoccupied	Takenaka	1992	1.63	1.68	1.83	1.93	1.90	1.71
RT, occupied	Takenaka	1992	1.63	1.57	1.65	1.80	1.74	1.58
EDT, unoccupied	Takenaka	1992	1.53	1.72	1.82	1.80	1.75	1.62
EDT, occupied	Takenaka	1992	1.51	1.63	1.64	1.76	1.65	1.50
$IACC_A$, unoccupied	Takenaka	1992	—	—	0.22	0.15	0.12	0.11
$IACC_A$, occupied	Takenaka	1992	—	—	0.25	0.09	0.12	0.14
$IACC_E$, unoccupied	Takenaka	1992	—	—	0.40	0.29	0.21	0.17
$IACC_E$, occupied	Takenaka	1992	—	—	0.34	0.18	0.22	0.22
$IACC_L$, unoccupied	Takenaka	1992	—	—	0.17	0.14	0.09	0.07
$IACC_L$, occupied	Takenaka	1992	—	—	0.23	0.07	0.07	0.07
C_{80}, dB, unoccupied	Takenaka	1992	−0.30	−1.90	−1.20	0.00	0.60	0.30
C_{80}, dB, occupied	Takenaka	1992	−1.70	−1.10	−0.80	0.10	0.60	1.10
G, dB, unoccupied	Takenaka	1992	7.50	7.60	9.80	10.00	10.80	11.30
G, dB, occupied	Takenaka	1992	4.30	4.40	6.00	7.20	8.70	11.40

78. TOKYO, METROPOLITAN ART SPACE (Opened 1990, 2,017 seats)

Attribute	Measured by	Year	125	250	500	1,000	2,000	4,000
RT, unoccupied	Nagata	1993	2.80	2.60	2.60	2.60	2.40	2.10
RT, occupied	Nagata, calculated	1993	2.60	2.30	2.10	2.10	2.00	1.70
	Matsuzawa	1993	2.50	2.23	2.17	2.19	2.08	1.91
EDT, unoccupied	Nagata	1993	2.80	2.70	2.60	2.50	2.40	2.00
$IACC_A$, unoccupied	Takenaka	1995	—	—	0.29	0.19	0.21	0.22
$IACC_E$, unoccupied	Takenaka	1995	—	—	0.48	0.37	0.37	0.35
$IACC_L$, unoccupied	Takenaka	1995	—	—	0.19	0.10	0.09	0.07
C_{80}, dB, unoccupied	Nagata	1993	−5.88	−3.55	−0.85	−1.50	−1.07	−0.30
G, dB, unoccupied	Takenaka	1995	4.50	4.20	3.90	4.50	4.50	4.90

Attribute	Measured by	Year of Data	Center Frequencies of Filter Bands					
			125	250	500	1,000	2,000	4,000
			Hertz					

79. TOKYO, NEW NATIONAL THEATRE (NNT) OPERA HOUSE (Opened 1997, 1,810 seats)

Attribute	Measured by	Year	125	250	500	1,000	2,000	4,000
RT, unoccupied	Takenaka	1997	1.65	1.66	1.73	1.85	1.83	1.60
RT, occupied	Takenaka	1997	1.62	1.59	1.49	1.49	1.42	1.32
EDT, unoccupied	Takenaka	1997	1.55	1.59	1.65	1.75	1.73	1.48
EDT, occupied	Takenaka	1997	1.53	1.50	1.36	1.28	1.22	1.10
$IACC_A$, unoccupied	Takenaka	1997	0.93	0.78	0.29	0.20	0.21	0.24
$IACC_A$, occupied	Takenaka	1997	0.89	0.78	0.43	0.36	0.32	0.31
$IACC_E$, unoccupied	Takenaka	1997	0.94	0.82	0.42	0.31	0.31	0.34
$IACC_E$, occupied	Takenaka	1997	0.92	0.83	0.54	0.45	0.39	0.38
$IACC_L$, unoccupied	Takenaka	1997	0.92	0.74	0.19	0.13	0.11	0.08
$IACC_L$, occupied	Takenaka	1997	0.87	0.72	0.25	0.16	0.12	0.10
C_{80}, dB, unoccupied	Takenaka	1997	−0.10	0.90	1.70	1.60	1.40	2.20
C_{80}, dB, occupied	Takenaka	1997	0.90	1.70	3.30	3.80	3.70	4.30
G, dB, unoccupied	Takenaka	1997	−0.40	0.20	1.20	2.20	2.40	−0.30
G, dB, occupied	Takenaka	1997	−0.80	−0.40	−0.50	0.40	0.10	−1.60

80. TOKYO, NHK HALL (Opened 1973, 3,677 seats)

Attribute	Measured by	Year	125	250	500	1,000	2,000	4,000
RT, unoccupied	NHK Laboratories	1973	2.30	1.90	1.90	2.00	1.90	1.60
	NHK Laboratories	1988	2.50	2.00	1.90	2.10	2.10	1.70
	Average		2.40	1.95	1.90	2.05	2.00	1.65
RT, occupied	Anonymous	1994	1.77	1.64	1.63	1.72	1.75	1.52
C_{80}, dB, unoccupied	NHK Laboratories	1973	−2.10	−0.10	0.00	0.00	0.00	0.00

81. TOKYO, ORCHARD HALL, BUNKAMURA (Opened 1989, 2,150 seats)

Attribute	Measured by	Year	125	250	500	1,000	2,000	4,000
RT, unoccupied	Ishi (22,500 m³, large orch.)	1989	2.29	2.25	2.22	2.27	2.28	2.04
	Ishi (18,490 m³, chamber)	1989	2.25	2.17	2.08	2.13	2.09	1.88
RT, occupied	Ishi (22,500 m³ w/65 players)	1989	1.90	2.01	1.88	1.92	1.91	1.69
	Anonymous (concert)	1994	1.96	1.95	1.83	1.77	1.64	1.42
C_{80}, dB, unocc.	Ishi (largest stage)	1989	—	—	−2.85	—	−1.58	—

82. TOKYO, SUNTORY HALL (Opened 1986, 2,006 seats)

Attribute	Measured by	Year	125	250	500	1,000	2,000	4,000
RT, unoccupied	Tachibana	1986	2.35	2.40	2.50	2.60	2.60	2.15
	Nagata	1986	2.40	2.60	2.60	2.60	2.60	2.40
RT, occupied	Tachibana	1986	2.20	2.10	2.00	2.00	1.90	1.75
	Anonymous	1994	2.14	2.08	1.95	2.03	2.00	1.77
EDT, unoccupied	Nagata	1986	2.30	2.40	2.30	2.60	2.50	1.90
$IACC_A$, unoccupied	Takenaka	1995	—	—	0.30	0.20	0.21	0.22
$IACC_E$, unoccupied	Takenaka	1995	—	—	0.53	0.45	0.42	0.39
$IACC_L$, unoccupied	Takenaka	1995	—	—	0.22	0.11	0.07	0.05
LF, unoccupied	Takenaka	1995	0.16	0.15	0.17	0.16	0.18	0.19
C_{80}, dB, unoccupied	Nagata	1986	−3.81	−2.68	−0.85	−0.91	−1.00	−0.31
G, dB, unoccupied	Takenaka	1995	3.30	3.80	4.60	5.30	5.40	5.60

Attribute	Measured by	Year of Data	Center Frequencies of Filter Bands					
			125	250	500	1,000	2,000	4,000
			Hertz					

83. TOKYO, TOKYO OPERA CITY, CONCERT HALL (Opened 1997, 1,636 seats)

Attribute	Measured by	Year of Data	125	250	500	1,000	2,000	4,000
RT, unoccupied	Takenaka	1997	2.16	2.51	2.72	2.88	2.98	2.72
RT, occupied	Takenaka	1997	2.07	2.03	1.99	1.93	1.84	1.66
EDT, unoccupied	Takenaka	1997	2.03	2.24	2.65	2.73	2.84	2.54
EDT, occupied	Takenaka	1997	1.76	1.77	1.84	1.81	1.73	1.51
$IACC_A$, unoccupied	Takenaka	1997	0.89	0.68	0.18	0.12	0.11	0.12
$IACC_A$, occupied	Takenaka	1997	0.86	0.66	0.26	0.19	0.17	0.21
$IACC_E$, unoccupied	Takenaka	1997	0.92	0.75	0.36	0.25	0.22	0.25
$IACC_E$, occupied	Takenaka	1997	0.88	0.66	0.33	0.29	0.26	0.31
$IACC_L$, unoccupied	Takenaka	1997	0.88	0.66	0.15	0.09	0.07	0.05
$IACC_L$, occupied	Takenaka	1997	0.85	0.69	0.23	0.13	0.08	0.06
C_{80}, dB, unoccupied	Takenaka	1997	−2.00	−3.50	−2.90	−2.70	−2.60	−2.20
C_{80}, dB, occupied	Takenaka	1997	−2.50	−1.60	−0.80	0.20	0.70	1.8
G, dB, unoccupied	Takenaka	1997	4.00	5.40	6.00	6.30	6.90	6.30
G, dB, occupied	Takenaka	1997	3.50	4.20	4.70	4.50	4.60	3.90

84. KUALA LUMPUR, DEWAN FILHARMONIK PETRONAS (Opened 1998, 850 seats)

Note: All data are from Kirkegaard & Assoc.

Attribute	Measured by	Year of Data	125	250	500	1,000	2,000	4,000
RT, unoccupied	Standard large stage hall	1998	2.75	2.37	2.30	2.24	2.00	1.75
	Half volume; 100% absorption	1998	2.32	1.90	1.63	1.67	1.55	1.40
	Minimal volume; 100% absorption	1998	2.00	1.67	1.50	1.42	1.37	1.23
	Full volume: no absorption	1998	2.90	2.50	2.46	2.42	2.17	1.84
	Half volume; no absorption	1998	2.70	2.24	1.97	1.96	1.80	1.55
	Minimal volume; no absorption	1998	2.20	1.77	1.63	1.65	1.55	1.30
RT, occupied	Est.: JASA **109**: 1028, std.	2001	2.50	2.20	2.05	1.90	1.70	1.50

85. MEXICO CITY, SALLA NEZAHUALCOYOTL (Opened 1976, 2,376 seats)

Attribute	Measured by	Year of Data	125	250	500	1,000	2,000	4,000
RT, unoccupied	Jaffe	1994	2.80	2.50	2.20	2.20	2.00	1.60
RT, occupied	Jaffe	1994	2.20	2.30	2.00	1.90	1.80	1.70

86. AMSTERDAM, CONCERTGEBOUW (Opened 1888, 2,037 seats)

Attribute	Measured by	Year of Data	125	250	500	1,000	2,000	4,000
RT, unoccupied	Takenaka	1993	2.68	2.53	2.59	2.63	2.43	2.05
	Bradley	1987	2.60	2.40	2.50	2.53	2.35	1.97
	Tachibana	1986	2.80	2.65	2.65	2.75	2.45	1.85
	Gade	1987	2.62	2.47	2.45	2.55	2.33	1.97
	Average		2.68	2.51	2.55	2.62	2.39	1.96
RT, occupied	Tachibana	1986	2.20	2.15	2.05	1.95	1.80	1.55

Attribute	Measured by	Year of Data	Center Frequencies of Filter Bands					
			125	250	500	1,000	2,000	4,000
			Hertz					

86. AMSTERDAM, CONCERTGEBOUW (Opened 1888, 2,037 seats) *continued*

Attribute	Measured by	Year	125	250	500	1,000	2,000	4,000
EDT, unoccupied	Takenaka	1993	2.51	2.47	2.58	2.64	2.44	1.98
	Bradley	1987	2.51	2.39	2.54	2.57	2.36	1.93
	Gade	1987	2.82	2.65	2.64	2.78	2.47	2.10
	Average		2.61	2.50	2.59	2.66	2.42	—
$IACC_A$	Takenaka	1993	0.91	0.69	0.21	0.17	0.27	0.28
$IACC_E$	Takenaka	1993	0.94	0.78	0.46	0.42	0.51	0.44
$IACC_L$	Takenaka	1993	0.90	0.66	0.15	0.10	0.07	0.06
LF, unoccupied	Bradley	1987	0.16	0.17	0.20	0.17	0.14	0.23
	Gade	1987	0.21	0.12	0.18	0.17	—	—
C_{80}, dB, unoccupied	Takenaka	1993	−5.20	−4.40	−3.90	−2.60	−1.70	−0.80
	Bradley	1987	−5.09	−4.80	−3.91	−2.60	−2.42	−1.38
	Gade	1987	−5.91	−4.80	−4.75	−4.02	−3.84	−2.22
	Average		−5.40	−4.67	−4.19	−3.07	−2.65	−1.47
G, dB, unoccupied	Takenaka	1993	5.80	5.90	6.20	6.50	5.80	4.90
	Bradley	1987	5.46	4.99	5.37	5.71	5.23	4.20
	Gade	1987	3.88	4.72	5.87	5.07	6.13	—

87. ROTTERDAM, DE DOELEN, CONCERTGEBOUW (Opened 1966, 2,242 seats)

Attribute	Measured by	Year	125	250	500	1,000	2,000	4,000
RT, unoccupied	Hak & Martin	1992	2.00	2.00	2.40	2.30	2.30	1.90
RT, occupied	Kimura	1992	1.90	2.00	2.00	2.10	2.00	1.85
EDT, unoccupied	Hak & Martin	1992	2.30	2.20	2.30	2.30	2.10	1.60
C_{80}, dB, unoccupied	Hak & Martin	1992	−6.50	−3.60	−2.90	−2.80	−2.70	−1.00

88. CHRISTCHURCH, TOWN HALL (Opened 1972, 2,662 seats)

Attribute	Measured by	Year	125	250	500	1,000	2,000	4,000
RT, unoccupied	Yamasaki	1992	2.60	2.20	2.35	2.35	2.20	1.70
	Marshall	1994	2.54	2.39	2.50	2.40	2.29	1.88
RT, occupied	Estimated (EDT-0.1 sec)	1994	2.10	1.70	1.80	1.80	1.70	1.40
EDT, unoccupied	Barron (w/overhung seats)	1983	1.88	1.72	1.94	2.07	2.06	—
	Marshall	1994	2.17	1.77	1.90	1.88	1.81	1.53
LF, unoccupied	Barron	1983	0.16	0.15	0.14	0.14	—	—
C_{80}, dB, unoccupied	Marshall	1994	−2.50	0.20	1.30	1.90	1.30	2.00

89. TRONDHEIM, OLAVSHALLEN (Opened 1989, 1,200 seats)

Attribute	Measured by	Year	125	250	500	1,000	2,000	4,000
RT, unoccupied	Strom Concert Hall	1991	1.90	1.90	1.80	1.80	1.70	1.40
	Holmefjord Concert Hall	1997	1.60	1.70	1.80	1.70	1.60	1.20
RT, occupied	Est.: JASA **109**: 1028	2001	1.65	1.65	1.65	1.65	1.50	1.20
RT, unoccupied	Strom Theater	1991	1.50	1.40	1.30	1.20	1.20	1.00
C_{80}, dB, unoccupied	Holmefjord Concert Hall	1997	1.70	0.70	1.70	4.10	5.60	6.70
LF, unoccupied	Holmefjord Concert Hall	1997	0.10	0.10	0.16	0.13	0.12	0.11

Attribute	Measured by	Year of Data	Center Frequencies of Filter Bands					
			125	250	500	1,000	2,000	4,000
			Hertz					

90. EDINBURGH, USHER HALL (Opened 1914; updated 2000; 2,502 seats)

Attribute	Measured by	Year	125	250	500	1,000	2,000	4,000
RT, unoccupied	Sandy Brown Associates	2001	2.40	2.20	2.55	2.35	2.00	1.55
RT, occupied	Sandy Brown, 90% occupied	2000	1.55	1.65	1.80	1.75	1.55	1.27

91. GLASGOW, ROYAL CONCERT HALL (Opened 1990, 2,459 seats) (Renovations in progress)

Attribute	Measured by	Year	125	250	500	1,000	2,000	4,000
RT, unoccupied	Barron	1990	2.32	2.15	1.99	1.87	1.71	—
RT, occupied	Sandy Brown Assoc.	1990	2.04	1.88	1.76	1.74	1.67	1.44
EDT, unoccupied	Barron	1990	2.27	1.97	1.78	1.67	1.53	—
LF, unoccupied	Barron	1990	0.20	0.28	0.24	0.19	—	—
C_{80}, dB, unoccupied	Barron	1990	−4.60	−1.30	0.70	1.10	1.40	—
G, dB, unoccupied	Barron	1990	1.30	1.60	2.40	0.80	1.70	—

92. MADRID, AUDITORIO NACIONAL DE MÚSICA (Opened 1988, 2,293 seats)

Attribute	Measured by	Year	125	250	500	1,000	2,000	4,000
RT, unoccupied	Garcia-BBM	1989	2.39	2.13	2.07	2.13	1.93	1.51
RT, occupied	Garcia-BBM	1989	2.10	2.02	1.85	1.62	1.50	1.46
EDT, unoccupied	Garcia-BBM	1989	2.18	2.19	2.03	2.08	1.85	1.35
LF, unoccupied	Garcia-BBM	1989	0.17	0.28	0.34	0.28	0.24	0.27
C_{80}, dB, unoccupied	Garcia-BBM	1989	−3.03	−1.97	−1.21	−0.06	0.06	2.30

93. VALENCIA, PALAU DE LA MÚSICA (Opened 1987, 1,790 seats)

Attribute	Measured by	Year	125	250	500	1,000	2,000	4,000
RT, unoccupied	Garcia-BBM	1987	3.05	3.6	3.35	3.00	2.60	2.20
RT, occupied	Garcia-BBM	1987	2.10	2.30	2.10	2.00	2.00	2.00
EDT, unoccupied	Garcia-BBM	1987	2.73	3.32	2.97	2.95	2.60	2.19
LF, unoccupied	Garcia-BBM	1987	0.22	0.27	0.36	0.33	0.29	0.25
C_{80}, dB, unoccupied	Garcia-BBM	1987	−4.98	−7.55	−5.45	−3.51	−2.12	0.98

94. GOTHENBURG, KONSERTHUS (Opened 1935; revised 2001; 1,286 seats)

Attribute	Measured by	Year	125	250	500	1,000	2,000	4,000
RT, unocccupied	Jorden and Rindell	2002	2.28	1.96	1.88	1.74	1.65	1.48
RT, occupied	Est.: JASA **109**: 1028	2002	2.10	1.85	1.70	1.60	1.50	1.35
EDT, unoccupied	Jorden and Rindell	2002	2.27	1.90	1.88	1.76	1.69	1.39
C_{80}, dB, unoccupied	Jorden and Rindell	2002	−1.86	−2.02	−1.31	0.16	0.95	1.25
G, dB, unoccupied	Jorden and Rindell	2002	10.8	8.2	3.6	3.1	3.6	3.1

95. BASEL, STADT-CASINO (Opened 1876, 1,448 seats)

Attribute	Measured by	Year	125	250	500	1,000	2,000	4,000
RT, unoccupied	Takenaka	1993	2.78	2.74	2.31	2.31	2.23	1.90
RT, occupied	Beranek	1965	2.20	2.00	1.80	1.75	1.60	1.50
EDT, unoccupied	Takenaka	1993	2.55	2.62	2.19	2.20	2.13	1.79
$IACC_A$, unoccupied	Takenaka	1993	0.90	0.72	0.22	0.13	0.17	0.18
$IACC_E$, unoccupied	Takenaka	1993	0.89	0.78	0.46	0.34	0.33	0.29
$IACC_L$, unoccupied	Takenaka	1993	0.90	0.69	0.17	0.09	0.07	0.06
C_{80}, dB, unoccupied	Takenaka	1993	−4.10	−4.50	−3.20	−2.00	−1.70	−0.70
G, dB, unoccupied	Takenaka	1993	9.10	8.90	7.90	8.30	7.70	7.20

Attribute	Measured by	Year of Data	Center Frequencies of Filter Bands					
			125	250	500	1,000	2,000	4,000
			Hertz					

96. LUCERNE, CULTURE AND CONGRESS CENTER, CONCERT HALL (Opened 1999, 1,892 seats)

Attribute	Measured by	Year	125	250	500	1,000	2,000	4,000
RT, occupied	(A) Doors open	1998	3.03	2.48	2.15	2.10	1.80	1.45
	(B) ¾ doors open	2002	2.65	2.45	2.15	2.00	1.70	1.50
	(A) Upper open, lower clsd.	1998	2.90	2.35	2.05	2.05	1.60	1.30
	(A) Doors closed	1998	2.20	2.05	1.95	1.70	1.50	1.40
	(B) Doors closed, Curtains	2002	2.30	1.90	1.60	1.60	1.40	1.30

(A) Data by ARTEC, (B) Data by Beranek (Curtains pulled over top layer of closed doors)

97. ZURICH, GROSSER TONHALLESAAL (Opened 1895, 1,546 seats)

Attribute	Measured by	Year	125	250	500	1,000	2,000	4,000
RT, unoccupied	Takaneka	1991	3.59	3.60	3.27	3.09	2.59	2.12
RT, occupied	Beranek	1965	2.50	2.40	2.15	1.95	1.75	1.62
EDT, unoccupied	Takenaka	1991	3.58	3.77	3.21	3.02	2.58	2.01
$IACC_A$, unoccupied	Takenaka	1991	0.89	0.68	0.21	0.11	0.15	0.15
$IACC_E$, unoccupied	Takenaka	1991	0.93	0.79	0.48	0.27	0.33	0.28
$IACC_L$, unoccupied	Takenaka	1991	0.89	0.66	0.16	0.09	0.09	0.06
C_{80}, dB, unoccupied	Takenaka	1991	−5.80	−6.80	−4.40	−3.60	−2.80	−1.50
G, db, unoccupied	Takenaka	1991	9.00	8.90	8.90	8.20	8.30	7.90

98. TAIPEI CULTURAL CENTRE, CONCERT HALL (Opened 1987, 2,074 seats)

Attribute	Measured by	Year	125	250	500	1,000	2,000	4,000
RT, unoccupied	Chung-Shan Institute	2002	2.71	2.42	2.46	2.55	2.50	2.19
RT, occupied	Beranek (s.c.)	1997	1.95	1.93	2.00	2.07	2.03	1.78
EDT, unoccupied	Chung-Shan Institute	2002	2.39	2.28	2.46	2.57	2.41	2.10
C_{80}, dB, unoccupied	Chung-Shan Institute	2002	−7.71	−5.87	−4.65	−3.28	−2.74	−2.70
$IACC_E$	Chung-Shan Institute	2002	0.94	0.70	0.20	0.15	0.14	0.19
LF_E	Chung-Shan Institute	2002	0.17	0.21	0.29	0.20	0.28	0.28

99. CARACAS, AULA MAGNA (Opened 1954, 2,660 seats)

Attribute	Measured by	Year	125	250	500	1,000	2,000	4,000
RT, unoccupied	M.A.A. (BBN)	1954	2.50	2.25	1.85	1.75	1.90	1.70
RT, occupied	M.A.A. (BBN)	1954	1.90	1.40	1.30	1.20	1.00	0.90

100. CARDIFF, ST. DAVID'S HALL (Opened 1982, 1,952 seats)

Attribute	Measured by	Year	125	250	500	1,000	2,000	4,000
RT, unoccupied	Barron	1982	1.83	1.98	2.07	2.15	2.10	—
	Gade	1986	1.95	2.00	2.10	2.25	2.15	1.75
RT, occupied	Sandy Brown Associates	1983	1.88	1.97	1.96	1.96	1.80	1.56
EDT, unoccupied	Barron	1982	1.90	2.13	2.11	2.14	2.11	—
	Gade	1986	1.93	2.10	2.01	2.03	2.01	1.68
LF, unoccupied	Barron	1982	0.17	0.14	0.19	0.16	—	—
	Gade	1986	0.15	0.19	0.18	0.15	—	—
C_{80}, dB, unoccupied	Barron	1982	−4.20	−1.50	−0.70	−0.50	−0.90	—
	Gade	1986	−2.31	−2.17	−1.04	−0.73	−0.88	−0.08
G, dB, unoccupied	Barron	1982	2.00	0.40	2.80	3.60	3.00	—
	Gade	1986	2.34	1.92	3.92	3.26	2.37	—

APPENDIX

3

Equations, Technical Data,
and Sound Absorption

$\mathscr{E}$QUATIONS FOR ACOUSTICAL PARAMETERS

THE INTERAURAL CROSS-CORRELATION FAMILY, IACF, IACC$_A$, IACC$_E$, AND IACC$_L$. A binaural measure of the difference in the sounds at the two ears produced by a sound source on the stage is the interaural cross-correlation function IACF$_t(\tau)$:

$$\text{IACF}_t(\tau) = \frac{\int_{t_1}^{t_2} p_L(t) p_R\ (t\ +\ \tau) dt}{\left(\int_{t_1}^{t_2} p_L^2\ dt \int_{t_1}^{t_2} p_R^2\ dt \right)^{1/2}} \tag{A3.1}$$

where L and R designate the entrances to the left and right ears, respectively. The maximum possible value of Eq. (A3.1) is unity. Time 0 is the time of arrival of the direct sound from the impulse radiated by the source. Integration from 0 to t_2 msec includes the energy of the direct sound and whatever early reflections and reverberant sounds fall within the t_2 time period.

Because the time it takes for a sound wave impinging perpendicular to one side of the head to travel to the other side is about 1 msec, it is customary to vary τ over the range of -1 to $+1$ msec. Further, to obtain a single number that measures the maximum similarity of all waves arriving at the two ears within the time integration limits and the range of τ, it is customary to select the maximum magnitude of Eq. (A3.1), which is then called the interaural cross-correlation coefficient (IACC):

$$\text{IACC}_t\ =\ |\text{IACF}_t(\tau)|\max \quad \text{for } -1 < \tau < +1. \tag{A3.2}$$

With different integration periods we have IACC$_A$ ($t_1 = 0$ to $t_2 = 1{,}000$ msec); IACC$_{E(\text{arly})}$ (0–80 msec); and IACC$_{L(\text{ate})}$ (80–1,000 msec). The E(arly) IACC is a measure of the apparent source width ASW and the L(ate) IACC is a (poor) measure of the listener envelopment LEV.

THE BINAURAL QUALITY INDEX, BQI. The Binaural Quality Index is given by the formula

$$BQI = [1 - IACC_{E3}] \qquad \text{(A3.2a)}$$

Where E indicates integration of the early sound (0–80 msec) and 3 indicates the average of the $IACC_E$ measured values in the three octave bands with center frequencies at 500, 1,000, and 2000 Hz.

THE CLARITY FACTOR, C_{80}. The clarity factor, (C_{80}), expressed in decibels, is the ratio of the early energy (0–80 msec) to the late (reverberant) energy (80–3,000 msec):

$$C_{80} = 10 \log \frac{\int_0^{.08} p^2(t)dt}{\int_{.08}^{\infty} p^2(t)dt} \text{ dB.} \qquad \text{(A3.3)}$$

THE LATERAL ENERGY FRACTION, LF. The lateral energy fraction (LF) is the ratio of the output of a figure-8 microphone (with its null direction aimed at the source) to the output of a non-directional microphone. The figure-8 microphone weights the non-direct energy by $\cos^2 \theta$, where $\theta = 90°$ is in the direction of the sound. LF is given by

$$LF = \frac{\int_{.005}^{.08} p_8^2(t)dt}{\int_0^{.08} p^2(t)dt}. \qquad \text{(A3.4)}$$

As shown, the time integration is usually performed over the interval of 5–80 msec for the figure-8 microphone and 0–80 msec for the omnidirectional microphone. The 5-msec value is introduced to make certain that the direct sound is eliminated.

THE STRENGTH FACTOR, G. The strength factor (G) is a measure of the sound-pressure level at a point in a hall, with an omnidirectional source on stage, minus the sound-pressure level that would be measured at a distance of 10 m from the same sound source operating at the same power level and located in an anechoic chamber. The equation is

$$G = 10 \log \frac{\int_0^{t_2} p^2(t)dt}{\int_0^{t_2} p_A^2(t)dt} \text{ dB,} \qquad (A3.5)$$

where $P_A(t)$ is the free-field sound pressure at a distance of 10 m.

THE BASS RATIO, RT. Bass ratio RT is the ratio of the low to mid-frequency reverberation times, given by

$$BR = \frac{RT_{125} + RT_{250}}{RT_{500} + RT_{1,000}}, \qquad (A3.6)$$

where the RT's are the reverberation times at the frequencies shown in the subscripts.

THE SUPPORT FACTOR, ST1. The support factor (ST1) is the difference in decibels between two measurements of sound-pressure level made on a stage or in a pit where the orchestra members play. A sound source (loudspeaker) emits an impulse and the microphone receives it at a point 1 m removed from the center of the source, which is omnidirectional. The first measurement is of the energy in the time interval from 0 to 10 msec and the second measurement is of the energy in the time interval from 20 to 100 msec. It is given by the equation

$$ST1 = 10 \log \left[\frac{\int_{.02}^{0.1} p^2(t)dt}{\int_0^{.01} p^2(t)dt} \right]. \qquad (A3.7)$$

THE LATE LATERAL ENERGY, LG. Late lateral energy (LG) is 10 log of the ratio of the delayed output of a figure-8 microphone (with its null direction aimed at the source) to the total output of a non-directional microphone, where the latter is measured at a distance of 10 m from the acoustical center of an omni-directional source in an anechoic chamber operating at the same power output. The time integration for the figure-8 microphone is performed from 80 msec to several seconds after the arrival of the impulse. It is given by the equation:

$$ LG = 10 \log \left[\frac{\int_{.08}^{\infty} p_8^2(t)dt}{\int_{0}^{\infty} p_A^2(t)dt} \right] \qquad (A3.8) $$

THE DISTINCTNESS (DEUTLICHKEIT) RATIO, D. The distinctness ratio (D) is the ratio of the sound in the first 50 msec after arrival of the direct sound to the total sound arriving. It is usually expressed as a percentage and is determined from the impulse response of the hall.

$$ D = 100 \frac{\int_{0}^{.05} p^2(t)dt}{\int_{0}^{\infty} p^2(t)dt} \ \% \ . \qquad (A3.9) $$

MEASURED ACOUSTICAL QUANTITIES

In Chapters 4 and 5 measured data are cited for a number of the acoustical attributes that are defined in Chapter 2 and Appendix 1. To facilitate research and to ease the search for data that are scattered throughout this book, a compilation of both acoustical and architectural information is presented in the next three tables.

Table A3.1 presents the available acoustical data for 86 concert halls, while Table A3.2 presents the same categories of available data for 34 opera houses. Architectural-type information is presented in Table A3.3. For example, the area allotted to each seat is shown by the quantity S_a/N and this varies from a low of 0.402 m² in Boston Symphony Hall to a high of 0.690 in the Munich, Philharmonie Am Gasteig. The term $[EDT_{unocc}/V] \times 10^6$, which appears in Figs. 4.8 and 4.9 and which is closely related to the strength of the sound G in a hall, is also presented.

$\mathcal{T}$ABLE A3.1. Basic material on concert halls relative to the studies in Chapter 4. Definitions of terms are given in Appendix 1 and formulas in Appendix 2. Opera houses are listed in Table A3.2.

Name of Hall	No. of Seats	Cubic Volume, m³	sec			BQI (3-Band) (Early)	dB		ST1 (4-Band)	LF (4-Band) (Early)
			RT_{mid} Occ.	RT_{mid} Unocc.	EDT_{mid} Unocc.		G_{mid} (Euro) Unocc.	G_{125} (Euro) Unocc.		
Amsterdam, Concertgebouw	2,037	18,780	2.00	2.59	2.63	0.54	5.4	4.7	−17.8	0.18
Aspen, Benedict Music Tent	2,050	19,830	3.50	—	—	—	—	—	—	—
Athens, Megaron, Concert Hall	1,962	19,100	1.85	2.30	—	—	3.1	—	—	—
Baden-Baden, Festspielhaus	2,300	20,100	1.92	2.09	1.82	—	3.4	−0.3	—	—
Baltimore, Meyerhoff Symphony Hall	2,467	21,530	2.00	2.34	2.31	0.56	3.6	3.0	−12.2	0.17
Basel, Stadt-Casino	1,448	10,500	1.78	2.31	2.20	0.62	6.9	7.9	−13.7	—
Bayreuth, Festspielhaus	1,800	10,308	1.65	—	—	—	—	—	—	—
Belfast, Waterfront Hall	2,250	30,800	2.15	2.35	—	—	—	—	—	—
Berlin, Kammermusiksaal (Philharmonie)	1,138	11,000	1.82	2.13	2.10	—	—	—	—	—
Berlin, Konzerthaus (Schauspielhaus)	1,575	15,000	2.00	2.47	2.43	0.64	5.7	7.1	−16.8	—
Berlin, Philharmonie	2,218	21,000	1.90	2.20	2.01	0.47	3.7	3.0	−16.8	—
Birmingham, Symphony Hall	2,211	25,000	1.85	2.52	2.00	—	—	—	—	—
Bonn, Beethovenhalle	1,407	15,730	1.65	1.80	—	—	—	—	—	—
Boston, Symphony Hall	2,625	8,750	1.90	2.52	2.37	0.61	4.2	2.9	−13.7	0.22
Brussels, Palais des Beaux-Arts	2,150	12,520	1.60	1.76	1.86	—	—	—	—	—

| Name of Hall | No. of Seats | Cubic Volume, m³ | sec | | | BQI (3-Band) (Early) | dB | | | LF |
			RT_{mid} Occ.	RT_{mid} Unocc.	EDT_{mid} Unocc.		G_{mid} (Euro) Unocc.	G₁₂₅ (Euro) Unocc.	ST1 (4-Band)	(4-Band) (Early)
Budapest, Pátria Hall	1,750	13,400	1.70	1.81	—	—	—	—	—	—
Buenos Aires, Teatro Colón	2,487	21,524	1.62	1.90	1.72	0.65	1.2	−0.8	—	—
Buffalo, Kleinhans Music Hall	2,839	18,240	1.50	1.82	1.61	0.41	2.7	3.9	−14.8	0.10
Caracas, Aula Magna	2,660	24,920	1.25	1.80	—	—	—	—	—	—
Cardiff, St. David's Hall	1,952	22,000	1.96	2.15	2.07	—	3.2	2.2	−16.6	0.17
Chicago, Orchestra Hall	2,530	27,000	1.72	1.90	—	—	—	—	—	—
Christchurch, Town Hall	2,662	20,500	1.90	2.34	1.89	—	—	—	—	0.15
Cleveland, Severance Hall	2,101	16,290	1.60	2.15	2.25	0.59	3.5	1.8	−14.8	—
Copenhagen, Radiohuset, Studio 1	1,081	11,900	1.50	1.96	1.96	—	−3.00	0.8	−14.5	0.16
Costa Mesa, Segerstrom Hall	2,903	27,800	1.60	2.40	2.18	0.61	4.2	1.3	−14.3	0.23
Dallas, Meyerson Symphony Center	2,065	23,900	2.80	2.90	1.90	—	—	—	—	—
Denver, Boettcher Hall	2,750	37,444	2.40	2.68	2.45	0.30	1.5	2.0	—	0.12
Edinburgh, Usher Hall	2,502	15,700	1.78	2.45	2.40	—	—	—	−16.3	—
Fort Worth, Bass Performance Hall	2,072	27,300	1.95	2.60	2.50	0.52	—	—	−16.3	—
Glasgow, Royal Concert Hall	2,457	22,700	1.75	1.95	1.72	—	2.00	2.00	—	—
Gothenberg, Konserthus	1,286	11,900	1.62	1.70	1.75	—	4.7	5.8	−14.3	0.10
Hong Kong, Cul. Ctr. Concert Hall	2,019	21,250	—	2.00	1.89	—	2.0	3.1	—	0.14
Jerusalem, Binyanei Ha'Oomah	3,142	24,700	1.75	1.80	1.80	0.55	—	—	—	—
Kuala Lumpur, Dewan Fil. Petronas	850	17,860	1.80	2.12	—	—	—	—	—	—
Kyoto, Concert Hall	1,840	20,000	1.95	2.25	2.10	—	—	—	—	—

Name of Hall	No. of Seats	Cubic Volume, m³	sec			BQI (3-Band) (Early)	dB			LF (4-Band) (Early)
			RT$_{mid}$ Occ.	RT$_{mid}$ Unocc.	EDT$_{mid}$ Unocc.		G$_{mid}$ (Euro) Unocc.	G$_{125}$ (Euro) Unocc.	ST1 (4-Band)	
Lahti, Sibelius/Talo	1,250	15,500	2.20	—	—	—	—	—	—	—
Leipzig, Gewandhaus	1,900	21,000	2.02	2.20	—	—	—	—	—	—
Lenox, MA, Seiji Ozawa Hall	1,180	11,610	1.66	2.25	—	—	—	—	—	—
Lenox, Tanglewood Music Shed	5,121	42,480	1.90	3.42	3.33	0.37	1.9	0.0	−12.8	—
Liverpool, Philharmonic Hall	1,803	13,560	1.50	1.72	1.79	—	3.4	1.2	—	0.17
London, Barbican, Concert Hall	2,026	17,750	1.68	1.90	1.91	—	4.0	2.0	—	0.16
London, Royal Albert Hall	5,222	86,650	2.41	3.08	2.65	—	−0.7	−1.7	—	0.14
London, Royal Festival Hall	2,901	21,950	1.46	1.64	1.42	—	1.9	0.2	−16.0	0.18
Lucerne, Cultural Ctr. Concert Hall	1,892	18,000	2.05	—	—	—	—	—	—	—
Madrid, Auditorio Nacional de Música	2,293	20,000	1.74	2.10	2.06	—	—	—	0.27	—
Manchester, Bridgewater Hall	2,357	25,000	2.00	2.44	2.40	—	3.2	5.0	0.26	—
Mexico City, Salla Nezahualcoyotl	2,376	30,640	1.95	2.20	—	—	—	—	—	—
Minneapolis, Minn. Orchestra Hall	2,450	18,975	1.85	2.28	—	—	—	—	—	—
Montreal, Salle Wilfrid-Pelletier	2,982	26,500	1.67	2.00	1.93	—	0.1	−0.4	—	0.12
Munich, Herkulessalle	1,287	13,590	1.80	2.30	—	—	—	—	—	—
Munich, Philharmonie Am Gasteig	2,487	29,700	1.80	2.22	2.12	—	1.9	0.6	—	0.12
New York, Avery Fisher Hall	2,742	20,400	1.76	2.24	1.96	—	—	—	—	0.12
New York, Carnegie Hall	2,804	24,270	1.79	2.05	—	—	—	—	—	—
Odense, Koncerthus, Nielsen Hall	1,320	14,000	—	2.25	2.25	—	3.9	1.8	−17.0	0.20
Osaka, Symphony Hall	1,702	17,800	1.80	2.20	2.10	0.59	—	—	—	—

Name of Hall	No. of Seats	Cubic Volume, m³	RT$_{mid}$ Occ.	RT$_{mid}$ Unocc.	EDT$_{mid}$ Unocc.	BQI (3-Band) (Early)	G$_{mid}$ (Euro) Unocc.	G$_{125}$ (Euro) Unocc.	ST1 (4-Band)	LF (4-Band) (Early)
			sec				dB			
Paris, Salle Pléyel	2,386	15,500	1.48	2.00	1.89	—	3.9	5.5	—	0.16
Philadelphia, Verizon Hall, Kimmel Center	2,519	23,520	1.92	—	1.72	—	—	—	—	—
Rochester, NY, Eastman Theatre	3,347	25,500	1.65	1.82	—	0.55	—	—	—	—
Rotterdam, De Doelen	2,242	24,070	2.05	2.35	2.30	—	—	—	—	—
Salt Lake City, Symphony Hall	2,812	19,500	1.70	2.03	2.08	0.59	1.4	0.7	− 12.9	—
Salzburg, Festspielhaus	2,158	15,500	1.50	1.94	1.87	—	3.8	1.8	− 15.8	0.16
San Francisco, Davies Hall	2,743	24,070	1.85	2.14	2.15	0.44	2.2	1.3	—	—
São Paulo, Sala São Paulo	1,610	20,000	2.05	—	—	—	—	—	—	—
Sapporo, Concert Hall	2,008	28,800	1.80	2.13	1.90	0.47	2.1	0.8	0.12	—
Seattle, Benaroya Hall	2,500	19,263	1.97	2.20	—	—	—	—	—	—
Shanghai, Grand Theatre	1,895	13,000	1.65	1.83	—	—	—	—	—	—
Stuttgart, Liederhalle, Grosser Saal	2,000	16,000	1.65	2.10	2.14	—	3.7	2.7	− 14.5	0.13
Sydney Opera House, Concert Hall	2,696	24,600	2.20	2.49	2.19	—	—	—	—	—
Taipei, Cultural Centre, Concert Hall	2,074	16,700	2.02	2.20	—	—	—	—	—	—
Tel Aviv, Fredric Mann Auditorium	2,715	21,240	1.50	1.67	1.70	0.47	—	—	—	—
Tokyo, Bunka Kaikan (Ueno)	2,327	17,300	1.52	1.96	—	0.59	3.1	3.3	—	0.18
Tokyo, Dai-ichi Seimei Hall	767	6,800	1.56	1.83	1.80	0.88	6.0	6.0	—	—
Tokyo, Hamarikyu Asahi Hall	522	5,800	1.72	1.88	1.86	0.70	8.7	6.3	—	—
Tokyo, Metropolitan Art Space	2,017	25,000	2.15	2.60	2.55	0.59	3.0	3.3	—	—
Tokyo, NHK Hall	3,677	25,200	1.69	1.98	—	—	—	—	—	—

Name of Hall	No. of Seats	Cubic Volume, m³	sec			BQI (3-Band) (Early)	dB			
			RT_{mid} Occ.	RT_{mid} Unocc.	EDT_{mid} Unocc.		G_{mid} (Euro) Unocc.	G_{125} (Euro) Unocc.	ST1 (4-Band)	LF (4-Band) (Early)
Tokyo, Orchard Hall	2,150	20,500	1.82	2.10	—	—	—	—	—	—
Tokyo, Suntory Hall	2,006	21,000	1.98	2.00	2.45	0.53	3.8	2.2	—	0.16
Tokyo, Opera City Concert Hall	1,636	15,300	1.96	2.80	2.69	0.72	5.0	2.8	—	—
Toronto, Roy Tompson Hall	2,613	24,500	1.80	—	—	—	—	—	—	—
Trondheim, Olavshallen	1,200	13,000	1.65	1.80	—	—	—	—	—	0.12
Valencia, Palau de la Música	1,790	15,400	2.05	3.17	2.96	—	—	—	—	0.30
Vienna, Grosser Musikvereinssaal	1,680	15,000	2.00	3.05	3.04	0.64	6.5	6.6	− 13.9	0.16
Vienna, Konzerthaus	1,865	16,600	1.88	2.19	2.12	0.66	3.8	6.6	—	—
Washington, DC, JFK Concert Hall	2,448	22,300	1.72	1.92	—	0.61	2.50	2.10	− 18.1	—
Worcester, Mechanics Hall	1,343	10,760	1.55	2.15	2.15	0.55	5.0	4.0	− 16.1	0.20
Zurich, Grosser Tonhallesaal	1,546	11,400	2.05	3.18	3.12	0.64	7.5	7.8	− 12.6	—
Average	2,163	20,379	1.84	2.20	2.14					

$\mathcal{T}$ABLE A3.2. Basic material on opera houses relative to the studies in Chapter 5. Definitions of terms are given in Appendix 1 and formulas in Appendix 2. Concert halls are listed in Table A3.1.

Name of Opera House	No. of Seats	Cubic Volume, m³	sec		BQI (3-Band) (Early)	G_{mid} (Jap.) Unocc. dB	ITDG, msec	Stage Set
			RT_{mid} Occ.	EDT_{mid} Unocc.				
Amsterdam, Music Theater	1,689	10,000	1.30	1.30	0.55	1.7	32	n
Athens, Megaron	1,700	15,000	1.60	1.70	—	3.5	40	—
Bayreuth, Festspielhaus	1,800	10,308	1.55	—	—	—	—	—
Berlin, Deutscheoper	1,900	10,800	1.36	1.60	0.39	1.2	33	n
Berlin, Komischeoper	1,222	7,000	1.25	1.23	0.62	6.0	20	y
Budapest, Magyar Allami Operahaz	1,277	8,900	1.34	1.37	0.65	4.4	15	y
Buenos Aires, Teatro Colón	2,487	20,570	1.56	1.72	0.65	2.4	18	y
Chicago, Civic Opera House	3,563	23,000	1.51	1.49	0.53	0.3	41	n
Dresden, Semperoper	1,284	12,480	1.60	1.83	0.72	2.7	20	n
Essen, Opera House	1,125	8,800	1.61	1.90	0.54	−0.4	16	n
Fort Worth, Bass Performance Hall	1,960	18,470	1.80	1.85	—	—	25	—
Glyndebourne, Opera House	1,243	7,790	1.25	1.23	—	3.6	—	—
Hamburg, Staatsoper	1,679	11,000	1.23	1.35	0.46	1.3	34	y
London, Royal Opera House	2,157	12,250	1.20	—	0.53	0.7	18	n
Milan, Teatro alla Scala	2,289	11,252	1.20	1.20	0.48	−0.3	16	y
Naples, Teatro di San Carlo	1,414	13,700	1.15	1.15	—	0.9	—	—
New York, Metropolitan Opera House	3,816	24,724	1.55	1.62	0.62	0.5	18	n
Paris, Opéra Bastille	2,700	21,000	1.57	1.55	—	—	—	—

Name of Opera House	No. of Seats	Cubic Volume, m³	sec		BQI (3-Band) (Early)	G$_{mid}$ (Jap.) Unocc. dB	ITDG, msec	Stage Set
			RT$_{mid}$ Occ.	EDT$_{mid}$ Unocc.				
Paris, Opéra Garnier	2,131	10,000	1.18	1.19	0.50	0.7	15	y
Philadelphia, Academy of Music	2,827	15,100	1.20	1.20	0.47	2.6	19	—
Prague, Staatsoper	1,554	8,000	1.23	1.17	0.64	2.2	16	y
Rochester, Eastman Teatre	3,347	23,970	1.65	1.9	0.54	3.6	22	y
Salzburg, Festspielhaus	2,158	14,020	1.50	1.80	0.40	1.2	27	n
San Francisco, War Mem. Opera House	3,252	20,900	1.50	—	—	—	—	—
Seattle, Opera House	3,099	22,000	2.02	2.50	0.48	2.7	25	n
Shanghai, Grand Theatre	1,676	13,000	1.30	1.37	—	—	—	—
Tokyo, Bunka Kaikan (Ueno)	2,303	16,250	1.51	1.75	0.56	0.3	14	n
Tokyo, New National Theatre Opera House	1,810	14,500	1.49	1.70	0.65	1.7	20	n
Tokyo, Nissei theater	1,340	7,500	1.11	1.06	0.58	5.3	17	y
Tokyo, NHK	3,677	25,200	1.60	1.70	—	—	23	—
Vienna, Staatsoper	1,709	10,665	1.30	1.42	0.60	2.8	17	y
Washington, DC, JFK Center, Opera House	2,142	13,027	1.28	1.27	0.53	3.1	15	y

ABLE A3.3. Technical details for the 100 halls of Chapter 3 in metric units.

Name of Hall	No. of Seats	m³		m²				V/S$_T$, m	EDT/V × 10⁶
		Volume	V/N	S$_A$	S$_o$, or S$_{pit}$	S$_T$	S$_a$/N		
Amsterdam, Concertgebouw	2,037	18,780	9.20	1,125	160	1,285	0.414	14.6	140
Aspen, Benedict Music Tent	2,050	19,830	9.67	1,197	487	1,377	0.470	14.4	—
Athens, Megaron Concert Hall	1,962	19,100	9.73	1,183	287	1,363	0.476	14.0	—
Baden-Baden, Festspielhaus	2,300	20,100	8.74	1,421	282	1,601	0.500	12.6	91
Baltimore, Meyerhoff Symphony Hall	2,467	21,530	8.73	1,487	229	1,667	0.485	12.9	109
Basel, Stadt-Casino	1,448	10,500	7.25	731	160	891	0.403	11.8	210
Bayreuth, Festspielhaus	1,800	10,308	5.72	845	35	1,032	0.419	10.0	—
Belfast, Waterfront Hall	2,250	30,800	13.70	1,301	200	1,481	0.475	20.8	—
Berlin, Kammermusiksaal (Philharmonie)	1,138	11,000	9.66	810	78	907	0.543	12.1	191
Berlin, Konzerthaus (Schauspielhaus)	1,575	15,000	9.53	943	158	1,101	0.498	13.6	162
Berlin, Philharmonie	2,218	21,000	9.00	1,385	172	1,558	0.455	13.5	93
Birmingham, Symphony Hall	2,211	25,000	11.30	1,320	279	1,500	0.466	16.7	80
Bonn, Beethovenhalle	1,407	15,730	11.20	1,115	205	1,295	0.614	12.1	—
Boston, Symphony Hall	2,625	18,750	7.14	1,370	152	1,522	0.402	12.3	126
Brussels, Palais des Beaux-Arts	2,150	12,520	5.83	1,300	186	1,486	0.474	8.4	148
Budapest, Magyar Allami Operahaz	1,277	8,900	6.97	644	58	861	0.405	10.3	—
Budapest, Pátria Hall	1,750	13,400	7.66	1,286	156	1,442	0.651	9.3	—
Buenos Aires, Teatro Colón	2,487	21,524	8.67	1,765	230	1,945	0.617	11.1	80
Buffalo, Kleinhans Music Hall	2,839	18,240	6.24	1,951	205	2,131	0.556	8.6	90
Caracas, Aula Magna	2,660	24,920	9.37	1,886	204	2,066	0.594	12.1	—

Name of Hall	No. of Seats	m³		m²				V/S$_T$, m	EDT/V × 10⁶
		Volume	V/N	S$_A$	S$_o$, or S$_{pit}$	S$_T$	S$_a$/N		
Cardiff, St. David's Hall	1,952	22,000	11.20	1,235	186	1,420	0.512	15.5	94
Chicago, Orchestra Hall	2,530	27,000	10.67	1,159	268	1,339	0.419	20.2	—
Christchurch, Town Hall	2,662	20,500	7.70	1,416	194	1,596	0.423	12.9	93
Cleveland, Severance Hall	2,101	16,290	7.75	1,210	215	1,390	0.443	11.7	138
Copenhagen, Radiohuset, Studio 1	1,081	11,900	11.00	721	288	901	0.660	13.2	165
Costa Mesa, Segerstrom Hall	2,903	27,800	9.58	1,742	223	1,922	0.518	14.5	78
Dallas, Meyerson Symphony Center	2,065	23,900	11.60	1,161	250	1,341	0.475	17.8	79
Denver, Boettcher Hall	2,750	37,444	13.60	1,628	238	1,808	0.457	20.7	65
Dresden, Semperoper	1,284	12,480	9.74	866	120	1,153	0.491	10.8	148
Edinburgh, Usher Hall	2,502	15,700	6.27	1,204	165	1,369	0.393	11.5	—
Fort Worth, Bass Performance Hall	2,072	27,300	13.18	1,222	304	1,402	0.456	19.5	92
Glasgow, Royal Concert Hall	2,457	22,950	9.35	1,365	218	1,545	0.466	14.9	76
Glyndebourne, Opera House	1,243	7,790	6.27	701	109	960	0.449	8.1	158
Gothenberg, Konserthus	1,286	11,900	9.25	666	170	836	0.455	14.2	146
Hong Kong, Cul. Ctr. Concert Hall	2,019	21,250	10.50	1,111	248	1,291	0.426	16.5	89
Jerusalem, Binyanei Ha'Oomah	3,142	24,700	7.90	2,137	260	2,317	0.532	10.7	75
Kuala Lumpur, Dewan Fil. Petronas	850	17,860	21.0	604	223	784	0.453	22.8	—
Kyoto, Concert Hall	1,840	20,000	10.90	1,162	237	1,342	0.482	14.9	105
Lahti, Sibelius/Talo	1,250	15,500	12.40	758	181	940	0.454	16.5	—
Leipzig, Gewandhaus	1,900	21,000	11.00	1,197	181	1,378	0.545	15.2	—
Lenox, MA., Seiji Ozawa Hall	1,180	11,610	9.83	739	202	919	0.420	12.6	—

Name of Hall	No. of Seats	m³		m²				V/S_T, m	EDT/V × 10⁶
		Volume	V/N	S_A	S_o, or S_{pit}	S_T	S_a/N		
Lenox, Tanglewood Music Shed	5,121	42,490	8.29	2,861	204	3,041	0.435	14.0	79
Liverpool, Philharmonic Hall	1,803	13,560	7.54	1,275	160	1,435	0.544	9.5	132
London, Barbican, Concert Hall	1,924	17,000	8.84	1,265	209	1,445	0.572	11.8	108
London, Royal Albert Hall	5,222	86,650	16.6	3,512	176	3,688	0.517	23.5	30
London, Royal Festival Hall	2,901	21,950	7.56	1,972	173	2,145	0.531	10.2	65
London, Royal Opera House	2,157	12,250	5.68	1,300	85	1,545	0.417	7.9	—
Lucerne, Cultural Center Concert Hall	1,892	17,823	9.42	1,256	242	1,436	0.456	12.4	—
Madrid, Auditorio Nacional de Música	2,293	20,000	8.72	1,705	278	1,885	0.611	10.6	103
Manchester, Bridgewater Hall	2,357	25,000	10.60	1,611	276	1,791	0.574	14.0	96
Mexico City, Salla Nezahualcoyotl	2,376	30,640	12.90	1,684	270	1,864	0.621	16.4	—
Milan, Teatro alla Scala	2,289	11,252	4.92	1,300	111	1,635	0.542	6.9	133
Minneapolis, Minn. Orchestra Hall	2,450	18,975	7.74	1,574	203	1,754	0.517	10.8	—
Montreal, Salle Wilfrid-Pelletier	2,982	26,500	8.90	1,767	172	1,939	0.520	13.7	73
Munich, Herkulessalle	1,287	13,590	10.60	674	168	842	0.455	16.1	—
Munich, Philharmonie Am Gasteig	2,487	29,700	12.45	1,639	230	1,818	0.557	16.4	71
Naples, Teatro di San Carlo	1,414	13,700	9.78	950	108	1,327	0.527	10.3	84
New York, Avery Fisher Hall	2,742	20,400	7.44	1,480	203	1,660	0.434	12.3	81
New York, Carnegie Hall	2,804	24,270	8.65	1,600	227	1,780	0.408	13.6	—
New York, Metropolitan Opera House	3,816	24,724	6.48	2,262	132	2,394	0.502	10.3	66

Name of Hall	No. of Seats	m³ Volume	V/N	m² S_A	S_o, or S_{pit}	S_T	S_a/N	V/S_T, m	EDT/V × 10⁶
Odense, Koncerthus, Nielsen Hall	1,320	14,000	10.60	651	176	827	0.415	16.9	161
Osaka, Symphony Hall	1,702	17,800	10.45	1,236	285	1,416	0.533	12.6	118
Paris, Opéra Bastille	2,700	21,000	7.80	1,522	186	1,951	0.470	10.8	—
Paris, Opéra Garnier	2,131	10,000	4.68	1,126	78	1,448	0.422	6.9	179
Paris, Salle Pléyel	2,386	15,500	6.50	1,058	242	1,238	0.327	12.5	122
Philadelphia, Academy of Music	2,827	15,100	5.34	1,460	59	1,740	0.445	8.7	79
Phildelphia, Verizon Hall, Kimmel Center	2,519	23,520	9.34	1,666	274	1,846	0.488	12.7	—
Rochester, NY, Eastman Theatre	3,347	25,500	7.62	2,021	204	2,201	0.472	11.6	—
Rotterdam, De Doelen	2,242	24,070	10.70	1,509	195	1,689	0.539	14.2	96
Salt Lake City, Symphony Hall	2,812	19,500	6.93	1,669	218	1,850	0.528	10.5	107
Salzburg, Festspielhaus	2,158	15,500	7.18	1,375	195	1,555	0.490	10.0	121
San Francisco, Davies Hall	2,743	24,070	8.78	1,562	200	1,742	0.442	13.8	89
San Francisco, War Mem. Opera House	3,252	20,900	6.43	1,973	71	2,276	0.471	9.2	—
São Paulo, Sala São Paulo	1,610	20,000	12.42	1,043	396	1,223	0.463	16.3	—
Sapporo, Concert Hall	2,008	28,800	14.34	1,438	240	1,618	0.504	17.8	66
Seattle, Benaroya Hall	2,500	19,263	7.70	1,452	216	1,632	0.471	11.8	—
Shanghai, Grand Theatre	1,895	13,000	6.86	842	85	1,175	0.387	11.1	105
Stuttgart, Liederhalle, Grosser Saal	2,000	16,000	8.00	1,300	176	1,533	0.500	10.4	134
Sydney Opera House, Concert Hall	2,679	24,600	9.18	1,563	181	1,743	0.508	14.0	89
Taipei, Cultural Centre, Concert Hall	2,074	16,700	8.00	1,261	269	1,441	0.492	11.6	—

Name of Hall	No. of Seats	m³ Volume	V/N	m² S$_A$	S$_o$, or S$_{pit}$	S$_T$	S$_a$/N	V/S$_T$, m	EDT/V × 10⁶
Tel Aviv, Fredric Mann Auditorium	2,715	21,240	6.76	1,700	195	1,880	0.497	11.0	80
Tokyo, Bunka Kaikan (Ueno)	2,327	17,300	7.40	1,301	241	1,481	0.422	11.2	—
Tokyo, Dai-ichi Seimei Hall	767	6,800	8.86	538	104	642	0.498	10.6	264
Tokyo, Hamarikyu Asahi Hall	552	5,800	10.51	395	73	468	0.513	11.4	320
Tokyo, Metropolitan Art Space	2,017	25,000	12.40	1,312	207	1,492	0.460	16.8	102
Tokyo, New National Theatre Opera House	1,810	14,500	8.00	1,153	102	1,460	0.492	9.9	117
Tokyo, NHK Hall	3,677	25,200	6.85	1,821	193	2,000	0.396	12.6	—
Tokyo, Orchard Hall	2,150	20,500	9.53	1,314	217	1,494	0.465	13.7	—
Tokyo, Suntory Hall	2,006	21,000	10.50	1,364	235	1,544	0.519	13.6	117
Tokyo, Opera City Concert Hall	1,636	15,300	9.40	1,052	168	1,220	0.485	12.5	176
Toronto, Roy Tompson Hall	2,812	28,300	10.06	1,681	222	1,861	0.498	15.2	67
Trondheim, Olavshallen	1,200	13,000	6.00	816	234	996	0.599	14.0	—
Valencia, Palau de la Música	1,790	15,400	8.60	812	155	967	0.400	15.9	192
Vienna, Grosser Musikvereinssaal	1,680	15,000	8.92	955	163	1,118	0.411	13.4	203
Vienna, Konzerthaus	1,865	16,600	8.90	881	137	1,018	0.350	16.3	128
Vienna, Staatsoper	1,709	10,665	6.24	1,194	107	1,460	0.544	7.3	133
Washington, DC, JFK Center, Concert Hall	2,448	22,300	9.11	1,425	228	1,605	0.439	13.9	—
Washington, DC, JFK Center, Opera House	2,142	13,027	6.08	1,289	109	1,595	0.529	8.2	97
Worcester, Mechanics Hall	1,343	10,760	8.01	701	154	855	0.402	12.5	—
Zurich, Grosser Tonhallesaal	1,546	11,400	7.37	877	145	1,022	0.454	11.2	274

$\mathscr{S}$ABINE EQUATION

$\mathbf{W}$allace Clement Sabine presented the acoustical world with the Sabine reverberation equation in 1898. The ability to calculate the reverberation time (RT) has been the cornerstone of acoustics since that date. The reverberation time is given by the formula

$$RT = 0.161 \frac{V}{A} \qquad (A3.10)$$

where V is the cubic volume of the room in m^3 measured as though there were no seats in the room, but with the volume occupied by balcony structures subtracted; and A is the total sound absorption in m^2 for both the surfaces of the room and losses in the air itself as the sound travels through it.

The total sound absorption can be subdivided,

$$A = S_T\alpha_T + S_R\alpha_R + S_{M1} + S_{M2} + \cdots + 4 \text{ mV} \quad m^2, \qquad (A3.11)$$

where S_T is the "acoustical area" over which the audience chairs sit, occupied or unoccupied, plus the stage area (not to exceed 180 m^2) when the orchestra is present (see Appendix 1 for a detailed definition of S_T); S_R is the actual area of all other surfaces in the room except the areas over which the audience and orchestra sit, including underbalcony soffits and all of the aisle areas; and S_{M1}, S_{M2}, S_{M3}, etc., are areas of special absorbing materials like rugs, draperies, and acoustical tiles. Each of the types of absorbing area S_T, S_R, and S_M has its associated absorption coefficient. The quantity $S_R\alpha_R$ is usually referred to as the "residual absorption" of the room.

All of the absorption coefficients presented later in this chapter are for use in the Sabine equation. Those that go with the "residual" area S_R have lumped in them the absorptions of ventilation openings, chandeliers, doors, and the like, and, therefore, may not agree with others in the literature.

The air absorption is equal to 4 mV, where V is the volume of the room in m^3, times the quantity 4 m (in m^{-1}), which is listed in Table A3.4 as a function of relative humidity assuming an ambient temperature of about 20°C. Air absorption is important only in the 2,000 and 4,000-Hz frequency bands, except in very large halls, where it begins to be important in the 1,000-Hz band.

It has been established that in a large hall for musical performances the absorbing power of a seated audience, chorus and orchestra, or empty upholstered

*T*ABLE A3.4. Air attenuation coefficient multiplied by 4, yielding 4 m in units of m^{-1} for an ambient temperature of about 20°C for four relative humidities (ISO, 1990).

Relative Humidity (%)	4 m (Frequency, Hz)			
	500	1,000	2,000	4,000
50	0.0024	0.0042	0.0089	0.0262
60	0.0025	0.0044	0.0085	0.0234
70	0.0025	0.0045	0.0081	0.0208
80	0.0025	0.0046	0.0082	0.0194

seats, increases in direct proportion to the floor area they occupy, almost independent of the number of seated persons or chairs in that area, provided the seats are nearly 100% occupied or unoccupied (Beranek and Hidaka, 1998; Barron and Coleman, 2001). This hypothesis is valid for seating densities (S_a/N) in the range of 0.40–0.65 m^2 of floor space per person and for halls with normally diffuse sound fields. No attempt has been made to extend the applicability of these data to small auditoriums, classrooms, or churches, where the state of sound diffusion in the room or the seating density may be substantially different.

*D*ERIVATIVES OF THE SABINE EQUATION

In the preliminary design of a hall it is desirable to have a simpler equation to use than Eqs. (A3.10) and (A3.11). A possible simplification is to assume that the total room absorption is attributable to the audience, and to restrict the calculation to the two mid-frequency bands at 500 and 1,000 Hz, in order to eliminate the air absorption term and to avoid the greater irregularities found in sound absorption at low frequencies. Also eliminated are any terms owing to special absorbing materials, such as carpets and absorbing means for echo control. The simplification leads to

$$RT_{mid} = 0.161 \ V/[S_T(\alpha_T + (S_R\alpha_R/S_T)] = K_1 \times (V/S_T) \quad (A3.12)$$

So that

$$K_1 = 0.161/[\alpha_T + (S_R\alpha_R/S_T)] \quad (A3.13)$$

Before proceeding further, let us look at three groups of halls to determine what values of K_1 are to be expected in concert halls and opera houses. Results using some data from Tables A3.1 to A3.3 are shown in Fig. A3.1. The coefficient $K_1 = RT/(V/S_T)$ is different for three types of halls. For the upper right-hand group of concert halls, $K_1 = 0.143$. The middle group of seven opera houses with higher RT's, would be better fit with a value of $K_1 = 0.145$. The lower group of four opera houses with low reverberation times would be better fit with a value of $K_1 = 0.17$. These latter two values were previously substantiated in Hidaka and Beranek (2000).

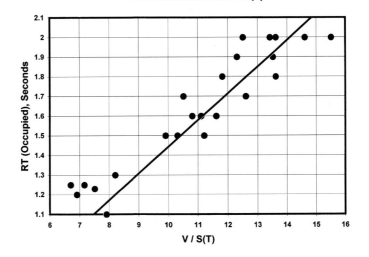

DETERMINATION OF K(1)

$\mathscr{F}$IGURE A3.1. Plot of mid-frequency reverberation time, RT, hall fully occupied, vs V/S_T, where V is the volume and S_T is the acoustical area (see Appendix 1 for definition of S_T). The upper right group is composed of concert halls; the middle group of 7 opera houses with high reverberation times; and the lower group of 6 opera houses with low reverberation times. $K_1 = RT/[V/S_T]$.

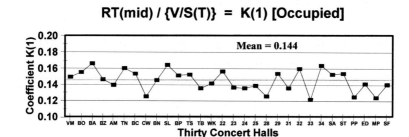

$\mathscr{F}$IGURE A3.2. Plot of K_1 = RT/[V/S_T] for 30 concert halls. The mean value is 0.144.

The accuracy of using this simplified formula and K_1 for 30 of the concert halls in Table 4.1 is illustrated in Fig. A3.2. A mean value of K_1 = 0.144 is indicated. Obviously, the value of K_1 is not highly accurate because the accuracy of the reverberation data, mostly from a few stop-chords at a limited number of seats, is unknown; and, perhaps more important, the halls have different audience absorptions owing to the type of seats, and $(S_R \alpha_R / S_T)$ varies from one hall to another. Thus, using K_1 = 0.143 is restricted to preliminary explorations in the design of halls.

The application of Eqs. (A3.12) and (A3.13) to the planning stage of a concert hall's design is given in Chapter 4 in the section titled, "Design procedures for preliminary determination of audience size and cubic volume of a hall given selected values for G_{mid} and RT_{occ}."

$\mathscr{C}$ALIBRATION OF THE DODECAHEDRAL SOUND SOURCE

The sound source used for the determination of the various objective parameters in halls for music is a 12-sided (dodecahedral) approximation to a sphere—an approximate omnidirectional source, ranging in diameter from 30 to 45 cm. In each of the 12 sides a cone loudspeaker is imbedded. For determination of the strength of the sound at seats in a hall, this source must be calibrated. Apparently, there are two methods of calibration in use, as evidenced by data from two different sources as reported in Chapter 4 that differ, on average, by about 1.2 dB. These calibration methods are referred to here as (1) reverberation chamber method and (2) field method.

The official definition of the strength factor in a hall, is $G = 10 \log [p^2/p_A^2]$, where p^2 is the mean-square sound pressure at the point (seat) where G is being measured and p_A^2 is the mean-square sound pressure measured in an anechoic chamber at point located a distance of 10 m from the acoustical center of the source.

The mean-square sound pressure p_r^2 at a distance r from a spherical source radiating an acoustical power measured in watts (W) in anechoic space is $p^2 = W\rho c/4\pi r^2$. The quantity $\rho c/4\pi$ equals 32.4 (mks units), so

$$p_r^2 = 32.4(W/r^2). \tag{A3.14}$$

If we measure at a distance of r = 10 m,

$$p_r^2 = p_A^2 = 0.324W \tag{A3.15}$$

If we divide both sides of the last equation by the standard reference quantities for sound pressure and sound power, i.e., so that the divisor equals $[10^{-12*}4^*10^{-10}]$, we can convert that equation to include the sound-pressure level SPL_A (at 10 m in an anechoic space) and sound-power level PWL. Thus, $SPL_A = PWL + 10 \log [(0.324/4)^*10^{-12*}10^{10}]$, and

$$SPL_A = PWL - 31 \text{ dB}. \tag{A3.16}$$

Let G indicate the strength at a seat and SPL be the measurement at a seat. Then, by definition of G,

$$G = SPL - SPL_A. \tag{A3.17}$$

The strength factor (G) at a seat in a hall is determined from Eq. (A3.17), using Eq. (A3.16) to obtain the SPL_A.

REVERBERATION CHAMBER METHOD OF CALIBRATION. For this method the source is calibrated by measuring its acoustical power output (PWL in decibels) in a reverberation chamber. International standard ISO-3382 is used for the procedure. The Takenaka reverberation chamber has seven walls with no parallel surfaces, a volume of 332.8 m², and 20 curved, randomly spaced diffusers, each with an area of 1.8 m². All of Takenaka calibrations are made in their reverberation chamber.

FIELD METHOD OF CALIBRATION. Another method for calibration is to set the omnidirectional source at its position on the stage in a hall and measure the sound-pressure level at 1 m from its acoustical center. The Takenaka group compared this method with the reverberation chamber method, taking data by the field method in six halls. The source is set 3 m from the front of the stage, on the centerline, and is supported on a tripod with the source's acoustical center 1.5 m above the floor. The acoustical center is assumed to be at the geometrical center. The 1-m measurement is made by two persons. One stands on the left side of the source holding the microphone with his arm stretched out as far as possible to avoid reflection from the body; the other does likewise on the right side. The sound-level meters are calibrated before and after the measurements with a pistonphone. The 12 loudspeakers were numbered and No. 1 always faced down the centerline of the hall.

Data taken in six halls in this manner are shown in Table A3.5 All halls were in Japan. The values for the field calibration are higher than for the reverberation calibration. This is to be expected, because the floor returns some energy to the microphones—more at the low frequencies than at the high. Note that the mid-frequency difference is 1.3 dB. The next question is, how does this affect the field values of G?

EFFECT OF CALIBRATION DIFFERENCE ON MEASURED G'S IN HALLS. For the reverberation chamber method of calibration: combine Eqs. (A3.16) and (A3.17), where G and SPL indicate measurements in the hall:

$$G = SPL - PWL + 31 \text{ dB.} \tag{A3.18}$$

For the field method:

$$SPL'_A = SPL_{10m} - 20 \log 10 + 1.3 \text{ dB.} \tag{A3.19}$$

Substituting (A3.19) in (A6.17) yields

$$G = SPL - SPL_A' = SPL - SPL_{10m} + 20 \log 10 - 1.3 \text{ dB} \tag{A3.20}$$

If there was no modification of G by +1.3 dB, $[-SPL_{10m} + 20 \log 10]$ would equal $[-PWL + 31 \text{ dB}]$. Substituting the latter in Eq. (A3.19) gives

$\mathscr{T}$ABLE A3.5. Calibration of the dodecahedral loudspeaker by two different methods: One method places the loudspeaker in a reverberation chamber where the total acoustic power radiated is determined, and the other places the loudspeaker on the stage of the hall and the sound pressure level is determined at one meter from the speaker's acoustical center. The differences in the calibrations in the low, middle and high frequency regions are determined.

		Frequency in Hz						
		63	125	250	500	1k	2k	4k
Reverberation chamber measurement	PWL re 10^{-12} watt based on ISO 3741, in dB	76.6	96.4	99.1	94.9	90.8	91.6	84.1
	SPL re $2*10^{-5}$ N/m² at 1 m (SPL = PWL − 11), in dB	65.6	85.4	88.1	83.9	79.8	80.6	73.1
Field measurement avg. of data in six halls	SPL re $2*10^{-5}$ N/m² measured at 1 m from source center, in dB	71.5	88.1	90.3	85.6	80.9	81.5	73.5
Difference	SPL (1 m, in field) minus SPL (1 m, by rev. chamber method)	5.9	2.7	2.2	1.7	1.1	0.9	0.4
Summary	Low (125, 250) = 2.4 dB; Mid (500, 1000) = 1.4 dB; High (2000, 4000) = 0.6 dB							

$$G_{field} = SPL - PWL + 31 - 1.3 \text{ dB, and} \qquad (A3.21)$$

$$G_{reverb} = SPL - PWL + 31 \text{ dB}. \qquad (A3.22)$$

This difference of 1.3 dB is to be compared to the difference between Figs. 4.8 and 4.9 of Chapter 4. The heavy line in Fig. 4.8 is about 1.2 dB higher than that in Fig. 4.9.

SOUND ABSORPTION COEFFICIENTS

Sound absorption coefficients for common building materials are presented in Table A3.6. These absorption coefficients were derived from data taken in halls on which the author has consulted. Other engineers may be using somewhat different coefficients based on experience with different halls. The reference is Beranek and Hidaka (1998).

$\mathcal{T}$ABLE A3.6. Sound absorption coefficients for building materials and audience areas. These coefficients must be used in the Sabine equation. The measurements on building materials were made in the types of diffuse sound fields found in concert halls without seats or audience. The absorptions by audience areas, with and without full occupancy, were determined from measurements made after the seats were installed in those halls.

Materials	Frequency, Hz						Mass, kg/m²
	125	250	500	1,000	2,000	4,000	
Gypsum, 2 layers, fiberglass reinforced, 25 mm w/lighting and ventilation	0.15	0.12	0.10	0.08	0.07	0.06	40
Note: Gypsum, plaster board, not reinforced, mass per m² equals [thickness in mm] $\times$ 1.0 kg/m², approximately							
Wood, ceiling, 2 layers, 28 mm w/lighting and ventilation	0.18	0.14	0.10	0.08	0.07	0.06	17
Wood, sidewalls, 1 layer, 20 mm w/doors and lighting	0.25	0.18	0.11	0.08	0.07	0.06	12
Wood, sidewalls, 1 layer, 12 mm w/doors and lighting	0.28	0.22	0.19	0.13	0.08	0.06	6.2
Wood, audience floor, 2 layers, 33 mm on sleepers over concrete	0.09	0.06	0.05	0.05	0.05	0.04	N/A
Wood, stage floor, 2 layers, 27 mm over airspace	0.10	0.07	0.06	0.06	0.06	0.06	17
Wood, 19 mm, over 25 mm compressed fiberglass, screwed to 150 mm concrete block w/doors and lighting	0.20	0.15	0.08	0.05	0.05	0.05	N/A
Plaster, ceiling, 60 mm w/lighting and ventilation	0.10	0.08	0.05	0.04	0.03	0.02	60
Plaster, ceiling, 30 mm w/lighting and ventilation	0.14	0.12	0.08	0.06	0.06	0.04	30
Plastic, fiberglass reinforced phenolic foam, filled with aluminum hydroxide, faced with very thin layer plywood, 8 mm (Tokyo, Hamarikyu-Asahi Concert Hall)	0.25	0.23	0.16	0.12	0.11	0.10	4

Materials	Frequency, Hz					
	125	250	500	1,000	2,000	4,000
Concrete floor, linoleum cemented to it	0.04	0.03	0.03	0.03	0.03	0.02
Concrete floor, woods boards, 19 mm, secured to it	0.10	0.08	0.07	0.06	0.06	0.06
Concrete block, plastered	0.06	0.05	0.05	0.04	0.04	0.04
Organ absorption, case opening 75 m² (Boston, behind grille)	41	26	19	15	11	11
Organ absorption, free standing (Tokyo, TOC Concert Hall)	65	44	35	33	32	31
Audience, seats fully occupied						
Heavily upholstered	0.72	0.80	0.86	0.89	0.90	0.90
Medium upholstered	0.62	0.72	0.80	0.83	0.84	0.85
Lightly upholstered	0.51	0.64	0.75	0.80	0.82	0.83
Seats unoccupied						
Heavily upholstered	0.70	0.76	0.81	0.84	0.84	0.81
Medium upholstered	0.54	0.62	0.68	0.70	0.68	0.66
Lightly upholstered	0.36	0.47	0.57	0.62	0.62	0.60
Absorption power of orchestra (m²), Tokyo, TOC Concert Hall and NNT Opera House						
Concert Hall (stage 170 m², vertical walls, sides (ends) splayed)						
13 string instruments	3	4	6	17	52	64
44 players (2 brass)	12	21	24	46	74	100
92 players (4 brass)	22	37	44	64	102	132
Opera House (pit opening 100 m²)						
40 players	10	13	17	41	50	57
80 players	12	17	23	56	67	71

Note: Surface density values do not include the mass of furring or wooden nailing strips

Note: The coefficients following were taken from the literature

Materials	125	250	500	1,000	2,000	4,000
Carpet, heavy, cemented to concrete	0.02	0.06	0.14	0.37	0.6	0.65
Carpet, heavy, over foamed rubber	0.08	0.24	0.57	0.69	0.71	0.73
Carpet, thin, cemented to concrete	0.02	0.04	0.08	0.2	0.35	0.4

Bibliography

Akaike, H. (1973) *Information Theory and an Extension of the Maximum Likelihood Principle*. Akademiai Kiado, Budapest.

Aoshima, N. (1981) Computer-generated pulse signal applied for sound measurement. J. Acoust. Soc. Am. **69**, 1484–1488.

Ando, Y. (1985) *Concert Hall Acoustics*. Springer-Verlag, Berlin.

Ando, Y. (1998) *Architectural Acoustics*. Springer-Verlag, New York.

Barron, M. (1971) The subjective effect of first reflections in concert halls—the need for lateral reflections. J. Sound Vib. **15**, 475–494.

Barron, M. and Marshall, A. H. (1981) Spacial impression due to early lateral reflections in concert halls: The derivation of a physical measure. J. of Sound and Vibration, **77**, 211–232.

Barron, M. (1988) Subjective study of British concert halls. Acustica **66**, 1–14.

Barron, M. and Lee, L.-J. (1988) Energy relation in concert auditoria. I. J. Acoust. Soc. Am. **84**, 618–628.

Barron, M. (1993) *Auditorium Acoustics and Architectural Design* E & FN Spon. Chapman & Hill, London & New York.

Barron, M. (1995) Balcony overhangs in concert auditoria, *J. Acoust. Soc. Am.* **98**, 2580–2589.

Barron, M. (1995) Interpretation of early decay times in concert auditoria. Acustica **81**, 320–331.

Barron, M. (1996) Loudness in concert halls. Acustica **82**, S21–S29.

Barron, M. (2000) Measured early lateral energy fractions in concert halls and opera houses. J. Sound Vib. **232**, 79–100.

Barron, M. (2001) Late lateral energy fractions and the envelopment question in concert halls. Applied Acoustics **62**, 185–202.

Barron, M. and Coleman, S. (2001) Measurements of the absorption by auditorium seating—a model study. J. Sound Vib. **239**, 573–587.

Beranek, L. L. (1962) *Music, Acoustics and Architecture*. Wiley, New York. In Japanese (1972) with modifications and added Appendix 4 by M. Nagatomo, Kajima Institute, Tokyo.

Beranek, L. L. (1992) Concert hall acoustics—1992. J. Acoust. Soc. Am. **92**, 1–39.

Beranek, L. L. (1996) *Concert and opera halls: How they sound*. Acoust. Soc. Am., Melville, New York.

Beranek, L. L. (2002). Concert hall acoustics: Addenda, 2001. J. Acoust Soc. Japan **58**, 61–71 (in Japanese).

Beranek, L. L. (1988) *Noise and Vibration*

Control. Institute of Noise Control Engineering, Saddle River, New Jersey.

Beranek, L. L. and Ver, I. L., (1992) *Noise and Vibration Control Engineering.* J. Wiley, New York.

Beranek, L. L. and Hidaka, T. (1998) Sound absorption in concert halls by seats, occupied and unoccupied, and by the hall's interior surfaces, J. Acoust. Soc. Am. **104**, 3169–3177.

Beranek, L. L., Hidaka T. and Masuda, S. (1998) Acoustical design of the drama and experimental theaters of the New National Theater, Tokyo, Japan. Proc. ICA **98**, 1793–1794.

Beranek, L. L., Hidaka, T. and Masuda, S. (2000) Acoustical design of the Opera House of the New National Theater (NNT), Tokyo, Japan. J. Acoust. Soc. Am., **107**, 355–367.

Blauert, J. (1983) *Spatial Hearing.* MIT Press, Cambridge, Massachusetts.

Blauret, J. and Lindermann, W. (1986) Auditory spaciousness: Some further psychoacoustic analyses. J. Acoust. Soc. Am. **80**, 533–542.

Blauert, J., Möbius, U. and Lindemann, W. (1986) Supplementary psychoacoustical results on auditory spaciousness. Acustica **59**, 292–293.

Bradley, J. S. (1991) A comparison of three classical concert halls. J. Acoust. Soc. Am. **89**, 1176–1192.

Bradley, J. S. (1992) Predicting chair absorption from reverberation chamber measurements. J. Acoust. Soc. Am. **91**, 1514–1524.

Bradley, J. S. (1994a) Reply to reports on [Bradley 1992]. J. Acoust. Soc. Am. **96**, 1155–1157.

Bradley, J. S. (1994b) Comparison of concert hall measurements of spatial impression. J. Acoust. Soc. Am. **96**, 3525–3535.

Bradley, J. S. (1994c) *Data from 13 North American Concert Halls.* Internal Report No. 668, National Research Council of Canada, Ottawa K1A OR6, July. (Partially funded by the Concert Hall Research Group.)

Bradley, J. S. and Halliwell, R. E. (1991) Ten years of newer auditorium acoustics measurements. J. Acoust. Soc. Am. **89**, 1856 (A).

Bradley, J. S. (1996) The sound absorption of occupied auditorium seating. J. Acoust. Soc. Am., **99**, 990ff.

Bradley, J. S. and Soulodre, G. (1995a) The influence of late arriving energy on spatial impression, J. Acoust. Soc. Am. **97**, 2263–2271.

Bradley, J. S. and Soulodre, G. (1995b) Objective measures of listener envelopment. J. Acoust. Soc. Am. **98**, 2590–2597.

Bradley, J. S. and Soulodre, G. (1997) Factors influencing the perception of bass. 133rd Meeting of the Acoustical Society of America. J. Acoust. Soc. Am. **101**, 3135 (A).

Bradley, J. S., Reich, R. D. and Norcross, S. G. (2000) On the combined effects of early- and late-arriving sound on spatial impression in concert halls. J. Acoust. Soc. Am. **108**, 651–661.

Cremer, L. and Müller, H.A. (1982) *Principles and Applications of Room Acoustics Vol. 1.* English translation with additions by T. J. Schultz (Applied Science Pub., Essex, England). In USA and Canada (Elsevier, New York). Originally published in German (1978) by Hirzel, Stuttgart.

Davies, W. J. and Orlowski, R. J. (1990) Methods of measuring acoustic absorption of auditorium seating. Proc. Institute of Acoustics, **12**, 299–306.

Davies, W. J., Orlowski, R. J. and Lam, Y. W. (1994) Measuring auditorium seat absorption. J. Acoust. Soc. Am. **96**, 879–888.

Egan, M. D. (1998) *Architectural Acoustics*. McGraw Hill, New York.

Fry, A. (1988) *Buildings for Music*. MIT Press, Cambridge, Massachusetts.

Fry, A. (1998) *Noise Control*. Pergamon, Oxford.

Fasold, W., Tennhardt, H. and Winkler, H. (1988) Ergänzende raumakustische Maßnahmen im Großen Konzertsaal des Schauspielhauses Berlin, Bauakademie der DDR.

Fukuchi, T. and Fujiwara, K. (1985) Sound absorption area per seat of upholstered chairs in a hall. J. Acoust. Soc. Jpn. (English) **6**, 271–279.

Gade, A. C. (1985) Objective measurements in Danish concert halls. Proc. Inst. Acoust. **7**, 9–16.

Gade, A. C. (1989a) Acoustical survey of eleven European concert halls, Report No. 44. The Acoustics Laboratory, Technical University of Denmark, Copenhagen.

Gade, A. C. (1989b) Investigations of musicians' room acoustic conditions in concert halls. Acustica **69**, 193–203 and 249–262.

Gade, A. C. (1991) Prediction of room acoustical parameters. J. Acoust. Soc. Am. **89**, 1857(A).

Griesinger, D. (1998) General overview of spatial impression, envelopment, localization, and externalization. Audio Engineering Society, 15th International Conference Proceedings, 136–149.

Griesinger, D. (1999) Objective measures of spaciousness and envelopment. Audio Engineering Society, 16th International Conference Proceedings, 27–41.

Haan, C. H. and Fricke, F. R. (1993) Surface diffusivity as a measure of the acoustic quality of concert halls. Proceedings of Conference of the Australia and New Zealand Architectural Science Association, Sydney, 81–90.

Haan, C. H. and Fricke, F. R. (1995) Musician and music critic responses to concert hall acoustics. Proceedings of the 15th International Congress on Acoustics, Trondheim, Norway, June.

Harris, C. (1991) *Handbook of Acoustical Measurements & Noise Control*. McGraw-Hill, New York, 3rd Ed.

Hawkes, R. J. and Douglas, H. (1971) Subjective acoustic experience in concert auditoria. Acustica **24**, 235–250.

Hidaka, T., Kageyama, K. and Masuda, S. (1988) Recording of anechoic orchestral music and measurement of its physical characteristics based on the auto-correlation function. Acustica **67**, 68–70.

Hidaka, T., Beranek, L. L. and Okano, T. (1995) Interaural cross-correlation IACC, lateral Fraction LF, and low- and high-frequency sound level G as measures of acoustical quality in concert halls. J. Acoust. Soc. Am. **98**, 988–1007.

Hidaka, T. and Beranek, L. L. (2000) Objective and subjective evaluations of twenty-three opera houses in Europe, Japan and the Americas. J. Acoust. Soc. Am. **107**, 368–383.

Hidaka, T., Beranek, L. L., Masuda, S., Nishihara, N. and Okano, T. (2000) Acoustical design of the Tokyo Opera City (TOC) Concert Hall, Japan. J. Acoust. Soc. Am. **107**, 340–354.

Hidaka, T., Nishihara, N. and Beranek, L. L. (2001) Relation of acoustical parameters with and without audiences in concert halls and a simple method for simulating the occupied state. J. Acoust. Soc. Am. **109**, 1028–1042.

Hidaka, T. and Nishihara, N. (2001) Objective evaluations of chamber music halls in Europe and Japan. Proceedings of 17th ICA meeting in Rome, Vol. 3, Sept. 3.

Hidaka, T. and Nishihara, N. (2002) On the objective measure of texture, Proc. of Forum Acusticum, Sevilla.

Hirata, Y. (1972) Measurement of absorption coefficient by electrically canceling method. J. Acoust. Soc. Jpn. **28**, 416 (Japanese).

ISO 3745 (1977) Acoustics—Determination of sound power levels of noise sources using pressure—Precision methods for anechoic and semi-anechoic rooms. Standards Secretariat, Acoustical Society of America, Melville, New York.

ISO 345 (1985) Measurement of sound absorption in a reverberation room. Standards Secretariat, Acoustical Society of America, Melville, New York.

ISO 3741 (1988) Determination of sound power levels of noise sources using pressure: Precision methods for reverberation rooms. Standards Secretariat, Acoustical Society of America, Melville, New York.

ISO 3382 (1997) Measurement of the reverberation time of rooms with reference to other acoustical parameters. Standards Secretariat, Acoustical Society of America, Melville, New York.

Jordan, V. L. (1980) *Acoustical Design of Concert Halls and Theaters*. Appl. Sci. Pub., London, p. 217.

Kath, U. and Kuhl, W. (1965) Messungen zur Schallabsorption von Polsterstuehlen mit und ohne Personen. Acustica **15**, 137–131.

Kirkegaard, D. L. (1996) Sound absorption of occupied chairs as a function of chair design and audience clothing. J. Acoust. Soc Am. **99**, 2458.

Kosten, C. W. (1965/66) New method for calculation of the reverberation time of

halls for public assembly. Acustica **16**, 325–330,

Kuhl, W. (1978) Räumlichkeit als komponente des raumeindrucks. Acustica **40**, 167–181.

Kuttruff, H. (1991) *Room Acoustics*. Elsevier Applied Science, Essex, England.

Kuttruff, H. (2000) *Room acoustics,* E & FN Spon, London, Chap. 8.

Litovsky, R. Y. and Colburn, H. S. (1999) The precedence effect. J. Acoust. Soc. Am. **106**, 1633–1654.

Marshall, A. H. (1968) Acoustical determinants for the architectural design of concert halls. Arch. Scio. Rev., Australia **11**, 81–87.

Marshall, A. H. and Barron, M. (2001) Spatial responsiveness in concert halls and the origins of spatial impression. Applied Acoustics **62**, 91–108

Marshall, A. H. and Hyde, J. R. (1980) Some practical considerations in the use of quadratic residue diffusing surfaces. Proceedings of 10th International Congress on Acoustics, Sydney, Australia.

Meyer, J. (1978) *Acoustics and the Performance of Music*. Verlag Das Musikinstrument, Frankfurt/Main.

Makita, Y. and Hidaka, T. (1987) Revision of the cos θ law of oblique incident sound energy and modification of the fundamental formulations in geometrical acoustics in accordance with the revised law. Acustica **63**, 163–173.

Morimoto, M. and Maekawa, Z. (1988) Effects of low frequency components on auditory spaciousness. Acustica **66**, 190–196.

Nakamura, S., Kan, S., and Nagatomo, M. (1991) Subjective evaluation of acoustics of hall stages by players of symphonic orchestras. (Personal communication.)

Nagata, M. (1991) *Acoustic Design of Architecture.* Ohmu-Sya, Tokyo, 215 (in Japanese).

Nishihara, N., Hidaka, T. and Beranek, L. (2001) Mechanism of sound absorption by seated audience in halls. J. Acoust. Soc. Am. **110**, 2398–2411.

Okano, T., Beranek, L. L. and Hidaka, T. (1998) Relations among interaural cross-correlation coefficient (IACC$_E$), lateral fraction (LF$_E$), and apparent source width (ASW) in concert halls. J. Acoust. Soc. Am. **104**, 255–265.

Parkin, P. H., Allen, W. A., Purkis, J., and Scholes, W. E. (1953) The acoustics of Royal Festival Hall, London. Acustica **3**, 1–21.

Potter, J. M., Bilsen, F. A. and Raatgever, J. (1995) Frequency dependence of spaciousness. Acta Acustica **3**, 417–427.

Potter, J. M., Raatgever, J. and Bilsen, F. A. (1995) Measures for spaciousness in room acoustics based on a binaural strategy. Acta Acustica **3**, 429–442.

Reichardt, W., Abdel, A., Alim O. and Schmidt, W. (1975) Definition und Messgrundlage eines objektiven Masses zur Ermittlung der Grenze zwischen brauchbarer und unbrauchbarer Derchsichtigkeit bet Musikdarbietung. Acustica **32**, 126–137.

Sabine, W. C. (1900) Architectural acoustics (published in seven parts in April, May and June, Am. Arch Building News **68**). Available in reprint book, 1992. *Collected Papers on Acoustics: Wallace Clement Sabine.* Peninsula Publishing, Los Altos, California.

Schroeder, M. R. (1954) Die statistischen parameter der frequenzkurven von grossen räumen. Acustica **4**, 594ff.

Schroeder, M. R., Gottlieb, D., and Siebrasse, K. F. (1974) Comparative study of European concert halls. Correlation of subjective preference with geometric and acoustic parameters. J. Acoust. Soc. Am. **56**, 1195–1201.

Schroeder, M. R. (1979) Binaural dissimilarity and optimum ceilings for concert halls: more lateral sound diffusion." J. Acoust. Soc. Am. **65**, 958–963.

Schultz, T. J. (1963) Problems in the measurement of reverberation time. J. Audio Eng. Soc. **11**, 307–317.

Schultz, T. J. (1980) Concert Hall Tour of North America. BBN Tech. Inf. Rep. No. 98.

Sotiropoulou, A. G. and Hawkes, R. J., and D. B. Fleming (1995) Concert hall acoustic evaluations by ordinary concert-goers: I, Multi-dimensional description of evaluations. Acustica **81**, 1–9.

Sotiropoulou, A. G. and Fleming, D. B. (1995) Concert hall acoustic evaluations by ordinary concert-goers: II, Physical room acoustic criteria subjectively significant. Acustica **81**, 10–19.

Tachibana, H., Yamasaki, Y., Morimoto, M., Hirasawa, Y., Maekawa, Z. and Poesselt, C. J. (1989) Acoustic survey of auditoriums in Europe and Japan, J. Acoust. Soc. Jpn. **10**, 73–85.

Toyota, Y., Oguchi, K., and Nagata, N. (1988) A study on the characteristics of early reflections in concert halls." J. Acoust. Soc. Am., (Suppl. 1) **84** (A) 130.

Veneklasen, P. S. and Christoff, J. P. (1964) Seattle opera house—acoustical design. J. Acoust. Soc. Am. **36**, 903ff.

Watanabe, K., Yoshihisa, K. and Tachibana, H (1985) Subjective evaluation of room acoustics by using impulsive sound source. Proc. of Autumn Meeting of Acoust. Soc. Jpn, 479–480 (in Japanese).

Name Index

Subject Index

A

Abravanel Hall, **133**, 592

Absorption by air, 631, 632

Absorption by materials, 429, 502, 503, 639, 640

Absorption coefficients, 501, 631, 638–640

Academy of Music, 2, 4, 5, 15, **119**, 591

Acoustical "clouds" ("saucers"), 499, 537,

Acoustical modeling, 545, 548, 549

Acoustical quality

  age, 498

  apparent source width, 519

  attack, 33

  audience capacity, 501

  balance, 32

  balconies, 545–548

  bass strength (*see* Bass)

  binaural quality index, *BQI*, 29, 506–510, 536, 558–560

  blend, 32

  brilliance, 32, 533, 534

  categories of, 495

  clarity (*see* Clarity)

  decay (*see* Reverberation)

  definition, (*see* Clarity)

  diffusion, (*see* Diffusion)

  distortion, 572

  dynamic range, 2, 33, 34

  early decay time, **EDT**, 505, 506, 536, 558, 577

  early sound, 23, 34

  echo (*see* Echo)

  ensemble (*see* Ensemble)

  envelopment (*see* Listener envelopment)

  fullness of tone (*see* Reverberation)

  glare, 31, 521, 528

  "good," 2

  hall uniformity, 35

  intimacy (*see* Intimacy)

  liveness, (*see* Reverberation)

  loudness (*see* Loudness)

  noise, 35, 534

  parameters, 577–580

  presence, 27

  ranking of

    concert halls, 496

    opera houses, 554–556

  response, 33

  reverberance (*see* Reverberation)

  shape, 493, 499

  size, 18, 493, 501

  spaciousness (*see* Spaciousness)

  stages, 541–545

  strength (*see* Strength)

  terminology, 577–580

  texture (*see* Texture)